ECOLOGY AND MANAGEMENT OF AQUATIC ENVIRONMENTS

ECOLOGY AND MANAGEMENT OF AQUATIC ENVIRONMENTS

Edited by:

Bruno Augusto Amato Borges

Delve
PUBLISHING

www.delvepublishing.com

Ecology and Management of Aquatic Environments

Bruno Augusto Amato Borges

Delve Publishing
2010 Winston Park Drive,
2nd Floor
Oakville, ON L6H 5R7
Canada
www.delvepublishing.com
Tel: 001-289-291-7705
001-905-616-2116
Fax: 001-289-291-7601
Email: orders@arclereducation.com

ISBN: 978-1-77361-042-9 (Hardcover)

DECLARATION

Some content or chapters in this book are open access copyright free published research work, which is published under Creative Commons License and are indicated with the citation. We are thankful to the publishers and authors of the content and chapters as without them this book wouldn't have been possible.

ABOUT THE EDITOR

Bruno Augusto Amato Borges

Bruno Augusto is Aquaculture Engineer from Federal University of Santa Catarina. His focus is on writing projects and development of new technologies in aquaculture. He is currently project service provider, consultant and develops research in the field of Biofloc Technology (BFT) for freshwater fish farming.

TABLE OF CONTENT

SECTION 5 – IMPACTS AND MANAGEMENT MEASURES

LIST OF CONTRIBUTORS

Bruno Augusto Amato Borges
Aquaculture Engineer, Florianopolis, Brazil

Guangyu Wang
Faculty of Forestry, University of British Columbia, 2424 Main Mall, Vancouver, BC V6T 1Z4, Canada

Shari Mang
Faculty of Forestry, University of British Columbia, 2424 Main Mall, Vancouver, BC V6T 1Z4, Canada

Haisheng Cai
Faculty of Forestry, University of British Columbia, 2424 Main Mall, Vancouver, BC V6T 1Z4, Canada
College of Land Resources and Environment, Jiangxi Agriculture University, 1225 Zhimin Ave, Qingshanhu, Nanchang 330029, People's Republic of China

Shirong Liu
Key Laboratory of Forest Ecology and Environment of State Forestry Administration, Institute of Forest Ecology, Environment and Protection, Chinese Academy of Forestry, Beijing 10091, People's Republic of China

Zhiqiang Zhang
Key Laboratory of Soil and Water Conservation and Desertification Combating, Ministry of Education, Beijing Forestry University, Beijing, People's Republic of China

Liguo Wang
College of Land Resources and Environment, Jiangxi Agriculture University, 1225 Zhimin Ave, Qingshanhu, Nanchang 330029, People's Republic of China

John L. Innes
Faculty of Forestry, University of British Columbia, 2424 Main Mall, Vancouver, BC V6T 1Z4, Canada

Rebecca M. Quinones
Center for Watershed Sciences, University of CaliforniaDavis, 1 Shields Ave, Davis, CA 95616, USA

Theodore E. Grantham
Center for Watershed Sciences, University of CaliforniaDavis, 1 Shields Ave, Davis, CA 95616, USA

Brett N. Harvey
California Department of Water Resources, 3500 Industrial Blvd, West Sacramento, CA 95691, USA

Joseph D. Kiernan
Southwest Fisheries Science Center, National Oceanic and Atmospheric Administration, La Jolla, CA 92037, USA

Mick Klasson
2814 Tiber Ave, Davis, CA 95616, USA

Alpa P. Wintzer
Center for Watershed Sciences, University of CaliforniaDavis, 1 Shields Ave, Davis, CA 95616, USA

Peter B. Moyle
Center for Watershed Sciences, University of CaliforniaDavis, 1 Shields Ave, Davis, CA 95616, USA

Trina Rytwinski
Fish Ecology and Conservation Physiology Laboratory, Department of Biology, Carleton University, Ottawa, ON, Canada
Canadian Centre for Evidence-Based Conservation and Environmental Management, Institute of Environmental Sciences, Carleton University, Ottawa, ON, Canada
Department of Biology and Institute of Environmental Science, Carleton University, Ottawa, ON, Canada

Dirk A. Algera
Fish Ecology and Conservation Physiology Laboratory, Department of Biology, Carleton University, Ottawa, ON, Canada
Canadian Centre for Evidence-Based Conservation and Environmental Management, Institute of Environmental Sciences, Carleton University, Ottawa, ON, Canada

Jessica J. Taylor
Fish Ecology and Conservation Physiology Laboratory, Department of Biology, Carleton University, Ottawa, ON, Canada

Canadian Centre for Evidence-Based Conservation and Environmental Management, Institute of Environmental Sciences, Carleton University, Ottawa, ON, Canada

Karen E. Smokorowski
Great Lakes Laboratory for Fisheries and Aquatic Sciences, Fisheries and Oceans Canada, Sault Ste. Marie, ON, Canada

Joseph R. Bennett
Canadian Centre for Evidence-Based Conservation and Environmental Management, Institute of Environmental Sciences, Carleton University, Ottawa, ON, Canada
Department of Biology and Institute of Environmental Science, Carleton University, Ottawa, ON, Canada

Philip M. Harrison
Fish Ecology and Conservation Physiology Laboratory, Department of Biology, Carleton University, Ottawa, ON, Canada
Department of Biology, University of Waterloo, Waterloo, ON, Canada.

Steven J. Cooke
Fish Ecology and Conservation Physiology Laboratory, Department of Biology, Carleton University, Ottawa, ON, Canada
Canadian Centre for Evidence-Based Conservation and Environmental Management, Institute of Environmental Sciences, Carleton University, Ottawa, ON, Canada

Jens Wollebæk
The Norwegian School of Veterinary Science, Dep. of Basic Sciences and Aquatic Medicine, Box 8146, Dep. 0033 Oslo, Norway
Telemark University College, Dep. of Environmental Sciences, Hallvard Eikas Plass, 3800 Bø i Telemark, Norway.

Jan Heggenes
Telemark University College, Dep. of Environmental Sciences, Hallvard Eikas Plass, 3800 Bø i Telemark, Norway.

Knut H Røed
The Norwegian School of Veterinary Science, Dep. of Basic Sciences and Aquatic Medicine, Box 8146, Dep. 0033 Oslo, Norway

Ki Won Jung
Pacific Northwest National Laboratory, Richland, WA 99352, USA

Z. Daniel Deng
Pacific Northwest National Laboratory, Richland, WA 99352, USA

Jayson J. Martinez
Pacific Northwest National Laboratory, Richland, WA 99352, USA

David R. Geist
Pacific Northwest National Laboratory, Richland, WA 99352, USA

Geoffrey A. McMichael
Pacific Northwest National Laboratory, Richland, WA 99352, USA

John R. Stephenson
Pacific Northwest National Laboratory, Richland, WA 99352, USA

Peter J. Graf
Grant County Public Utility District, Ephrata, WA, USA.

Gulab D Khedkar
Paul Hebert Centre for DNA Barcoding and Biodiversity Studies, Dr. Babasaheb Ambedkar Marathwada University, Aurangabad 431004, India
Department of Cell and Molecular Biology, University of Hawaii, 1960 East-West Rd, Honolulu, HI 96822, USA

Rahul Jamdade
Paul Hebert Centre for DNA Barcoding and Biodiversity Studies, Dr. Babasaheb Ambedkar Marathwada University, Aurangabad 431004, India

Amol Kalyankar
Paul Hebert Centre for DNA Barcoding and Biodiversity Studies, Dr. Babasaheb Ambedkar Marathwada University, Aurangabad 431004, India

Anita Tiknaik
Paul Hebert Centre for DNA Barcoding and Biodiversity Studies, Dr. Babasaheb Ambedkar Marathwada University, Aurangabad 431004, India

Tetsuzan Benny Ron
College of Tropical Agriculture and Human Resources, Department of Human Nutrition, Food and Animal Sciences, University of Hawaii, Honolulu, HI, USA

David Haymer
Department of Cell and Molecular Biology, University of Hawaii, 1960 East-West Rd, Honolulu, HI 96822, USA

Kari Stange
Environmental Policy Group, Wageningen University & Research, Postbus 81306700, EW, Wageningen, The Netherlands

Judith van Leeuwen
Environmental Policy Group, Wageningen University & Research, Postbus 81306700, EW, Wageningen, The Netherlands

Jan van Tatenhove
Environmental Policy Group, Wageningen University & Research, Postbus 81306700, EW, Wageningen, The Netherlands

Lina M. Saavedra-Díaz
Department of Biology, University of Magdalena, Carrera 32 No. 22 – 08, Santa Marta, Magdalena, Colombia

Robert Pomeroy
University of Connecticut-Avery Point, Agricultural and Resource Economics, Room 380 - Marine Science Building, 1080 Shennecossett Road, Groton, Connecticut 06340-6048, USA

Andrew A. Rosenberg
Center for Science & Democracy, Union of Concerned Scientists, 2 Brattle Square, Cambridge, Massachusetts 02138, USA

Tatenda Dalu
Zoology and Entomology, Rhodes University, Grahamstown 6140, South Africa

P William Froneman
Zoology and Entomology, Rhodes University, Grahamstown 6140, South Africa

Magdalena Bełdowska
Institute of Oceanography, University of Gdansk, Pilsudskiego 46, 81-378 Gdynia, Poland

Agnieszka Jędruch
Institute of Oceanography, University of Gdansk, Pilsudskiego 46, 81-378 Gdynia, Poland

Joanna Słupkowska
Institute of Oceanography, University of Gdansk, Pilsudskiego 46, 81-378 Gdynia, Poland

Dominka Saniewska
Institute of Oceanography, University of Gdansk, Pilsudskiego 46, 81-378 Gdynia, Poland

Michał Saniewski
Institute of Meteorology and Water Management - National Research Institute, Maritime Branch, Waszyngtona 42, 81-342 Gdynia, Poland

Thaıs Dalzochio
Graduate Program in Environmental Quality, Feevale University, RS 239, 2755, Novo Hamburgo, RS CEP 93352-000, Brazil

Gabriela Zimmermann Prado Rodrigues
Graduate Program in Environmental Quality, Feevale University, RS 239, 2755, Novo Hamburgo, RS CEP 93352-000, Brazil

Ismael Evandro Petry
Laboratory of Comparative Histology, Feevale University, RS 239, 2755, Novo Hamburgo, RS CEP 93352-000, Brazil

Gunther Gehlen
Graduate Program in Environmental Quality, Feevale University, RS 239, 2755, Novo Hamburgo, RS CEP 93352-000, Brazil

Luciano Basso da Silva
Graduate Program in Environmental Quality, Feevale University, RS 239, 2755, Novo Hamburgo, RS CEP 93352-000, Brazil

Jonathan D Midwood
Fish Ecology and Conservation Physiology Laboratory, Department of Biology, Carleton University, 1125 Colonel By Dr., Ottawa, ON, Canada

Jason D Thiem
Fish Ecology and Conservation Physiology Laboratory, Department of Biology, Carleton University, 1125 Colonel By Dr., Ottawa, ON, Canada

Peter Klimley
Biotelemetry Laboratory, Department of Fish, Wildlife and Conservation Biology, University of California at Davis, Davis, CA, USA

Martyn C Lucas
Aquatic Animal Ecology Research Group, School of Biological and Biomedical Sciences, Durham University, Durham, UK

Eva B Thorstad
Norwegian Institute for Nature Research, Trondheim, Norway

John Eiler
Auke Bay Laboratories, Alaska Fisheries Science Center, National Marine Fisheries Service, National Ocean and Atmospheric Administration, Juneau, AK, USA

Chris Holbrook
Hammond Bay Biological Station, United States Geological Survey, Millersburg, MI, USA

Brendan C Ebner
Tropical Landscapes Joint Venture, CSIRO Ecosystem Sciences & TropWATER, James Cook University, Atherton, QLD, Australia

Steven Emmerson Orchard
Sustainability Research Institute, School of Earth and Environment, University of Leeds, Leeds, West Yorkshire LS2 9JT, UK

Lindsay Carman Stringer
Sustainability Research Institute, School of Earth and Environment, University of Leeds, Leeds, West Yorkshire LS2 9JT, UK

Claire Helen Quinn
Sustainability Research Institute, School of Earth and Environment, University of Leeds, Leeds, West Yorkshire LS2 9JT, UK

Roger J. H. Herbert
Faculty of Science and Technology, Department of Life and Environmental Sciences, Bournemouth University, Talbot Campus, Fern Barrow, Poole, Dorset BH12 5BB, UK

John Humphreys
Faculty of Science and Technology, Department of Life and Environmental Sciences, Bournemouth University, Talbot Campus, Fern Barrow, Poole, Dorset BH12 5BB, UK

Clare. J. Davies
Faculty of Science and Technology, Department of Life and Environmental Sciences, Bournemouth University, Talbot Campus, Fern Barrow, Poole, Dorset BH12 5BB, UK

Caroline Roberts
ABP Marine Environmental Research Ltd., Quayside Suite, Medina Chambers, Town Quay, Southampton, Hampshire SO14 2AQ, UK

Steve Fletcher
Centre for Marine and Coastal Policy Research, Plymouth University, Plymouth PL4 8AA, UK
United Nations Environment Programme-World Conservation Monitoring Centre, Huntingdon Road, Cambridge CB3 0DL, UK

Tasman. P. Crowe
Earth Institute and School of Biology and Environmental Science, Science West, University College Dublin, Belfield, Dublin 4, Ireland

Hui Liu
Yellow Sea Fisheries Research Institute, Chinese Academy of Fishery Sciences, Qingdao, China

Jilan Su
Second Institute of Oceanography, State Oceanic Administration, Hangzhou, China

PREFACE

Aquatic environments have an important role in the maintenance of life on Earth, occupying approximately 2/3 of the land area and represent large portion on primary productivity of the planet.

Currently, according to the FAO (2016), the total production of aquatic organisms by-catch (including fish, molluscs, crustaceans, etc.) represents 83.5 million tonnes in the year 2014, with a stable production for some years. On the other hand, aquaculture has grown enough in recent years. In the year 2006 the total output of aquatic organisms through cultivation was approximately 45 million tonnes in 2014 the production over 73 million tonnes.

In this sense, aquaculture is becoming increasingly important and critical to maintaining the productivity required by the demand to supply the world market, in addition to serve as a mechanism for the production of organisms affected by overfishing. It is clearly now that the worldwide demand for seafood will continue to rise. Increased production from aquaculture may meet some of this demand (Kennelly and Broadhurst 2002).

Several improvements and has been employed in aquaculture, seeking to reduce the environmental impacts that the activity generates. The use of water and pollution are among the main problems that activity has been showing from an environmental point of view. In this way, it is extremely important to employ management techniques that minimize such impacts while maintaining productivity. To develop an efficient production of food is essential to the understanding of the processes of assimilation and nutritional demand, in addition to handling and appropriate use.

So, it is necessary to search for increasingly apply management techniques that integrate different species with productive potential and bioremediation of environment.

Editor

SECTION I

ECOLOGY AND MANAGEMENT OF AQUATIC ENVIRONMENTS

CHAPTER

1

ECOLOGY AND MANAGEMENT OF AQUATIC ENVIRONMENTS

Bruno Augusto Amato Borges

Aquaculture Engineer, Florianopolis, Brazil

MANAGEMENT, MONITORING AND CONSERVATION OF AQUATIC ENVIRONMENTS: AN ECOLOGICAL APPROACH

Aquatic ecosystem is everyone who presents for biotope some body of water such as oceans, rivers, oceans, lakes, swamps, etc. Aquatic environments have an important role in the maintenance of life on Earth, occupying approximately 2/3 of the land area and represent large portion on primary productivity of the planet. It is extremely important to seek alternatives to maintenance and conservation of these ecosystems, which are severely affected by pollution from human activities.

To establish policies and methodologies of environmental management and conservation is essential to monitor local conditions so that they can be drawn the appropriate strategies. Environmental monitoring involves a process of data collection, study and continuous and systematic monitoring of environmental variables, with the objective of identifying and assessing qualitatively and quantitatively the conditions of natural resources at any given time, as well as trends over time.

In addition to environmental conservation, the preservation of aquatic ecosystems is extremely important to ensure the economy and survival of families and towns that depend on riparian activities to survive.

The recovery of lakes, reservoirs, rivers and flooded areas demand a set of integrated actions involving the watershed, the aquatic ecosystem and its components, physical, chemical and biological. The exploitation of water resources for energy, biomass production and irrigation, water supply to the major urban centers demand a strong articulation between research and scientific knowledge accumulated, and the actions of management and engineering.

FISHERY MANAGEMENT

Currently, according to the FAO (2016), the total production of aquatic organisms by-catch (including fish, molluscs, crustaceans, etc.) represents 83.5 million tonnes in the year 2014, with a stable production for some years. On the other hand, aquaculture has grown enough in recent years. In the year 2006 the total output of aquatic organisms through cultivation was approximately 45 million tonnes in 2014 the production over 73 million tonnes.

Thus, it is easy to observe that fish stocks are collapsing, because over time has been increased fishing efforts getting the same productivity. In addition, over the last century there was a great development in the technologies employed in the processes, which possibly caused worldwide overfishing.

Several works coming fishing management failures pointing caused the crisis in which the industry is currently inserted. According to Beddington et al., (2007), the main issue to be worked to restore fishing stocks affected is the management. It is important to adopt an approach in which the authorities provide incentives for conservation-based fisheries, in addition to the establishment of harvest strategies.

In this sense, aquaculture is becoming increasingly important and critical to maintaining the productivity required by the demand to supply the world market, in addition to serve as a mechanism for the production of organisms affected by overfishing.

It's a big responsibility, and should be crafted in a responsible way and planned to not repeat the same mistakes of yesteryear. It is clearly visible that the population generally comes looking and investigating increasingly on environmental and social aspects that involve several processes in the industry. I believe this is the way for the re-establishment of the fishing industry.

In research conducted by Birkeland & Dayton (2005) proposed greater focus on the protection of the larger or older individuals of a population, seeking to ensure reproductive success in spawning and subsequent recruitment. However, you can secure the growth and conservation of these individuals, Berkeley et al. (2004) concluded that booking areas should be established, with the aim to conserve the populations inhabiting the particular ecosystem.

The main current discussion with regard to the development and production of fisheries is the concern with the environment coupled with the search for new technologies. So grows the need to discuss the situation and seek improvements aimed at the sustainable development of the activity and prosperity for future generations. This way starts a different approach regarding the ecosystem to fisheries production.

The term Ecosystem approach to fisheries (EAF) is presented by Ward et al. (2002) as "an extension of conventional fisheries management recognizing more explicitly the interdependence between human well-being and ecosystem health and the need to maintain ecosystems productivity for present and future generations, conserving critical habitats, reducing pollution and degradation, minimizing waste, protecting endangered species". This management approach seeks to understand the existing relations and their characteristics within the environment, with the objective of improving the management and use of resources by activity. In addition it is extremely important to consider the impact that fisheries production has been producing in each ecosystem and globally, so that decisions involving changes in management can be evaluated.

Currently, we can observe numerous advances in fishing technology. This new technology makes possible greater efficiencies, mainly for commercial trawlers. The increase in productivity reduces the price of fishing fish on the market, generating later less profit for the fisherman. That way, the search for increased productivity becomes an endless cycle.

It is clearly now that the worldwide demand for seafood will continue to rise. Increased production from aquaculture may meet some of this demand (Kennelly and Broadhurst 2002).

According to Agboola (2014), is extremely important to understand how technological advances affect natural aquatic resources and how ecosystems have been modified over time. That way, several strategies have been considered in the context of environmental change at the global level.

So, come show my appeal related to good and responsible management strategies involving fisheries stocks. Evaluate possible ways to minimize the impacts on wild populations is critical to maintaining the environment and restoration of stocks already achieved by human action.

ECOLOGY AND MANAGEMENT OF MIGRATORY FISH IN DAMS

With the development and acquisition of energy through the construction of hydroelectric plants in the last century, there was a big problem regarding the recruitment and maintenance of the fish species that inhabit these rivers. The main affected species are the migratory fish's that depend on large stretches of rapids present for reproductive success.

The reproductive strategy of migratory fish migration involves ascending and descending along the river, and various habitats to get success, since spawning sites to power homes to larvae. With the intensification of construction of dams such environments become increasingly restricted. To minimize this impact, mechanisms were created by the dam fish passage, among them ladders, stairs, doors, elevators, among others. But your real efficiency is still doubtful, because only solved the migration upstream of the dam, without worrying about the downward migration of eggs and larvae, which is extremely important for success in spawning and subsequent recruitment. Some studies have been made in the search for understanding of the real needs of the fish ladders. For example, on a stretch of the river that does not have good locations to spawn upstream of the dam, it doesn't make any sense to build such structures. Therefore, before constructing a fish passage structure must be analyzed several factors to ensure your efficiency.

In addition, these mechanisms of fish passage are extremely selective when it comes to success in the process of ascent or descent. Agostinho et al. (2007) conclude in a study that the selectivity present in these fish passage systems is the major bottleneck in the process of translocation of fish in order to preserve natural stocks and biodiversity in rivers where dams are present.

A lot of studies have been conducted to evaluate recruitment and reproductive success of migratory species (Agostinho et al., 2002; Esguícero & Arcifa, 2010; Pelicice & Agostinho, 2008; Agostinho et al. 2007). Evaluation studies of the effectiveness of fish ladders still limited to upward migration, checking to see if she is being used and by whom. The next step is to analyze the viability of eggs and survival of larvae after the down ward migration to verify the need for mechanisms to aid the descent of eggs and larvae. To work around this problem many studies still need to be made because the subject is extremely complex and depends on many variables for it to work. The design of buses must be preceded by an accurate survey on the existing species and habitats available on-site, such as feeding and breeding sites.

Currently, stocking programs has been developed for several species, from the reproduction of individuals bred in captivity and release of the offspring in the natural environment. But it's not as simple as it seems, to get success in restocking and population growth of the species several factors should be analyzed in addition to the reproduction and release of juveniles. Bell (2004) relates that when fetching start restocking programs, not just possess production technology and release the juveniles in the environment.

These programs create large amount of offspring from crossing a few breeders, causing significant population level effects related to diversity and genetic variability of the species. According to Rourke,et al. (2009) there are several genetic implications involved in stocking programs of freshwater fish, mainly related to the reduction of the effective population size. In work done by Hansen et al. (2000), were analyzed different wild populations of brown trout submitted to stocking programs. The results demonstrated a significant reduction of effective population size in two of the three populations analyzed.

In addition, it is essential to the understanding of the reproductive biology of the species for which they are intended for fish passages.

Evaluate the homes of spawning and nursery is necessary to protect the important sites for the recruitment success. Still, it must be considered that the fish can seek alternative routes in the downstream migration, seeking different locations to spawn. In work done by Angelo et al. (2007), it was concluded that for the maintenance of migratory species which occupy with presence of buses, should be studied and preserved of reproduction and spawning areas.

The creation of a reservoir has as a consequence the appearance of a lotic stretches on ecosystem, which changes over time, with gradual replacement of species. At a certain age that environment will be colonized by aquatic macrophytes to a greater or lesser extent.

Therefore, it is extremely important to adopt a plan for the use and operations of the reservoirs, in order to minimum harm the local and adjacent ecosystems, in addition to the organisms that inhabit it. Moreover, it is essential to the establishment of policies that assess the implementation of these projects from the point of view of social, environmental and economic.

MANAGEMENT OF MACROPHYTES

Reservoir construction causes changes in bodies of water, among reduction in current velocity and increases in shoreline development, water transparency, and sediment stability (Tundisi *et al.*, 1993). These changes can lead to the rapid development of different species and ecological types of aquatic macrophytes (Pieterse & Murphy, 1990).

Aquatic macrophytes are photosynthetic vegetables visible to the naked eye, entirely or partly submerged in fresh or brackish water for several months every year, or still floating in the same (Irgang & Gastal Jr.1996). The sharp growth of aquatic macrophytes in reservoirs comes increasingly interesting to the scientific community and to the hydroelectric projects to seek appropriate solutions to the damage caused by them on the environment.

In General, for lakes, reservoirs or ponds for aquaculture, there are management features when the objective is the control of macrophytes, which are the control of nutrients or the aquatic plants.

The control of nutrients (especially phosphorus) has the goal reduce algal growth, and can also control vascular plants, but not controlling plants with roots. Nutrients can be controlled at the level of river basin or lake itself, but if not done periodically a control at the level of the basin, the control in the Lake will bring only temporary solutions. Direct control of aquatic plants is normally run as maintenance, but some cases are more severe and persistent.

There are different management techniques that are used to reduce and/or remove the entire biomass of macrophytes of water bodies: controlling point sources of pollution, transshipment and dilution techniques, precipitation of phosphorus, artificial circulation and aeration, bacterial additives, dredging, among others. The mentioned methods have been applied successfully in various aquatic ecosystems of reduced size, but your use in large environments still remains a challenge. In addition, there are the biological and chemical methods to remove macrophyte from water, beyond herbicides use or insert of predators, respectively.

According to Nichols (1991), understanding macrophyte is important to solve problems of managing aquatic plants and aquatic ecosystems, and a thorough knowledge of macrophyte biology makes the development of new management techniques, the efficacy of present techniques, and the assessment of environmental impacts more efficient.

Knowing this can be seen clearly that the issues related to management and control of aquatic plants involve more than just management. Must be invested efforts to better understand the biology of species and the mechanisms for removing from the environment. It's not a simple problem to solve, but currently showing high demand knowledge comes in this area of study.

INTEGRATED MULTI-TROPHIC AQUACULTURE (IMTA): A SUSTAINABLE MANAGEMENT OF AQUATIC ENVIRONMENTS

Currently, there is a growing concern related to conservation and responsible use of natural resources. In General, the search for environmental conservation comes followed by the evolution of the knowledge about the importance of the environment for the development and welfare of mankind. Thus, various research has been developed in order to use resources with minimal waste as possible and by reusing and cycling of nutrients involved in aquaculture processes.

The term sustainability is gaining enough space over the last decade and has been inserted into some new ventures in a variety of ways; claims for the use of engineering tools to development of a more possible sustainable aquaculture have increased over the past few years (Chopin et al., 2008).

Several improvements and has been employed in aquaculture, in seeking to reduce the environmental impacts that the activity generates. The use of water and pollution are among the main problems that activity has been showing from an environmental point of view. In this way, it is extremely important to employ management techniques that minimize such impacts while maintaining productivity.

Currently, there is a growing quest for more sustainable crops that uses the available resources in a responsible manner with minimal waste. Various researches has been made in the pursuit of knowledge and improvements regarding the integrated more than one specie on cultivation, named Integrated Multi-Trophic Aquaculture (IMTA), system that reduces the environmental impact and waste of nutrients from feed and feces (Samocha et al., 2015; Matos et al., 2005; Hadley et al., 2015; Abreu et al., 2013; Pereira & Abreu, 2012). The main proposal of this cultivation system is bio mitigation and diversification of products, seeking benefits in relation to the environment, society and the economy.

The functioning of this cultivation system is based on the use of individuals who meet a certain role in the maintenance and quality of the environment. In most cases, there are used species that require external feed supply (fish, shrimps) together with filtrating (shellfish, marine plants) species that use these organic compounds for growth. According to Troell et al. (2005), the main purpose of IMTA systems is increasing long-term sustainability and profitability of aquaculture, as the wastes of one crop (fed animals) are converted into particulate matter – macro and microalgae – that can be used to feed low trophic level animals that can in turn be sold on the market. The concept of IMTA has been applied throughout human history. In China, microalgae have been used for millennia in integrated systems, alongside with macroalgae and cyanobacteria. In such systems, rice is cultured with fish and algae (Neori et al., 2004), potentially increasing production by 8 to 47% (Kangmin, 1988). According to Bosma & Verdegem (2011), the most effective solution is to intensify ecological and technological research on aquaculture for sustainable development, through optimizing aquaculture structure and practicing IMTA system to reduce organic and inorganic pollution. Barrington (2009) related that this combination provides a sustainable environmental balance and biomitigation, economic stability due to diversification of produced species, and has high social acceptability, mainly related to good management practices.

A few years here, has been discussed and increasingly used the concept of best management practices (BMP) in aquaculture. BMP consists in the realization of the various production processes in a responsible manner taking into account all aspects. Best management practices in feeding, in the treatment of effluents, in the choice of the area, in the management of harvest, in the settlement, are some examples of processes that can be improved and standardized in search of a responsible and sustainable aquaculture.

According to Chopin et al. (2001), in its search for best management practices, the aquaculture industry should develop innovative and responsible practices that improve its efficiency and provides diversification, while ensuring the remediation of the consequences of its activities to keep in the health of coastal waters. Therefore, a different approach to aquaculture has been developed, with a focus on ecosystem. Ecosystem approach to aquaculture was defined by FAO (FAO, 2008) during your workshop in 2007 as "a strategy for the integration of the activity within the wider ecosystem such that it promotes sustainable development, equity, and resilience of interlinked social-ecological systems". This concept is increasingly widespread and applied by producers and researchers, mainly because of the current need to protect the ecosystems affected by industrialization and human action.

Some study comes showing the effective potential of some species of seaweed in the nitrogen reduction, biofiltration, among others uses. Abreu et al. (2009) reported in your work several benefits of *Gracilaria* in IMTA systems, with the ability to reduce the input of inorganic nitrogen from a 1500 tonnes of salmon farming for each 100 hectare of *G. chilensis*.

Marinho et al. (2013) demonstrated in a work that the incorporation of 10% of the macroalgae *Ulva* spp. from IMTA systems on nile tilapia dietary didn't compromise the growth performance. In this way, in addition to bioremediation of the environment in which it is inserted, the macroalgae may return as food for species cultivated commercially.

According to Troell et al. (2009), must be invested efforts in search of different extractive species with potential for use in IMTA ventures, so this method of cultivation could be adapt to different regions.

To develop an efficient production of food is essential to the understanding of the processes of assimilation and nutritional demand, in addition to handling and appropriate use. Some performance indicators are used to evaluate these systems of production, including measurement of species diversity, abundance, colonization, and growth rate as a function of ecosystem and nutrient cycling, interactions among species, among others (Chopin et al., 2012).

So, it is necessary to search for increasingly apply management techniques that integrate different species with productive potential and bioremediation of environment. Due to the numerous benefits that the activity presents, I believe is the way to a responsible cultivation will allow evolution of activity without compromising other factors related to production.

AQUACULTURE: ZONING AND AREA MANAGEMENT

With the growth of the cultivation of aquatic organisms across the world, the need to establish suitable sites for the production that impact the environment as little as possible while providing good production indices. For that, most producing countries around the world start to be implemented zoning policies and management of areas for deployment of aquaculture enterprises.

In most cases, are taken into account various factors that may or may not define the success of the production, mainly related to biological, social and economic issues for the practice of aquaculture. In addition, evaluate the areas with the greatest potential for development of activity where it is still non-existent. That way, the zoning of aquatic environments can be used to several purposes, including prevent and control environmental deterioration at the farm; implement biosecurity measures; reduce adverse social and environmental interactions; estimates of environmental capacity, among others.

The main benefit of zoning is the control and knowledge about the production in a certain area, subject to supervision and monitoring. This is the way when fetching the evolution and development of the activity, and not a growth without control and knowledge related to the operation and environmental dynamics.

Knowing the importance of aquaculture in the worldwide production scenario of aquatic organisms, zoning and correct choice of the areas to be implemented production systems must be taken seriously, because it is the main factor that will define the success or not of an enterprise, in addition to giving due attention to environmental and social issues related to the site to be chosen.

REFERENCES

1. Abreu, M.H., Varela, D.A., Henríquez, L., et al. (2009). Traditional vs. Integrated Multi-Trophic aquaculture of Gracilaria chilensis C.J. Bird, J. McLachlan & E.C. Oliveira: Productivity and physiological performance. Aquaculture, v. 293, issues 3-4, 211-220. doi: 10.1016/j.aquaculture.2009.03.043.
2. Abreu, M.H. et al. (2013). New developments of IMTA in Europe: Fucusing on seaweed. 21st International Seaweed Symposium. Seaweed Science for Sustainable Prosperity. April, 2013. Bali, Indonesia.
3. Agboola, J.I. (2014). Technological Innovation and Developmental Strategies for Sustainable Management of Aquatic Resources in Developing Countries. Environmental Management (2014) 54:1237–1248.
4. Agostinho, A.A., Domingo, L.C.S., Fernandez, R., Suzuki, H.I. (2002). Efficiency of fish ladders for neotropical ichthyofauna. River Res. Applic., 18: 299-306.
5. Agostinho, A.A., Marques, E.E., Agostinho, C.S. et al. (2007). Fish ladder of Lajeado Dam: migrations on one-way routes? Neotropical Ichtiology, 5 (2): 121-130.
6. Agostinho, C.S., Agostinho, A.A., Pelicice, F. et al. (2007). Selectivity of fish ladders: a bottleneck in tropical fish movement. Neotropical Ichtiology, 5 (2): 205-213.
7. Antonio, R.R., Agostinho, A.A., Pelicice, F.M. et al. (2007). Blockage of migration routes by dam construction:can migratory fish find alternative routes? Neotropical Ichthyology, 5(2):177-184,

8. Barrington, K., Chopin, T., Robinson, S. (2009). Integrated multi-trophic aquaculture (IMTA) in marine temperate waters. In D. Soto (ed.). Integrated mariculture: a global review. FAO Fisheries and Aquaculture Technical Paper. No. 529. Rome, FAO. pp. 7–46.
9. Beddington, J.R., Agnew, D.J., Clark, C.W. (2007). Current Problems in the Management of Marine Fisheries *Science* 316 (5832), 1713-1716. [doi: 10.1126/science.1137362]
10. Bell, J. (2004). Management of restocking and stock enhancement programs: the need fordifferent approaches. Chapter 18. Book: Stock enhancement and sea ranching: developments, pitfalls and opportunities, p. 213. Edited by Leber,K.M., Kitada, S., Blankenship,H.L., Svasand,T. Second edition, Blackwell publishing.
11. Berkeley, S.A. et al. (2004) Fisheries sustainability via protection of age structure and spatial distribution of fish populations. Fisheries 29, 23–32.
12. Birkeland, C. & Dayton, P.K. (2005). The importance on fishery management of leaving the big ones. Trends in ecology & evolution, v.20, Issue 7, 356-358.
13. Bosma, R.H., Verdegem, M.C.J. (2011). Sustainable aquaculture in ponds: principles, practices and limits. Livest Sci 139(1–2):58–68.
14. Chopin, T. et al. (2001). Integrating seaweeds into marine aquaculture systems: a key toward sustainability. Journal of Phycology, v. 37, Issue 6, 975-986. DOI: 10.1046/j.1529-8817.2001.01137.x.
15. Chopin, T., Robinson, S.M.C., Troell, M., Neori, A., Buschmann, A.H., Fang, J., (2008). Multitrophic integration for sustainable marine aquaculture. Ecological Engineering. Vol. 3. Encyclopedia of Ecology, vol. 5. Elsevier, Oxford, pp. 2463–2475.
16. Chopin, T., Cooper, J.A., Reid, G., Cross, S., Moore, C. (2012). Open-water integrated multi-trophic aquaculture: environmental biomitigation and economic diversification of fed aquaculture by extractive aquaculture. Reviews in Aquaculture, 4, 209–220. doi: 10.1111/j.1753-5131.2012.01074.x
17. Esguícero, A.L.H. & Arcifa, M.S. (2010). Fragmentation of a neotropical migratory fish population by a century-old dam. Hydrobiologia, 638: 41-53.
18. FAO (2008). Human dimensions of the ecosystem to fisheries: an

overview of context, concepts, tools and methods. FAO Fisheries Technical Paper. No. 489, FAO, Rome (2008) 152 pp.

19. FAO. 2016. FAO yearbook. Fishery and Aquaculture Statistics. 2014. Rome, Italy.
20. Garcia, S.M., Zerbi, A., Aliaume, C., Do Chi, T., Lasserre, G. The ecosystem approach to fisheries. Issues, terminology, principles, institutional foundations, implementation and outlook. FAO Fisheries Technical Paper. No. 443. Rome, FAO. 2003. 71 p.
21. Hadley, S., Wild-Allen, K., Johnson, C., Macleod, C. (2015). Modeling macroalgae growth and nutrient dynamics for integrated multi-trophic aquaculture. Journal of Applied Phycology, 27: 901-916.
22. Hansen, M.M., Nielsen, E.E., Ruzzante, D.E., Bouza, C. Mensberg, K.D. (2000). Genetic monitoring of supportive breeding in brown trout (Salmo trutta L.), using microsatellite DNA markers. Canadian Journal of Fisheries and Aquatic Sciences, 57(10): 2130-2139, 10.1139/f00-187
23. Irgang, B.E. & Gastal JR, C.V.S. (1996). Macrófitas aquáticas da planície costeira do Rio Grande do Sul. Porto Alegre: Botânica/ UFRGS, 290 pg.
24. Kangmin, L. (1988). Rice-fish culture in China: a review, Aquaculture 71 (3), 173–186.
25. Kennelly, S.J., Broadhurst, M.K. (2002). Bye-catch begone: changes in the philosophy of fishing technology. Fish Fish 3:340–355
26. Marinho, G., Nunes, C., Sousa-Pinto, I. et al. (2013). The IMTA-cultivated Chlorophyta *Ulva* spp. as a sustainable ingredient in Nile tilapia (*Oreochromis niloticus*) diets. J Appl Phycol, 25: 1359. doi:10.1007/s10811-012-9965-3.
27. Matos, J., Costa, S., Rodrigues, A., Pereira, R., Sousa Pinto, I. (2005). Experimental integrated aquaculture of fish and red seaweeds in Northern Portugal. Aquaculture, 252, 31-42.
28. Neori, A., Chopin, T., Troell, M., Buschmann, A.H., Kraemer, G.P., Halling, C. et al. (2004). Integrated aquaculture: rationale, evolution and state of the art emphasizing seaweed biofiltration in modern mariculture, Aquaculture 231 (1–4).361–391.

29. Nichols, S.A. (1991). The interaction between biology and the management of aquatic macrophytes. Aquatic Botany, 41: 225-252.
30. Pelicice, F.M. & Agostinho, A.A. (2008). Fish-passage facilities as ecological traps in large neotropical Rivers. Conservation biology, v. 22, n.1, 180-188.
31. Pereira, R., Abreu, H. (2012). Seaweed aquaculture in on-shore IMTA systems: some limitation and future perspectives. Inter-regional network to promote sustainable development in the marine algal industry. November, 2012. San Sebastian, Spain.
32. Pieterse, A. H. & Murphy, K. (1990), *Aquatic Weeds. The Ecology and Management of Nuisance Aquatic Vegetation.* Oxford Science Publications, New York.
33. Rourke, M.L., McPartlan, H.C., Ingram, B.A., Taylor, A.C. (2009). Polygamy and low effective population size in a captive Murray cod (*Maccullochella peelii peelii*) population: genetic implications for wild restocking programs. Marine and freshwater research, 60 (8): 873-883.
34. Samocha, T.M., Fricker, J., Ali, A.M., Shpigel, M., Neori, A. (2015). Growth and nutrient uptake of the macroalga *Gracilaria tikvahiae* cultured with the shrimp *Litopenaeus vannamei* in an Integrated Multi-Trophic Aquaculture (IMTA) system. Aquaculture, 446, 263-271.
35. Troell, M., Neori, A., Chopin, T., Buschmann, A.H. (2005). Biological wastewater treatment in aquaculture more than just bacteria, WORLD Aquac. 36 (1), 27–29.
36. Troell, M., Joyce, A., Chopin, T. et al. (2009). Ecological engineering in aquaculture – Potential for integrated multi-trophic aquaculture (IMTA) in marine offshore systems. Aquaculture, 297: 1-9.
37. Tundisi, J. G.; Matsumura-Tundisi, T. & Calijuri, M. C. (1993), Limnology and management of reservoirs in Brazil. In - *Comparative reservoir limnology and water quality management*,eds. M., Straškraba; Tundisi, J. G. & Duncan, A. Kluger Academic Publishers, Dordrecht, pp.25-55
38. Ward, T., Tarte, D., Hegerl, E. & Short, K. (2002). Ecosystem-based management of marine capture fisheries. World Wide Fund for Nature Australia, 80 pp.

SECTION II

WATERSHED MANAGEMENT

CHAPTER

2

INTEGRATED WATERSHED MANAGEMENT: EVOLUTION, DEVELOPMENT AND EMERGING TRENDS

Guangyu Wang[1], Shari Mang[1], Haisheng Cai[1,2], Shirong Liu[3], Zhiqiang Zhang[4], Liguo Wang[2], John L. Innes[1]

[1] Faculty of Forestry, University of British Columbia, 2424 Main Mall, Vancouver, BC V6T 1Z4, Canada

[2] College of Land Resources and Environment, Jiangxi Agriculture University, 1225 Zhimin Ave, Qingshanhu, Nanchang 330029, People's Republic of China

[3] Key Laboratory of Forest Ecology and Environment of State Forestry Administration, Institute of Forest Ecology, Environment and Protection, Chinese Academy of Forestry, Beijing 10091, People's Republic of China

[4] Key Laboratory of Soil and Water Conservation and Desertification Combating, Ministry of Education, Beijing Forestry University, Beijing, People's Republic of China

ABSTRACT

Watershed management is an ever-evolving practice involving the management of land, water, biota, and other resources in a defined area for ecological, social, and economic purposes. In this paper, we explore the following questions: How has watershed management evolved? What new tools are available and how can they be integrated into sustainable watershed management? To address these questions, we discuss the process of developing integrated watershed management strategies for sustainable management through the incorporation of adaptive management techniques and traditional ecological knowledge. We address the numerous benefits from integration across disciplines and jurisdictional boundaries, as well as the incorporation of technological advancements, such as remote sensing, GIS, big data, and multi-level social-ecological systems analysis, into watershed management strategies. We use three case studies from China, Europe, and Canada to review the success and failure of integrated watershed management in addressing different ecological, social, and economic dilemmas in geographically diverse locations. Although progress has been made in watershed management strategies, there are still numerous issues impeding successful management outcomes; many of which can be remedied through holistic management approaches, incorporation of cutting-edge science and technology, and cross-jurisdictional coordination. We conclude by highlighting that future watershed management will need to account for climate change impacts by employing technological advancements and holistic, cross-disciplinary approaches to ensure watersheds continue to serve their ecological, social, and economic functions. We present three case studies in this paper as a valuable resource for scientists, resource managers, government agencies, and other stakeholders aiming to improve integrated watershed management strategies and more efficiently and successfully achieve ecological and socio-economic management objectives.

Keywords- Adaptive, management, Climate, change, impacts, Social-ecological, systems, analysis, Traditional, ecological, knowledge, Watershed, management

INTRODUCTION

A watershed is a topographically delineated area that is drained by a stream system—it is the total area above some point on a stream or river that drains past that point. A watershed is also a hydrological response unit, a biophysical unit, and a holistic ecosystem in terms of the materials, energy, and information that flow through it. Therefore, as well as being a useful unit for physical analyses, it can also be a suitable socioeconomic-political unit for management planning and implementation. Watersheds can vary in size from thousands of square kilometres to a small area drained by a freshet.

Watershed management is the process of organizing and guiding land, water, and other natural resources used in a watershed to provide the appropriate goods and services while mitigating the impact on the soil and watershed resources. It involves socio-economic, human-institutional, and biophysical inter-relationships among soil, water, and land use and the connection between upland and downstream areas (Ffolliott et al. 2002). In essence, it is resource management with the watershed as the basic organizing unit.

The concept of watershed management dates back to 2000 BC (Zheng 2004; Chen 2007), and it has continuously evolved and improved over time. Watershed management can broadly be defined as "the study of the relevant characteristics of a watershed aimed at the sustainable distribution of its resources and the process of creating and implementing plans, programs, and projects to sustain and enhance watershed function that affect the plant, animal, and human communities within a watershed boundary" (California Department of Conservation 2015). Through the evolution of watershed management, the practice of integrated watershed management has now become more prominent. Integrated watershed management builds upon the foundational principles of watershed management to integrate various social, technical, and institutional dimensions, as well as conservation, social, and economic objectives (German et al. 2007). This integration generates "An adaptive, comprehensive, integrated multi-resource management planning process that seeks to balance healthy ecological, economic, and cultural/social conditions within a watershed. It serves to integrate planning for land and water; it takes into account both ground and surface water flow, recognizing and planning for the interaction of water, plants, animals, and human land use found within the physical boundaries of a watershed" (Red Deer River Watershed Alliance 2015).

The integrated watershed management approach exemplifies the importance of looking at multiple uses of watershed resources, rather than simply the hydrology. It attempts to balance human and environmental needs, while simultaneously guarding ecosystem services and biodiversity (Bakker 2012). Managing watersheds in this manner allows the needs of society and the environment to be accounted for, even with increasing population pressures and demand for higher productivity and multiple uses of forests and related landscapes (Dortignac 1967). For the purpose of this paper, we define integrated watershed management as an adaptive, integrated, and multidisciplinary systems approach to management that aims to preserve productivity and ecosystem integrity regarding the water, soil, plants, and animals within a watershed, thereby protecting and restoring ecosystem services for environmental, social, and economic benefit.

Improvements to this integrative approach over the last few decades are largely accredited to the rapid development of computer science and geo-spatial technology. The integration of remote sensing imagery, geographical information systems, global positioning systems, meta-analysis approaches, and computer simulation models, as well as access to large databases have provided explicit interfaces for decision makers, communities, public interest groups, and other stakeholders to interact with each other. The development of the democratic process, as well as public participatory and outreach programs integrated with web-based technologies, have improved watershed management quality and led to sustainable watershed management.

This paper focuses on the development of watershed management, potential uses of new technologies, current issues, and the future direction of watershed management and research. It also examines three case studies from China, Europe, and Canada to evaluate their major management issues and the strategies and technologies used to overcome them.

EVOLUTION OF WATERSHED MANAGEMENT

The concept of watershed management has existed for millennia. The Atharva Veda text from 800 BC contains what may well be the first written reference to watershed management. Atharva Veda verse 19, 2.1 states that: "one should take proper managerial action to use and conserve water from mountains, wells, rivers and also rainwater for use in drinking, agriculture, industries" (Chandra 1990). In the West, Benjamin Franklin recognized the need for watershed management as early as 1790. However, watershed management as a holistic concept was not defined until the mid-20th century.

By the late twentieth century, population growth in many areas was resulting in increasing constraints on the availability of land, water, and other natural resources. Scarcity of fresh water supply, contamination of agricultural land, and polluted streams were affecting millions of lives. Currently, almost half of the countries in the world have low to very low fresh water availability. The importance of watershed mismanagement can be illustrated by the history of the Aral Sea Basin (Aladin and Potts 1992; Glantz 1999; Cai et al. 2002). Stream diversions for cotton cultivation starting in the 1950s hindered the ability of the world's fourth largest lake to keep up with the 33–36 km^3 of annual water loss through evaporation. The result has been the progressive shrinkage of the Aral Sea, with a 74 % decrease in area and 90 % decrease in volume, isolation of coastal villages, destruction of local economies, extinction of 20 fish species, and a 30- to 60-fold increase in human kidney, liver, arthritic, and bronchial diseases (Jensen et al. 1997; Nearly 2000; Small et al. 2001; Micklin 2007). This situation increased recognition of the significance of and need for a holistic, ecosystem-based, multiple-use approach to land stewardship.

Watershed management has evolved from a focus on water resource management and the hydrological cycle to the current integrated approach of managing the biological, physical, and social elements in a landscape within a watershed's boundaries (Ffolliott et al. 2002). A strong global consensus is emerging around the notion that watersheds are the best units for the management of not only water resources, but also ecosystems in general (Montgomery et al. 1995). The World Bank uses watershed management assessment approaches as the key to identifying the linkages between landscape improvements, productivity increases, and attainment of true natural resource sustainability. Their approach to watershed management extends well beyond hydrological considerations—it aims to utilize the land and resources within a watershed to obtain the desired goods and services without harming the soil and water, while recognizing the links between upstream and downstream areas (Nearly 2000). Brooks et al. (2013) adopted a similar definition that emphasized that by having a good perspective of how a watershed functions and a clear understanding of the linkages between the uplands and downstream areas, a watershed manager should be able to design long-term, sustainable solutions to human natural resource problems, and avoid disasters that can cause human suffering due to lack of water or water pollution (Nearly 2000).

These ideas have been expanded in studies of the nature of the relationships between human health and the sustainability of natural ecosystems, particularly as they relate to watersheds (Gleick 2000). However, watershed management for human health and well-being requires the ability to move beyond typical reductionist approaches towards more holistic methods (Bunch et al. 2014). Understanding these complex relationships has required the development of interdisciplinary studies of catchments to resolve complex problems (Rapport et al. 1998) with an emphasis on the links between land-use change and hydrological systems, ecosystems and human health, as well as scientific and political aspects of watershed management (Bakker 2012) and how these all relate to socio-economic development (Witten et al. 2000). Such ideologies and relationships can be extended to an examination of the links between natural resource management, rural and community development, and public and environmental health (Parkes and Panelli 2001).

In summary, integrated watershed management is the process of creating and implementing plans, programs, and projects to sustain and enhance watershed functions that provide the goods, services, and values desired by the community affected by conditions within a watershed boundary. The management is integrated and complex, including components inside (e.g., upstream, midstream, downstream) and outside the watershed, affecting both man-made and natural factors.

CONCEPTS OF WATERSHED MANAGEMENT

Many countries are now trying to place watershed management within the context of natural and human systems (Bunch et al. 2014; Sanchez et al. 2014). The World Bank and the United Nations Food and Agriculture Organization (FAO), amongst others, use watershed management approaches to assess the environmental benefits of development projects and aim to improve the provision of goods and services from watersheds including agriculture, forestry, and fisheries in a sustainable manner (Kumar et al. 2014; Sha et al. 2014). Components of watersheds such as low-lying lakes can serve as indicators for overall ecosystem health and change as they aggregate materials from the water and air (Hampton 2013) and thus are an appropriate means to assess the larger system. Watershed management recognizes that people are affected by the interaction of water with other resources and that people can influence the nature and magnitude of those interactions (Manuelli et al. 2014).

Watersheds for Sustainable Resource Management

Watersheds are natural environmental and land management units that determine the health of a nation. Poor ecosystem management within watersheds has and will result in the impaired functioning of the watershed, which in fragile environments can lead to ecosystem collapse (Eswaran and Samra 1997). Watershed management has moved from a focus on physical water and soil utilization and conservation to the integration of social, economic, and environmental development.

As part of these changes, traditional watershed management must shift its conventional focus away from wildlands to include the urban fringe and urbanized areas if it is to satisfy society's needs (Nearly 2000). Watershed management must also be capable of providing solutions to watershed problems such as plans for water augmentation or watershed restoration. That being said, watershed management can be difficult to regulate; while regulations can prevent people from undertaking harmful actions, the regulations often provide no incentive for people to take beneficial actions. As such, watershed management requires the application of the three "Ps": planning, partnership, and participation by stakeholders (Blomquist and Schlager 2005). Public–private partnerships (PPPs) are becoming a more prevalent way to improve management success; PPPs join these two sectors to work towards a common goal and enables each to benefit from the other's strengths and resources to more effectively meet management objectives (PPPIRC 2016).

Additionally, watershed management relies heavily on the science of watershed (forest/range/wildland/land use) hydrology, a branch of hydrology that addresses the effects of vegetation and land management on water quality, erosion, and sedimentation. Embedded in both watershed hydrology and management is the acknowledgment of the linkages between upstream and downstream areas and interrelationships among land use, soil, and water. With the increasing awareness that land management decisions cannot be made in isolation, the principles of watershed management are being used as the basis for many environmental and natural resource management decisions (Hebin and Ueta 2012; Karcher et al. 2013).

Means of Achieving Sustainable Watershed Management

Integrated watershed management grapples with the complexity of interactions between ecosystems and socio-economic systems, and aims to restore and sustain the health, productivity, and biodiversity of ecosystems through strategies that integrate the needs of society and the economy (Szaro et al. 1998; Einar 2010; Qi and Altinakar 2013). The concept of integrated watershed management is fairly easy to envisage, but the practice of integration is complicated. It is a dynamic process that crosses temporal-spatial spectra, jurisdictional boundaries, and social, cultural, economic, and environmental systems. In order to succeed, it must be participatory, adaptive, and experimental, involve all pertinent stakeholders, identify an appropriate balance between development and protection, and integrate all relevant scientific knowledge and user-supplied information about the social, economic, and environmental processes affecting natural resources within the watershed (Calder 1999; Yang et al. 2006; VanHouten 2014). Additionally, effective integration is largely dependent on current regional development levels and the aims of future development (Thorburn 2012; Mutekanga et al. 2013; Ozturk et al. 2013). Through monitoring of and research on ecological and socio-economic interactions and processes, integrated watershed management can remain adaptable and help to develop strategies to sustain ecosystem composition, structure, and function through policies, protocols, and practices that are based on sound research.

Adaptive management (AM) can be coupled with integrated management and can improve the ability to cope with the inherent uncertainties of managing complex, dynamic systems such as watersheds by learning from the outcomes of management implementation and adjusting future approaches accordingly (Allan et al. 2008; Porzecanski et al. 2012). For AM to be successful, it requires set timeframes for management evaluation where failures and inadequacies can be openly discussed and dealt with. Management strategies need to be continuously improved by learning from the implemented policies and remaining flexible (Raadgever et al. 2008). This requires financial support as well as the dedication of time and resolution of those involved to critically and objectively evaluate their own work (Allan et al. 2008). Without the dedication and means to identify mistakes and rectify them, AM will not be successful.

The use of Traditional Ecological Knowledge (TEK) is also a key component in successful watershed management and can be incorporated into adaptive and integrated management techniques. It is the accumulation of knowledge, experience, and values regarding the local ecosystem held by communities with a history of subsistence living that addresses the interactions among and between organisms (including humans) and their environment (Berkes et al. 2000; Olsson and Folke 2001; Ellis 2005). Traditional knowledge approaches are holistic and adaptive by nature, acknowledge ever changing environmental conditions, and are often derived with the purpose of maintaining ecosystem structure and function while providing resources to the local community—a goal similar to those of today's watershed management strategies. There are several components of TEK that may be complementary to adaptive management techniques, such as management regulations that are locally developed and enforced by the users; the flexible use of resources through rotations; innate flexibility within the approach of decision-makers to respond to environmental feedbacks; the use of an array of resources to minimize environmental risk; and the use of qualitative measurements to evaluate the direction of future management strategies (Berkes et al. 2000). However, there are several barriers to the successful incorporation of TEK into modern management strategies. These include conceptual barriers due to differing values, communication barriers arising from different languages, and political barriers resulting from the unwillingness to acknowledge traditional knowledge that opposes political or industry agendas (Ellis 2005). For instance, certain traditional practices may be preferentially selected because they conform more to Western-style governance and values. To integrate TEK in watershed management more successfully, there needs to be a shift from forcing selected elements of traditional practices into conventional strategies to making current methods more malleable so that they may be integrated with TEK.

In general, the watershed management process has 6 steps (Figure 1): (1) Survey the status of the watershed and identify its situation; (2) identify stakeholders; (3) identify interests and objectives; (4) determine the target and plan; (5) implement the plan; and (6) evaluate management success and failures, reassess objectives, and adjust the plan to improve management success. In the first step, the implementing agency, whether it is private, governmental, academic, etc., needs to survey indicators of ecosystem health to evaluate ecosystem function, such as physico-chemical (e.g., pH, water temperature, concentration of nutrients), biological (e.g., flora and fauna biodiversity, algal growth), habitat (e.g., riparian habitat species composition, degree of bank erosion), and water flow indicators (e.g., peak flow, base flow) (Queensland Government 2016). This step will help identify any ecological issues and trends, which will help inform the decision-making and planning process. Next, stakeholders need to be identified; this includes communities, people, and organizations that have an interest in the watershed and who will be influenced by management decisions and outcomes. The third step is to identify stakeholders' interests and management objectives. This information can be obtained through surveying and engagement with stakeholders and from the initial ecosystem surveys. The next step is to determine the management target and plan based on the information gathered during the ecological survey and the concerns and needs of stakeholders. For the fifth step, the management plan needs to be implemented by the responsible agency; they should oversee implementation, monitoring, and enforcement of management strategies. Lastly, throughout the life of the management plan, the implementing agency needs to assess and improve their management plan to account for changes in the environment and stakeholder needs and address management activities that are insufficiently meeting management objectives. This self-assessment and evaluation involves reflection and evaluation of the effectiveness of the management plan in meeting its objectives; development of solutions to improve management in areas where objectives are not being met; identification of new social or ecological issues related to the watershed; adjustment of management objectives where needed to more effectively meet existing and new management issues; and redevelopment of the management action plan to reflect the new and improved objectives that account for lessons learned in the self-assessment of management success. Developing site-specific management plans based on these generalized steps can lead to watershed management strategies that achieve the goals of sustainability in relation to land resource use, the ecosystem, the ecological economy, and human health and well-being.

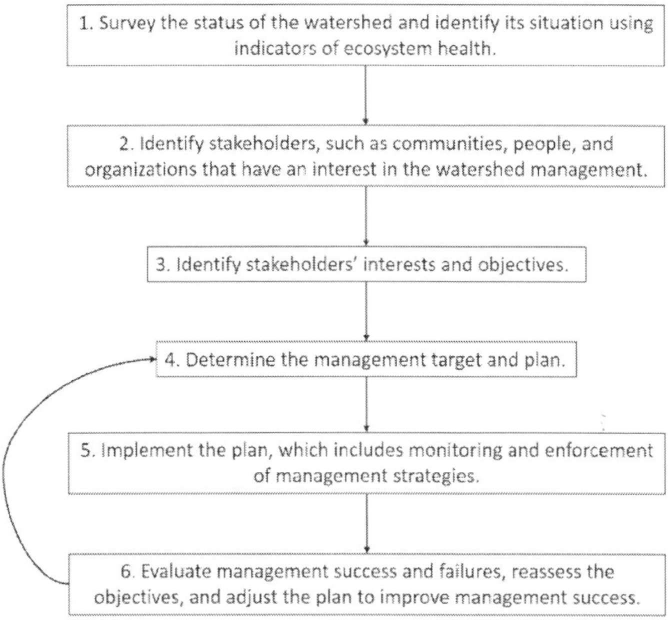

Figure 1: Conceptual model for developing an integrated watershed management plan.

In summary, watershed management provides a framework to integrate decision-making to help assess the nature and status of the watershed; identify watershed issues; define and re-evaluate short- and long-term objectives, actions and goals; assess benefits and costs; and implement and evaluate actions. Sustainable utilization and management of resources and the environment is the key to watershed management (Wagner et al. 2002; Sungjun 2007; Davenport and Seekamp 2013). Additionally, an equitable partnership between stakeholders is a crucial component to watershed management (Davenport and Seekamp 2013; Mutekanga et al. 2013; Manuelli et al. 2014). Now, sustainable watershed management has become a common process, and sustainability guidelines on watershed management are available for industrialized, newly industrialized, and developing countries to generate better, more locally specific, watershed management strategies.

TECHNOLOGIES AND TOOLS FOR MANAGEMENT OF WATERSHEDS

One of the underlying principles of watershed management is the recognition of the interrelationships among land use, soil, and water, and the linkages between upland and downstream areas. Physical changes in a watershed can result in a variety of responses ranging from short-term events, such as flooding, landslides, and point-source pollution, to long-term processes, such as soil degradation, water depletion, and non-point-source pollution. The development and assessment of watershed management, therefore, requires the integration of a vast array of spatial information and temporal data.

Remote Sensing

Remote sensing (RS) technology utilizes electromagnetic radiation reflected or emitted from Earth's surface to derive information and images about Earth's land and water surfaces (Campbell 2002; Brooks et al. 2013). This accurate and real time data source provides a means of surveying, identifying, classifying, and monitoring various components within a watershed, such as land use/cover, physiography, soil distribution, and drainage characteristics (Singh and Woolhiser 2002; Pandey et al. 2007), as well as lake temperature, water level, and algal dynamics (Hampton 2013). RS can be used alone or in combination with geographic information systems (GIS) to provide spatial input data for watershed management models. This is especially useful for remote, inaccessible areas within a watershed (Pandey et al. 2007). Additionally, utilizing RS as model input data can generate real-time information, such as seasonal snow melt, evolution of watershed management strategies for conservation planning, environmental impact assessments of water resource projects, flood damage and drought assessments, and many others (Singh and Woolhiser 2002). This technology can also assist policy makers to produce watershed management plans efficiently through the development of alternative management scenarios (Pandey et al. 2007). Overall, using remotely sensed data such as satellite imagery enables analysis across a broader spatial extent than could otherwise be achieved using in situ data alone. However, it is not a replacement for in situ measurements, as RS cannot (so far) provide the necessary detailed species data, and the relationship between satellite-derived and in situ data still requires further investigation (Hampton 2013).

Geographic Information Systems

GIS are a critical tool for watershed management, as they can be used to assess watershed conditions through modelling impacts of human activities, as well as to visualize the impacts of alternative management scenarios. These modelling and visualization capabilities are fundamental tools to understand the processes and dynamics that shape the physical, biological, and chemical environment of watersheds (Singh and Woolhiser 2002; Tim and Mallavaram 2003). Incorporating GIS into hydrological simulation models generates more spatial detail that most other hydrological models and leads to the ability to analyze combinations of slope, aspect, and hydrological-response units in the simulation (Brooks et al. 2013). As well, GIS enables the watershed to be subdivided into more discrete units vertically and horizontally, which enhances the resultant model outputs (Singh and Woolhiser 2002). GIS coupled with the Internet's increased accessibility of data and information have provided a means of overcoming the limitations of computer-based models regarding data preparation and visualization (Choi et al. 2005). There are some publicly available tools utilizing GIS such as the understanding your watershed program created by the institute of water research of Michigan State University. This tool is a web-based mapping program that can be applied to any watershed within the state (Shi et al. 2004; Michigan State University 2014). GIS's high-quality outputs, easy updating capabilities, and potential for testing management options make it a useful tool for providing management information to decision-makers (Pandey et al. 2007).

Global Positioning System

The Global Positioning System (GPS) is a freely accessible space-based global-navigational system that provides users with their precise location and time for almost anywhere on Earth (Brooks et al. 2013). The precision of data can be increased by using GPS to track the exact location of gauges, monitoring devices, study plots, etc. This information can be incorporated into database management systems or GIS to improve data quality and accuracy (Brooks et al. 2013). GPS combined with radio-tracking has proved invaluable for tracking the movement of animals, birds, and fish within watersheds, enabling detailed studies to be made of where and when particular habitats are being used.

Internet

Several government organizations, institutions, and universities have made tools, data, and educational information regarding watersheds easily accessible through the Internet. The US environmental protection agency (EPA) website (EPA 2013) has abundant information that can be used for managing watersheds or organising stakeholders on a watershed basis. They also have numerous online tools to address all stages of watershed management and planning including education and outreach tools, data collection and technical tools, resources for mapping and accessing data and models, as well as planning and management tools (EPA 2015a). Additionally, the EPA offers training for all the tools they offer (EPA 2015b) and Online Training in Watershed Management that consists of a variety of self-paced training modules for a basic to broad introduction to the field of watershed management (EPA 2015c). The Institute of Water Research offers the watershed comprehensive assessment tool (Institute of Water Research 2014), making it easy and accessible to understand the current conditions of a watershed of interest, even for those with little experience in running models and analyzing data related to watersheds. All the major components for watershed management are now offered online, which should facilitate well developed watershed management strategies.

Big Data

Big data is a popular term used to describe the exponential growth and availability of data, both structured and unstructured; it has five basic characteristics: volume, velocity, variety, variability, and complexity (Davenport and Dyché 2013; SAS 2014). Big data is beneficial as it can dramatically reduce the cost and time dedicated to computing tasks (Davenport and Dyché 2013). Large volumes of data can provide more accurate analyses that may lead to more confident and well-informed decision-making, subsequently facilitating greater operational efficiencies, cost reductions, and reduced risk. Time should be devoted to collecting detailed data that can be used to model and test alternative management scenarios of watersheds. This is not only important locally for the watershed of interest, but it is globally important that watersheds be managed sustainably and to the best of our abilities (Gizjen 2013). Data mining is a commonly used technique to make sense of big data. It uses pattern recognition technology with statistical and mathematical techniques to sift through large repositories of data to discover new correlations, patterns, and trends, as well as classify and cluster information into groups (Loucks et al. 2005; Larose 2014).

Meta-Analysis

Meta-analysis is a statistical technique for integrating quantitative findings from several studies and distilling them into broad conclusions (Joshi et al. 2005; Locatelli and Vignola 2009). Using this approach, one can collate studies that may themselves be inconclusive and generate a statistically conclusive synthesis. Although it is not a common approach for evaluating hydrology, an analysis of its usefulness indicates that it is a promising approach for integrating results from several studies (Locatelli and Vignola 2009). This comprehensive amalgamation of information can be beneficial for policy and decision-makers who are dealing with an abundance of information, often undertaken as individual studies and reviews and some of which may be conflicting. However, for meta-analysis to be successful, there needs to be sufficient data and research available to synthesize, which may be a limiting factor when using this technique for certain elements of watersheds.

Multi-Level Social-Ecological System Analysis

There have been several advances in analytical models for watershed management. These include: system dynamic modelling, watershed simulation modelling, watershed decision-making systems, stakeholder analysis modelling, and so on (e.g., Westervelt 2001; Anil et al. 2003; Davenport and Seekamp 2013; Mutekanga et al. 2013; Qi and Altinakar 2013). Additionally, watershed management models embrace the rapid advances occurring in remote sensing and satellite technology, GIS, database management systems, error analysis, risk and reliability analysis, and expert systems (Singh and Frevert 2006).

System dynamic (SD) modelling was first introduced by Forrester (1961) and is a concept based on systems thinking whereby dynamic interactions between the elements of the system are considered to reflect the behavior of the system as a whole. As the name suggests, the behavior of the system is monitored over time and is thus dynamic. The main idea of SD modelling is to understand the behavior of the system by the use of simple mathematical structures. According to Anil et al. (2003), SD concepts can help describe the system; understand the system; develop quantitative and qualitative models; identify how information feedback governs the behavior of the system; and develop control polices for better management of the system. This type of analysis is crucial for watershed management issues as a variety of components are interrelated with each other leading to complicated systems that are dynamic, interactive, and uncertain (Qi and Altinakar 2013).

Watershed simulation modelling, or hydrologic simulation, is a useful tool to achieve optimal management strategies that balance several benefits of land and water resources in a watershed. This is done through the analysis of watershed processes and their interactions and the development and assessment of management scenarios (He 2003) that simultaneously consider upstream soil conservation, midstream land use, and downstream reservoir level sediment control (Lee et al. 2013; Qi and Altinakar 2013). These models simulate the dynamic behavior of flow and storage processes and generate water balance information (quantity and associated hydraulic characteristics, source and pathway, residence time, etc.) for past, present, and future streamflow regimes (Brooks et al. 2013). Simulation modelling for watershed management allows users to generate models that are site-specific and tailored to the characteristics of the environment and anticipated changes or management strategies that will be applied over time (Westervelt 2001; Brooks et al. 2013). This makes them more useful for managers than model results published in previous research, which provide insight and information into watershed dynamics, but may not be relevant outside the context of the watershed in which those research models were run (Westervelt 2001). These types of models often need to be combined with GIS or remote sensing technologies to develop input parameters and to analyse and visualize simulation results (He 2003).

Recently, there have been many studies of watershed decision-making support systems designed for various purposes, such as water supply (Koutsoyiannis et al. 2003; Ghahraman and Sepaskhah 2004; Chung et al. 2008), soil conservation (Rahman et al. 2009; Markose and Jayappa 2016; Rejani et al. 2016), pollution (Djodjic et al. 2002; Santhi et al. 2006; Ouyang et al. 2007), sustainable resource development (Smith et al. 2003; Prodanovic and Simonovic 2010; Weng et al. 2010; Mocanu et al. 2013), the impact of land-use change (Engel et al. 2003; Mango et al. 2011), and stakeholder analysis in integrated watershed management (Luyet et al. 2012; Mutekanga et al. 2013). User-friendly decision support systems (DSS) are needed to help watershed managers and planners develop, understand, and evaluate alternative watershed management strategies, while accounting for the interests and goals of several stakeholders (Loucks et al. 2005). The DSS should integrate computer programs with components of database management systems (DBMS), GIS, simulation models, decision models, and easy-to-understand user interfaces (Miller et al. 2004). One such DSS is the multiple-criteria decision-making (MCDM) technique, whereby multiple discrete alternatives are evaluated against specified criteria to generate management strategies that meet multiple objectives (Brooks et al. 2013). However, the success of DSS developed thus far remains inconclusive, as they are often found to be poorly suited for real-world issues and face general problems such as data availability. Further development of these tools will be beneficial, as they have the potential to generate better understanding of environmental feedbacks and make use of the increased availability of watershed simulation models (Mysiak et al. 2005).

Stakeholder analysis (SA) is a holistic approach used to understand a system and assess the impacts of changes to that system through the identification of key stakeholders and their interests in the system (Mutekanga et al. 2013). SA is recognized as a suitable tool to avoid inflaming conflicts, to represent diverse interests, and to identify key stakeholders and assess their respective interests in the system. The steps involved in SA for sustainable watershed management include: (1) identification of natural resource management problems and stakeholders involved; (2) selection of key stakeholders to be involved in decision-making for integrated watershed management; and (3) workshops at community and watershed levels to formulate concrete action and work plans (Mutekanga et al. 2013).

There has been increasing recognition that public participation can lead to better management of common resources. Benefits include a better-informed public, reduced conflict amongst different users, greater democracy through greater involvement of people in decision-making, and more effective implementation of conservation measures. Public participation is a vitally important aspect of planning watershed management, but it needs to be conducted in an appropriate way to be successful (Konisky and Beierle 2001; Webler and Tuler 2001). A management plan requires the active involvement of all interested parties in developing the best approach to achieve its objectives. However, involving the public and stakeholders in decision-making requires forethought and planning as to what their involvement will be so that the interaction and outcomes are positive and beneficial. Nine steps are recommended in the development of a general public participation plan: (1) identify the watershed problem(s); (2) set project goals and objectives; (3) define the study area and pilot projects to be completed; (4) identify objectives for public involvement; (5) identify the stakeholders and interest group; (6) outline the benefits of and obstacles to public participation; (7) outline methods of public participation; (8) establish an action plan; and (9) put the plans into action. Another means of enhancing community involvement in watershed management is through the multilevel community capacity model (MCCM) for sustainable watershed management developed by Davenport and Seekamp (2013). This model addresses community conditions, characteristics, and interactions that are essential for sustainable watershed management and provides a framework for resource managers and decision makers to understand, evaluate, and build community capacity for responding to natural resource stressors or problems. It is organized in four levels: member engagement, relational networks, organizational development, and programmatic coordination.

Overall, multi-level social-ecological system analyses are required for effective watershed management and involve responsible government agencies, locally led partnerships, corporations and other institutions, and all other stakeholders. The goals and objectives of sustainable watershed management will be realized through planning, partnership, and participation by stakeholders.

CASE STUDIES: WATERSHED MANAGEMENT IN CHINA, EUROPE, AND CANADA

We have selected three watershed management case studies for the purpose of this paper—Poyang Lake basin, China; Rhine River basin, Europe; and Fraser River basin, Canada. We use these case studies to highlight the success and failure of various management strategies and different approaches to integrated watershed management under vastly different ecological, social, and political contexts. Each watershed is also ecologically and socio-economically important to the local and national economies. In each region, there were numerous ecological and social issues that needed to be addressed, and although not all have been fully resolved, these examples showcase progress through integrated watershed management to improve the quality of the ecological and social components of the watershed.

Poyang Lake Basin

Setting and Issues

Poyang Lake is situated on the south bank of the mid-lower reaches of the Yangtze River with its watershed being entirely contained within Jiangxi province. Located in a 162,250 km^2catchment, it is the largest freshwater lake in China (Figure 2). Poyang Lake has five main tributaries, Gan, Fu, Xin, Rao, and Xiu Rivers, and releases water into the Yangtze River (Shen and Wu 2004). The geographic alignment of the watershed and Jiangxi provincial boundary in principle enables more effective management to be applied to the entire basin as it eases administrative functions and conflict resolution between economic development and ecological protection and allows for greater control over industrialization and zoning throughout the watershed (Chen et al. 2011).

Poyang Lake lies within a migratory corridor for waterfowl known as the East Asian Flyway (Takekawa et al. 2010) and provides wintering grounds for 19 species of birds listed as threatened by IUCN (Ji et al. 2007). As such, the area is extremely important for migrating birds, with 98 % of the world population of the critically endangered Siberian crane (*Grus leucogeranus*) over-wintering there (Global Nature Fund 2016). Due to its importance in supporting several rare and critically endangered bird species, Poyang Lake and surrounding area is designated as a Wetland of International Importance (Global Nature Fund 2016).

Figure 2: Poyang Lake Basin, China.

Degradation in the Poyang Lake basin is linked to the rapid increase in population, which has led to the conversion of forest to grain production, land reclamation from the lake, sand-dredging, pollution, ship traffic, and over-fishing. In the early 1980s, the area impacted by water and soil erosion in the upper reaches of the Gan River reached 17,732 km^2 (Gong et al. 2006), while forest cover was reduced to 31.5 % of its original extent in the catchment (Shen and Wu 2004). Furthermore, impoundment of the Three Gorges Dam (TGD) in 2003 altered the interaction between the lake and the Yangtze River as the dam reduced river flow causing increased outflow from Poyang Lake, reducing the lake's volume, altering hydrological processes, and impacting water resources (Guo et al. 2012; Liu et al. 2013; Lai et al. 2014). As a result of this and several other factors, the surface area of Poyang Lake has been drastically reduced over the last few decades (Liu et al. 2011), its ecological functions have become compromised, and floods have occurred regularly (Liu et al. 2015). There have been several adverse effects on the ecosystem, including loss of biodiversity, loss of wetland habitat, spread of schistosomiasis (a disease caused by parasitic worms), and degradation of water quality. In addition, the degradation of the ecosystem has been accompanied by increasing poverty that has proven extremely difficult to eradicate because of the connections between the environmental state of the watershed and its economy (Huang et al. 2012).

Remediation and Management

Although management strategies for Poyang Lake have yet to resolve several issues related to environmental degradation, they provide a unique example of collaboration among local government, local communities, and international partners to develop comprehensive monitoring and research in the watershed that can be used to improve economic and ecological conditions. Several large-scale projects have been implemented in Poyang Lake basin over the last 30 years that aim to sustainably manage water resources, preserve ecosystem function, and support economic development through a holistic approach. One such program is the Mountain–River–Lake (MRL) program, developed and implemented by the Jiangxi provincial government in 1983 (Shen and Wu 2004). The MRL program aims to promote the sustainable development of the region through environmentally sound policies, integrated regional management, cooperation between agencies and organizations located along the upper and lower reaches of the watershed, and the protection of water resource based on extensive research that identifies the principal problems and cause-effect links for those problems in the watershed (Shen and Wu 2004). The MRL management strategy emphasizes the inter-dependency among the surrounding mountains, lake, tributaries, and human populations in maintaining productivity and quality of the watershed. Numerous collaborations between local universities and international organizations have been carried out to improve research and monitoring of conditions within the basin (MRL 2006) [e.g., elimination of schistosomiasis carried out with the World Bank (Yuan et al. 2000; Xianyi et al. 2005)].

MRL's research approach, level of planning, and intensive implementation led to the program being selected as a key Chinese project presented at the technical fair associated with the 1992 UN Conference on Environment and Development in Rio de Janeiro, Brazil. It was also featured at the Hannover World Expo in 2000 and at the sustainable development summit in Johannesburg, South Africa in 2002 (Shen and Wu 2004). Although there have been some setbacks in the program, such as the failure of two thirds of pilot site experiments due to the mismatch between technology and local circumstance (Shen and Wu 2004), there has also been substantial progress; the MRL had successfully increased forest cover from 33 % in the 1980s to 63 % in 2010 (Statistics Bureau of Jiangxi 2014) and improved the livelihood of about 4 million poor farmers in the area (Shen and Wu 2004). Furthermore, between 1996 and 2012, fifteen regulations and acts were implemented by the government, such as the Poyang Lake wetland protection act and Poyang Lake environmental protection act to protect water ways, wetlands, migratory birds, and biodiversity and reduce pollution in the basin.

The Poyang Lake eco-economic zone is a project developed out of the MRL and was approved by the state council in 2012. The program acts as a natural demonstration site for the development of lakes in China (Cao et al. 2012). It aims to increase the economic well-being of residents of Jiangxi, one of China's poorest provinces, while improving wetland conservation, pollution prevention, and schistosomiasis control (Adameit 2010; Cao et al. 2012). The World Bank has also developed a 5-year project in conjunction with MRL's Eco-Economic Zone program focusing on ecological economic development in small towns; this project will be carried out until 2018 (The World Bank 2016).

There are 237 natural reserves in the Poyang Lake basin operating at various levels (i.e., national, provincial, and county) that cover 1,208,581 ha (State Forestry Administration 2014). The Poyang Lake national nature reserve (PLNR), established in 1983, has made significant contributions to environmental protection and watershed management through research and management activities (Global Nature Fund 2016). In 1992, the 22,400 ha reserve was designated a Ramsar site by the Chinese government (Finlayson et al. 2010). PLNR won Best Protection Practice of 11th World Living Lake Convention in 2006 (Global Nature Fund 2016) and is the most important reserve in Poyang Lake Basin for cranes and other migratory birds. Scientific research and monitoring is carried out in the PLNR in partnership with numerous research institutions and universities in China and abroad including research related to wintering migratory birds, aquatic plants, fish populations, hydrology, and meteorology of the watershed (Global Nature Fund 2016).

Currently there are fifteen protected areas for waterfowl (Finlayson et al. 2010) and 77 wetland parks (State Forestry Administration 2014) in the basin, which reflects the importance of waterfowl protection in management activities. In 1998, the International crane foundation and PLNR initiated the Siberian Crane wildlife project, a long term study that aims to improve ecological knowledge, particularly in regards to plants and wintering waterfowl, to improve the design of conservation programs for the entire watershed and better management strategies in the basin's protected areas (Siberian Crane Wetland Project 2011). In 2010, the Poyang Lake area wintering bird and wetland joint protection committee was established to supervise and assess wintering migratory birds and wetland protection (Global Nature Fund 2016).

Management strategies have also focused on ecosystem restoration. The returning land from farming to forest program increased the forest area by 623,333 ha between 2001 and 2008 through replanting of agricultural land and afforestation of bare land (State Forestry Administration 2008). This initiative increased forest landscape connectivity, and decreased ecological risk in the project areas, which was previously high due to landscape fragmentation and exploitation associated with conversion to agricultural land (Xie et al. 2013). Reforestation has also significantly delayed the average timing of flow and reduced the duration and magnitude of flow during high flow periods, countering the effects of deforestation in the previous decades that had increased magnitude, return period, and timing of flow (Liu et al. 2015).

However, despite these efforts, there are still numerous issues in the watershed such as biodiversity loss and habitat degradation. The situation of decreasing lake area and volume and worsening water quality has yet to be successfully managed. Since the early 2000s, Poyang Lake has experienced continuous extreme low water levels and an earlier onset of the dry season (Du et al. 2014; Mei et al. 2015). This has resulted in negative social and environmental consequences, such as water shortages for irrigation and domestic use, reduction in suitable habitat for wintering migratory birds, depletion of fisheries resources, and deterioration of water quality (Finlayson et al. 2010). In 1999, the majority (58.33 %) of Poyang Lake's water ranked Grade II on a five-class grade scale. By

2009, the best quality water was Grade III (41.67 %) with 16.67 % of the water classified as Grade V, the lowest quality water class (Wu et al. 2011). Although ecosystem quality has improved in some areas of the watershed, an analysis of ecological risk across the basin revealed that the overall environmental quality of the basin has still declined, with the proportion of high level ecological risk areas increasing from 4.46 % in 1995 to 18.2 % in 2005 (Xie et al. 2013). Furthermore, although forest areas increased, grassland and wetland habitat decreased by 44.4 and 39.0 %, respectively, from 1989 to 2000 (Chen et al. 2006). Although overall vegetation area expanded from 1995 to 2013, the expansion was towards to the lake centre as a result of a reduction in lake size (Han et al. 2015). The decrease in water level has impacted wetland biodiversity, particularly aquatic plants and waterfowl, the numbers of which have dropped significantly (Finlayson et al. 2010).

A serious challenge to managing the Poyang Lake watershed is the altered flow dynamics between Poyang Lake and the Yangtze River as a result of the TGD reducing river flow in the Yangtze. Poyang Lake has naturally high seasonal variability in lake volume and area, with the outflow from Poyang Lake being controlled by the interaction with the Yangtze River (Liu et al. 2013). The reduced streamflow in the upper Yangtze River caused by the impoundment of TGD in 2003 has profoundly changed the hydrology of the lake (Lai et al. 2014); outflow from Poyang Lake to the Yangtze has increased and resulted in significant changes to Poyang Lake such as an earlier start to the dry-season (Feng et al. 2013; Lai et al. 2014; Dia et al. 2015) and reduction in lake volume and area (Liu et al. 2013; Zhang et al. 2014, 2015). Although changes in climate, such as increasing temperature and evapotranspiration, have contributed to the decrease in lake volume and area (Feng et al. 2013; Liu et al. 2013), the modifications caused by the TGD have had a much greater impact on the lake's seasonal dryness (Zhang et al. 2014). The degradation of water quality mentioned above is also partially linked to the TGD, as there was a drastic decline in quality after 2003 (Wu et al. 2011). The issues caused by the impoundment of the TGD are challenging to overcome as they result from an external factor; nonetheless these issues need to be addressed as they may lead to severe and irreversible environmental degradation.

The provincial government has proposed the Poyang Lake hydraulic project, which would involve damming the river linking Poyang Lake to the Yangtze River to control lake area and volume, increase development and management of water resources, and conserve protected areas in the watershed (Finlayson et al. 2010). The initial proposal put forward in 2002 was met with opposition and debate amongst local scientists and the international community. As a result, it was restructured in 2008 with less control of the waterway during the flooding season so that the river and lake would naturally be connected, with control implemented in the dry season to establish the lake as a water reserve (BaiduBaike 2015). However, there has still been concern related to the ecological consequences of manipulating water levels (Finlayson et al. 2010; Harris and Hao 2010; Jiang et al. 2014) in an ecosystem adapted to natural seasonal fluctuations (Han et al. 2015). The water depth is a critical component of the Poyang Lake ecosystem, as submerged vegetation needs shallow water to receive adequate sunlight and birds that feed on this vegetation, including the endangered Siberian Crane, need to access it (Barzen et al. 2009). In the most recent proposal, the controlled water depths closely mimic the natural seasonal fluctuations in Poyang Lake (Lai et al. 2015). Waterflow will be controlled starting September 1st to maintain a depth of 15.5 m until September 30th (Lai et al. 2015), which is similar to historical depths ranging from 1814 m (Zhang et al. 2015). From October 1st to the end of November, the lake level will decrease from 14 to 11 m, again tracking similar natural progressions, and from December to the end of March, the water level will be allowed to fluctuate between 10 and 11 m depending on the ecological needs of the area, but will not exceed 11 m (Lai et al. 2015; Zhang et al. 2015). Modelling shows that controlling lake level at 11 m will actually uplift the average water level from December to April (Lai et al. 2015). From the end of March to the end of August, water level will not be controlled using the dam, allowing natural interaction between Poyang Lake and the Yangtze River during the high water period (Lai et al. 2015). The proposed low water period from October to the end of March corresponds to the lake's natural low water period when birds migrate to Poyang Lake to overwinter (Zhang et al. 2015). The maximum winter water depth is recommended to be 12 m, the historic average depth (Harris and Hao 2010), as this is the optimal depth for Siberian Cranes and would prevent a large decrease in wetland habitat (Barzen et al. 2009; Jiang et al. 2014). By the end of

October the water depth would be 11.5 m and then remain no higher than 11 m until the end of March, which is below the recommended maximum of 12 m and thus should have minimal impacts on waterfowl habitat. There is still concern, however, regarding the reduction in the current rapid water exchange rate between the lake and the Yangtze River that helps maintain relatively good water quality (Harris and Hao 2010). As the lake already receives high nutrient inputs (Finlayson et al. 2010), the increased residence time during dam closure may increase turbidity and phytoplankton concentrations and cause a decline in water quality (Lai et al. 2015). This would adversely affect sensitive aquatic vegetation and the species dependent on it (Harris and Hao 2010) as well as the surrounding human population. Further investigation into the effects of the dam on factors such as water quality, biodiversity, and disease spread are essential. The Chinese government's plans to assess the dam's impact on these factors, as well as reviewing the engineering proposal for the project during the next 5-year plan (2016–2020). This will then inform a decision on whether or not to proceed with the project (BaiduBaike 2015).

Rhine River Basin

Location and Issues

The Rhine River flows from its source in Switzerland 1320 km through France, Germany, and the Netherlands to the North Sea, encompassing a catchment area of 170,000 km^2 that includes parts of Italy, Austria, Liechtenstein, Luxembourg, and Belgium (Dieperink 2000; Frijters and Leentvaar 2003) (Figure 3). As Europe's most densely navigated shipping route, it is lined with many densely populated urban areas, and several major industrial complexes and chemical production plants border it (Frijters and Leentvaar 2003). The Rhine serves a multitude of purposes, such as industrial, agricultural, waste disposal, energy generation, recreation, and drinking water, all of which differ in importance between countries. This can lead to numerous conflicts and concerns, such as problems with water quality, river ecology, and native habitat loss (Frijters and Leentvaar 2003).

Figure 3: Rhine River Basin, Europe.

Pollution in the Rhine River became increasingly problematic after the industrialization and population growth that occurred from 1850 onwards. The river was contaminated by heavy metals, pesticides, hydrocarbons, organic chlorine compounds, as well as wastewater discharge by households, industries, and agriculture. This led to severe toxicological problems in the ecosystem (Frijters and Leentvaar 2003). By the 1970s, it had become “a dead river” (Dieperink 2000, p 353). The issue of water quality was especially pertinent to the Netherlands, being the most downstream point, so an international agreement with upstream states was essential for the Netherlands to achieve its desired water quality standards (Dieperink 2000).

The Rhine basin states have dealt with several conflicts between ecological concerns and increasing development (Frijters and Leentvaar 2003). Despite a salmon treaty having been in force amongst all riparian states since 1885, the Rhine’s salmon population crashed in the 1950s due to overfishing, declining water quality (Dieperink 2000), and the emphasis on socio-economic gains through activities such as navigation and hydropower generation (Frijters and Leentvaar 2003). The prioritization of these industries increased the development of dams and weirs, preventing salmon from reaching their spawning grounds, as well as harming the ecosystem through altered river velocity and sedimentation conditions in spawning areas (Frijters and Leentvaar 2003). Moreover, the river passes through a diverse array of ecosystems, such as alluvial forests, reed plains, wooded fringes, and swamps, and development along the river has resulted in damage to some regions and habitats. Several dams and dikes have been constructed and some channels have been completely cut off along the river to control flooding, reducing its length by about 100 km (Frijters and Leentvaar 2003; Loucks et al. 2005). The reduced inundation has enabled agriculture to develop along the river, but it has also resulted in an 85 % reduction in the natural alluvial floodplains area over the last two centuries, and thus reduction in habitat for animals and plants dependent on those floodplains (Frijters and Leentvaar 2003).

Remediation and Management

An essential component to sustainably manage the ecological and socio-economic concerns in the Rhine basin has been the collaboration between upstream and downstream states. The establishment of organizations at the catchment level enabled research, information sharing, policy development, and initiatives to be cross-jurisdictional, which was necessary to address the issues encompassing the entire watershed (Dieperink 2000). There are several organizations focused on watershed and resource management related to the Rhine basin. Some of the organizations and strategies that are in place are detailed below.

The International commission for the hydrology of the Rhine basin (CHR) is a permanent, autonomous, international commission that was developed in 1970 to promote international cooperation among countries associated with the Rhine. It comprises representatives from each member state who are responsible for involving their nations' public and private sector research organizations in watershed management. It encourages increased hydrological knowledge through joint research, as well as exchange of data, methods, and information. It aims to solve cross-border issues through the management and use of information systems such as GIS and models such as those for hydrology and resource management.

This international collaboration has generated complex, uniform databases and projects that can relate to the entire Rhine basin, rather than discrete river segments. Projects have included: hydrological interests in water management and flood control; sedimentation management; hydrological forecasts and models; research into climate change; and investigations of the interactive relationships between various factors influencing hydrology within the Rhine basin (Frijters and Leentvaar 2003).

The International commission for the protection of the Rhine (ICPR) is responsible for investigating the type, source, and extent of Rhine pollution, recommending measures to reduce it, and preparing agreements between involved countries (Dieperink 2000; Frijters and Leentvaar 2003; Raadgever et al. 2008). Additionally, ICPR is responsible for any task that states jointly agree upon, such as the rehabilitation of ecosystems in 1987, and managing flood problems in 1995 (Frijters and Leentvaar 2003). However, the ICPR is only a negotiation platform and an advisor to basin governments; the implementation and funding of research and projects is the responsibility of individual states. That being said, the ICPR has been an integral part of the development and exchange of knowledge as well as negotiations between the Rhine states, which has led to the successful management of the Rhine watershed (Dieperink 2000). ICPR encourages member states to share data, cooperate in research, and exchange information, and implemented legal obligations for states to make information available at some levels, with the ICPR helping to disseminate that knowledge through its website (Raadgever et al. 2008).

After water quality in the Rhine reached an all-time low in 1971, the conventions on the protection of the Rhine against chemical pollution and chloride pollution were established by ICPR and ratified by all parties in 1976. They established legally binding threshold values for the discharge of toxic substances (Frijters and Leentvaar 2003; Dieperink 2000). Since their implementation, oxygen levels in the downstream region have rebounded, most chemicals have declined to appropriate levels, the salinity has returned to a more natural state, and some of the biodiversity has returned (Frijters and Leentvaar 2003; Dieperink 2000). They have been largely successful, although some targets have not been met. An international target of 50 % reduction of surface water phosphorous and nitrogen inputs was set; however, by the early 2000s the reduction target had not been met for N (reduced by 23 %) in the Swiss region of the Rhine (Prasuhn and Sieber 2005).

The Rhine action plan (RAP) is a flexible and adaptable plan that was introduced in 1987 (Raadgever et al. 2008) to encourage industry to continually improve technology to avoid environmental pollution and to promote ecosystem restoration (Dieperink 2000; Frijters and Leentvaar 2003; Loucks et al. 2005). Although it is not legally binding, it has been highly successful in achieving its goals. RAP included initiatives to improve the Rhine ecosystem to encourage the return of migratory and native species. Habitat for some endangered species, such as the European otter (*Lutra lutra*) and European beaver (*Castor fiber*), has been restored. The Ecological Master Plan, referred to as the "Salmon 2000″ Action Plan was implemented in 1991 to re-establish the Rhine's Atlantic salmon (*Salmo salar*) populations by the end of the decade (Dieperink 2000; Frijters and Leentvaar 2003). Fish passages were built to bypass physical barriers, while action was taken to restore the alluvial areas along the Rhine, improve the habitat in spawning grounds, reduce chemical discharge, and improve the Rhine's hydrology (Dieperink 2000; Frijters and Leentvaar 2003). Since the salmon population was almost unviable, young fish were reared and released into the river, and by the end of the twentieth century the populations of salmon and sea trout (*Salmo trutta*) had begun to rebound (Frijters and Leentvaar 2003).

The ICPR developed the "Action Plan on Flood Defense" following devastating floods of riparian cities in 1995 and 1997 (Frijters and Leentvaar 2003). Ironically, a few months after the floods in 1995, there was barely enough water for river navigation (Loucks et al. 2005). The Rhine states acknowledged the need for action, and they recognized how upstream action will impact downstream communities and agreed upon interdisciplinary and transboundary cooperation. The action plan was implemented in 1998 and established clear targets to integrate ecological improvements with greater protection of society and development against flood risk (Frijters and Leentvaar 2003).

The use of modelling has increased the understanding of how urbanization and various land-use scenarios around the Rhine influence river runoff (Hundecha and Bárdossy 2004), which compliments the development of flood management strategies. Surface runoff contributes to point-source pollution, particularly after a storm, as well as diffuse pollution from industries such as agriculture. Models, such as SWAT (soil water assessment tool) developed by Abbaspour et al. (2007), can be used to simulate the flow and transport of pollution, as well as physical characteristics such as river discharge and sediment concentration at the watershed scale. These types of tools can increase the understanding of the relationship between land use and runoff (Hundecha and Bárdossy 2004) and enhance initiatives aimed at managing the impacts of different land uses on the watershed.

The Rhine states have shown that successful international watershed management must involve moving from an "upstream–downstream" mindset to a holistic ecosystem protection approach, the use of binding yet flexible agreements, basin-wide research and data sharing, and the even, systematic distribution of information and resources to ensure that agreed measures are implemented (Bernauer 2002).

Fraser River Basin

Location and Issues

The Fraser River basin catchment covers 234,000 km^2 or 25 % of British Columbia (BC), Canada, and supports a population of over 2.9 million people (FBC 2015a) (Figure 4). Its headwaters lie in the province's central interior Rocky Mountains, where it flows 1400 km to the southernmost part of the province where it discharges into the Pacific Ocean via the Strait of Georgia (Watson 2004; Ferguson et al. 2011; FBC 2015a). The major environmental issues within the basin include ecosystem degradation and water pollution from agriculture, mining, forestry, and industrial and human development; water depletion; and biodiversity loss from habitat destruction and environmental degradation (Gustavson et al. 1999). More recently, issues related to climate change such as sea level rise, increased risk of flooding (Barron et al. 2012), rising summer water temperatures, and the potential decline in salmon stocks (Reed et al. 2011) have become major concerns.

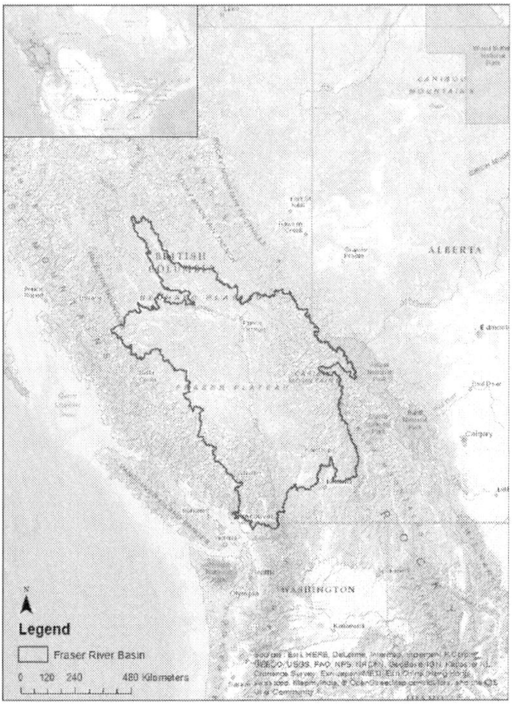

Figure 4: Fraser River Basin, Canada.

The Fraser River is the greatest salmonid-producing system in the world, with seven salmonid species living in the basin—sockeye (*Oncorhynchus nerka*), pink (*Oncorhynchus gorbuscha*), chum (*Oncorhynchus keta*), coho (*Oncorhynchus kisutch*), and Chinook salmon (*Oncorhynchus tshawytscha*); and steelhead (*Oncorhynchus mykiss*) and cutthroat trout (*Oncorhynchus clarki*) (FBC 2015a; Ferguson et al. 2011). The five salmon species are commercially harvested with an economic value of C$41.7 million annually (Ferguson et al. 2011), although this has fluctuated markedly in recent years because of government restrictions on catch limits. Additionally, there are several Indigenous communities within the basin that depend on salmon for non-economic purposes such as cultural and spiritual ceremonies and subsistence needs (Ferguson et al. 2011). Most of the salmonid species migrate up the Fraser River to reach their spawning grounds, which have been severely altered due to human development and increasing population. The gold rush of the mid-to-late ninth century led to the establishment of mines and dams, such as that in Quesnel, which decimated a run of 10 million sockeye salmon

(Ferguson et al. 2011). Dams were also constructed throughout the basin for log storage in the early twentieth century, which blocked migratory routes and destroyed eggs when the log booms were floated downstream. Despite the popularity of hydropower dams in British Columbia, there are no dams along the main stem of the Fraser River (Ferguson et al. 2011). There are however seven dams along its tributaries that have altered the downstream flows along the migratory routes of all five salmon species (Ferguson et al. 2011). One of the most devastating events was a rockslide in 1914 in Hell's Gate Canyon, triggered by railroad construction, which blocked passage to the spawning grounds of most of the sockeye salmon, as well as some Chinook and coho. This event, in combination with heavy commercial fishing pressure from the US and Canada, led to the collapse of the sockeye population around 1917. Pink salmon also declined by 75 % at this time due to fishing pressure (Ferguson et al. 2011).

Urban development and increased agricultural production along the Fraser River have also led to issues for terrestrial and aquatic wildlife in the basin. These activities have eliminated natural habitat, altered waterways, reduced diversity of shelter and food, as well as increased pollution of waterways from nutrient and pesticide runoff (Environment Canada 1998). Nutrient loading has led to eutrophication of river segments and its tributaries causing them to become temporarily anoxic, conditions that are lethal to any oxygen dependent species. Some chemicals in pesticides are taken up by aquatic species and bioaccumulate through the food chain (Environment Canada 1998), which can then carry these damaging chemicals throughout the environment as individuals move or migrate. It is not only the damage to the waterways themselves that are of concern, but also the loss and degradation of riparian areas, as these provide food, shelter, and nesting sites for many terrestrial vertebrates, and are important stopovers for migratory birds. However, the expansion of urban areas and agricultural lands, particularly in the Lower Fraser Valley portion of the basin, has eliminated most natural wetland habitat (Environment Canada 1998).

Remediation and Management

In the latter half of the twentieth century, a more integrative approach to watershed management in the Fraser River basin began to be adopted. This triggered institutional reform within BC towards the establishment of numerous organizations and management initiatives that provide opportunities for local government, First Nations, and environmental groups to participate in decision-making (Watson 2004). The federal government implemented the Fraser River action plan (FRAP) in 1991 to reduce pollution and restore habitat and fish stocks (Watson 2004). In that same year, they also proposed an integrated management approach for the Fraser River, which led to the development of the Fraser Basin management board (FBMB). Through public consultation, auditing of management activities, and review reports, they identified issues within the Basin and changes to policy and practice that were needed to develop a collaborative program for sustainable development in the watershed (Watson 2004). This eventually led to the Charter for Sustainability that outlined twelve guiding principles for economic, social, and environmental sustainability in the Fraser Basin. Sustainability principles for watershed management were also developed by the Canadian water resources association (CWRA) between 1992 and 1994 that focused on the need for integrated watershed management, consensus through negotiations, and increased public participation in decision-making (Watson 2004), while staying committed to ecological integrity and biological diversity, a dynamic economy, and social equity for present and future generations (CWRA 2015).

An essential component to the integrative watershed management approach for the Fraser River is the Fraser Basin council (FBC). The FBC is a non-profit organization established in 1997 to facilitate sustainable development, healthy waterways and watersheds, action plans on climate change and air quality, and to develop strong, resilient communities in the Fraser River Basin. It is a collaborative effort involving four orders of government—federal, provincial, local, and First Nations—along with the private sector and civil society to develop action plans and solutions collectively to issues in the Basin that account for economic, social, and environmental dimensions (Watson 2004; FBC 2015a). The FBC has played a key role in facilitating collaborative initiatives on flood management, invasive plant species, sustainable fisheries, and economic diversification in rural areas, and operating the provincial information exchange on climate change (Watson 2004; FBC 2015a). In order to monitor and evaluate the outcomes of their initiatives, the FBC developed a set of indicators, covering social, economic, and environmental dimensions, to identify areas where progress had been made and where conditions were deteriorating (Watson 2004). The results are presented in annual reports—*Sustainability Snapshot Indicators Report* (FBC 2010)—and are invaluable tools in evaluating current trends in the watershed, for examining the impact of policy, and for identifying where more attention is needed.

One of the first management initiatives in the Fraser River basin was in 1937 after the collapse of the sockeye and pink salmon populations. The International Pacific Salmon Fisheries Commission (IPSFC) was established; they proposed the development of fish passages at Hell's Gate to enable sockeye to circumvent the blockage to reach their spawning grounds. As a result, sockeye salmon began to slowly increase in the late twentieth century (Ferguson et al. 2011). Carrying on this effort to restore fish populations, the Fraser Salmon and Watershed Program (FSWP) was an initiative co-managed by the Pacific Salmon Foundation (PSF) and FBC between 2006 and 2012. It focused on watershed planning and governance, habitat restoration and stewardship, sustainable fisheries management, and education and engagement of regions and sectors throughout BC to improve the health and sustainability of wild salmon (FBC 2015b). Over 300 projects were carried out over the 6-year period, and FBC has committed to continue funding projects with a similar focus (FBC 2015b, c). Although the populations of Fraser River salmon species have risen and fallen over time, the Fraser remains of the most productive fishery systems in the world (Ferguson et al. 2011).

There has been considerable involvement by First Nations' communities in the Fraser River Basin in management and decision-making. The Fraser salmon management council (FSMC) was established to increase First Nations' involvement in decisions regarding Fraser salmon. These communities negotiate with the Canadian department of fisheries and oceans to build management agreements that consider First Nations' cultural and spiritual connections to salmon, while ensuring preservation of the fish populations (Fraser River Aboriginal Fisheries Secretariat 2015). The BC Ministry of environment also established the roundtable on the environment and the economy in 2014. It aims to bring together representatives from First Nations, communities, industry, and the environmental sector to develop environmental policy that will balance sustainable resource development and protection of the environment and human health (Watson 2004; BC Ministry of the Environment 2015). These types of organizations allow traditionally underrepresented members of the community to be involved in the management of this ecosystem.

The Fraser River Action Plan (FRAP) is a multifaceted approach to deal with numerous issues within the Fraser River Basin, including agriculture, forestry, and urban development through partnerships with local organizations and communities, and education of the citizens and industry. Several programs have been developed to address the loss of riparian and wetland habitat due to agriculture and urban expansion, as well as the pollution and contamination associated with these activities. In 1991, they partnered with local farmers, residents, and environmental organizations in Delta to promote winter cover crops to provide food for birds while reducing soil erosion. Since then, every winter thousands of hectares of farmland are planted with winter crops. The Interior wetlands program was launched in the middle and upper regions of the Fraser in 1992 in partnership with three provincial ministries and Ducks unlimited Canada—a charity for wetland conservation (Ducks Unlimited Canada 2015). The program focuses on habitat conservation, improving water quality and quantity, and sustainable agriculture through the distribution of information, workshops, and working with local landowners to improve irrigation and reduce habitat destruction by livestock (Environment Canada 1998).

In 2009, the Canadian government established the *Commission of Inquiry into the Decline of Sockeye Salmon in the Fraser River*, or the *Cohen Commission,* to identify factors contributing to the recent declines in population and productivity of Fraser River sockeye salmon. The report, published in 2012, investigated both natural and anthropogenic causes of population decline, and the current state and long-term projections for the population. Although Canada adopted the Wild Salmon Policy (WSP) in 2005 to sustainably manage fisheries and maintain genetic diversity, habitat, and ecosystem integrity for wild Pacific Salmon, the report found that little had been done to implement the policy. Furthermore, inconsistencies were found in management statistics, monitoring, and reporting between various sectors that may have impacted the accuracy of fishery management practices and statistics. Thus, the report recommended changes to the management of the sockeye salmon fishery, and suggested several procedures and programs to improve fisheries statistics. However, the government's Department of Fisheries and Oceans was under no obligation to act on the recommendations, and there has been a lack of serious action on the Commission's recommendations (Cohen Commission 2015).

Dealing with climate change related issues is a high priority in the Fraser River Basin. As such, the BC Regional Adaptation Collaborative (BC RAC) program was implemented as part of a national strategy for adapting communities to climate change. Its partnership with FBC and the BC Ministry of the Environment is focused on developing tools and resources to plan for climate change adaptation, securing water resources, minimizing water-related risks, and generating opportunities to collaborate with First Nations communities, researchers, government agencies, and resource planners. Climate and hydrological modelling are essential tools for the development of climate change mitigation and adaptation strategies. The Government of Canada and the BC Ministry of Environment have both developed hydraulic models for the lower Fraser Valley to determine changes to extreme flooding events for both coastal and inland communities. This type of information has helped communities develop long-term flood management strategies, improve the understanding of the risks of climate change, and update infrastructure such as dikes to deal with potential inundation (BC Ministry of Forests, Lands and Natural Resources Operations 2014; FBC 2015d). Models can also be used to understand how climate change will impact salmon migration and spawning (Rand et al. 2006; Macdonald et al. 2010). Understanding how the salmon populations will fair under a warmer climate is critical for decisions regarding annual allowable catch and development of strategies to facilitate salmon spawning and survival.

Although not all management initiatives in the Fraser River Basin have been successful, this system serves as an excellent example of the success of integrative management strategies and engagement of local communities and interest groups. Several initiatives are thought to provide an excellent framework for engagement of First Nations, local communities and industry in decision-making and resource management (Watson 2004; Blomquist et al. 2005; Mitchell 2005).

CONCLUSIONS FROM THE CASE STUDIES

The three case studies described here should not be viewed as unqualified successes. Each has numerous strong points, but each has also failed at some point, so there is no room for complacency. However, it is evident that success emerges when integrated watershed management is utilized to address the complexity of the social, economic, environmental, and cross-jurisdictional issues through the involvement of industry, local communities, and environmental organizations with government decision-making bodies. As well, models and technological advances can better the knowledge of the system and allow for more beneficial long-term strategies to be developed.

ISSUES IN WATERSHED MANAGEMENT

Land-use and water resource concerns are often as diverse as the different countries' cultures, economies, and stage of technical development in which they arise. To successfully develop and implement an integrated watershed management strategy under various conditions, there needs to be clear goals and objectives based on sound science that consider all elements within the watershed and how they will change temporally, while accounting for the needs and opinions of the public and stakeholders. Obstacles to this process mainly arise from the following:

LACK OF UNIFIED ORGANIZATION FOR MANAGEMENT AND CLEAR BOUNDARIES

Management authority is often decentralized with different managers controlling different areas, especially between upstream and downstream regions and between regions inside and outside the watershed. Defining the watershed can be challenging, as there is often tension between the natural contours of a watershed and social and political boundaries (Blomquist and Schlager 2005). In addition, some components of a watershed may lie within national parks or protected areas, but their effectiveness may be limited if the headwaters lie outside their boundaries (Soares-Filho et al. 2006). These factors can seriously affect the effectiveness of integrated management. Watersheds may also be nested within one another; a major watershed may contain hundreds of moderate- to small-sized watersheds that can generate issues when trying to determine at which scale the integrated watershed management should be organized, particularly if the major watershed crosses regional or national boundaries. Managing small, nested watersheds independently discourages integration, but management at a larger scale may result in localized problems being neglected with management strategies being unable to account for all the issues and interactions that occur in such a complex system (Blomquist and Schlager 2005). Small-scale management may sometimes be necessary, however, depending on restrictions imposed by watersheds crossing jurisdictions where management strategies, priorities, and objectives may greatly differ making watershed assessment and management planning difficult (Wickham et al. 2011).

Lack of Appropriate Big Data and Sharing Mechanisms for Watershed Research

The lack of standardized and available data makes it difficult to compare data from different locations (Hampton 2013), or for the same location from different sources. Statistical frameworks for integrating these data exist, but they are often unfamiliar to environmental scientists. There needs to be increased availability of adequate computational infrastructure for data sharing, analysis, and archiving, as well as training for researchers on its use (Hampton 2013). In addition, there is often a lack of control data, which can hinder the ability of researchers to classify and interpret the results of the watershed management strategies. Additionally, one cannot assume that there were similar conditions within the watershed pre- and post-implementation of management strategies, so the necessary data must be available for impact evaluations. Post-implementation satellite images should be compared to pre-implementation images from the same time of year, as even a minor difference in date can have a dramatic impact of the greenness of vegetation, especially at the beginning or end of a season (Kumar et al. 2014). Overall, adequate data are necessary to identify the main issues and establish an effective management plan. However, to use the data to their full capacity, expertise is needed in database management systems, GIS, computer operating systems, RS, data collection from the Internet, and graphics, as well as knowledge of watersheds and the processes that affect them. Few professionals have all these skills, much less the typical watershed stakeholder, so the development of these skills among management team members is critical.

Additionally, the required information is often held in many different databases by various agencies, not all of which are openly accessible (Environment Canada 2010). This system needs to be improved so that decision makers can access the information necessary and collaborate with people/institutions to properly analyze the data to study the watershed of interest and develop the most appropriate management strategies.

Lack of Interagency Communication and Cooperation and Tools for the Development and Application of Sustainable Watershed Management

Due to watersheds potentially spanning multiple regional and international jurisdictions, cooperation and coordination of management strategies can be difficult to achieve. Improving watershed management under these conditions requires better data and information-sharing among parties, an integrated management approach, greater transparency and accountability, stakeholder involvement, and clear goals. The development of a legal framework between all parties involved will help to ensure cooperation and progress towards the outlined goals and strategies (Environment Canada 2010). In addition, there has been little development regarding a set of criteria and indicators on which to base a plan for sustainable development. Some progress was made by De Steiguer et al. (2003), who proposed an Analytical Hierarchy Process (AHP), which is a Multi-Attribute Decision Method (MADM). It provides a systematic process for decision-making by comparing multiple competing criteria and alternatives, including situations involving subjective judgments, to select the best watershed plan. It provides numerical weights to alternatives, which can aid in group-decision making and allow easy comparison between competing factors.

Lack of a Long-Term Perspective and Quantitative Goals for the Whole Watershed over the Next 50 Years

There needs to be more use of GIS and simulation models in long-term research and decision-making. It is particularly important to account for the potential impacts of climate change in the long-term management of watersheds, and as such this aspect needs to be integrated into models and decision support tools.

Lack of Consideration of Public Opinion and Positive Public Participation in Watershed Management

Often, there is little involvement with the public or the main stakeholders in the watershed and little encouragement for cooperation between the different agencies. Decision makers are often evaluating several alternative management plans to determine which best serve the watershed's social and environmental goals (Blomquist and Schlager 2005). Public participation can help clarify the social goals and how they may change over time. Even in complex modelling exercises, public participation is possible but requires an informed and skilled public. The exact processes taken to include the public and develop management activities are context-dependent. However, it is increasingly clear that unilateral processes that exclude the public will fail (Heathcote 2009).

Moving Forward With Integrated Watershed Management

Watershed management needs to be a standard component in development programs that focus on water resources, forestry, agriculture, and related land and resource use. To be effective, land-use administrators, water resource managers, foresters, and agriculturalists, along with professional watershed managers, must all be involved. Integrated watershed management neither seeks nor needs a cure-all watershed model. It needs to be based on the interdisciplinary integration of the physical, biological, chemical, social, economic, and political sciences as they relate to the watershed of interest. Successful future integrated watershed management plans need to incorporate a holistic approach that accounts for environmental as well as socio-economic elements of the system, that embraces the advances in technology and modelling, and that are based on clear legal obligations that are understood by all parties involved. The future development of watershed management will be focused on the following aspects.

Holistic Sustainable Watershed Management

Integrated watershed management is a holistic approach that regards a watershed as a holistic system where social, cultural, economic, and environmental components interact and interweave together, and the principles of sustainable development are used to guide watershed management (Muschett and Campbell 1997). A key component of integrated watershed management is to balance development and protection in such a way that it is consistent with local social, economic, and environmental needs (Heathcote 2009). The integration needs to be broad and should include all aspects of watershed resources (natural, human, and political resources; science and technology) and watershed issues (economic development, water shortages, natural disasters, biodiversity, soil erosion, sedimentation, resource depletion, poverty), as well as involving multiple agencies and jurisdictions and local communities (CCICED 2005; Smit 2005; Yang et al. 2006; Heathcote 2009). To achieve this, integrated watershed management needs to evaluate and consider several elements, such as the spatial and temporal scale, the relationship between ecosystem function and structure, diversity and integrity of the system, ecosystem dynamics in space and time, and natural resource utilization and management by stakeholders. It needs to be adaptive and responsive to changing environmental, social, economic, and political aspects. This will involve revisiting the initially outlined strategies to determine if they need to be improved, and continuously incorporating new technology and knowledge into the management framework (Heathcote 2009).

Multilevel Objectives Social-Ecological System Project

The social-ecological system of a watershed has several functions, and different regions have different requirements. Those involved in watershed management need to identify the various regions and functions of the watershed in order to develop the appropriate strategies for each region. First, there is a need to try to understand the watershed ecosystem; then, via a bottom-up approach in which the local people are truly involved, alternative livelihoods that conflict as little as possible with nature should be identified. Market mechanisms could be combined with the financial leverage of government so that watershed management can better balance the benefits of all stakeholders and thus gain their support (Smit 2005, p 35). Governance approaches should include enhancing water use efficiency, environmental laws and regulation enforcement, the use of smart economics and market mechanisms, and improving public involvement (Gleick 2008). Additionally, sustainable watershed management must consider the implementation of zoning and projects for different goals and objectives. The zoning includes: land-use functional zoning, ecological function regionalization, water resource function zoning, ecological protection zones, and so on. This approach will help to achieve the goals for watershed sustainable management that include: (a) clean water and sufficient water flows and water reserves for environmental, industrial, agricultural, and urban use; (b) profitable land-using enterprises which provide regional development and food production and support resilient communities and economies; (c) terrestrial and aquatic biodiversity protection; and (d) resources that meet aesthetic, recreational, and spiritual needs.

Integrated Up-To-Date Science and Technologies

Watershed managers need to utilize a multidisciplinary approach by bringing together the developments of computer science, geo-spatial technologies, the support of big data, simulation modelling techniques, as well as optimized decision making approaches to systematically integrate various dimensions of data to develop a comprehensive management tool. This will enable managers and stakeholders to work together effectively and generate the best possible solutions to watershed management problems. To achieve this, more interdisciplinary research is needed to solve the complex integration of population, resources, environment, and development.

Innovative Watershed Management Legal Systems and Institution Structures

There is a need for an international convention promoting the development of watershed management that legally requires the cooperation between jurisdictions, inter-governmental agencies, and upland and downstream regions, with the coordinated organizations having full authority to exercise the duty to facilitate, implement, and monitor the watershed development. Such conventions have been created and implemented for specific rivers, such as the Rhine, but there are many other watersheds encompassing multiple nations that do not have this level of cooperation. A convention would provide the necessary framework for such specific agreements. There is a need to classify agency and institutional jurisdictions over watershed management activities and improve the coordination between the agencies and institutions. An improvement in inter-governmental agency communication and in the communication between stakeholders throughout the entire watershed is necessary. Future watershed management must consist of more and improved regional cooperation, community empowerment, active democracy, as well as political decision-making that involves effective public participation to solve poverty problems and ensure equity and justice for all stakeholders. As evident from the MRL program in China, each stakeholder group needs clear responsibilities, with the government adopting a leadership role, individual departments fulfilling their statutory responsibilities, adequate supervision by the environmental agency, appropriate treatment of effluents by local enterprises, and surveillance and participation by the public.

Widely Expand Public Participation

Key factors that determine the success of watershed management include the degree of public input, contribution of local knowledge, democratic decision-making processes, and whether all the stakeholders can be brought together in planning and implementing watershed development strategies. The experiences from Poyang Lake (CCICED 2005), from the International Rhine Commission (Smit 2005), Tennessee Valley Authority (USA) (Tan and Wan 2002; Yang et al. 2006; Heathcote 2009), the Fraser Basin Council (Blomquist and Schlager 2005), and from many others (USEPA 1997; Mody 2004) suggest that most problems have stemmed from centralized management approaches that failed to take into account the emphasis placed by local stakeholders on rapid economic growth (Wang 1999; Yang et al. 2006). In order to incorporate ever-changing social values and priorities, these components initially have to be known, and the plan must adapt to accommodate them over time. To achieve this, all major public and stakeholder viewpoints need to be considered, with the views of the majority reflected in the management strategies. The exact form of public involvement will depend on the current issues and the community and its values (Heathcote 2009). There must be efforts to continuously enhance public environmental education and to encourage public participation in watershed planning and monitoring (Yang et al. 2006), as well as improve the stakeholders' awareness about the importance of sustainability and to fairly distribute the outcomes associated with upland natural resource development in order to ensure the success of management strategies.

FUTURE RESEARCH

Efforts are needed in research focusing on the growing concerns over several water-related human and ecosystem vulnerabilities, such as threats to drinking water supply, loss of water-related ecosystem services due to pollution and increased water consumption, and increasing hydrological variability associated with climate change (Bakker 2012). There needs to be emphasis on research that examines watersheds across the broad range of interconnected socio-economic and environmental components. Researchers need to make use of the large data sets and technology available and do what they can to make their own data and technological advancements available to their fellow colleagues. Based on big data and advanced technologies and from macro- and micro-perspectives, comprehensive studies of watershed management need to focus on the optimization of management strategies. Research needs to be presented in a straightforward manner so that policy makers and managers can integrate knowledge into practical applications.

Accounting for the Impacts of Climate Change in Large-Scale Watershed

A key component of future watershed management will be to account for the potential impacts of climate change. Milly et al. (2008) concept that "stationarity is dead" due to anthropogenic climate change suggests that the environment is moving away from familiar patterns of hydrology and ecosystem dynamics and that previous methods of understanding ecosystems and developing long-term plans will be ineffective. Climate change may significantly affect the ecological and socio-economic components of watershed management through altered precipitation patterns, increased risk of drought/flooding, changes to volume and timing of peak flows, and other changes (Brooks et al. 2013; Khedum and Singh 2014; Singh et al. 2014). However, there is uncertainty associated with predictions of how climate change will affect the hydrology of various regions. In order to achieve integrated watershed management in the long-term, management needs to account for and be adaptable to climatic variability and weather extremes (Brooks et al. 2013). This is particularly important for developing nations and regions that are already struggling to manage water resources, as they are most vulnerable to climate change (Vörösmarty et al. 2000; Tompkins and Adger 2004; Khedum and Singh 2014; Singh et al. 2014). Developing long-term management strategies based on past climatic conditions is not sufficient (Milly et al. 2008) and will undoubtedly result in social, economic, and environmental losses due to unsuitable management strategies that could potentially lead to conflicts over water resources (Bakker 2012; Khedum and Singh 2014).

The use of tools and technology that can predict local future changes in climate is essential, as the impacts of climate change on the hydrologic system will vary across the globe and within a country (Milly et al. 2008; Singh et al. 2014). Tools such as hydrological simulation models, stochastic models (Brooks et al. 2013), GIS, and high-resolution regional climate models (Christensen and Christensen 2003) should be used to guide the development of long-term watershed management plans. Within the context of global climate change, the big data network system could lead to better analyses of the changing ecological and environmental dynamics of watersheds. It could also enable the evaluation of the influence of human and natural factors, enabling more objective evaluations of watershed condition and the implementation of more scientific management (Gizjen 2013). It is important to utilize technology and models that are downscaled (Milly et al. 2008) and regionally specific, as this enables management strategies to be tailored to the specific ecological and socio-economic needs of the watershed of interest to ensure they remain viable and contribute to mitigating climate change. However, the current tools and technologies need to be improved upon to generate more accurate projections of how the future climate will influence watershed dynamics. Some countries such as China have already dedicated themselves to increasing their capacity to cope with climate change—expanding technology research and development and improving adaptive capacity of key sectors such as water resources to climate change—and other countries need to follow suite.

A recent study by Zhou et al. (2015) highlights future research opportunities in studying how climate and watershed characteristics interactively affect hydrological responses within a watershed. It was found that when $m < 2$, where m represents watershed characteristics in Fuh's model (Fuh 1981) and the watershed is thus smaller or more simplistic, watershed characteristics can play a larger role in hydrological responses than climate change. Watershed characteristics (*m*) are positively related to forest cover and watershed areas and negatively to watershed slope. As such, land cover and land-use change should be incorporated into climate change scenarios to increase the accuracy of predicting the response of water resources to climate change (Zhou et al. 2015). If watershed characteristics are excluded, the models may miss the most important factors influencing the hydrology of the watershed, leading to inaccurate predictions and thus ineffective watershed management strategies.

There are several other strategies to better integrate future climatic variability into watershed management. These include improving watershed resilience (social and ecological); improving and sharing knowledge about climate variability and changes to water resources; incorporating climate variability into planning efforts; and developing practices that protect, maintain, and restore watershed processes and services (Tompkins and Adger 2004; Brooks et al. 2013). Flexibility in management strategies is also crucial due to the uncertainty associated with climate predictions. In addition, the environmental, social, and economic costs associated with climate change and the interactions between them need to be considered for successful future management (Brooks et al. 2013). This reiterates the importance of a holistic approach to watershed management that encompasses managing the total environment (Singh et al. 2014) and considers the interactions among land, water, natural resources, and the community and industry dependent on them (Vörösmarty et al. 2000; Brooks et al. 2013).

CONCLUSION

It is evident that watershed management has transitioned to a more holistic resource management approach, employing integrated and adaptive management strategies to account for biological, physical, and social elements within the landscape. Technological advancements have significantly contributed to this well-rounded approach. Information is now more easily shared within and across disciplines, there is improved accuracy in data collection techniques and model development, and it has become possible to develop integrated, multi-level analysis that generates more complete information about the watershed system both socially and ecologically. These improvements in watershed management and technology provide more comprehensive and multi-dimensional information for decision makers to assess the status of a watershed and implement necessary regulations. As evident in the case studies, social, economic, and environmental issues of concern are unique to each watershed. However, the use of integrated management strategies, local knowledge, cross-jurisdictional cooperation and information sharing, advanced data collection and analysis methods, and the consideration of both ecological and socio-economic concerns can remedy social issues, environmental degradation, and improve the health and management of a watershed. That being said, there is still room for improvement in watershed management strategies and research. Future watershed management should involve integrated watershed management techniques based on the latest science and technologies, as well as local knowledge and stakeholder input. Strategies should account for social and ecological needs and the potential changes associated with climate change. By developing management strategies in this way and continuing to improve research techniques and technology, environmental and social situations within watershed should continue to improve and thrive.

This paper has highlighted the importance of employing integrated watershed management strategies and has outlined numerous methods for improving management strategies, such as incorporation of new technologies and models, inclusion of holistic and adaptive management strategies, stakeholder participation, cross-jurisdictional partnership, and the consideration of climate change impacts in management planning. This evaluation of integrated watershed management and presentation of tools and strategies to improve sustainable watershed management represents an important synthesis of knowledge for scientists, stakeholders, resource managers, and government agencies. Our work can be used to improve agencies' management strategies and better the ecological and socio-economic conditions within watersheds of concern. By employing the steps outlined for developing and maintaining watershed management strategies, integrating appropriate technologies and resources, and gleaning lessons from the case studies presented, future watershed management applied to any circumstances and location may be improved and more successful achieve ecological, social, and economic management objectives.

ACKNOWLEDGMENTS

We thank the anonymous reviewers for their comments on the previous version of the manuscript, which helped us to greatly improve the quality and readability of the paper. We also would like to thank Dr. Thomas D Dahmer for his invaluable comments. This study is supported by Canadian SSHRC Standard Research Grant project entitled "Application of sustainable forest management in a culturally modified landscape" and the project "Adaptation of Asia–Pacific Forests to Climate Change" which is funded by Asia–Pacific Forests Network (APFNET/2010/FPF/001).

REFERENCES

1. Abbaspour KC, Yang J, Maximov I, Siber R, Bogner K, Mieleitner J, Zobrist J, Srinivasan R (2007) Modelling hydrology and water quality in the pre-alpine/alpine Thur watershed using SWAT. J Hydrol 333:413–430
2. Adameit M (2010) How can management influence industry and urbanization? Poyang Lake Eco-economic Zone of Jiangxi Province, China. https://www.globalnature.org/bausteine.net/file/showfile.aspx?downdaid=7226&domid=1011&fd=0. Accessed 20 Apr 2016
3. Aladin NV, Potts WTW (1992) Changes in the Aral Sea ecosystems during the period 1960–1990. Hydrobiologia 237:67–79
4. Allan C, Curtis A, Stankey G, Shindler B (2008) Adaptive management and watersheds: a social science perspective. J Am Water Resour Assoc 44(1):166–174
5. Anil K, Tangirala Ramesh SV, Teegavarapu Lindell O (2003) Modeling adaptive water quality management strategies using system dynamics simulation. Environ Inform Arch 1:245–253
6. BaiduBaike (2015) Poyang Lake water conservancy hub project. http://baike.baidu.com/subview/3313644/3313644.htm. Accessed 02 May 2016
7. Bakker K (2012) Water security: research challenges and opportunities. Science 337:914–915
8. Barron S, Canete G, Carmichael J, Flanders D, Pond E, Sheppard S, Tatebe K (2012) A climate change adaptation planning process for

low-lying communities vulnerable to sea level rise. Sustainability 4:2176–2208

9. Barzen J, Engels M, Burnham J, Harris J, Wu G (2009) Potential impacts of water control structure on the abundance and distribution of wintering waterbirds at Poyang Lake. Unpublished report submitted to hydro-ecology institute of the Yangtze water resources commission. International Crane Foundation, Baraboo, p 54
10. BC Ministry of Forests, Lands, and Natural Resource Operations (2014) Simulating the effects of sea level rise and climate change on fraser river flood scenarios—final report. http://www.env.gov.bc.ca/wsd/public_safety/flood/pdfs_word/Simulating_Effects_of_Sea_Level_Rise_and_Climate_Change_on_Fraser_Flood_Scenarios_Final_Report_May-2014.pdf. Accessed 02 March 2015
11. BC ministry of the environment (2015) Roundtable on the environment and the economy. http://www.env.gov.bc.ca/main/roundtable/. Accessed 04 March 2015
12. Berkes F, Colding J, Folke C (2000) Rediscovery of traditional ecological knowledge as adaptive management. Ecol Appl 10(5):1251–1262
13. Bernauer T (2002) Explaining success and failure in international river management. Aquat Sci 64:1–19
14. Blomquist W, Schlager E (2005) Political pitfalls of integrated watershed management. Soc Nat Resour 18:101–117
15. Blomquist W, Dinar A, Kemper K (2005) Comparison of institutional arrangements for river basin management in eight basins. World bank policy research working paper 3636, June 2005, p 46
16. Brooks KN, Ffolliott PF, Magner JA (2013) Hydrology and the management of watersheds, 4th edn. Wily-Blackwell, Ames, p 552
17. Bunch MJ, Parkes M, Zubrycki K, Venema H, Hallstrom L (2014) Watershed management and public health: an exploration of the intersection of two fields as reported in the literature from 2000 to 2010. Environ Manag 54:240–254
18. Cai X, McKinney DC, Rosegrant MW (2002) Sustainability analysis for irrigation water management in the Aral Sea region. Agric Syst 76:1043–1066

19. Calder IR (1999) The Blue Revolution: land use and integrated water resource management, 1st edn. Earthscan Publications Ltd, London, p 376
20. California Department of Conservation (2015). Watershed Program. http://www.conservation.ca.gov/dlrp/wp/Pages/Index.aspx. Accessed 23 Feb 2016
21. Campbell JB (2002) Introduction to remote sensing, 3rd edn. The Guilford Press, New York, p 621
22. Canadian water resources association (CWRA) (2015). Policies and principles. https://www.cwra.org/en/about/policies-and-principles. Accessed 02 March 2015
23. Cao L, Gemmer M, Jiang T (2012) Adaptation to climate change in China: policy, action, and progress. In: Weiguang W, Zheng G, Pan J, Luo Y, Chen Y, Chen H (eds) China's climate change policies. Social Sciences Academic Press, China, p 292
24. Chandra S (1990) Hydrology in ancient India. National Institute of Hydrology, Rookee, p 106
25. Chen SJ (2007) History of China water resource management. China Water Resources Management Publishing House, Beijing, p 134
26. Chen X, Li H, Bao Z, Wu Z, Fu W, Cai X, Zhao H, Guo P (2006) RS and GIS based study on landscape pattern change in the Poyang Lake wetland area, China. In: geoinformatics 2006: remotely sensed data and information, pp 64192E–64192E
27. Chen M, Wei X, Huang H, Lu T (2011) Poyang Lake basin: a successful, large-scale integrated basin management model for developing countries. Water Sci Technol 63(9):1899–1905
28. China council for international cooperation on environment and development (CCICED) (2005) Lessons learned for integrated River Basin management: proceedings of international symposium on integrated River Basin management. CCICED task force task on integrated River Management. China Environmental Science Press, Beijing, p 328
29. Choi JY, Engel BA, Farnsworth RL (2005) Web-based GIS and spatial decision support system for watershed management. J Hydroinform 7(3):165–174

30. Christensen JH, Christensen OB (2003) Climate modelling: severe summertime flooding in Europe. Nature 421:805–806

31. Chung G, Lansey K, Blowers P, Brooks P, Ela W, Stewart S, Wilson P (2008) A general water supply planning model: evaluation of decentralized treatment. Environ Model Softw 23(7):893–905

32. Cohen Commission (2015) Cohen commission on salmon stocks. http://cohencommission.ca/about/. Accessed 29 Sep 2015

33. Davenport TH, Dyché J (2013) Big data in big companies. Ambio 31(1):2–13

34. Davenport MA, Seekamp E (2013) A multilevel community capacity model for sustainable watershed management. Soc Nat Resour Int J 26(9):1101–1111

35. De Steiguer JE, Duberstein J, Lopes V (2003) The analytic hierarchy process as a means for integrated watershed management. In: Renard KG (eds) First interagency conference on research on the watersheds, pp 739–740

36. Dia X, Wan R, Yang G (2015) Non-stationary water-level fluctuation in China's Poyang Lake and its interactions with Yangtze River. J Geog Sci 25(3):274–288

37. Dieperink C (2000) Successful international cooperation in the Rhine catchment area. Water Int 25(3):347–355

38. Djodjic F, Montas H, Shirmohammadi A, Bergström L, Ulén B (2002) A Decision support system for phosphorus management at a watershed scale. J Environ Qual 31:937–945

39. Dortignac EJ. 1967. Forest water yield management opportunities, Pg. 579–592. In: Sopper WE, Lull HW (eds) Forest hydrology: proceedings of a National science foundation advanced seminar, The Pennsylvania State University, University Park, August 29-September 1965, Pergamon Press, Oxford, p 813

40. Du Y, Zhou H, Peng W, Liu X, Wang S, Yin S (2014) Modeling the impacts of the change of river-lake relationship on the hydrodynamic and water quality revolution in Poyang Lake. Acta Sci Circumst 35(5):1278–1284

41. Ducks Unlimited Canada. 2015. Who we are. http://www.ducks.ca/who-we-are/. Accessed 05 March 2015

42. Einar B (2010) Integrated watershed management: perspectives and problems. Springer Capital Publishing Company, New Delhi, p 273
43. Ellis SC (2005) Meaningful Consideration? A review of traditional knowledge in environmental decision making. Arctic Inst N Am 58(1):66–77
44. Engel BA, Choi JY, Harbor J, Pandey S (2003) Web-based DSS for hydrologic impact evaluation of small watershed land use changes. Comput Electron Agric 39(3):241–249
45. Environment Canada (1998) Fraser River action plan: agriculture. http://research.rem.sfu.ca/frap/agrie.pdf. Accessed 03 March 2015
46. EnvironmentCanada(2010)Integratedwatershedmanagement.http://www.ec.gc.ca/eau-water/default.asp?lang=En&n=13D23813-1. Accessed 23 Feb 2015
47. Environmental Protection Agency (EPA) (2013) Water: watersheds. http://www.epa.gov/owow/watershed/. Accessed 16 Feb 2015
48. Environmental Protection Agency (EPA) (2015a) Resource listing: epa internet-based tools for watershed management. http://water.epa.gov/learn/training/wacademy/upload/ch99_resource_listing.pdf. Accessed 26 Feb 2015
49. Environmental Protection Agency (EPA) (2015b) Key EPA internet tools for watershed management: a hands-on computer training course. http://water.epa.gov/learn/training/wacademy/epatools_index.cfm. Accessed 26 Feb 2015
50. Environmental Protection Agency (EPA) (2015c) Watershed academy web: online training in watershed management. http://cfpub.epa.gov/watertrain/index.cfm. Accessed 26 Feb 2015
51. Eswaran H, Samra JS (1997) Challenges in ecosystem management in a management in a watershed context in Asia. In: Lal R (ed) Integrated watershed management in the global ecosystem. CPC Press, Boca Raton, pp 19–32
52. Feng L, Hu C, Chen X, Zhao X (2013) Dramatic inundation changes of China's two largest freshwater lakes linked to the Three Gorges Dam. Environ Sci Technol 47:9628–9634
53. Ferguson JW, Healey M, Dugan P, Barlow C (2011) Potential effects of dams on migratory fish in the Mekong River: lessons from salmon

in the Fraser and Columbia Rivers. Environ Manag 47:141–159

54. Ffolliott PF, Baker MB, Edminster CB, Dillon MC, Kora KL (eds) (2002) Land stewardship through watershed management, perspective for 21st Century. Kluwer Academic/Plenum Publishers, New York, p 137
55. Finlayson M, Harris J, McCartney M, Young L, Chen Z (2010) Report on Ramsar visit to Poyang Lake Ramsar site, P.R. China, 12–17 April 2010. Report prepared on behalf of the Secretariat of the Ramsar Convention, p 34
56. Forrester JW (1961) Industrial Dynamics. Productivity Press, Cambridge, p 482
57. Fraser Basin Council (FBC) (2010) Sustainability snapshot 2010: working together in the lower Mainland. http://www.fraserbasin.bc.ca/_Library/Comm_Indicators/report_sslm_2010.pdf. Accessed 05 March 2015
58. Fraser Basin council (FBC) (2015a) About the Basin. http://www.fraserbasin.bc.ca/about_fraser_basin.html. Accessed 04 March 2015
59. Fraser Basin Council (FBC) (2015b) Fraser salmon and watersheds. http://www.fraserbasin.bc.ca/water_fswp.html. Accessed 04 March 2015
60. Fraser Basin council (FBC) (2015c) FSWP highlights. http://www.fraserbasin.bc.ca/water_fswp_highlights.html. Accessed 04 March 2015
61. Fraser Basin council (FBC) (2015d) Flood projects. http://www.fraserbasin.bc.ca/water_flood_projects.html. Accessed 04 March 2015
62. Fraser River Aboriginal Fisheries Secretariat (2015) Fraser salmon management council (FSMC). http://www.frafs.ca/node/52. Accessed 05 March 2015
63. Frijters ID, Leentvaar J (2003) Rhine case study. UNESCO-IHP, technical documents. In: hydrology, project PCCP series no. 17. http://unesdoc.unesco.org/images/0013/001333/133303e.pdf
64. Fuh BP (1981) On the calculation of the evaporation from land surface. Sci Atmos Sin 5:23–31

65. German L, Mansoor H, Alemu G, Mazengia W, Amede T, Stroud A (2007) Participatory integrated watershed management: evolution of concepts and methods in an ecoregional program of the eastern African highlands. Agric Syst 94:189–204
66. Ghahraman B, Sepaskhah AR (2004) Linear and non-linear optimization models for allocation of a limited water supply. Irrig Drain 53(1):39–54
67. Gizjen H (2013) Big data for a sustainable future. Nature Corresp 502:38
68. Glantz MH (1999) Creeping environmental problems and sustainable development in the Aral Sea Basin. Cambridge University Press, Cambridge, p 304
69. Gleick PH (2000) The changing water paradigm: a look at twenty-first century water resources development. Water Int 25:127–138
70. Gleick PH (2008) The world water 2008–2009: the biennial report on freshwater resources. Island Press, Washington, D.C., p 432
71. Global Nature Fund (2016) The introduction of Jiangxi Poyang Lake National nature reserve. http://www.globalnature.org/33539/Nature-Reserve/02_vorlage.asp. Accessed 20 Apr 2016
72. Gong XM, Li K, Liu YQ, Fu CJ (2006) Soil erosion and development trends in the Gan River watershed. J Yangtze 37(8):48–51
73. Guo H, Hu Q, Zhang Q, Feng S (2012) Effects of the Three Gorges Dam on Yangtze River flow and river interactions with Poyang Lake, China: 2003–2008. J Hydrol 416–417:19–27
74. Gustavson KA, Lonergan SC, Ruitenbeek HJ (1999) Selection and modeling of sustainable development indicators: a case study for the Fraser River Basin, British Columbia. Ecol Econ 28:117–132
75. Hampton SE (2013) Understanding lakes near and far. Science 342:815–816
76. Han X, Chen C, Feng L (2015) Four decades of winter wetland changes in Poyang Lake based on Landsat observations between 1973 and 2013. Remote Sens Environ 156:426–437
77. Harris J, Hao Z (2010) An ecosystem approach to resolving conflicts among ecological and economic priorities for Poyang Lake wetlands. Gland, Switzerland: wetlands International—IUCN

SSC Crane Specialists Group, 40 ppHe C. 2003. Integration of geographic information systems and simulation model for watershed management. Environ Model Softw 18:809–813

78. He C (2003) Integrated watershed management—principles and practice, 2nd edn. Wiley, New Jersey
79. Heathcote IW (2009) Integrated watershed management—principles and practice, 2nd edn. Wiley, New Jersey, p 464
80. Hebin L, Ueta K (2012) Lake watershed management: services, monitoring, funding and governance. Lakes Reserv Res Manag 17:207–223
81. Huang L, Shao Q, Liu J (2012) Forest restoration to achieve both ecological and economic progress, Poyang Lake basin, China. Ecol Eng 44:53–60
82. Hundecha Y, Bárdossy A (2004) Modeling of the effect of land use changes on the runoff generation of a river basin through parameter regionalization of a watershed model. J Hydrol 292:281–295
83. Institute of water research (2014) Watershed comprehensive assessment tool. http://www.iwr.msu.edu/WCAT/. Accessed 26 Feb 2015
84. Jensen S, Mazhitova Z, Zetterström R (1997) Environmental pollution and child health in the Aral Sea region in Kazakhstan. Sci Total Environ 206:187–193
85. Ji W, Zeng N, Wang Y, Gong P, Xu B, Bao S (2007) Analysis on the waterbirds community survey of Poyang Lake in winter. Geograph Inf Sci 13(1–2):51–64
86. Jiang J, Qi S, Liao F, Ding M, Wang Y (2014) Vulnerability of siberian crane habitat to water level in Poyang Lake wetland. Chin Giscience Remote Sens 51(6):662–676
87. Joshi PK, Jha AK, Wani SP, Joshi L, Shiyani RL (2005) Meta-analysis to assess impact of watershed program and people's participation. Comprehensive assessment research report 8, Colombo: comprehensive assessment secretariat. http://www.iwmi.cgiar.org/assessment/files_new/publications/CA%20Research%20Reports/CARR8_001.pdf
88. Karcher SK, VanBriesen JM, Nietch CT (2013) Alternative land-

use method for spatially informed watershed management decision making using SWAT. J Environ Eng 39(12):1413–1423

89. Khedum CP, Singh VP (2014) Climate change, water, and health: a review of regional challenges. Water Qual Expo Health 6:7–17
90. Konisky DM, Beierle TC (2001) Innovations in public participation and environmental decision making: examples from the Great Lakes Region. Soc Nat Resour 14(9):815–826
91. Koutsoyiannis D, Karavokiros G, Efstratiadis A, Mamassis N, Koukouvinos A, Christofides A (2003) A decision support system for the management of the water resource system of Athens. Phys Chem Earth A/B/C 28(14–15):599–609
92. Kumar G, Sena DR, Kurothe RS, Pande VC, Rao BK, Vishwakarma AK, Bagdi GL, Mishra PK (2014) Watershed impact evaluation using remote sensing. Curr Sci 106(10):1369–1379
93. Lai X, Liang Q, Jiang J, Huang Q (2014) Impoundment effects of the Three-Gorges-Dam on flow regimes in Two China's Largest Freshwater Lakes. Water Resour Manag 28(5):5111–5124
94. Lai G, Wang P, Huang X, Xiong J, Liu Y, Zeng F (2015) A simulation research of impacts of the Lake Poyang hydraulic project on hydrology and hydrodynamics. J Lake Sci 27(1):128–140
95. Larose DT (2014) Discovering knowledge in data: an introduction to data mining, 2nd edn. Wiley, Hoboken, p 240
96. Lee V, Yoon T, Shah FA (2013) Optimal watershed management for reservoir sustainability: economic appraisal. J Water Resour Plan Manag 139(2):129–138
97. Liu Y, Song P, Peng J, Fu Q, Dou C (2011) Recent increased frequency of drought events in Poyang Lake basin, China: climate change or anthropogenic effects?. IAHS-AISH publication, Vienna, pp 99–104
98. Liu Y, Wu G, Zhao X (2013) Recent declines in China's largest freshwater lake: trend or regime shift? Environ Res Lett 8:9
99. Liu W, Wei X, Fan H, Guo X, Liu Y, Zhang M, Li Q (2015) Response of flow regimes to deforestation and reforestation in a rain-dominated large watershed of subtropical China. Hydrol Process 29:5003–5015

100. Locatelli B, Vignola R (2009) Managing watershed services of tropical forests and plantations: can meta-analyses help? For Ecol Manag 258:1864–1870

101. Loucks DP, van Beek E, Stedinger JR, Dijkam JPM, Villars MT (2005) Water resources systems planning and management: an introduction to methods, models, and applications. United Nations Educational, Scientific and Cultural Organization (UNESCO), Italy, p 680

102. Luyet V, Schleapfer R, Parlange MB, Buttler A (2012) A framework to implement stakeholder participation in environmental projects. J Environ Manag 111:213–219

103. Macdonald JS, Patterson DA, Hague MJ, Guthrie IC (2010) Modeling the influence of environmental factors on spawning migration mortality for sockeye salmon fisheries management in the Fraser River, British Columbia. Trans Am Fish Soc 139:768–782

104. Mango LM, Melesse AM, McClain ME, Gann D, Setegn SG (2011) Land use and climate change impacts on the hydrology of the upper Mara River Basin, Kenya: results of a modeling study to support better resource management. Hydrol Earth Syst Sci 15:2245–2258

105. Manuelli S, Hofer T, Vita A (2014) FAO's work on sustainable mountain development and watershed management. Mt Res Dev 34(1):66–70

106. Markose VJ, Jayappa KS (2016) Soil loss estimation and prioritization of sub-watersheds of Kali River basin, Karnataka, India, using RUSLE and GIS. Environ Monit Assess 188(4):225–241

107. Mei X, Dai Z, Du J, Chen J (2015) Linkage between Three Gorges Cam impacts and the dramatic recessions in China's largest freshwater lake, Poyang Lake. Nature 5:18197

108. Michigan State University (2014) Understanding your watershed interactive mapping program. http://msustatewide.msu.edu/Programs/Details/2083. Accessed 26 Feb 2015

109. Micklin P (2007) The Aral Sea disaster. Annu Rev Earth Planet Sci 35:47–72

110. Miller RC, Guertin DP, Heilman P (2004) Information technology in watershed management decision making. J Am Water Resour

Assoc 40:347–357

111. Milly PCD, Betancourt J, Falkenmark M, Hirsch RM, Kundzewicz ZW, Lettenmaier DP, Stouffer RJ (2008) Stationarity is dead: whither water management? Science 319:573–574
112. Mitchell B (2005) Integrated water resource management, institutional arrangements, and land-use planning. Environ plan 37:1335–1352
113. Mocanu M, Vacariu L, Drobot R, Muste M (2013) Information-centric systems for supporting decision-making in watershed recourse development. 19th international conference on control systems and computer science, Bucharest, pp 611–616
114. Mody J (2004) Achieve accountability through decentralization: lessons for integrated river basin management. World Bank Policy Research Working Paper 3346, June 2004, p 52
115. Montgomery DR, Grant GE, Sullivan K (1995) Watershed analysis as a framework for implementing ecosystem management. Water Resour Bull 31:369–386
116. MRL (2006) Integrated management of Poyang Lake Basin, Jiangxi, P.R. China. Presented by The Mountain–River–Lake program office, Nanchang City, China
117. Muschett FD, Campbell CL (1997) Principles of sustainable development. CRC Press, Boca Raton, p 176
118. Mutekanga FP, Kessler A, Leber K, Visser S (2013) The use of stakeholder analysis in integrated watershed management. Mt Res Dev 33(2):122–131
119. Mysiak J, Giupponi C, Rosato P (2005) Towards the development of a decision support system for water resource management. Environ Model Softw 20:203–214
120. Nearly DG (2000) Changing perceptions of watershed management from a retrospective viewpoint. USDA For Serv Process RMRS 13:167–176
121. Olsson P, Folke C (2001) Local ecological knowledge and institutional dynamics for ecosystem management: a study of Lake Racken Watershed, Sweden. Ecosystems 4:85–104
122. Ouyang W, Hao FH, Wang X (2007) Regional non-point source

organic pollution modeling and critical area identification for watershed best environmental management. Water Air Soil Pollut 187(1):251–261

123. Ozturk I, Erturk A, Ekdal A, Gurel M, Cokgor E, Insel G, Pehlivanoglu-Mantas E, Ozabali A, Tanik A (2013) Integrated watershed management efforts: case study from Melen Watershed experiencing interbasin water transfer. Water Sci Technol 13(5):1272–1280
124. Pandey A, Chowdary VM, Mal BC (2007) Identification of critical erosion prone areas in the small agricultural watershed using USLE, GIS and remote sensing. Water Resour Manag 21(4):729–746
125. Parkes M, Panelli R (2001) Integrating catchment ecosystems and community health: the value of participatory action research. Ecosyst Health 7(2):85–106
126. Porzecanski I, Saunders LV, Brown MT (2012) Adaptive management fitness of watersheds. Ecol Soc 17(3):29–43
127. Prasuhn V, Sieber U (2005) Changes in diffuse phosphorous and nitrogen inputs into surface waters in the Rhine watershed in Switzerland. Aquat Sci 67(3):363–371
128. Prodanovic P, Simonovic SP (2010) An operational model for support of integrated watershed management. Water Resour Manag 24(6):1161–1194
129. Public-private partnership in infrastructure resource center (PPPIRC) (2016) What are Public Private Partnerships? http://ppp.worldbank.org/public-private-partnership/overview/what-are-public-private-partnerships. Accessed 09 May 2016
130. Qi H, Altinakar MS (2011) A conceptual framework of agricultural land use planning with BMP for integrated watershed management. J Environ Manag 92(1):149–155
131. Qi H, Altinakar M (2013) Integrated watershed management with multi-objective land-use optimizations under uncertainty. J Irrig Drain Eng 139(3):239–243
132. Queensland Government: Department of Environment and Heritage Protection (2016) Ecosystem health indicators. https://www.ehp.qld.gov.au/water/monitoring/assessment/water_quality_indicators.html. Accessed 23 Feb 2016

133. Raadgever GT, Mostert E, Kranz N, Interwies E, Timmerman JG (2008) Assessing management regimes in transboundary river basins: do they support adaptive management? Ecol Soc 13(1):14

134. Rahman RM, Shi ZH, Chongfa C (2009) Soil erosion hazard evaluation—an integrated use of remote sensing, GIS and statistical approaches with biophysical parameters towards management strategies. Ecol Model 220(13–14):1724–1734

135. Rand PS, Hinch SG, Morrison J, Foreman MGG, MacNutt MJ, Macdonald JS, Healey MC, Farrell AP, Higgs DA (2006) Effects of river discharge, temperature, and future climates on energetics and mortality of adult migrating Fraser River sockeye salmon. Trans Am Fish Soc 135:655–667

136. Rapport DJ, Costanza R, McMichael AJ (1998) Assessing ecosystem health. Trends Ecol Evol 13:397–402

137. Red Deer River Watershed Alliance (2015) What is watershed management? http://www.rdrwa.ca/node/27. Accessed 23 Feb 2016

138. Reed TE, Schindler DE, Hague MJ, Patterson DA, Meir E, Waples RS, Hinch SG (2011) Time to evolve? Potential evolutionary responses of Fraser River sockeye salmon to climate change and effects of persistence. PLoS One 6(6):1–13

139. Rejani R, Rao KV, Osman M, Rao CS, Reddy KS, Chary GR, Pushpanjali Samuel J (2016) Spatial and temporal estimation of soil loss for the sustainable management of a wet semi-arid watershed cluster. Environ Monit Assess 188(3):143–159

140. Sanchez GM, Nejadhashemi AP, Zhang Z, Woznicki SA, Habron G, Marquart-Pratt HS, Shortridge A (2014) Development of a socio-ecological environmental justice model for watershed-based management. J Hydrol 518:162–177

141. Santhi C, Srinivasan R, Arnold JG, Williams JR (2006) A modeling approach to evaluate the impacts of water quality management plans implemented in a watershed in Texas. Environ Mod Softw 21(8):1141–1157

142. SAS (2014) Big data: what it is and why it matters. http://www.sas.com/en_us/insights/big-data/what-is-big-data.html. Accessed 26 Feb 2015

143. Sha J, Li Z, Swaney DP, Hong B, Wang W, Wang Y (2014)

Application of a bayesian watershed model linking multivariate statistical analysis to support watershed-Scale nitrogen management in China. Water Resour Manag 28:3681–3695

144. Shen D, Wu J (2004) Mountain-river-lake integrated water resources development program, Jiangxi, People's Republic of China. Water and poverty: the realities—experience from the field. Asian Development Bank, Mandaluyong, pp 129–138
145. Shi Y, Asher J, Bartholic J, Choi JY, Engel B, Farnsworth R (2004) An online WebGIS-based hierarchical watershed decision support system for United States. Environ Inform Arch 2:838–845
146. Siberian Crane Wetland Project (2011). Poyang Lake Ecology Study. http://www.scwp.info/documents/Poyang%20Lake%20Ecology%20Study_A4.pdf. Accessed 16 Apr 2016
147. Singh VP, Frevert DK (2006) Watershed models. Taylor and Francis Group, Boca Raton, p 680
148. Singh VP, Woolhiser DA (2002) Mathematical modeling of watershed hydrology. J Hydrol Eng 7:270–292
149. Singh VP, Khedun CP, Mishra AK (2014) Water, environment, energy, and population growth: implications for water sustainability under climate change. J Hydrol Eng 19:667–673
150. Small I, Vander Meer J, Upshoor REJ (2001) Acting on an environmental health disaster: the case of the Aral Sea. Environ Health Perspect 109:547–549
151. Smit AJM (2005) Evolution of river basin management in the Rhine River Basin and lessons for China in its quest to develop a sustainable land use and water management strategy. In: CCICED task force task on integrated river management (ed) Lessons learned from integrated River Basin management: proceedings of international symposium on integrated River Basin management. China Environmental Science Press, Beijing, pp 75–87
152. Smith DW, Prepas EE, Putz G, Burke JM, Meyer WL, Whitson I (2003) The forest watershed and riparian disturbance study: a multi-discipline initiative to evaluate and manage watershed disturbance on the Boreal Plain of Canada. J Environ Eng Sci 2:S1–S13
153. Soares-Filho BS, Nepstad DC, Curran LM, Cerequeira GC, Garcia RA, Ramos CA, Voll E, McDonald A, Lefebvre P, Schlesinger

P (2006) Modelling conservation in the Amazon basin. Nature 440:520–523

154. State Forestry Administration (2008) Jiangxi province forest statistical yearbook. China Forestry Publishing House, Beijing, p 356
155. State Forestry Administration (2014) Jiangxi province forest statistical yearbook. China Forestry Publishing House, Beijing, p 320
156. Statistical Bureau of Jiangxi (2014) Jiangxi statistical yearbook. China Statistics Press, Beijing, p 211
157. Sungjun M (2007) Toward sustainable watershed management: a case study of San Francisquito creek watershed. Int Rev Public Adm 12(1):149–159
158. Szaro RC, Sexton WT, Malone CR (1998) The emergence of ecosystem management as a tool for meeting people's needs and sustaining ecosystems. Landsc Urban Plan 40:1–7
159. Takekawa JY, Newman SH, Xiao X, Prosser DJ, Spragens KA, Palm EC, Yan B, Li T, Lei F, Zhao D, Douglas DC, Muzaffar SB, Ji W (2010) Migration of waterfowl in the East Asian flyway and spatial relationship to HPA1 H5N1 outbreaks. Avian Dis 54:466–476
160. Tan GL, Wan J (2002) River basin management in tennessee valley, USA. Chin Water Resour 10:157–159
161. The World Bank (2016). Cn-jiangxi Poyang Lake basin and ecological economic zone small town development project. http://www.worldbank.org/projects/P126856/cn-jiangxi-poyang-lake-basin-ecological-economic-zone-small-town-development-demonstration-project?lang=en. Accessed 18 Apr 2016
162. Thorburn C (2012) Farmer field schools for integrated watershed management. Dev Pract 22(1):3–175
163. Tim US, Mallavaram S (2003) Application of gis technology in watershed-based management and decision making. Watershed Update 1(5):1–6
164. Tompkins EL, Adger WN (2004) Does adaptive management of natural resources enhance resilience to climate change? Ecol Soc 9(2):10

165. USEPA (1997) Top 10 watershed lessons learned. http://www.epa.gov/owow/lessons/lessonspdfs/top10.pdf. Accessed 20 May 2009
166. VanHouten JW (2014) Large watershed management and restoration: dioxin sediment remediation case study. Int J Environ Stud 71(4):570–577
167. Vörösmarty CJ, Green P, Salisbury J, Lammers RB (2000) Global water resources: vulnerability from climate change and population growth. Science 289:284–288
168. Wagner W, Gawel J, Furuma H, De Souza MP, Teixeira D, Rios L, Ohgaki S, Zehnder AJ, Hemond HF (2002) Sustainable watershed management: an international multi-watershed case study. Ambio 31(1):2–13
169. Wang LX (1999) Watershed management. Chin Forestry Publishing House, Beijing, p 325
170. Watson N (2004) Integrated river basin management: a case for collaboration. Int J River Basin Manag 2(4):243–257
171. Webler T, Tuler S (2001) Public participation in watershed management planning: views on process from people in the field. Hum Ecol Rev 8(2):29–39
172. Weng SQ, Huang GH, Li YP (2010) An integrated scenario-based multi-criteria decision support system for water resources management and planning—a case study in the Haihe River Basin. Exp Syst Appl 37(12):8242–8254
173. Westervelt J (2001) Challenges in management and modeling. Simulation modeling for watershed management. Springer, New York, pp 9–26
174. Wickham JD, Wade TG, Riitters KH (2011) An environmental assessment of United States drinking water watersheds. Landsc Ecol 26:605–616
175. Witten K, Parkes M, Ramasubramanian L (2000) Participatory environmental health research in Aotearoa/New Zealand: constraints and opportunities. Health Educ Behav 27(3):371–384
176. Wu L, Li M, Guo Y, Yang X (2011) Influence of three gorges project on water quality of Poyang Lake. Proced Environ Sci 10:1496–1501
177. Xianyi C, Liying W, Jiming C, Xiaonong Z, Jiang Z, Jiangang G,

Xiaohua W, Engels D, Minnang C (2005) Schistosomiasis control in China: the impact of a 10-year World Bank Loan Project (1992–2001). Bull World Health Org 83:43–48

178. Xie H, Wang P, Huang H (2013) Ecological risk assessment of land use change in the Poyang Lake eco-economic zone, China. Int J Environ Res Public Health 10:328–346
179. Yang GS, Yu XP, Li HP, Gao JF (2006) Introduction to integrated watershed management. Science Press, Beijing, p 238
180. Yuan H, Jiagang G, Bergquist R, Tanner M, Xianyi C, Huanzeng W (2000) The 1992–1999 World Bank schistosomiasis research initiative in China: outcomes and perspectives. Parasitol Int 49:195–207
181. Zhang Q, Ye X, Werner AD, Li Y, Yao J, Li X, Xu C (2014) An investigation of enhances recessions in Poyang Lake: comparison of Yangtze River and local catchment impacts. J Hydrol 517:425–434
182. Zhang Z, Chen X, Xu C, Hong Y, Hardy J, Sun Z (2015) Examining the influence of river-lake interaction on the drought and water resources in the Poyang Lake basin. J Hydrol 522:510–521
183. Zheng LD (2004) History of watershed resource management. China Water Resources Management Publishing House, Beijing, p 305
184. Zhou G, Wei X, Chen X, Zhou P, Liu X, Xiao Y, Sun G, Scott DF, Zhou S, Han L, Su Y (2015) Global pattern for the effect of climate and land cover on water yield. Nat Commun 6:1–9

CHAPTER

3

DAM REMOVAL AND ANADROMOUS SALMONID (ONCORHYNCHUS SPP.) CONSERVATION IN CALIFORNIA

Rebecca M. Quinones[1], Theodore E. Grantham[1], Brett N. Harvey[2], Joseph D. Kiernan[3], Mick Klasson[4], Alpa P. Wintzer[1], Peter B. Moyle[1]

[1] Center for Watershed Sciences, University of CaliforniaDavis, 1 Shields Ave, Davis, CA 95616, USA

[2] California Department of Water Resources, 3500 Industrial Blvd, West Sacramento, CA 95691, USA

[3] Southwest Fisheries Science Center, National Oceanic and Atmospheric Administration, La Jolla, CA 92037, USA

[4] 2814 Tiber Ave, Davis, CA 95616, USA

ABSTRACT

Dam removal is often proposed for restoration of anadromous salmonid populations, which are in serious decline in California. However, the benefits of dam removal vary due to differences in affected populations and potential for environmental impacts. Here, we develop an assessment method to examine the relationship between dam removal and salmonid conservation, focusing on dams that act as complete migration barriers.

Specifically, we (1) review the effects of dams on anadromous salmonids, (2) describe factors specific to dam removal in California, (3) propose a method to evaluate dam removal effects on salmonids, (4) apply this method to evaluate 24 dams, and (5) discuss potential effects of removing four dams on the Klamath River. Our flexible rating system can rapidly assess the likely effects of dam removal, as a first step in the prioritization of multiple dam removals. We rated eight dams proposed for removal and compared them with another 16 dams, which are not candidates for removal. Twelve of the 24 dams evaluated had scores that indicated at least a moderate benefit to salmonids following removal. In particular, scores indicated that removal of the four dams on the Klamath River is warranted for salmonid conservation. Ultimately, all dams will be abandoned, removed, or rebuilt even if the timespan is hundreds of years. Thus, periodic evaluation of the environmental benefits of dam removal is needed using criteria such as those presented in this paper.

Keywords: Dam effects, Pacific salmon, Steelhead, Climate change, Klamath River, Mediterranean environments

INTRODUCTION

Dams are present in nearly all California watersheds. Whether constructed for flood control, irrigation, urban water use, or hydroelectric power generation, dams alter the movement of water, sediment, nutrients and organisms in riverine ecosystems (Magilligan and Nislow 2005; Graf 2006). This disruption of longitudinal connectivity changes the structure of biotic communities and the function of ecosystems (Ward and Stanford 1983; Knighton 1998; Wohl 2012), frequently with adverse consequences for native species (Marchetti and Moyle 2001).

Anadromous Pacific salmon and trout (salmonids; *Oncorhynchus* spp.) are particularly vulnerable to impacts of dams, especially dams that prevent migration between spawning and rearing habitats (Hanak et al. 2011). Salmonids in California are blocked from about 45 % of historical habitats in major mainstem rivers but percentages vary by location. In California's Central Valley, for example, dams have blocked access to more than 80 % of historical salmonid spawning areas (Yoshiyama et al. 1998; Lindley et al. 2006). Recent status assessments (Moyle et al. 2008;

Katz et al. 2013) have highlighted that nearly all of California's native anadromous salmonids are in serious decline (Table 1). Dams negatively affect all of the state's anadromous salmonids (Moyle et al. 2008) and are significant contributors to the decline of 43 % of the taxa (Katz et al. 2013).

Table 1: Anadromous salmonid taxa found in California, their status and level of impact sustained from dams

Species	Taxa	Federal/state listing	Status	Impact from dams
Chinook salmon, *Oncorhynchus tshawytscha*	Upper Klamath-Trinity spring	–	E	High
	Upper Klamath-Trinity fall	–	V	Critical
	Southern Oregon Northern California Coast fall	–	WL	Moderate
	California Coast fall	T/–	V	Moderate
	Central Valley winter	E/E	V	Critical
	Central Valley spring	T/T	V	Critical
	Central Valley fall	–	V-E	High
	Northern California Coast	T/–	WL-E	Moderate
Steelhead, *Oncorhynchus mykiss*	Klamath Mountains Province winter	–	WL	Moderate
	Klamath Mountains Province summer	–	E	High
	Central California Coast winter	T/–	V	Moderate
	Central Valley	T/–	V	High
	South Central California Coast	T/–	V	Moderate
	Southern California	E/–	E	High
Coho salmon, *Oncorhynchus kisutch*	Central Coast	E/T	E	Moderate
	Southern Oregon Northern California Coast	E/T	E	Moderate
Chum salmon, *Oncorhynchus keta*	N/A	–	E	Moderate
Pink salmon, *Oncorhynchus gorbuscha*	N/A	–	E	Low

Coastal cut-throat trout, *Oncorhynchus clarki clarki*	N/A	–	WL	Low

Most taxa have formal recognition as distinct management units by state and federal agencies. We provide both federal listing status and status from Katz et al. (2013) with *T* threatened, *E* endangered, *V* vulnerable and *WL* watchlist. "Impact from dams" measures the magnitude of adverse impacts on specific runs.

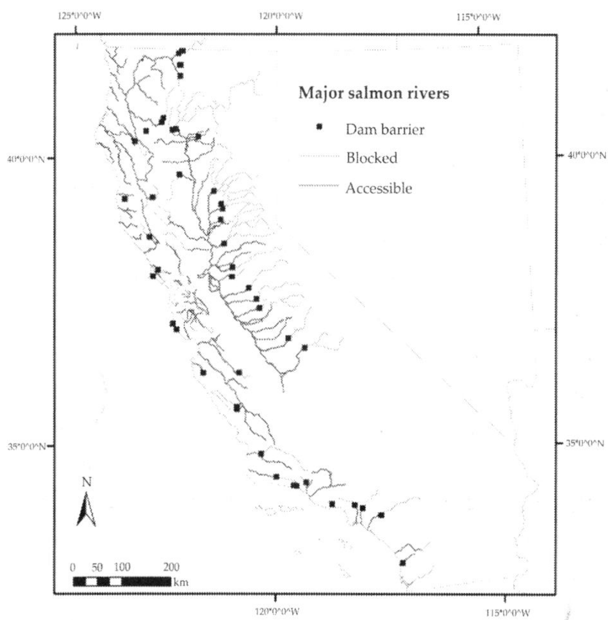

Figure 1: Anadromous salmonid habitat, accessible and blocked by dams, in major California rivers (modified from Hanak et al. 2011).

Since the 1850s, declining numbers of California's anadromous salmonids have been attributed, in part, to dams that blocked migration (Groot and Margolis 1991; Yoshiyama et al. 1998; Figure 1). Consequently, there have been many attempts to mitigate for loss of salmon and steelhead *Oncorhynchus mykiss* trout due to dams and their operations. Starting in 1870, the California State Legislature passed a series of clearly-stated laws to protect fish from the impacts of dams (Börk et al. 2012). The present manifestation of these laws is Fish and Game Code § 5937, which states, "the owner of any dam shall allow sufficient water at all times to pass over, around, or through the dam, to keep in good condition any fish that may be planted or exist below the dam." Despite this mandate, the majority of dams constructed in California during the last century were designed and operated with little consideration for their impact on fishes (Börk et al. 2012). In particular, most large dams do not provide structures for fish passage. While restoration efforts for salmonid habitat below dams typically focus on reoperation of dams to produce more suitable flows, dam removal is increasingly an option, especially for dams that no longer serve their original purposes well (Kiernan et al. 2012).

Here, we explore the relationship between dam removal and salmonid conservation in California, focusing on dams that act as complete barriers to anadromy (as in Sheer and Steel 2006). Specifically, we (1) review potential positive and negative effects of dam removal on anadromous salmonids, (2) describe factors which must be considered for dam removal in California, (3) propose a method to evaluate benefits and detriments that dam removal poses to salmonids, (4) apply this method to evaluate potential benefits/detriments of removing 24 dams, and (5) discuss the potential impacts of removing four dams on the Klamath River. We recognize there are also socioeconomic effects (e.g., Born et al. 1998; Johnson and Graber 2002; Robbins and Lewis 2008) of dam removal and salmonid conservation (e.g., to commercial and tribal fisheries) but do not discuss them here.

Review: Effects of Dams on Anadromous Salmonids

Dams disrupt movement of water, sediment, nutrients and biota between upstream and downstream habitats (Knighton 1998; Poff et al. 2007; Arthington 2012), thus altering many key processes in riverine ecosystems that are important to salmonids (Kondolf 1997; Poff et al. 1997; Poff and Hart 2002). The magnitude of these effects on anadromous salmonids depends largely on dam size, location, and operation (e.g., flow releases), and whether the dam is passable to salmonid migration (Collins 1976; Raymond 1988). Here, we briefly review major effects dams have through the creation of reservoirs and downstream changes in habitat, as well as impacts of hatcheries that are created to mitigate for loss of habitat above dams. Effects that are rare in California are not discussed (e.g., gas supersaturation; Beeman and Maule 2006).

Reservoirs

Although some salmonids have adapted successfully to use reservoirs to complete their life cycle (e.g. Chinook salmon in the Snake River; Connor et al. 2005), changes in stream flow and velocity associated with reservoirs can interfere with migration patterns (Tiffan et al. 2009; Pavlov et al. 2008). Adults migrating upstream through reservoirs can have difficulty finding spawning streams, while downstream migrating juveniles often encounter high predation and limited food resources (Rieman et al. 1991; Beauchamp et al. 1999; Carey et al. 2011). The impact of reservoirs on salmonids differs by location and in some cases they do not impede migration (USDOI and USDOC-NMFS 2012). However, most large dams in California do not have fish passage over them. As an alternative to fish passage, mitigation hatcheries were built downstream of many large dams in California, and at least six species of salmonids have been planted in reservoirs to support in-reservoir fisheries (Moyle 2002).

Downstream Effects

The reduction of peak flows commonly observed below dams has a broad range of effects on downstream habitat that adversely affect salmonids and other stream biota. Reductions in peak (geomorphic) flows and resultant encroachment by riparian vegetation can narrow channels and diminish floodplain connectivity (Kloehn et al. 2008). Isolation from floodplains reduces ecosystem productivity by limiting riparian vegetation regeneration (Jacobson et al. 2011; Benjankar et al. 2012) and the lateral movement of nutrients and organic material that support fish growth (Henery et al. 2010; Jeffres et al. 2008). Diminished peak flows limit recruitment of large wood, an important component for habitat complexity, nutrient retention, and channel stability (Bilby and Ward 1991; Naiman et al. 2002). Because sediment from upstream sources is retained by dams, streambeds downstream become coarser or armored, hindering excavation of redds by spawning salmonids (Knighton 1998; Poff and Hart 2002; Morley et al. 2008). In an embedded reach, spawning success is further reduced by accumulation of fine sediment in gravel that is normally flushed out by more frequent and larger flows (Louhi et al. 2011; Jensen et al. 2009). These substrate changes can reduce the abundance of macroinvertebrate prey for juvenile salmonids (Marchetti et al. 2011; Carlisle et al. 2010). In contrast, large flow releases from dams can result in channel incision in which substrates are scoured and sediment recruitment is disrupted by the dam (Kondolf 1997). Large fluctuations in dam releases over short time frames can increase fish mortality through stranding when high flows drop faster than salmonids are able to move (reviewed in Young et al. 2011).

Dam operations often reduce downstream habitat suitability by shifting water temperatures. Reservoirs create a large exposed surface area that absorbs solar radiation. If warm surface water from the reservoir is released, dams will increase downstream water temperatures (Risley et al. 2010), particularly in summer, when flows are lowest. Lower base flows and warm-water releases can reduce the amount of available habitat, increase the metabolic demands of fishes, and disrupt fish migration patterns (reviewed in Bednarek 2001; Olden and Naiman 2010). Warm water can also facilitate the spread of disease (Okamura et al. 2011; Kocan et al. 2009). Some larger dams, however, release cold water from the bottom of reservoirs. Cold-water releases that maintain or increase downstream base flows will usually reduce water temperatures in summer and fall (Huang et al. 2011; Yates et al. 2008), effectively shifting cold-water rearing habitat for juvenile anadromous salmonids from headwaters to below reservoirs (Ward and Stanford 1983). Cold-water releases are often crucial for sustaining remnant salmonid populations. For example, endangered winter run Chinook salmon *O. tshawytscha* in the Sacramento River are maintained entirely by cold-water flows from Shasta Dam, which prevents access to their former habitats (Moyle 2002). However, reliance on cold-water releases to protect salmon can be a problem if there is insufficient cold water in the reservoir to keep temperatures cool during late summer or during periods of drought (e.g., Thompson et al. 2011). Cooler temperatures can also delay juvenile migration cues and slow juvenile growth (Xu et al. 2010; Moyle and Cech 2004).

Hatcheries

Hatcheries are often constructed to mitigate for loss of upstream spawning and rearing habitat when dams obstruct migration or otherwise reduce salmonid populations. Hatcheries have often successfully maintained salmon fisheries for long periods of time, but at a cost of reduced genetic diversity, domestication of hatchery stocks, and major impacts on wild populations. Hatchery supplementation can result in significant differences between hatchery-produced and wild salmonids with respect to evolutionary fitness (Araki et al. 2008; Reisenbichler and Rubin 1999), phenotype (Hjort and Schreck 1982; Knudsen et al. 2006; Kostow 2004), behavior (Dittman et al. 2010; Melnychuk et al. 2010), and physiology (Shrimpton et al. 1994; Chittenden et al. 2008). Abundant hatchery stocks can further impact wild populations by supporting elevated predation (Nickelson 2003) and fishing pressure (Hard et al. 2008). Hatchery operations in California have only partially compensated for the loss of habitat upstream dams and in some cases have contributed to the decline of wild salmonids (Moyle et al. 2008). For example, hatchery practices can contribute (e.g., Klamath River; Hamilton et al. 2011) or introduce disease into the wild (reviewed in Katz et al. 2013).

Review: Effects of Dam Removal

The main goals of dam removal from a salmonid perspective are to restore natural river processes below dams and restore access to upstream habitats (Roni et al. 2002). However, dam removal will likely have short-term adverse effects; the return to geomorphic and biological processes that existed prior to dam construction will not take place immediately but at different temporal scales, assuming other upstream dams and diversions have not also altered the hydrologic regime. In the short term, stream channels are expected to narrow in the location of former reservoirs but widen below the dam as accumulated fine sediment and bedload are redistributed (Pearson et al. 2011). Salmon have spawned in newly deposited gravel just months after dam removal (e.g., Chinook salmon in the Rogue River; Vial 2012), so short-term effects may be minimal. Short-term improvements to benthic habitat in the former reservoir reach will result in recovery of macroinvertebrate abundance and biomass, but standing stocks similar to those prior to dam construction may take many years to achieve (Orr et al. 2008; Hansen and Hayes 2011).

Over the long-term (decades), stream substrate composition will become more heterogeneous (Kloehn et al. 2008). Bed loads of natural materials will encourage lateral migration of the stream channel, increasing habitat diversity and connectivity to the river floodplain (Kloehn et al.

2008), assuming downstream habitats are not constrained by levees and bank armoring. Restoration of naturally dynamic flows can increase native fish abundance where non-native fishes are few or removed (Marks et al. 2010). Upon dam removal, anadromous salmonids will likely recolonize newly accessible habitats (Anderson and Quinn 2007; Brenkman et al. 2008b; Nicole 2012; Engle et al. 2013), facilitating the return of locally extripated species. Dam removal also may increase life history diversity as run-types such as spring run Chinook salmon recolonize upstream reaches or populations isolated above the dams reestablish anadromy (Brenkman et al. 2008b). Life history diversification coupled with increased carrying capacity (Pess et al. 2008) is likely to build more resilient populations with higher fisheries productivity (Burroughs et al. 2010). Restored connectivity and fish migration to diverse habitats, including spring-fed habitats, have the potential to contribute to more resilient populations and buffer the impacts of climate change (Hamilton et al. 2011).

Nonetheless, dam removal also represents a significant disturbance that can adversely affect the ecology of rivers and their watersheds (Catalano et al. 2007; Online Resource 1). Immediately after dam removal, large quantities of impounded fine sediments may mobilize (Mussman et al. 2008) and deposit downstream (Chang 2008). High turbidity from entrained fine sediments can disrupt fish migrations and fill interstitial spaces in the substrate to the detriment of embryos incubating in the gravel (Pess et al. 2008). Moreover, increased fine sediment deposition can decrease benthic invertebrate production in the short term (Orr et al. 2008). Toxic materials (e.g., mercury) locked in reservoir sediments may become suddenly available to contaminate food webs and inhibit nutrient cycling (as in Colas et al. 2013).

Where reservoir releases support cold-water fisheries, dam removal may reduce or eliminate spawning and rearing habitat immediately downstream (e.g., winter Chinook salmon habitat downstream Shasta Dam; Moyle et al. 2008). The draining of a reservoirs results in the loss of lentic habitat and may decrease overall species richness and diversity of the river system (Catalano et al. 2007), although in California most reservoir fishes and many invertebrates are alien species (Moyle 2002). Removal of migration barriers in some situations has the potential to spread disease through interactions of fishes in stream reaches previously separated by dams (Brenkman et al. 2008a) and increase the vulnerability of populations to previously segregated invaders and predators (Stanley et al. 2007).

Review: Factors Affecting Dam Removal in California

General frameworks exist to assess the ecological benefits and detriments of dam removal (Hart et al. 2002; Whitelaw and MacMullan 2002; Yang et al. 2011; Zheng and Hobbs 2013). However, not all elements presented in these studies are applicable to dam removal in areas with Mediterranean climate such as California. Here, we discuss the distinct suite of factors that are particularly relevant for evaluating dam removal in California and similar areas: water infrastructure, sediment accumulation, mercury deposition, and climate change effects. The effects from these factors can be significant stressors to salmonid conservation and may be augmented by dam removal.

Water Infrastructure

California has been called the 'hydraulic society' due to its vast water infrastructure, designed to move water from places and times of abundance to those of scarcity (Hundley 2001). Most large watersheds contain multiple dams that range in size from small dirt dams on headwater streams to major edifices on mainstem river. When multiple dams occur on a single river, consideration must be given to the location of each dam relative to other dams in order to evaluate removal benefits, because remaining structures may continue to degrade habitats and restrict connectivity to upstream areas (Musil et al. 2012; Grantham and Moyle 2013).

A particular problem in California is transfer of water to other basins via aqueducts. These transfers dramatically alter flow regimes of both diverted and receiving streams. In source basins (e.g., Trinity River), up to 90 % of stream flow is diverted, resulting in hydrographs that are dramatically smaller and less variable relative to historical (natural) conditions (Hanak et al. 2011). In receiving basins (e.g., Sacramento River from the Trinity River), flows are augmented in summer, thus improving water quality for salmonids. Therefore, dam removal may directly benefit salmonids via increased stream flow in the source river, while paradoxically impacting salmonids by removal of water from the receiving basin. Thus, on California's north coast, water diverted from the Eel River to the Russian River is associated with declines of salmonids in the Eel River but increases in the Russian River (Yoshiyama and Moyle 2010).

Accumulated Sediment

Reservoirs behind dams frequently contain large amounts of accumulated sediment and this problem is particularly acute in California. While high sedimentation rates in southern California are a natural consequence of the area's erosive geology, in northern California, hydraulic mining, used extensively during the Gold Rush, delivered massive sediment loads to aquatic habitats for nearly 50 years. San Francisco Bay, for example, experienced an order of magnitude increase in sediment transport from 1.5 to 14×10^6 m^3/year (van Geen and Luoma 1999). The excess sediment flooded towns and agricultural fields and impeded river navigation (James 2005). While hydraulic mining was eventually banned, many reservoirs in northern California continue to retain mining sediment from the 19th and early 20th centuries. In addition, natural processes continue to re-work and release mine tailings into streams, adding to the sediment load (Childs et al. 2003).

Given the large quantities of sediment trapped behind many dams, mobilization of this material following dam removal has the potential to be devastating to downstream habitat and all life stages of anadromous salmon, at least in the short-term (Kemp et al. 2011). During migrations, high loads of suspended fine sediment can damage gills, increase physiological stress, and significantly alter behavior patterns (e.g., migration timing; Bjornn and Reiser 1991; reviewed in Newcombe and MacDonald 1991). During egg incubation, fine sediment incorporated in redds decreases substrate permeability resulting in reduced growth, disrupted development, and mortality of embryos (reviewed in Jensen et al. 2009). Moreover, fine sediments can cap redds and prevent fry from emerging. During rearing, juveniles feeding in turbid waters are often smaller than conspecifics in less turbid conditions due to decreased prey capture rates (reviewed in Kemp et al. 2011). Gill abrasion from suspended sediment during juvenile development can result in reduced ability to osmoregulate (Shrimpton et al. 2007). Indirect effects of high fine sediment concentrations (e.g., via loss of riparian vegetation, or filling of interstitial spaces) can further impair juvenile survival and growth by increasing the risk of predation due to loss of cover and altering of food web structure and productivity; even modest increases in streambed fine sediment (i.e. 20 %) are harmful to juvenile salmonids rearing in rivers (Suttle et al. 2004). However, in some situations, populations have recovered quickly from catastrophic fine sediment input (e.g., steelhead in Mount St. Helens streams after volcanic eruption; Bisson et al. 1988) and salmonids have been shown to avoid high turbidity (e.g., Bisson and Bilby 1982).

Mercury Deposition

Mobilization of mercury that has accumulated in reservoir sediments is a major environmental concern when considering dam removal in California. Although high mercury levels occur naturally in some California watersheds, an estimated 4.5 million kilograms of mercury were released into California streams as a by-product of hydraulic mining during the California Gold Rush (Alpers et al. 2005). Conversion of inorganic mercury to methylmercury, a biological neurotoxin, occurs at particularly high rates in low-oxygen habitats where dissolved organic carbon is available, such as reservoirs (Domagalski 1998). Methylmercury bioaccumulates in tissues of animals (James 2005) and poses a serious health risk to organisms at higher trophic levels. Some dams were built for the express purpose of catching mining debris and have accumulated large amounts of mercury-tainted sediment in their reservoirs (e.g., Englebright Dam). Therefore, the current methyl mercury load as well as conditions that might promote either mercury methylation or de-methylation need assessment prior to dam removal in California (James 2005).

Climate Change

California marks the southern end of the geographic range of six species of anadromous salmonids (Table 1), likely exposing them to conditions (e.g., high summer water temperatures) near the limits of environmental tolerances. In California, climate change is degrading aquatic habitats already affected by other stressors (Battin et al. 2007) through increases in temperature, decreases in snow retention, changes in precipitation patterns, and increases in the magnitude of extreme events, such as floods and droughts (Field et al. 1999; Null et al. 2012). These changes threaten the persistence of anadromous salmonid populations (Moyle et al. 2013) and, in some cases, will strongly influence the relative benefits and detriments of dam removal for salmonids.

Although water temperatures under future climate change scenarios may exceed salmonid tolerance limits in the lower reaches of streams during the warmer months of the year, mainstem habitats will continue to serve as migration corridors during winter, with cooler headwater streams offering the only suitable habitat in the system (Crozier et al. 2008). Consequently, habitat connectivity will become as important as habitat quantity or quality (Isaak et al. 2007) in conserving California salmonids. In some instances, dam removal may be the only option to provide anadromous salmonids with access to suitable cold-water habitats including high-elevation tributaries and spring-fed streams that would be buffered from warming impacts of climate change.

Conversely, water releases from cold, thermally-stratified depths present in some existing reservoirs may provide cold-water habitat downstream of dams and mitigate the effect of elevated air temperatures (Poff and Hart 2002). Such water releases can favor cold-water species immediately below the dam (Poff and Hart 2002) and affect riverine community composition for many kilometers downstream (reviewed in Haxton and Findlay 2008). However, cooling effects from dam releases differ by reservoir size and location (i.e. elevation and latitude; Null et al. 2014). Summer water releases can also augment seasonal low flows in areas where groundwater pumping lowers water tables or dries stream reaches (Constantz and Essaid 2007). Augmentation of flows may be particularly important for providing suitable salmonid habitat as water demands increase with human population growth (Tanaka et al. 2006). However, if available cold water cannot simultaneously meet human water demands and maintain stream flows in particularly dry years, salmonid populations may be extirpated (e.g., adult spring-run Chinook in Butte Creek; Thompson et al. 2011). Under such conditions, cold water combined with favorable habitat conditions may create an 'ecological trap' in which juvenile salmonids are attracted to cold flows during rearing and later die when cold-water releases are curtailed and temperatures become too warm (Jeffres and Moyle 2012).

METHODS

Rating the Effects of Dam Removal

We developed a simple, flexible rating system to rapidly assess the relative magnitude of impacts from dam removal (Table 2), considering both positive and negative impacts that dam removal may have on anadromous salmonids over different spatial and temporal scales (see Online Resource 1). We intend this rating system to serve as a first step in a dam removal evaluation process and as a way to prioritize dams for removal. The system scores nine metrics relating to the characteristics of salmonid populations, habitat quality, and stressors in the watershed of interest. Benefits from dam removal were considered to be none or limited (scores = 0–9), moderate (scores = 10–18), or high (scores = 19–27) based on the sum of metric scores. The system was designed to encompass issues that are generally applicable to most dams that prevent fish passage in Mediterranean-climate rivers.

Table 2: Scoring rubric for evaluating benefits of dam removal for salmonid conservation

	Score
1. Characteristics of salmonid taxa	
a. Number of taxa that may benefit or number of endemics	
None	0
1 taxon/≥1 also found in other locations	1
2 taxa/taxa in only one other location	2
3+ taxa/only found in removal watershed	3
b. Status of source population(s)	
Extirpated	0
Most small and or declining	1
Stable or increasing	2
Abundant	3
c. Life history diversity of source population(s)	
Extirpated	0
Single life history pattern dominates	1
Some runs/patterns lost	2
All historic life histories represented	3

2. Quality and quantity of aquatic habitat	
a. Migration barriers below dam	
Barriers exist near or downstream dam	0
Multiple permanent barriers in watershed	1
Seasonal or partial barriers exist	2
No additional barriers	3
b. Habitat suitability above dams	
No suitable habitat exists	0
Limited or degraded habitat exists	1
Suitable but slightly degraded habitat	2
Habitat in good condition	3
c. Amount of spawning and rearing habitat above dam	
None	0
<30 % of former habitat	1
30–50 % of former habitat	2
>50 % of former habitat	3
3. Stressors to salmonid recovery	
a. Number of additional limiting stressors (excluding hatcheries)	
Stressors cannot be overcome	0
>5 additional stressors	1
2–5 additional stressors	2
0 or 1 additional stressor	3
b. Duration of adverse impacts from dam removal	
Indefinite or long-term	0
Longer than 2 years	1
1–2 years	2
<1 year	3
c. Extent of hatchery supplementation	
Entirely reliant on hatcheries	0
>70 % of taxa supplemented	1
30–70 % of taxa supplemented	2
<30 % of taxa supplemented	3

Characteristics of Salmonid Taxa

The first metric in Table 2 considers either the level of endemism or number of native salmonid taxa that could benefit from dam removal. If protection or restoration of endemic salmonids is a desirable goal, this metric scores 0 for no existing endemics, 1 if one or more endemics exist but are found in multiple other drainages in the region, 2 if the endemics are only found in one other drainage, and 3 if the endemics occur only in the dam-removal watershed. If protection of salmonid diversity is the desirable goal, the metric scores 0 for no taxa present, 1 for one taxa, 2 for two taxa, and 3 for three or more taxa. Flexibility in the first metric allows equal consideration between areas characterized by few but highly endemic taxa and areas containing diverse assemblages. The second metric scores the ability of likely source populations to re-colonize the newly accessible areas. A source population is defined as the population that would most likely recolonize available habitats upon dam removal. Extirpated populations score a 0, small populations or those with low densities score a 1, stable or increasing populations score a 2, and abundant or highly dense populations score a 3. Status for California salmonids is evaluated in Moyle et al. (2008, 2011) and Katz et al. (2013). For the third metric, life history diversity, we assumed taxa with multiple life histories (i.e. life history portfolios as in Carlson and Satterthwaite 2011), are more adaptable to changing environmental conditions than those with extirpated or hatchery-dominated runs (as in Greene et al. 2010). Extirpated populations score 0, populations with a single life history score 1, populations with potential to regain lost life histories score a 2, and populations with all life histories intact score a 3.

Habitat Quality and Quantity

For metrics pertaining to aquatic habitat, we considered accessibility, condition, and quantity of habitat upstream and downstream of dam sites. For the first metric, users must determine if other barriers are present in the system. Removal of a single dam within a basin containing multiple barriers may result in limited benefits unless a large area of habitat is made accessible and all salmonid life-stages benefit. Streams with other barriers immediately adjacent to the dam under consideration, such that little to no additional habitat will be made accessible, score 0; streams with multiple remaining barriers within the range of anadromy, but not immediately adjacent, score a 1; streams with additional seasonal or temporary barriers score a 2; streams with no additional barriers score a 3. For the second metric, the quality of accessible habitat above the dam is considered (Roni et al. 2002). Dam removal may only have limited benefits if the watershed upstream has been irreversibly altered (e.g., by water diversion or urbanization). For example, streams upstream of the dam site with unscreened water diversions can result in high fish mortality due to entrainment or stranding of fish (as in Carlson and Rahel 2007). If a watershed requires extensive habitat restoration beyond dam removal, there will likely be limited benefits. Systems with no suitable habitat made available score a 0, while systems with habitat requiring extensive restoration score a 1, if dam removal provides access to habitat that is suitable most of the year score 2, or habitat in good condition year round (including cold-water springs) score a 3. The third metric scores the proportion of former spawning and rearing habitat reopened to anadromous salmonids. Habitats that function primarily as adult migration corridors are not considered here. In cases where multiple salmon taxa are present, the amount of habitat made available to the furthest migrant is estimated, usually for steelhead in California. Sheer and Steel (2006) used stream gradients to define general and optimal habitat for Chinook salmon and steelhead; Chinook salmon generally will use gradients from 0 to 7 % (optimal at 1–2 %), while steelhead use gradients 0–12 % (optimal at 1–5 %). Coho salmon *O. kisutch* use shallower gradients than Chinook because they do not often migrate as far inland, especially in California (Groot and Margolis 1991). Sheer and Steel (2006) determined that barriers had large, moderate, and limited impacts on steelhead and Chinook salmon distribution in the Willamette and Lower Columbia River basins, Oregon, when they blocked >50, 30–50, and <30 % of optimal habitat, respectively. We therefore used stream gradients above dams for estimating the proportion of high quality habitat accessible to different taxa upon dam removal. We used the thresholds determined by Sheer and Steel (2006) in our rating system because salmonids in Oregon and California have overlapping habitat requirements (Groot and Margolis 1991). Systems with no spawning or rearing habitat upstream of the dam score a 0, systems with <30 % of former habitat score a 1, systems with 30–50 % of former habitat score a 2, and systems with >50 % of former habitat score a 3.

Stressors

In addition to current habitat impairment, watersheds in California are subject to ongoing anthropogenic stressors that may limit the potential for salmonid populations to recover following dam removal [e.g., stream sedimentation, mercury accumulation, estuarine alteration, etc.; see Katz et al. (2013) for a list of 15 stressors evaluated]. Our stressor metric scores a 0 if one or more stressors clearly prevent habitat recolonization (e.g., chronic source of pollution such as acid mine drainage) or population expansion, scores a 1 where greater than five stressors were identified in the watershed, a 2 for 2–5 stressors, and a 3 for one or no stressors.

A particular stressor that we considered separately is the dam removal process itself which can represent a large disturbance (Stanley and Doyle 2003). Dam removal that is likely to result in permanent unfavorable downstream changes (e.g., scour the stream channel to bedrock) would score a 0. Metric scores increase with decrease in time to recovery. Adverse impacts lasting more than 2 years but expected to disappear over time score a 1, impacts lasting 1–2 years score a 2, and impacts expected to last <1 year score a 3.

The third metric concerns the degree of influence of hatcheries on salmonid populations in the system. We address this stressor separately because hatchery production can alter population characteristics intrinsically (e.g., genetics or fitness) as well as extrinsically (competition with and predation on wild fish). Several studies (reviewed in Katz et al. 2013) have shown that species persistence can be threatened at different levels based on the magnitude of hatchery supplementation (including juvenile releases and adult straying) in a system. In systems where salmon runs are constantly supplemented by hatcheries, dam removal may only have limited initial benefits because fishes may already have lost key adaptations to former habitat (e.g., run timing). In such situations, redevelopment of local adaptations is likely to take multiple generations. Accordingly, benefits may be intermediate in systems with partial hatchery supplementation but high where hatchery supplementation is absent or supplementation has been managed in small conservation hatcheries over short duration (Lackey et al. 2006). Systems with taxa largely reliant on hatchery supplementation score a 0, while systems with >70, 30–70, <30 % hatchery-supplemented populations score a 1, 2, or 3.

Certainty

A score of certainty is also given to each metric. A score of 1 represents limited knowledge based primarily on expert opinion. A score of 2 reflects moderate knowledge supported with a few papers and reports, usually from gray literature. High degree of certainty, a score of 3, is largely based on peer-reviewed publications (e.g., Yoshiyama et al. 2001) and high familiarity of the system by the scorers. An overall certainty score for each dam removal is calculated as the average of the certainty scores for all metrics.

Application of the Rating System

We first selected eight dams as case studies of dam removal for salmon conservation, based on recommendations by Hanak et al. (2011; Table 3) and the personal knowledge of one or more of the authors with these sites. These dams have all been considered for removal at one time or another. The eight dams studied in detail were Scott, Englebright, Matilija, Rindge, and the lower four dams on the Klamath River (Iron Gate, Copco 1, Copco 2, and J.C. Boyle dams; Figure 2). Although J.C. Boyle is located in Oregon, just north of the California border, we include it here because it is part of a complex of four dams formally proposed for removal on the Klamath River. The context of each of the case studies is described below. An additional 16 large dams (see Online Resource 2 for characteristics) which are complete migration barriers to salmonids, but are not currently candidates for removal, were also rated (Online Resource 3). These dams were chosen on the basis of location, size, and their consideration for reoperation to improve habitat conditions for salmonids (i.e. environmental flows; T. Grantham, unpublished data).

Table 3: Evaluation of potential benefits of proposed dam removal in five California watersheds

Dam	**Scott**	**Englebright**	**Rindge**	**Matilija**	**Iron Gate + 3**[a]
Watershed	**Eel River**	**Yuba River**	**Malibu Ck.**	**Matilija Ck.**	**Klamath River**
Metric					
1a	3	3	1	1	3
1b	1	1	1	1	2
1c	2	2	1	1	2

2a	2	2	3	2	2
2b	2	2	2	2	2
2c	1	3	3	3	2
3a	2	3	2	2	2
3b	1	0	3	1	2
3c	3	1	3	3	1
Score	17	17	19	16	20
Category	M	M	H	M	H
Certainty	3	3	3	3	3

For metrics, see Table 2. Categories of benefits from dam removal: *H* high (scores = 19–27), *M* moderate (scores = 10–18), *L* none or limited (scores = 0–9)

[a]Copco 1, Copco 2, J.C. Boyle dams

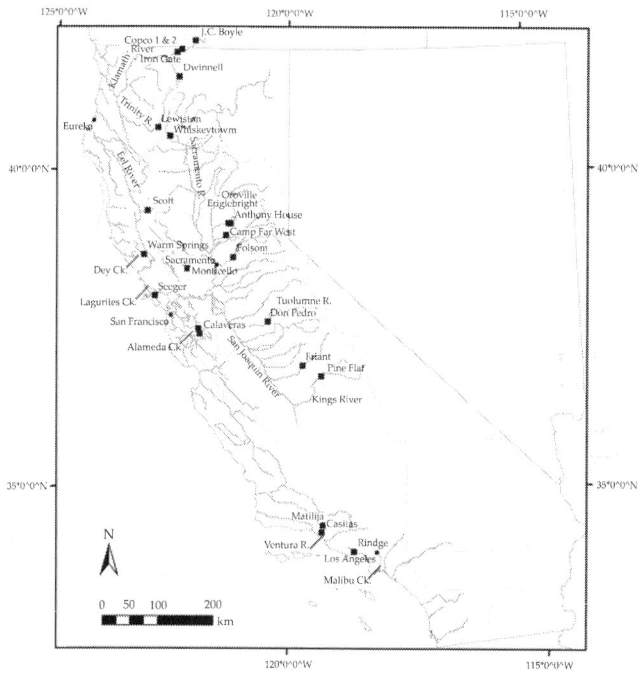

Figure 2: Dam removal sites evaluated for effects on anadromous salmonid conservation in California.

Scott Dam

Cape Horn Dam was built in Mendocino County as part of the Potter Valley Project on the Eel River, approximately 290 km from the ocean, to divert water into the Russian River. The reservoir quickly filled with sediment and, in 1921, Scott Dam (Lake County) was constructed upstream of Cape Horn Dam to increase storage capacity and provide hydropower (Pejchar and Warner 2001). Reduced flows in the mainstem river, coupled with widespread habitat degradation by logging, flooding, and mining (Pejchar and Warner 2001) and the introduction of the predatory Sacramento pikeminnow *Ptychocheilus grandis* resulted in a greater than 99 % reduction in historical salmonid populations (Yoshiyama and Moyle 2010). While Cape Horn Dam has a functional fish ladder, Scott Dam does not and it prevents runs of Chinook salmon, coho salmon and steelhead trout from reaching ~93 km of potential habitat (Yoshiyama and Moyle 2010). The amount of accessible habitat, however, is lower during periods of low flow when functionality of the ladder at Cape Horn dam is compromised (Yoshiyama and Moyle 2010). The inter-dam reach (between Scott and Cape Horn dams) is an important rearing area for steelhead even at low flows because cold-water releases from Scott Dam are present all summer (Yoshiyama and Moyle 2010).

Englebright Dam

Englebright Dam, located in Yuba and Nevada counties, was built in 1941 on the Yuba River, just downstream of the point where the North, Middle, and South Yuba rivers join together to form the main river. It is located approximately 314 km upstream from the Pacific Ocean. The dam was constructed for hydropower generation and mining debris containment, although hydraulic mining ceased upstream shortly after construction. Today, its reservoir is 25 % filled with mercury-tainted sediment that continues to accumulate in the reservoir (Childs et al. 2003). Another debris dam, Daguerre Point Dam, 16 km downstream, has a fish ladder allowing fish passage up to Englebright Dam. The absence of fish ladders on Englebright Dam makes it a barrier to anadromous fishes; blocking access to ~160 km of potential spawning and rearing habitat (USGS 2004). Yoshiyama et al. (2001) indicate that 70 % (90 km) of historically available spawning habitat for Chinook salmon and steelhead has been cut off by the dam.

Matilija Dam

Matilija Creek, located in Ventura County approximately 27 km from the ocean, is a major tributary to the Ventura River. Matilija Dam was built in 1947 on Matilija Creek for water storage and flood control. However, due to the high natural erosion rates of the region, the dam has trapped more than 2.5×10^6 m^3 of sediment, completely filling the reservoir (Slagel and Griggs 2008). Trapped sediments were once important for creating and maintaining channel structure and fish spawning habitat downstream (Jenkin 2009). Additionally, use of construction materials of poor quality resulted in structural weakening of the dam (Minear 2003). The Ventura River watershed historically supported large runs of southern steelhead (Allen et al. 2007), but prime spawning and rearing habitat became severely limited following construction of three large dams: Matilija, Casitas, and Robles Diversion dams. Furthermore, altered flow and sediment transport regimes have substantially degraded the quality of accessible habitat (Becker et al. 2010). Matilija Dam is currently the upstream-most barrier to habitats that historically provided more than half of the spawning and rearing habitat in the system (P. Moyle and R. Quiñones, personal observations).

Rindge Dam

Rindge dam was built in 1926 on Malibu Creek in Los Angeles County for private water supply; its reservoir was completely filled with sediment by 1955 (Courter 2002). The estimated $0.6–1.2 \times 10^6$ m^3 of sediment currently trapped behind the dam historically balanced shoreline erosion at the creek's mouth, which supports an estuary that provides habitat for several listed species (e.g., tidewater goby *Eucyclogobius newberryi*, willow flycatcher *Empidonax traillii*; M. Capelli, NMFS, unpublished observations). Rindge Dam is located just 4.4 km inland from the coast, leaving little spawning and rearing habitat accessible to endangered southern steelhead populations in Malibu Creek. Removal of the dam would about triple the amount of quality habitat available to steelhead (Abramson and Grimmer 2005).

Iron Gate, Copco 1, Copco 2, and J.C. Boyle Dams

Between 1911 and 1962, six dams were built on the mainstem Klamath River: three in Klamath County, Oregon (Link River, Keno, and J.C. Boyle) and three in Siskiyou County, California (Copco 1, Copco 2, and Iron Gate). The downstream-most dam, Iron Gate Dam, is approximately 308 km from the ocean. The primary functions of these dams are to impound and divert water (Link River and Keno dams), or to generate hydropower (J.C. Boyle, Copco 1, Copco 2, Iron Gate dams). Only Link River and Keno dams, the two most upstream structures, have fishways that facilitate salmonid passage. As a result, approximately 675 km of potential migration, rearing and spawning habitat are currently inaccessible to anadromous salmonids in the basin (Hamilton et al. 2011), although not all this habitat is likely suitable (P. Moyle and R. Quiñones, personal observations). As a condition for a hydropower relicensing, the Departments of Interior and Commerce prescribed mandatory fishways at the lower four dams to provide upstream access to migrating anadromous fishes. This has resulted in agreements (USDOI and USDOC-NMFS 2012) to decommission the lower four dams because of the high cost of building fishways.

RESULTS AND DISCUSSION

Using the rating system, we evaluated the benefits and detriments of removing 24 dams, including potential negative impacts of dam removal not always considered in previous assessments. None of the dams are currently passable for salmon. In our case, the goal was to prioritize dam removal for the conservation of salmonid diversity, as it pertains to metric 1a. Removal of twelve (50 %) of the dams evaluated seems to result in at least a moderate advance in salmonid conservation due to the large number of salmonids that would benefit or the amount of habitat that would become accessible. All eight of the dams with proposals for removal show moderate to high benefits for salmonids. For example, the moderate benefits of removing Scott Dam (score = 17) were largely based on the number of salmonid taxa found in the Eel River (1a), the quantity (2c) of habitat upstream of the dam, and the few negative effects expected from dam removal (3b). Benefits were diminished by the status of declining salmonid populations; for instance, runs of spawning Chinook salmon in the entire Eel River basin have decreased from an estimated 56,000 to 1,000 individuals (Yoshiyama and Moyle 2010). Sacramento pikeminnow are also a major stressor that may inhibit the recovery of anadromous salmonids in the Eel River (Yoshiyama and Moyle 2010). Although several salmonids species exist in the Yuba River (1a), benefits from removing Englebright Dam would likely be moderate (score = 17) due to the status of existing populations (1b; declining), high concentrations of fine sediment already in the mainstem Yuba River (2b) as well as by the large volume of mercury-tainted sediment (James et al. 2009) that would be released upon removal (3b). Additional barriers (2a) and hatchery supplementation (3c) also likely decrease the ability of salmonids to recover. However, large portions of formerly occupied habitat could be made accessible to some taxa upon dam removal (e.g., >80 % for spring-run Chinook salmon; NMFS 2012).

The high and moderate benefits of removing Rindge (score = 19) and Matilija (score = 16) dams were due to the quantity (2c) of habitat upstream of the dams, the limited duration of adverse effects expected from dam removal (3b), usually 1–5 years (as in Hamilton et al. 2011), and the lack of hatchery supplementation in these systems (3c). Removal of the four Klamath dams (score = 20) would likely have high benefits to all existing salmonid taxa (1a), all with intact life histories (1c) due to the short duration of adverse impacts from dam removal (3b; Hamilton et al. 2011). However, net benefits in the Klamath River were diminished by the magnitude of other stressors in the system (3a), including additional barriers (2a; i.e. seasonally poor water conditions in reaches near Kendo dam) and land use practices that decrease the suitability of habitat (2b; discussed below). The degree of hatchery supplementation (3c) for Chinook and coho salmon and steelhead trout also threaten the recovery capability of salmonids in the basin (Quiñones et al. 2013).

Although the eight dams discussed above have been proposed for removal at one time or another, the other 16 dams we rate have not. We evaluate them here for comparison, since all are also large dams that block salmonid migration. The removal of two dams (Seeger, score = 20; Warm Springs, score = 22) could yield high benefits to salmonid conservation in California, while the removal of four (Whiskeytown, score = 6; Monticello, score = 7, Don Pedro Main, score = 8; Folsom score = 8) appear unlikely to have benefits. The other 10 dams were thought to advance salmon conservation at least moderately.

The high benefits of removing Seeger and Warm Springs are largely due to the protected status of these watersheds as important sources of domestic water to Marin County and the city of Santa Rosa. For this same reason, however, their removal is unlikely. The negligible benefits of removing Whiskeytown, Monticello, Folsom, and Don Pedro Main dams are a result of their location. Whiskeytown is located high in the watershed; its removal would afford little new salmonid habitat (2c). Monticello, Folsom, and Don Pedro Main dams are located near other permanent barriers (i.e. Putah Creek diversion, Nimbus dam, La Grange dam) so other dams would continue to degrade habitat and disrupt salmonid migration (2a) upon their removal. However, the likelihood of removing additional dams within the same management complex may increase once operational links between dams (e.g., Putah Creek diversion dam with Monticello dam, Folsom dam with Nimbus dam) are lost. All eight case studies received high certainty scores (3; Table 3) due to the abundance of literature available for each and the familiarity of at least some of the authors with these systems. Most (10 of 16) of the other 16 dams evaluated also received high certainty scores but knowledge of six was considered to be of moderate certainty. Overall, we concluded we had sufficient data to knowingly evaluate all 24 dams.

Of the dams evaluated, the Klamath dams are the only ones being formally proposed for removal at the present time. Consequently, the benefits and detriments of their removal are discussed in additional detail to illustrate the complexities of dam removal even when salmon conservation is the primary consideration. Full evaluation of Klamath dam removal effects are discussed in Hamilton et al. (2011) and USDOI and USDOC-NMFS (2012).

Klamath River Case Study

The Klamath River basin drains approximately 40,000 km^2 in southern Oregon and northern California (Figure 2). The Klamath basin once produced an estimated 650,000–1 million adult salmonids per year (Gresh et al. 2000) distributed across multiple species: Chinook (fall and spring runs), coho, chum *O. keta* and pink salmon *O. gorbuscha* and steelhead. However, abundances of runs of wild salmonids have been declining (e.g., spring-run Chinook; Nehlsen et al. 1991; NRC 2004; USDOI and USDOC-NMFS 2012).

Conditions without Dam Removal

In the absence of dam removal, freshwater habitats above Iron Gate Dam will remain degraded and inaccessible to anadromous fishes, unless major investment is made in fish passage facilities. Stream reaches between Keno Reservoir and Iron Gate Dam (Figure 2) are characterized by highly-simplified flow regimes (Bartholow et al. 2004) and low water quality (e.g., low dissolved oxygen, algal blooms, high water temperature; reviewed in Hamilton et al. 2011). Furthermore, the sheer volume of water and warm temperatures in reservoirs prevent tributaries from contributing cold water to mainstem habitats (Hamilton et al. 2011). Unmodified, future operation of these dams will continue to limit the viability of fishes in areas upstream of Iron Gate Dam (Hamilton et al. 2011).

Below Iron Gate Dam, the natural flow regime and riverine processes will remain disrupted (reviewed in Hamilton et al. 2011). Construction and operation of the dams has changed the timing and magnitude of peak and base flows in the Klamath River mainstem (Stillwater Sciences 2009; Quiñones 2011), disrupting riparian plant succession, channel formation, and spawning gravel recruitment (KRBFTF 1991; FERC 2007; Varyu and Greimann 2010). These and other dam-related changes to riverine processes have contributed to conditions that foster the spread of fish disease below Iron Gate Dam (Stanford et al. 2011).

Conditions with Dam Removal

Our rating indicated that removal of these dams will yield benefits, supporting the conclusion of Stanford et al. (2011) that removal of the dams would be "ecologically beneficial to the Klamath River ecosystem (p. 157)." Our rating is independent of extensive restoration activities proposed concurrent with dam removal that will further benefit fisheries while sustaining agriculture consistent with environmental laws [as in Klamath Basin Restoration Agreement 2010 (KBRA)]. These actions are expected to greatly improve the resilience of Klamath salmonid fisheries over the next few decades (Hamilton et al. 2011). The removal of the four dams is expected to substantially improve water quality and habitat conditions in the upper basin (reviewed in Hamilton et al. 2011 and Stanford et al. 2011). Flow and temperature regimes are expected to more closely reflect historical conditions. Consequently, water that currently evaporates from reservoirs should augment flows after dams are removed (Hamilton et al. 2011). Cyanobacteria blooms, acidity, and problems associated with low dissolved oxygen are expected to diminish because nutrient cycling, uptake and transport rates are greater in lotic systems than in reservoirs (Asarian et al. 2009; Stanford et al. 2011). Water temperatures are also expected to track air temperatures more closely, so will be slightly warmer in summer but cooler in fall and winter than with dams intact (Stanford et al. 2011). There may also be some seasonal or regional cooling due to reduction in hydraulic residence time and increased riparian shading along restored river reaches (Hamilton et al. 2011). The reach that contains Iron Gate and the two Copco dams should provide high quality habitat for holding adult and rearing juvenile salmonids because several large springs provide a continuous source of cool water (Hamilton et al. 2011). Restoration of more natural flow and sediment transport regimes is expected to improve habitat diversity by moving sediment, building bars, eroding banks and toppling large woody debris into the stream channel (NRC 2007). Removal of the dams, in conjunction with KBRA, would provide anadromous salmonids access to approximately 675 km of additional habitat and improve the viability of all species in the system (Hamilton et al. 2011). The suitability of most of this habitat for spawning and rearing has not been closely evaluated so the amount of high quality habitat may be considerably less than the 675 km of accessible habitat. However, even a fraction of the newly accessible habitat could still provide significantly more spawning and rearing habitat, particularly to anadromous fishes eradicated from upstream reaches.

Removal of Klamath dams can also have negative effects on fish, mostly through short-term adverse effects on downstream aquatic habitats (Stillwater Sciences 2011; Stanford et al. 2011). Dam removal may reduce dissolved oxygen and increase turbidity levels in the short term, due to high levels of total suspended organic sediment. The worst-case scenario for salmon exposed to high fine sediment loads is of a one-time loss of production from redds in the mainstem river, potentially resulting in a loss of approximately one-third of adult fall-run Chinook spawners 3 and 4 years after dam removal (Stillwater Sciences 2008). However, the higher than normal fine sediment concentrations are not expected to impact spawning habitats beyond one year, because high flows will clean spawning gravels and remove excess organic sediments from the system. Also, formerly disrupted bedload transport will resume upon dam removal, creating and enhancing spawning habitat in former reservoir and downstream reaches (USDOI and USDOC-NMFS 2012). Over the long-term, benefits of dam removal are expected to outweigh adverse effects such that impacted populations of fall Chinook salmon are predicted to fully recover within 5 years of dam removal (Hamilton et al. 2011).

CONCLUSIONS

Dam removal is likely to improve ecosystem function and condition in most cases. However, dam removal is unlikely to restore rivers systems to pre-dam conditions because the effects of dams are amplified through time and dam removal itself is a disturbance. On the other hand, benefits from dam removal can be enhanced by additional habitat restoration and modification of existing resource management strategies. For instance, the magnitude of benefit from the removal of Englebright Dam can be greatly increased if a viable method is found to deal with the mercury-laden sediment stored in its reservoir prior to dam removal and if releases from upstream dams can be adjusted to benefit the restored salmonid populations.

Overall, we found that most dam removals we investigated would result in at least moderate benefits to salmonid conservation. Net positive effects are most likely in two situations. The first is where salmonid populations below the dams are still large (e.g., fall-run Chinook salmon in the Klamath River), making them more resilient to the temporary negative impacts of dam removal such as sedimentation. The second is where populations are imperiled but adverse effects of dam removal are few (e.g., southern steelhead in Malibu Creek). The magnitude of the benefit from dam removal is therefore expected to be modulated by the conservation status of the impacted species. We assumed that large populations are more stable and able to withstand significant adverse impacts in comparison to small, imperiled populations which are likely more vulnerable. Furthermore, dam removal is likely to produce no or only marginal benefit to salmonids if additional stressors are major causes of a population's decline (e.g., hatchery supplementation, water diversions) or are preventing recovery. The four dams on the Klamath

River were given special attention since they are the only ones formally being considered for removal in California at the present time. Our analysis concludes that removal of these dams can be warranted for salmonid conservation.

Our method provides a simple initial assessment of whether dam removal will benefit salmonids. However, dam removal may present benefits to other taxa and river ecosystems as a whole. For example, Bushaw-Newton et al. (2002) found that communities of aquatic insects became more diverse upon dam removal in Manatawny Creek. Freshwater mussels were also expected to benefit from dam removal through the recolonization of upstream habitats by host fishes. In the Carmel River, removal of San Clemente Dam is expected to benefit the imperiled red-legged frog *Rana draytonii* (CDWR 2012). Native riparian vegetation may also benefit from dam removal (Poff et al. 1997), although benefits may only manifest after long periods and in absence of invasive non-native plants (reviewed in Shafroth et al. 2002). River ecosystems overall are expected to become more resilient upon dam removal if they are not so altered as to inhibit restoration of natural conditions (e.g., Hamilton et al. 2011). However, dam removal can also present detriments to river ecosystems. In the Klamath River, dam removal is expected to increase high water temperatures in spring and early summer to harmful levels for rearing juvenile salmonids (Bartholow et al. 2004). Short-term but significant impacts from the release of sediments trapped behind dams may present the largest detriment from dam removal; trapped sediments can degrade water quality and river morphology, and negatively impact salmonids and other aquatic life for several years after dam removal (USDOI and USDOC-NMFS 2012). Consequently, the benefits and detriments of dam removal must be carefully weighed prior to implementation.

While our protocol was developed for anadromous salmonids in California's Mediterranean climate, it could easily be modified for use in other regions and for other taxa, such as freshwater mussels or other anadromous and resident fishes, bearing in mind that the characteristics of the taxa of interest, quality of aquatic habitat, and additional stressors specific to each region and river system must be considered. We can envision such species-based protocols being incorporated into a broader systematic strategy for evaluating dams as operation licenses are evaluated for renewal. Such a strategy would likely include alternative criteria, such as economic considerations, that may reach conclusions contrary to assessments focused on fish.

We recognize that economic benefits and other reasons for maintaining dams are often a priority. Yet when salmon and steelhead conservation are considered in conjunction with other potential benefits, the case for dam removal can out-weigh reasons for maintaining a dam. For example, both Matilija and Rindge dams store large amounts of sediment that, if flushed downstream, would help restore important recreational ocean beaches near their mouths. In exchange for the short-term negative impacts of sediment flushing, there could be long-term benefits to both endangered southern steelhead and to local beach-based economies.

As a final thought, we note that no dam lasts forever and most gradually become reduced in function as the reservoir fills with sediment. Ultimately, all dams will be abandoned, removed, or rebuilt, even if the time span is hundreds of years. Thus, periodic evaluation using criteria such as we have presented in conjunction with economic evaluation is a worthwhile exercise. As dams decline in economic value, the relative benefits of removal will increase.

ACKNOWLEDGMENTS

This work was partially funded by the California Landscape Conservation Cooperative.

REFERENCES

1. Abramson M, Grimmer M (2005) Fish migration severity and steelhead habitat quality in the Malibu Creek watershed. California State Coastal Conservancy and California Department of Parks and Recreation, Oakland
2. Allen MA, Riley SC, Gast T (2007) Steelhead population and habitat assessment in the Ventura River/Matilija Creek basin 2006 final report. Thomas R. Payne and Associates, Arcata
3. Alpers CN, Hunerlach MP, May JT, Hothern RL (2005) Mercury contamination from historical gold mining in California. US Geol Surv Fact Sheet 2005–3014:1–6

4. Anderson JH, Quinn TP (2007) Movements of adult coho salmon (*Oncorhynchus kisutch*) during colonization of newly accessible habitat. Can J Fish Aquat Sci 64:1143–1154
5. Araki H, Berejikian BA, Ford MJ, Blouin MS (2008) Fitness of hatchery-reared salmonids in the wild. Evol Appl 1:342–355
6. Arthington AH (2012) Environmental flows: saving rivers in the third millennium. University of California Press, Berkeley
7. Asarian E, Kann J, Walker WW (2009) Multi-year nutrient budget dynamics for Iron Gate and Copco reservoirs, California. Karuk Tribe Department of Natural Resources, Eureka
8. Bartholow JM, Campbell SG, Flug M (2004) Predicting the thermal effects of dam removal on the Klamath River. Environ Manag 34:856–874
9. Battin J, Wiley MW, Ruckelshaus MH, Palmer RN, Korb E, Bartz KK, Imaki H (2007) Projected impacts of climate change on salmon habitat restoration. PNAS 104:6720–6725
10. Beauchamp DA, Baldwin CM, Vogel JL, Gubala CP (1999) Estimating diel, depth-specific foraging opportunities with a visual encounter rate model for pelagic piscivores. Can J Fish Aquat Sci 56:128–139
11. Becker GS, Smetak KM, Asbury DA (2010) Southern steelhead resources evaluation: identifying promising locations for steelhead restoration in watersheds south of the Golden Gate. Center for Ecosystem Management and Restoration, Oakland
12. Bednarek AT (2001) Undamming rivers: a review of the ecological impacts of dam removal. Environ Manag 27:803–814
13. Beeman JW, Maule AG (2006) Migration depths of juvenile Chinook salmon and steelhead relative to total dissolved gas supersaturation in a Columbia River reservoir. Trans Am Fish Soc 135:584–594
14. Benjankar R, Klaus J, Yager EM, Egger G, Goodwin P, Glenn NF (2012) The impact of river modification and dam operation on floodplain vegetation succession trends in the Kootenai River, USA. Ecol Eng 46:88–97
15. Bilby RE, Ward JW (1991) Characteristics and function of large woody debris in streams draining old-growth, clear-cut, and second-

growth forests in Southwestern Washington. Can J Fish Aquat Sci 48:2499–2508

16. Bisson PA, Bilby RE (1982) Avoidance of suspended sediment by juvenile coho salmon. N Am J Fish Manag 2:371–374
17. Bisson PA, Nielsen JL, Ward JW (1988) Summer production of coho salmon stocked in Mount St. Helens streams 3–6 years after the 1980 eruption. Trans Am Fish Soc 117:322–335
18. Bjornn TC, Reiser DW (1991) Habitat requirements of salmonids in streams. In: Meehan WR (ed) Influences of forest and rangeland management on salmonid fishes and their habitats. American Fisheries Society, Bethesda, pp 83–138
19. Börk KS, Krovoza JF, Katz JV, Moyle PB (2012) The rebirth of California fish and Game Code section 5937: water for fish. UCDL Rev 45:809–913
20. Born SM, Genskow KD, Filbert TL, Hernandez-Mora N, Keefer ML, White KA (1998) Socioeconomic and institutional dimensions of dam removal: the Wisconsin experience. Environ Manag 22:359–370
21. Brenkman SJ, Mumford SL, House M, Patterson C (2008a) Establishing baseline information on the geographic distribution of fish pathogens endemic in Pacific salmonids prior to dam removal and subsequent recolonization by anadromous fish in the Elwha River, Washington. Northwest Sci 82:142–152
22. Brenkman SJ, Pess GR, Torgersen CE, Kloehn KK, Duda JJ, Corbett SC (2008b) Predicting recolonization patterns and interactions between potamodromous and anadromous salmonids in response to dam removal in the Elwha River, Washington state, USA. Northwest Sci 82:91–106
23. Burroughs BA, Hayes DB, Klomp KD, Hansen JF, Mistak J (2010) The effects of the Stronach dam removal on fish in the Pine River, Manistee County, Michigan. Trans Am Fish Soc 139:1595–1613
24. Bushaw-Newton KL, Hart DD, Pizzuto JE, Thomson JR, Egan J, Ashley JT, Johnson TE, Horwitz RJ, Keeley M, Lawrence J, Charles D, Gatenby C, Kreeger DA, Nightengale T, Thomas RL, Velinsky DJ (2002) An integrative approach towards understanding ecological responses to dam removal: the Manatawny Creek study.

J Am Water Resour Assoc 38:1581–1600

25. California Department of Water Resources (CDWR) (2012) Final supplement to the San Clemente Dam seismic safety project final environmental impact report/environmental impact statement. Department of Water Resources, Sacramento
26. Carey MP, Sanderson BL, Friesen TA, Barnas KA, Olden JD (2011) Smallmouth bass in the Pacific Northwest: a threat to native species; a benefit for anglers. Rev Fish Sci 19:305–315
27. Carlisle DM, Wolock DM, Meador MR (2010) Alteration of streamflow magnitudes and potential ecological consequences: a multiregional assessment. Front Ecol Environ 9:264–270
28. Carlson AJ, Rahel FJ (2007) A basinwide perspective on entrainment of fish in irrigation canals. Trans Am Fish Soc 136:1335–1343
29. Carlson SM, Satterthwaite WH (2011) Weakened portfolio effect in a collapsed salmon population complex. Can J Fish Aquat Sci 68:1579–1589
30. Catalano MJ, Bozek MA, Pellett TD (2007) Effects of dam removal on fish assemblage structure and spatial distributions in the Baraboo River, Wisconsin. N Am J Fish Manag 27:519–530
31. Chang HH (2008) Case study of fluvial modeling of river responses to dam removal. J Hydraul Eng 134:295–302
32. Childs JR, Snyder NP, Hampton MA (2003) Bathymetric and geophysical surveys of Englebright Lake, Yuba-Nevada Counties, California. US Geological Survey Open File Report 03-383:1–20, Menlo Park
33. Chittenden CM, Sura S, Butterworth KG, Cubitt KF, Manel-La NP, Balfry S, Okland F, McKinley RS (2008) Riverine, estuarine and marine migratory behaviour and physiology of wild and hatchery-reared coho salmon *Oncorhynchus kisutch* (Walbaum) smolts descending the Campbell River, BC, Canada. J Fish Biol 72(3):614–628
34. Colas F, Baudoin J, Danger M, Usseglio-Polatera P, Wagner P, Devin S (2013) Synergistic impacts of sediment contamination and dam presence on river functioning. Freshw Biol 58:320–336
35. Collins GB (1976) Effects of dams on Pacific salmon and steelhead

trout. Mar Fish Rev 1222:39–46

36. Connor WP, Sneva JG, Tiffan KF, Steinhorst RK, Ross D (2005) Two alternative life history types for fall Chinook salmon in the Snake River basin. Trans Am Fish Soc 134:291–304
37. Constantz J, Essaid H (2007) Influence of groundwater pumping on streamflow restoration following dam removal. Hydrol Process 21:2823–2834
38. Courter C (2002) Trout coalition call for removal of Rindge Dam. Malibu Times, Malibu
39. Crozier LG, Hendry AP, Lawson PW, Quinn TP, Mantua NJ, Battin J, Shaw RG, Huey RB (2008) Potential responses to climate change in organisms with complex life histories: evolution and plasticity in Pacific salmon. Evol Appl 1:252–270
40. Dittman AH, May D, Larsen DA, Moser ML, Johnston M, Fast D (2010) Homing and spawning site selection by supplemented hatchery- and natural-origin Yakima River spring Chinook salmon. Trans Am Fish Soc 139:1014–1028
41. Domagalski J (1998) Occurrence and transport of total mercury and methyl mercury in the Sacramento River basin, California. J Geochem Explor 64:277–291
42. Engle R, Skalicky J, Poirier J (2013) Translocation of lower Columbia River fall Chinook salmon (*Oncorhynchus tshawystcha*) in the year of Condit dam removal and year one post-removal assessments. Report to US Fish and Wildlife Service, Vancouver
43. Federal Energy Regulatory Commission (FERC) (2007) Final environmental impact statement for relicensing of the Klamath hydroelectric Project No. 2082-027, Washington, DC
44. Field CB, Daily GC, Davis FW, Gaines S, Matson PA, Melack J, Miller NL (1999) Confronting climate change in California: ecological impacts on the golden state. Union of Concerned Scientists, Cambridge
45. Graf WL (2006) Downstream hydrologic and geomorphic effects of large dams on American rivers. Geomorphology 79:336–360
46. Grantham TE, Moyle PB (2013) Assessing flows for fish below dams: a systematic approach to evaluate compliance of California's

dams with fish and Game Code Section 5937. Center for Watershed Sciences, Davis

47. Greene CM, Hall JE, Guilbault KR, Quinn TP (2010) Improved viability of populations with diverse life-history portfolios. Biol Lett 6:382–386
48. Gresh T, Lichatowich JA, Schoonmaker P (2000) An estimation of historic and current levels of salmon production in the Northeast Pacific ecosystem: evidence of a nutrient deficit in the freshwater systems of the Pacific Northwest. Fisheries 25(1):15–21
49. Groot C, Margolis L (eds) (1991) Pacific salmon: life histories. UBC Press, Vancouver
50. Hamilton J, Rondorf D, Hampton M, Quiñones RM, Simondet J, Smith T (2011) Synthesis of the effects to fish species of two management scenarios for the secretarial determination on removal of the lower four dams on the Klamath River. US fish and Wildlife Service, Yreka
51. Hanak E, Lund J, Dinar A, Gray B, Howitt R, Mount J, Moyle PB, Thompson B (2011) Managing California's water: from conflict to reconciliation. Public Policy Institute of California, San Francisco
52. Hansen JF, Hayes DB (2011) Long-term implications of dam removal for macroinvertebrate communities in Michigan and Wisconsin rivers, United States. River Res Appl 28:1540–1550
53. Hard JJ, Gross MR, Heino M, Hilborn R, Kope RG, Law R, Reynolds JD (2008) Evolutionary consequences of fishing and their implications for salmon. Evol Appl 1(2):388–408
54. Hart DD, Johnson TE, Bushaw-Newton KL, Horwitz RJ, Bednarek AT, Charles DF, Kreeger DA, Velinsky DJ (2002) Dam removal: challenges and opportunities for ecological research and river restoration. Bioscience 52(8):669–681
55. Haxton TJ, Findlay CS (2008) Meta-analysis of the impacts of water management on aquatic communities. Can J Fish Aquat Sci 65:437–447
56. Henery RE, Sommer TR, Goldman CR (2010) Growth and methylmercury accumulation in Juvenile Chinook salmon in the Sacramento River and its floodplain, the Yolo Bypass. Trans Am Fish Soc 139:550–563

57. Hjort RC, Schreck CB (1982) Phenotypic differences among stocks of hatchery and wild coho salmon, *Oncorhynchus kisutch*, in Oregon, Washington and California. Fish Bull 80:105–119
58. Huang B, Langpap C, Adams RM (2011) Using instream water temperature forecasts for fisheries management: an application in the Pacific Northwest. J Am Water Resour As 47:861–876
59. Hundley JN (2001) The great thirst: Californians and water: a history. University of California Press, Berkeley
60. Isaak DJ, Thurow RF, Rieman BE, Dunham JB (2007) Chinook salmon use of spawning patches: relative roles of habitat quality, size, and connectivity. Ecol Appl 17:352–364
61. Jacobson R, Janke T, Skold J (2011) Hydrologic and geomorphic considerations in restoration of river-floodplain connectivity in a highly altered river system, Lower Missouri River, USA. Wetl Ecol Manag 19:295–316
62. James LA (2005) Sediment from hydraulic mining detained by Englebright and small dams in the Yuba basin. Geomorphology 71:202–226
63. James LA, Singer MB, Ghoshal S, Megison M (2009) Historical channel changes in the lower Yuba and Feather Rivers, California: long-term effects of contrasting river-management strategies. Geol Soc Am Spec Paper 45:57–81
64. Jeffres C, Moyle PB (2012) When good fish make bad decisions: coho salmon in an ecological trap. N Am J Fish Manag 32:87–92
65. Jeffres C, Opperman J, Moyle PB (2008) Ephemeral floodplain habitats provide best growth conditions for juvenile Chinook salmon in a California river. Environ Biol Fish 83:449–458
66. Jenkin P (2009) Managed shoreline retreat and dam removal: ecosystem restoration to benefit coastal resources. Reef J 1:140–152
67. Jensen DW, Steel EA, Fullerton AH, Pess GR (2009) Impact of fine sediment on egg-to-fry survival of Pacific salmon: a meta-analysis of published studies. Rev Fish Sci 17:348–359
68. Johnson SE, Graber BE (2002) Enlisting the social sciences in decisions about dam removal. Bioscience 52:731–738
69. Katz JV, Moyle PB, Quiñones RM, Israel JA, Purdy SE (2013)

Impending extinction of salmon, steelhead, and trout (Salmonidae) in California. Environ Biol Fish 96:1169–1186

70. Kemp P, Sear D, Collins A, Naden P, Jones I (2011) The impacts of fine sediment on riverine fish. Hydrol Process 25:1800–1821
71. Kiernan JD, Moyle PB, Crain PK (2012) Restoring native fish assemblages to a regulated California stream using the natural flow regime concept. Ecol Appl 22:1472–1482
72. Klamath River Basin Fisheries Task Force (KRBFTF) (1991) Long-range plan for the Klamath River basin conservation area fishery restoration program. US Fish and Wildlife Service, Yreka
73. Kloehn KK, Beechie TJ, Morley SA, Coe HJ, Duda JJ (2008) Influence of dams on river-floodplain dynamics in the Elwha River, Washington. Northwest Sci 82:224–235
74. Knighton D (1998) Fluvial forms and processes: a new perspective. Oxford University Press, New York
75. Knudsen CM, Schroder SL, Busack CA, Johnston MV, Pearsons TN, Bosch WJ, Fast DE (2006) Comparison of life history traits between first-generation hatchery and wild Upper Yakima river Spring Chinook salmon. Trans Am Fish Soc 135:1130–1144
76. Kocan R, Hershberger P, Sanders G, Winton J (2009) Effects of temperature on disease progression and swimming stamina in Ichthyophonus-infected rainbow trout, *Oncorhynchus mykiss* (Walbaum). J Fish Dis 32:835–843
77. Kondolf GM (1997) Hungry water: effects of dams and gravel mining on river channels. J Environ Manag 21:533–551
78. Kostow KE (2004) Differences in juvenile phenotypes and survival between hatchery stocks and a natural population provide evidence for modified selection due to captive breeding. Can J Fish Aquat Sci 61:577–589
79. Lackey RT, Lach DH, Duncan SL (eds) (2006) Salmon 2100: the future of wild Pacific salmon. American Fisheries Society, Bethesda
80. Lindley ST, Schick RS, Agrawal A, Goslin M, Pearson TE, Mora E, Anderson JJ, May B, Greene S, Hanson C, Low A, McEwan D, MacFarlane RB, Swanson C, Williams JG (2006) Historical population structure of Central Valley steelhead and its alterations

by dams. San Francisco Estuary Watershed Sci 4:1–19

81. Louhi P, Ovaska M, Mäki-Petäys A, Erkinaro J, Muotka T (2011) Does fine sediment constrain salmonid alevin development and survival? Can J Fish Aquat Sci 68:1819–1826
82. Magilligan FJ, Nislow KH (2005) Changes in hydrologic regimes by dams. Geomorphology 71:61–78
83. Marchetti MP, Moyle PB (2001) Effects of flow regime on fish assemblages in a regulated California stream. Ecol Appl 11:530–539
84. Marchetti MP, Esteban E, Smith ANH, Pickard D, Richards AB, Slusark J (2011) Measuring the ecological impact of long-term flow disturbance on the macroinvertebrate community in a large Mediterranean climate river. J Freshw Ecol 26:459–480
85. Marks JC, Haden GA, O'Neill M, Pace C (2010) Effects of flow restoration and exotic species removal on recovery of native fish: lessons from a dam decommissioning. Restor Ecol 18:934–943
86. Melnychuk MC, Welch DW, Walters CJ (2010) Spatio-temporal migration patterns of Pacific salmon smolts in rivers and coastal marine waters. PLoS One 5:e12916
87. Minear JT (2003) Sediment dynamics and steelhead habitat (*Oncorhynchus mykiss*) in the Matilija Creek watershed, southern California. Thesis, University of California
88. Morley SA, Duda JJ, Coe HJ, Kloehn KK, McHenry ML (2008) Benthic invertebrates and periphyton in the Elwha River basin: current conditions and predicted response to dam removal. Northwest Sci 82:179–196
89. Moyle PB (2002) Inland fishes of California. University of California Press, Berkeley
90. Moyle PB, Cech JJ (2004) Fishes: an introduction to ichthyology. Prentice-Hall Inc., Upper Saddle River
91. Moyle PB, Israel JA, Purdy SE (2008) Salmon, steelhead, and trout in California: status of an emblematic fauna. Caltrout, San Francisco
92. Moyle PB, Katz JV, Quiñones RM (2011) Rapid decline of California's native inland fishes: a status assessment. Biol Conserv 144:2414–2423
93. Moyle PB, Kiernan JD, Crain PK, Quiñones RM (2013) Climate

change vulnerability of native and alien freshwater fishes of California: a systematic assessment approach. PLoS One 8:e63883

94. Musil J, Horky P, Slavik O, Zboril A, Horka P (2012) The response of the young of the year fish to river obstacles: functional and numerical linkages between dams, weirs, fish habitat guilds and biotic integrity across large spatial scales. Ecol Indic 23:634–640
95. Mussman EK, Zabowski D, Acker SA (2008) Predicting secondary reservoir sediment erosion and stabilization following dam removal. Northwest Sci 82:236–245
96. Naiman RJ, Bilby RE, Schindler DE, Helfield JM (2002) Pacific salmon, nutrients, and the dynamics of freshwater and riparian ecosystems. Ecosystems 5:399–417
97. National Marine Fisheries Service (NMFS) (2012) Biological opinion for the U.S. Army Corps of Engineers' operation and maintenance of Englebright and Daguerre Point dams and Englebright Reservoir on the Yuba River. National Marine Fisheries Service, Sacramento
98. National Research Council (NRC) (2004) Endangered and threatened fishes in the Klamath River basin: causes of decline and strategies for recovery. The National Academies Press, Washington, DC
99. Nehlsen W, Williams JE, Lichatowich JA (1991) Pacific salmon at the crossroads: stocks at risk from California, Oregon, Idaho, and Washington. Fisheries 16:4–21
100. Newcombe CP, MacDonald DD (1991) Effects of suspended sediments on aquatic ecosystems. N Am J Fish Manag 11:72–82
101. Nickelson T (2003) The influence of hatchery coho salmon (*Oncorhynchus kisutch*) on the productivity of wild coho salmon populations in Oregon coastal basins. Can J Fish Aquat Sci 60:1050–1056
102. Nicole W (2012) Lessons of the Elwha River: managing health hazards during dam removal. Environ Health Perspect 120:a430–a435
103. NRC (2007) Hydrology, ecology, and fishes of the Klamath River basin. National Academies Press, Washington, DC
104. Null SE, Viers JH, Deas ML, Tanaka SK, Mount JF (2012) Stream

temperature sensitivity to climate warming in California's Sierra Nevada: impacts to coldwater habitat. Clim Change. doi:10.1007/s10584-012-0459-8

105. Null SE, Ligare ST, Viers JH (2014) A method to consider whether dams mitigate climate change effects on stream temperatures. J Am Water Resour Assoc 49:1456–1472

106. Okamura B, Hartikainen H, Schmidt-Posthaus H, Wahli T (2011) Life cycle complexity, environmental change and the emerging status of salmonid proliferative kidney disease. Freshw Biol 56(4):735–753

107. Olden JD, Naiman RJ (2010) Incorporating thermal regimes into environmental assessments: modifying dam operations to restore freshwater ecosystem integrity. Freshw Biol 55:86–107

108. Orr CH, Kroiss SJ, Rogers KL, Stanley EH (2008) Downstream benthic responses to small dam removal in a coldwater stream. River Res Appl 24:804–822

109. Pavlov D, Mikheev V, Lupandin A, Skorobogatov M (2008) Ecological and behavioural influences on juvenile fish migrations in regulated rivers: a review of experimental and field studies. Hydrobiologia 609(1):125–138

110. Pearson AJ, Snyder NP, Collins MJ (2011) Rates and processes of channel response to dam removal with a sand-filled impoundment. Water Resour Res 47(8):W08504

111. Pejchar L, Warner K (2001) A river might run through it again: criteria for consideration of dam removal and interim lessons from California. Environ Manag 28:561–575

112. Pess GR, McHenry ML, Beechie TJ, Davies J (2008) Biological impacts of the Elwha River dams and potential salmonid responses to dam removal. Northwest Sci 82:72–90

113. Poff NL, Hart DD (2002) How dams vary and why it matters for the emerging science of dam removal. BioScience 52:659–668

114. Poff NL, Allan JD, Bain MB, Karr JR, Prestegaard KL, Richter BD, Sparks RE, Stromberg JC (1997) The natural flow regime: a paradigm for river conservation and restoration. Bioscience 47:769–784

115. Poff NL, Olden JD, Merritt DM, Pepin DM (2007) Homogenization of regional river dynamics by dams and global biodiversity implications. PNAS 104:5732–5737
116. Quiñones RM (2011) Recovery of Pacific salmonids (*Oncorhynchus* spp.) in the face of climate change: a case study of the Klamath River basin, California. Dissertation, University of California
117. Quiñones RM, Johnson ML, Moyle PB (2013) Hatchery practices may result in replacement of wild salmonids: adult trends in the Klamath River basin, California. Environ Biol Fish. doi:10.1007/s10641-013-0146-2
118. Raymond HL (1988) Effects of hydroelectric development and fisheries enhancement on spring and summer Chinook salmon and steelhead in the Columbia River basin. N J Fish Manag 8:1–24
119. Reisenbichler RR, Rubin SP (1999) Genetic changes from artificial propagation of Pacific salmon affect the productivity and viability of supplemented populations. ICES J Mar Sci 56:459–466
120. Rieman BE, Beamesderfer RC, Vigg S, Poe TP (1991) Estimated loss of juvenile salmonids to predation by northern squawfish, walleyes, and smallmouth bass in John Day reservoir, Columbia River. Trans Am Fish Soc 120:448–458
121. Risley JC, Constantz J, Essaid H, Rounds S (2010) Effects of upstream dams versus groundwater pumping on stream temperature under varying climate conditions. Water Resour Res 46:W06517
122. Robbins JL, Lewis LY (2008) Demolish it and they will come: estimating the economic impacts of restoring a recreational fishery. J Am Water Resour 44:1488–1499
123. Roni P, Beechie TJ, Bilby RE, Leonetti FE, Pollock MM, Pess GR (2002) A review of stream restoration techniques and a hierarchical strategy for prioritizing restoration in Pacific Northwest watersheds. N Am J Fish Manag 22:1–20
124. Shafroth PB, Friedman JM, Auble GT, Scott ML, Braatne JH (2002) Potential response of riparian vegetation to dam removal. Bioscience 52:703–713
125. Sheer MB, Steel EA (2006) Lost watersheds: barriers, aquatic habitat connectivity, and salmon persistence in the Willamette and lower Columbia River basins. Trans Am Fish Soc 135:1654–1669

126. Shrimpton JM, Bernier NJ, Randall DJ (1994) Changes in cortisol dynamics in wild and hatchery-reared juvenile coho salmon (*Oncorhynchus kisutch)* during smoltification. Can J Fish Aquat Sci 51:2179–2187

127. Shrimpton JM, Zydlewski JD, Heath JW (2007) Effect of daily oscillation in temperature and increased suspended sediment on growth and smolting in juvenile Chinook salmon, *Oncorhynchus tshawytscha*. Aquaculture 273:269–276

128. Slagel MJ, Griggs GB (2008) Cumulative losses of sand to the California coast by dam impoundment. J Coast Res 243:571–584

129. Stanford J, Duffy W, Asarian E, Cluer B, Detrich P, Eberle L, Edmondson S, Foott S, Hampton M, Kann J, Malone K, Moyle PB (2011) Conceptual model for restoration of the Klamath River. In: Thorsteinson L, VanderKooi S, Duffy W (eds). Proceedings of the Klamath Basin science conference, Medford, pp 151–184

130. Stanley EH, Doyle MW (2003) Trading off: the ecological effects of dam removal. Front Ecol Environ 1:15–22

131. Stanley EH, Catalano MJ, Mercado-Silva N, Orr CH (2007) Effects of dam removal on brook trout in a Wisconsin stream. River Res Appl 23:792–798

132. Stillwater Sciences (2008) Klamath River dam removal study: sediment transport DREAM-1 simulation. Technical report. California Coastal Conservancy, Oakland

133. Stillwater Sciences (2009) Dam removal and Klamath River water quality: a synthesis of the current conceptual understanding and a assessment of data gaps. Technical Report, Oakland

134. Stillwater Sciences (2011) Klamath dam removal drawdown scenario 8: potential impacts of suspended sediments on focal fish species with and without mechanical sediment removal, Berkeley

135. Suttle KB, Power ME, Levine JM, McNeely C (2004) How fine sediment in riverbeds impairs growth and survival of juvenile salmonids. Ecol Appl 14:969–974

136. Tanaka SK, Zhu T, Lund JR, Howitt RE, Jenkins MW, Pulido MA, Tauber M, Ritzema RS, Ferreira IC (2006) Climate warming and water management adaptation for California. Clim Change 76:361–387

137. Thompson LC, Escobar MI, Mosser CM, Purkey DR, Yates D, Moyle PB (2011) Water management adaptations to prevent loss of spring-run Chinook salmon in California under climate change. J Water Res Plan Manag. doi:10.1061/(ASCE)WR.1943-5452

138. Tiffan KF, Kock TJ, Haskell CA, Connor WP, Steinhorst RK (2009) Water velocity, turbulence, and migration rate of subyearling fall Chinook salmon in the free-flowing and impounded Snake River. Trans Am Fish Soc 138:373–384

139. US Department of Interior and US Department of Commerce-National Marine Fisheries Service (USDOI and USDOC-NMFS) (2012) Klamath dam removal overview report for the secretary of the interior: an assessment of science and technical information. US Geological Survey, Portland

140. US Geological Survey (USGS) (2004) Should Englebright dam be removed? Pacific Coastal and Marine Science Center, Santa Cruz

141. van Geen A, Luoma SN (1999) The impact of human activities on sediments of San Francisco Bay, California: an overview. Mar Chem 64:1–6

142. Varyu DR, Greimann BP (2010) Sediment mobilization analysis at Little Bogus Creek and Beaver Creek for Klamath dam removal studies. Bureau of Reclamation, Sacramento

143. Vial J (2012) Gold Ray Dam removal monitoring. Open River Initiative Report no. 0348–0044. NOAA, White City

144. Ward JV, Stanford JA (1983) Serial discontinuity concept of lotic ecosystems. In: Bartell SM, Fontaine TD (eds) Dynamics of lotic systems. Ann Arbor Science, Ann Arbor

145. Whitelaw E, MacMullan E (2002) A framework for estimating the costs and benefits of dam removal. Bioscience 52:724–730

146. Wohl E (2012) Identifying and mitigating dam-induced declines in river health: three case studies from western United States. Int J Sediment Res 27:271–287

147. Xu C, Letcher BH, Nislow KH (2010) Context-specific influence of water temperature on brook trout growth rates in the field. Freshw Biol 55:2253–2264

148. Yang M, Qian X, Zhang YH, Sheng JB, Shen DL, Ge Y (2011)

Spatial multicriteria decision analysis of flood risks in aging-dam management in china: a framework and case study. Int J Environ Res 8:1368–1387

149. Yates D, Galbraith H, Purkey D, Huber-Lee A, Sieber J, West J, Herrod-Julius S, Joyce B (2008) Climate warming, water storage, and Chinook salmon in California's Sacramento Valley. Clim Change 91:335–350
150. Young PS, Cech Jr JJ, Thompson LC (2011) Hydropower-related pulse-flow impacts on stream fishes: a brief review, conceptual model, knowledge gaps, and research needs. Rev Fish Biol Fish 21:713–731
151. Yoshiyama RM, Moyle PB (2010) Historical review of Eel River anadromous salmonids, with emphasis on Chinook salmon, coho salmon and steelhead. University of California, Davis
152. Yoshiyama RM, Fisher FW, Moyle PB (1998) Historical abundance and decline of Chinook salmon in the central valley region of California. N Am J Fish Manag 18:487–521
153. Yoshiyama RM, Gerstung ER, Fisher FW, Moyle PB (2001) Historical and present distribution of Chinook salmon in the Central Valley drainage of California. Calif Dep Fish Game Fish Bull 179:71–176
154. Zheng P, Hobbs B (2013) Multiobjective portfolio analysis of dam removals addressing dam safety, fish populations, and cost. J Water Res Plan 139:65–75

CHAPTER

4

WHAT ARE THE CONSEQUENCES OF FISH ENTRAINMENT AND IMPINGEMENT ASSOCIATED WITH HYDROELECTRIC DAMS ON FISH PRODUCTIVITY? A SYSTEMATIC REVIEW PROTOCOL

Trina Rytwinski[1,2,3], Dirk A. Algera[1,2], Jessica J. Taylor[1,2], Karen E. Smokorowski[4] , Joseph R. Bennett[2,3], Philip M. Harrison[1,5] and Steven J. Cooke[1,2]

[1] Fish Ecology and Conservation Physiology Laboratory, Department of Biology, Carleton University, Ottawa, ON, Canada

[2] Canadian Centre for Evidence-Based Conservation and Environmental Management, Institute of Environmental Sciences, Carleton University, Ottawa, ON, Canada

[3] Department of Biology and Institute of Environmental Science, Carleton University, Ottawa, ON, Canada

[4] Great Lakes Laboratory for Fisheries and Aquatic Sciences, Fisheries and Oceans Canada, Sault Ste. Marie, ON, Canada

[5] Department of Biology, University of Waterloo, Waterloo, ON, Canada.

ABSTRACT

Background

This systematic review will address the need for a better understanding of the impacts of fish entrainment and impingement associated with hydroelectric dams on fish productivity in freshwater temperate environments. As the number of dams continues to increase worldwide, so too has concerns for their effects on fish populations. Fish injury and mortality at hydroelectric facilities may have serious consequences for fish populations, which are generally the result of three main sources: (1) fish passage through hydroelectric facilities (i.e., turbines, spillways, sluiceways, and other passage routes) during downstream migration for migratory fish; (2) the entrainment of resident fish; and (3) the impingement of fish (migratory or resident) against screens/trash racks. Most studies on the impacts of entrainment and impingement at hydroelectric facilities on fish have primarily focused on: (1) how fish injury and mortality occurs; and (2) evaluations of the effectiveness of various management strategies used to mitigate harm during downstream passage. Given the contributions of migratory and resident adults and juveniles to fish production, a necessary extension is to evaluate the impacts of fish injury and mortality from hydropower dam entrainment and impingement on fish productivity. Therefore, to ensure the sustainability of fishes dependent on our freshwater ecosystems, a better understanding of the impacts of fish entrainment and impingement associated with hydroelectric dams on fish productivity is needed.

Methods

This systematic review will search for, compile, summarize and synthesize evidence on the impacts of fish entrainment and impingement associated with hydroelectric dams on fish productivity in freshwater temperate environments. Considered studies will include (but not be limited to): (1) those that report a metric related to mortality and injury as an indication of the effect on fish productivity; (2) the change in a metric related to mortality and injury relative to an appropriate control; and (3) articles that scale-up the evaluation to include some estimate of a change in a component of fish productivity (e.g., articles that include an estimate of fish loss from the population due to entrainment/impingement by comparing a metric related to mortality or injury to an estimate of population size or biomass). Only studies where the causal relationship between intervention and outcome is made clear to allow for the effects of entrainment and impingement to be isolated from other potential impacts of hydroelectric power production (e.g., barriers to migration and/or habitat degradation), will be included. The review will use public search engines and specialist websites, and will include both primary and grey literature. Potential effect modifiers will be identified to obtain a better understanding of the factors that are associated with variation in effects among studies, given differences in: (1) site-specific factors (e.g., turbine type, size, power output); (2) methodologies and study designs used to assess impacts; and (3) biological factors (e.g., fish life history stage, body size and morphology). Study quality will be assessed to allow for critical evaluation, including study design, confounding factors and statistical analysis. Data will be compiled into a narrative synthesis and a meta-analysis will be conducted where data availability and quality allow.

Keywords: Dam, Evidence-based policy, Fish biomass, Fish abundance, Fish density, Fish migration, Fish survival, Hydropower, Reservoirs

BACKGROUND

Worldwide over 58,000 dams (>15 m height) have been constructed for various uses including irrigation, flood control, navigation, and hydroelectric power generation [1]. As the number of dams continues to increase worldwide, so too has concerns for their effects on fish populations. Dams can act as a barrier to migratory (i.e., anadromous, catadromous, potamodromous) and resident fish (i.e., those that complete their life cycle within a reservoir or section of the river), fragmenting rivers and degrading habitats. The negative impacts of dams on upstream migration of diadromous fish are widely acknowledged, and the installation of various types of fishways to facilitate upstream passage are commonplace [2]. However, downstream migration of fish at dams remains a challenge [3, 4]. Depending upon the natural history of a given migratory fish, mature adults seeking spawning grounds (e.g., catadromous species) or juveniles or post-spawn adults (for iteroparous species) seeking rearing and feeding habitats (e.g., anadromous species) may all need to move downstream past dams. Resident species may also move considerable distances throughout a riverine system for reproduction, rearing, and foraging (e.g., Kokanee salmon *Oncorhynchus nerka*; white sucker *Catostomus commersonii*; walleye *Sander vitreus*) or simply move throughout reservoirs where they may traverse forebay areas.

Fish injury and mortality resulting from entrainment or impingement associated with hydroelectric facilities may have serious consequences for fish populations [5, 6]. Sources of entrainment or impingement-related injury or mortality include the following: (1) fish passage through hydroelectric facilities (i.e., turbines, spillways, sluiceways, and other passage routes) during downstream migration for migratory fish; (2) the entrainment of resident fish; and (3) the impingement of adult or large fish (migratory or resident) against screens/trash racks. Migrating fish will use existing dam structures such as spillways and outlet works, used to release and regulate water flow, for downstream passage. When no spills occur owing to low reservoir water levels, both resident and facultative migrant fish can be attracted to the turbine intake tunnels, often the only other source of downstream flow present in the forebay area of the dam. Entrainment, occurring when fish travel through a hydro dam to the tailraces, can result in physical injury and mortality from fish passing through turbines and associated components [6, 7]. Injury and mortality can occur through several means from hydroelectric components. Freefall from passing over a spillway, abrasions, scrapes, and mechanical strikes from turbine blades are well known causes of physical injury and mortality (reviewed in [6, 7, 8]). Injuries from turbulence and shear owing to different water velocity experienced across the body length, occurs when passing over a spillway or through turbine components [9]. Water pressure associated injuries and mortality can occur from low pressure, rapid changes in pressure, shear stress, turbulence, cavitation (extremely low water pressures that cause the formation of bubbles which subsequently collapse violently), strike, or grinding when fish become entrained in turbine components [5, 10, 11]. Injury and mortality can also occur from fish being impinged against screens or trash racks that are intended to prevent debris, or in some cases fish, from being drawn into water intakes [12].

Since downstream migrants are not often observed (e.g., juvenile fish), historically far less consideration has been afforded to downstream passage, such that management strategies and/or structures specifically designed to accommodate downstream passage were not implemented nearly as frequently [13]. To date, literature on downstream passage largely focuses on juvenile survival, particularly in Pacific salmonids (*Oncorhynchus* spp.), popular commercial and recreational species in which the adults senesce after spawning. Minimal research exists on downstream passage and entrainment risk of resident fish species [8]. However, research on adult downstream passage in migratory fish is growing in popularity in temperate Europe and North America, particularly for species of conservation interest such as eels (*Anguilla* spp., [14, 15, 16, 17, 18, 19]) and sturgeons (*Acipenser* spp., [20, 21, 22]). To enhance downstream passage and reduce mortality, management strategies have included selectively timing spills to aid juvenile fish, the installation of "fish friendly" by-pass systems and screens directing fish to these systems, and retrofitting dams with low-volume surface flow outlets [23] or removable spillway structures designed to minimize fish harm [24]. The use of light, sound, bubble curtains, and electrical currents to act as repellent from harmful paths or potentially an attractant to more desirable (fish friendly) paths have been explored [25]. Given that the timing of downstream migration differs among life stages and is species-dependent [8], mitigating injury and mortality during downstream passage in a multispecies system could prove challenging and disruptive to power generation operations. Furthermore, operational strategies can be complicated by environmental regulations such as water quality requirements.

From a fish productivity perspective, minimizing impacts during downstream passage for migratory fish, unintended entrainment of resident species, and/or fish impingement, is an integral part of managing fish productivity. Downstream passage mortality from a single hydropower dam may appear low (i.e., 5–10%), but system-wide cumulative mortalities may be considerable in systems greatly fragmented by multiple dams [26]. Adult survival affects population dynamics (e.g., effective population size), and thus fisheries yields (e.g., sustainable yield, maximum sustainable yield). Juvenile survival affects recruitment (i.e., fish reaching an age class considered part of a fishery), ultimately contributing to fisheries productivity. Literature reviews and technical reports compiled to date primarily focus on how fish injury and mortality occurs, and/or evaluate the effectiveness of various management strategies used to mitigate harm during downstream passage [6, 7, 8]. Given the contributions of migratory and resident adults and juveniles to fish production, a natural extension would be evaluating the impacts of fish injury and mortality from hydropower dam entrainment and impingement on fish productivity. Systematic reviews and meta-analyses can be valuable tools to (1) reveal generalized relationships between the impacts of fish entrainment and impingement and fish productivity; (2) summarize the various research methods and study designs for studying such relationships; and (3) identify knowledge gaps and future research directions.

ADVISORY TEAM INPUT

During the formulation of the question for this review, an Advisory Team made up of stakeholders and experts was established and consulted. This team included academics, staff from the Oak Ridge National Laboratory (U.S. Department of Energy), and staff from Fisheries and Oceans Canada (DFO), specifically the Fisheries Protection Program (FPP), and DFO Science Branch. The Advisory Team guided the focus of this review to ensure the primary question was both answerable and relevant, and suggested search terms to capture the relevant literature. The Advisory Team was also consulted in the development of the inclusion criteria for article screening and the list of specialist websites for searches.

OBJECTIVE OF THE REVIEW

The objective of the systematic review is to evaluate the existing literature base to determine the consequences of fish entrainment and impingement associated with hydroelectric dams on fish productivity in freshwater temperate environments.

Primary Question

What are the consequences of fish entrainment and impingement associated with hydroelectric dams on freshwater fish productivity in temperate regions?

Components of the Primary Question

The primary study question can be broken down into the study components:

Population: Freshwater fish, including diadromous species, in temperate regions.

Intervention: Infrastructure associated with hydroelectric facilities (i.e., turbines, spillways, sluiceways, outlet works, screens, tailraces, water bypasses, tailwaters, penstocks, trash racks, draft tubes).

Comparator: No intervention or modification to intervention.

Outcomes: Change in a component of fish productivity (broadly defined in terms of: mortality, injury, biomass, yield, abundance, diversity, growth, survival, individual performance, migration, reproduction, population sustainability, and population viability).

METHODS

Searches

Search Terms

A list of relevant search terms was generated by the Advisory Team, and divided into three components: the population (subject and environment type), intervention and outcome, and will be combined using Boolean operators "AND" and/or "OR" (Table 1).

Table 1: Proposed search string for the execution of the search strategy

Component	Search string
Population terms	(Fish* AND (Reservoir$ OR Impoundment$ OR Dam$ OR "Hydro electric*" OR Hydroelectric* OR "Hydro dam*" OR Hydrodam* OR "Hydro power" OR Hydropower OR "Hydro"))
	AND
Intervention terms	(Turbine$ OR Spill* OR Outlet* OR Overflow* OR Screen$ OR Tailrace$ OR "Tail race" OR Diversion OR Bypass* OR Tailwater$ OR Penstock$ OR Entrain* OR Imping* OR Blade$ OR Intake$ OR "Trash rack$" OR "Draft tube$")
	AND
Outcome terms	(Productivity OR Growth OR Performance OR Surviv* OR Success OR Migrat* OR Passag* OR Reproduc* OR Biomass OR Stress* OR Mortalit* OR Abundance$ OR Densit* OR Yield$ OR Injur* OR Viability OR Sustainability OR "Vital rate$" OR Persistence OR "Trauma")

The asterisk (*) is a wildcard and represents any characters, including no character (e.g., Spill* includes spill, spilling, spills, spilled) while the dollar sign ($) includes zero or one character (e.g., Reservoir$ includes reservoir or reservoirs). Since it was not feasible to include every possible relevant species name, the broad search term "fish" was included in an attempt to capture studies of every relevant species. While this could potentially exclude articles that do not contain the word 'fish' in the title/abstract/keywords, we were confident that our searches would still find relevant articles because some of the databases/search engines (e.g., Science.gov and Google scholar) search the full record for terms (not optional), increasing the chance that 'fish' would be selected. Furthermore, additional targeted searches will be conducted to locate articles that have not been found using this search strategy, including: (1) website searches, (2) hand searching reference sections of included articles and reviews to evaluate relevant titles, symposium papers and other articles, and (3) evidence calls to the broader scientific community via social media and email to help locate research articles that are difficult to obtain, or for suggestions of articles to include.

Abbreviated Search

When a complex search string is not accepted by the search engine, the help menu will be consulted and the search terms will be modified. The search terms will be recorded in the article databases in order to preserve all metadata associated with the search.

Article Type

The search will include a variety of article types, including primary literature in peer-reviewed journals and grey literature. The search strategy will strive to minimize publication biases by focusing efforts equally on each article type, and all articles will be equally critically appraised to ensure validity.

Document/File Formats

The search will not have any document type restrictions (e.g., PDF vs. PowerPoint vs. MS-Word). All formats will be acquired and if specialized software is required, alternative formats will be requested for ease of file transferability. Where books are identified, digital copies will be sought (either through internet searches for availability or requests to authors) in order to ensure that all obtainable records are made available as an output from this review.

Computer Settings

The browsing history and cookies will be disabled on all computers used to conduct the search. The members of the Review Team will not access any electronic accounts (e.g., email, website) during the search period and will use "private mode" (Safari) for web browsers to reduce the possibility of user-specific search results.

Language

English search terms will be used to conduct all searches in all databases. All references that are returned will be included in the database. When articles in other languages are returned using the search strategy, those records will be reported in the database.

Publication Databases

The following online databases will be searched.

- Waves (Fisheries and Oceans Canada)—Canadian government books, reports, government documents, theses, conference proceedings and journal titles
- ProQuest Dissertations & Theses Global—Collection of dissertations and theses from around the world, spanning from 1743 to the present
- Science.gov—U.S. Federal Science
- ISI Web of Science Core Collection—Multidisciplinary research topics including journals, books, proceedings, published data sets and patents
- Scopus—Abstract and citation database of peer-reviewed literature including journals, books and conference proceedings.

Search Engines

Search terms will be entered into Google Scholar and the first 500 hits (sorted by relevance) will be screened for the appropriate fit for the review questions.

Specialist Websites

Specialist organization websites listed below will be searched using connectors created in Import.io and abbreviated search terms [i.e., search strings (1) fish AND hydro AND entrainment; (2) fish AND hydro AND impingement; (3) fish AND hydro AND mortality; and (4) fish AND hydro AND injury]. Page data from the first 20 search results for each search string will be extracted, screened for relevance, and searched for links or references to relevant publications and data and grey literature. The list of websites was narrowed to the following 29 organizations after consulting with our Advisory Team for relevance.

1. Alberta Hydro
2. British Columbia Hydro
3. Centre for Ecology and Hydrology
4. Centre for Environment, Fisheries and Aquaculture Science
5. Commonwealth Scientific and Industrial Research Organisation (CSIRO)
6. Electric Power Research Institute
7. EU water framework directive
8. Federal Energy Regulatory Commission
9. Fisheries and Oceans Canada
10. Fisheries Research Service
11. Food and Agriculture Organization of the United Nations
12. Hydro Québec
13. Land and Water Australia
14. Manitoba Hydro
15. Ministry of Natural Resources and Environment of the Russian Federation
16. Ministry of the Environment New Zealand

17. National Institute of Water and Atmospheric Research New Zealand
18. Natural Resources Canada
19. Natural Resources Wales
20. Newfoundland and Labrador Hydro
21. National Institute of Water and Atmospheric Research
22. Northern Ireland Environment Agency
23. Office of Scientific and Technical Information (U.S. Department of Energy)
24. Pacific Fisheries Environmental Laboratory
25. Parks Canada
26. The Nature Conservancy
27. Trout Unlimited
28. United Nations Environment Programme
29. US Fish and Wildlife Service

Other Literature Searches

Reference sections of accepted articles will be hand searched to evaluate relevant titles, symposium papers and other articles that have not been found using the search strategy. Authors of any unpublished references will be contacted to request access to the full article. Stakeholders will be consulted for insight and advice for new sources of information.

The Review Team will contact authors of unobtainable articles in an attempt to gain access to the full article. We will also use social media and email to alert the community of this systematic review and to reach out to area experts for research articles that are difficult to obtain, or for suggestions of articles to include. Any article provided will also be used to test the comprehensiveness of our search strategy and, where appropriate, adjustments will be made to the search strategy to ensure it is comprehensive and inclusive. Any changes made to the search strategy will be justified and documented in the final review document.

Search Record Database

All articles generated by each of the search strategies will be exported into separate Zotero databases. After all searches have been completed and references found using each different strategy have been compiled, the individual databases will be exported into EPPI-reviewer as one database. Duplicates will be identified and merged. All references regardless of their perceived relevance to this systematic review will be included in the database. This database will act as the archive and will remain unchanged throughout the review process, since it is the direct product of the search strategy and will be useful in the future when updating the systematic review archive (general updating timeframe is currently every 5 years).

Article Screening and Study Inclusion Criteria

Screening Process and Inclusion Criteria

Articles found using the search criteria will be screened in two distinct stages: (1) title and abstract, and (2) full text. Before the screening of title and abstracts begins, two reviewers using a subset of 10% of all articles or 100 abstracts (whichever is bigger) will undertake consistency checks to ensure consistent and repeatable decisions are being made in regards to which articles get screened out and which go on in the process to be further reviewed. The two reviewers will use a Kappa test to determine consistencies in screening decisions. A Kappa score of $\geq$0.6 indicates substantial agreement between reviewers and will be required to be achieved before any further screening is conducted for the review. The results from the consistency check will be discussed and discrepancies will be reviewed by both reviewers to understand why the choice was made to include/exclude the article. This same process will be repeated prior to screening articles at full text i.e., two reviewers using a subset of 10% of all articles that were included at title and abstract, will undertake consistency checks and a Kappa score of $\geq$0.6 will be required before any further screening is conducted.

All article screening decisions will be included in the database, so it will be clear at what level any article was excluded. If the decision to include or exclude a specific article is unclear, that article will be retained and will go on to the next level of screening. If there is further doubt, the Review Team will discuss those articles as a group to come up with a decision. Any articles that do not have abstracts (as is the case for some grey literature), those articles will automatically be screened at the full text level. Justification of the reason for inclusion or exclusion of an article will be explained and recorded using EPPI reviewer, and all articles excluded at the full text level will be included with the review, in compliance with CEE guidelines [27]. Only English-language literature will be included during the screening stage.

Articles will be excluded based on the following pre-defined inclusion criteria developed in consultation with the Advisory Team:

Relevant Subjects

Any fish species, including diadromous species, in North (23.5°N–66.5°N) or South (23.5°S–66.5°S) temperate regions. Only studies located in freshwater ecosystems, including lakes, rivers, and streams that contain fish species that are associated with a hydroelectric dam system will be included.

Relevant Interventions

Articles that describe infrastructure associated with hydroelectric facilities that may cause fish to be entrained or impinged (i.e., turbines, spillways, sluiceways, outlet works, screens, tailraces, water bypasses, tailwaters, penstocks, trash racks, draft tubes etc.). Only articles that describe water that moves via gravity will be included. Studies will be excluded where water is actively pumped for: (1) power generation (e.g., storage ponds: Robbins 1976); (2) irrigation; or (3) cooling-water intake structures for thermoelectric power plants. Other studies excluded describe infrastructure associated with other operations: (1) nuclear facilities; (2) dams without hydro; (3) hydrokinetic systems (i.e., energy from waves/currents); or (4) general water withdrawal systems (e.g., for municipal drinking, recreation, etc.).

Relevant Comparators

Relevant comparators will include: (1) before intervention data within same waterbody (i.e., pre-installation/modification); (2) no intervention (e.g., control experiments whereby each phase of a test procedure is examined for sources of mortality/injury other than passage through infrastructure e.g., mortality attributable to upstream introduction and/or downstream recovery apparatus while turbine is operating but no fish passage through turbine); (3) an unmodified version of the intervention on the same or different study waterbody, or (4) controlled flume study.

Relevant Outcomes

Most metrics used to evaluate consequences of fish entrainment and impingement will be related to fish mortality and injury. Any article that uses a metric related to: (1) lethal impact: direct fish mortality or indirect mortality (e.g., fish are disoriented after passage through hydroelectric dam and then predated upon), and (2) sublethal impacts: external and/or internal injury assessments (e.g., signs of scale loss, barotrauma, blade strike, etc.)—will be included. These metrics can include, but not limited to, reported mortality rate (%), survival rate (%), injury rate (% of population) with particular types of injuries (e.g., signs of blade strike) or all injury types combined.

To evaluate the impact of fish mortality/injury from entrainment and impingement for fish productivity, the reported measured effect should indicate some change in a component of fish productivity. We will use a broad definition of fish productivity to include any measurement related to: mortality, biomass, abundance, yield, diversity, growth, survival, migration, reproduction, population viability, sustainability, persistence, stress, injury, trauma, or surrogate thereof. In many situations, the effect for fish productivity will be the reported metric related to mortality and injury, or the change in these metrics relative to the control. However, we will also include articles that scale-up the evaluation to include some estimate of a change in a component of fish productivity. For example, articles that include an estimate of fish loss from the population due to entrainment/impingement by comparing a metric related to mortality or injury to an estimate of population size, will be included. Therefore, if the article evaluates mortality/injury (e.g., % mortality) and also scales-up the evaluation by using either (1) population estimates upstream of the dam based on field collected data (e.g., hydroacoustic sampling, electrofishing, etc.), or (2) some form of modeling (e.g., individual-based simulation models or population viability analysis), these will be included.

Furthermore, the causal relationship between intervention and outcome should be made clear to allow for the effects of fish mortality/injury from entrainment and impingement to be isolated from other potential impacts of hydroelectric power production such as barriers to migration and/or habitat degradation. Studies where no clear mechanism is identified (e.g., if fish density is surveyed up-and down-stream of a hydro dam but any difference or change in fish density cannot be clearly linked to impingement or entrainment in isolation of other effects) will be excluded.

Relevant Types of Study Design

Study designs with appropriate comparators including Before/After (BA), Control/Impact (CI), as well as studies combining these types of comparisons, Before/After/Control/Impact (BACI) and Randomized Controlled Trials (RCT) will be included. Review papers and policy discussions will be excluded.

Study Quality Assessment

Each of the studies that make it to the full text screening level will be classified and coded in the article database using a number of parameters including (but not limited to):

- Study setting—field or Lab
- Study design—BA/CI/BACI/RCT
- Temporal extent of study—before monitoring, after monitoring, and total duration
- Replication—replicated or unreplicated (i.e., in time and/or space)
- Randomization—randomized or not (i.e., outcome assessed using randomized methods)
- Confounding factors—present, not present, unclear, and whether they were accounted for in the study (e.g., mortality from turbines vs. mortality from recovery nets)

- Clarity of objectives in relation to methods used
- Use of (and number of) controls
- Effort devoted to assessing a change in fish productivity
- Statistical methods used in assessment of impacts of impingement/entrainment mortality on fish productivity (e.g., were results analysed statistically?)
- Statistical power of methods reported (i.e., was a power analysis conducted and results thereof reported?)
- Accounting for and/or identifying potential effect modifiers (see list in following sections).

Bilotta et al. [28] have outlined criteria for the assessment of the internal validity of a study. Their assessment criteria have been adapted from the Cochrane Collaboration's Risk of Bias Tool [29] for use in the field of environmental science. The assessment criteria include assessing selection bias, attrition bias, reporting bias, as well as criteria relevant to our study (e.g., duration of monitoring, endpoints). The criteria outlined in Bilotta et al. [28] will be used by the Review Team for this review and included in the reference database. The information for each article retrieved using the search strategy will be uniquely coded based on the criteria (generally categorised as "low risk", "high risk", or "unclear risk") to help assess the quality of each article, and to provide insight into any potential risk of bias present in each of the studies. This information will be instrumental in helping to determine reliability of the evidence base available for potentially conducting a meta-analysis on the impacts of turbine and spillway mortality on fish productivity.

Data Extraction Strategy

Meta-data will be extracted from the included studies by the Review Team and will be recorded in a MS-Excel database that will be made available with the published systematic review article, as additional supporting files. The extracted information will be used to assess the overall effect of fish entrainment and impingement associated with hydroelectric dams on fish productivity, and when sufficient, good quality data exist, the information will be used in a meta-analysis. Some of the outcome data that will be recorded will include: outcome means, measures of variation (e.g., standard deviation, standard error, confidence intervals), and sample sizes. When data are presented in tables or graphs, all information will be extracted and recorded. If it is not possible to decipher information from graphs, the main contact author for the article will be contacted (via email or phone) by the Review Team to request the information. During that request, the Review Team will also solicit the author to suggest any grey literature that they may know of related to the systematic review topic. Where only raw data are provided in the article, the Review Team will calculate summary statistics. In those instances, we will record how the calculations were conducted and with what information. To ensure that data is being extracted in a consistent and repeatable manner, two reviewers will extract information from 10 of the same articles. Afterwards, the information will be compared. Any inconsistencies will be discussed amongst the Review Team members, and if any disagreement occurs, they will be discussed with the entire Review Team to ensure all reviewers are extracting and interpreting data in the same manner.

Potential Effect Modifiers and Reasons for Heterogeneity

Potential effect modifiers will be identified to obtain a better understanding of the factors that are associated with variation in effects among studies, given different: (1) site-specific factors (e.g., turbine type, size, power output); (2) methodologies and study designs used to assess impacts, and (3) biological factors (e.g., fish life history stage, body size and morphology). The Review Team will extract data on potential effect modifiers from articles that are included at the full-text level of screening. All information will be recorded in the MS-Excel database. The following list is not exhaustive, but includes potential effect modifiers and reasons for heterogeneity which will be recorded where available. Further factors will be identified, defined and included throughout the process, through consultation with the Advisory Team.

- Study location
- Environment type
- Lowest taxonomical level of subject
- Life history stage and strategy
- Fish body size/morphology
- Study design

- Monitoring duration
- Intervention type
- Turbine type/size
- Spillway type
- Comparator type
- Sampling method

Data Synthesis and Presentation

A narrative synthesis of data from all articles included in the systematic review will be generated. The synthesis will aim to be as visual as possible, summarizing information in tables and figures. The ultimate goal of this review is to reveal generalized relationships between the impacts of fish entrainment and impingement and fish productivity and to identify study design and methodological factors associated with these relationships. The findings of this systematic review will help to inform evidence-based management and conservation activities for resource managers that deal with maintaining fish productivity in freshwater ecosystems. All efforts will be made to provide quantitative assessments and meta-analysis of the articles included in this review, when the study designs and evidence-base allow. In the case that meta-analysis is possible (given a sufficient sample size of studies), study effect sizes will be standardized and weighted appropriately, and analysis will take the form of random-effects models. Depending on the availability of the data, meta-regressions or subgroup analyses of categories of studies will also be performed. Publication-bias and sensitivity analysis will be carried out where possible. Overall effects will be presented visually in plots of mean effect sizes and variance.

AUTHORS' CONTRIBUTIONS

The manuscript was drafted by TR and DAA. JJT, KES, JRB, PMH, and SJC provided comments and revisions. All authors read and approved the final manuscript.

ACKNOWLEDGEMENTS

The authors would like to thank several reviewers and collaborators who provided valuable insights to strengthen this review protocol including: Mike Power (University of Waterloo, CA), Mark Bevelhimer (Oak Ridge National Laboratory, USA), and Fisheries and Oceans Canada staff including Richard Janusz, Keith Clarke, Eva Enders, David Gibson. We also thank the reviewers for their constructive comments.

REFERENCES

1. International Commission on Large Dams (ICOLD). Register of Dams—general synthesis. (2015). http://www.icold-cigb.net/GB/World_register/general_synthesis.asp. Accessed 24 Nov 2016.
2. Bunt CM, Castro-Santos T, Haro R. Performance of fish passage structures at upstream barriers to migration. River Res Appl. 2012;28:457–78.
3. Calles O, Karlsson S, Hebrand M, Comoglio C. Evaluating technical improvements for downstream migrating diadromous fish at a hydroelectric plant. Ecol Eng. 2012;48:30–7.
4. Buysse D, Mouton AM, Baeyens R, Coeck J. Evaluation of downstream migration mitigation actions for eel at an Archimedes screw pump pumping station. Fish Manag Ecol. 2015;22:286–94.
5. Čada G. The development of advanced hydroelectric turbines to improve fish passage survival. Fisheries. 2001;26:14–23.
6. Electric Power Research Institute (EPRI). Fish passage through turbines: application of conventional hydropower data to hydrokinetic technologies. Final Rep. 2011;1024638:1–40.
7. Čada GF, Coutant CC, Whitney RR. Development of biological criteria for the design of advanced hydropower turbines. Washington: EERE Publication and Product Library Department of Energy; 1997. p. 1–85.
8. Larinier M, Travade F. Downstream migration: problems and facilities. Bull Fr Pêche Piscic. 2002;364(suppl):181–207.
9. Čada G, Loar J, Garrison L, Fisher R Jr, Neitzel D. Efforts to reduce mortality to hydroelectric turbine-passed fish: locating and

quantifying damaging shear stress. Environ Manag. 2006;37:898–906.

10. Čada G, Ahlgrimm J, Bahleda M, Bigford T, Stavrakas SD, Hall D, Moursund R, Sale M. Potential impacts of hydrokinetic and wave energy conversion technologies on aquatic environments. Fisheries. 2007;32:174–81.

11. Brown RS, Colotelo AH, Pflugrath BC, Boys CA, Baumgartner LJ, Deng D, Silva LGM, Brauner CJ, Mallen-Cooper M, Phonekhampeng O, Thorncraft G, Singhanouvong D. Understanding barotrauma in fish passing hydro structures: a global strategy for sustainable development of water resources. Fisheries. 2014;39:108–22.

12. Barnthouse LW. Impacts of entrainment and impingement on fish populations: a review of the scientific evidence. Environ Sci Policy. 2013;31:149–56.

13. Katopodis C, Williams JG. The development of fish passage research in a historical context. Ecol Eng. 2012;48:8–18.

14. Jansen HM, Winter HV, Bruijs MCM, Polman HJG. Just go with the flow? Route selection and mortality during downstream migration of silver eels in relation to river discharge. ICES J Mar Sci. 2007;64:1437–43.

15. Carr JW, Whoriskey FG. Migration of silver American eels past a hydroelectric dam and through a coastal zone. Fish Manag Ecol. 2008;15:393–400.

16. Travade F, Larinier M, Subra S, Gomes P, De-Oliveira E. Behaviour and passage of European silver eels (Anguilla anguilla) at a small hydropower plant during their downstream migration. Knowl Manag Aquat Ecosyst. 2010;398(01):1–19.

17. Besson ML, Trancart T, Acou A, Charrier F, Mazel V, Legault A, Feunteun E. Disrupted downstream migration behaviour of European silver eels (Anguilla anguilla, L.) in an obstructed river. Environ Biol Fish. 2016;99:779–91.

18. Eyler SM, Welsh SA, Smith DR, Rockey MM. Downstream passage and impact of turbine shutdowns on survival of silver American eels at five hydroelectric dams on the Shenandoah River. Trans Am Fish Soc. 2016;145:964–76.

19. Haro A, Watten B, Noreika J. Passage of downstream migrant

American eels through an airlift-assisted deep bypass. Ecol Eng. 2016;91:545–52.

20. Acolas ML, Rochard E, Le Pichon C, Rouleau E. Downstream migration patterns of one-year-old hatchery-reared European sturgeon (Acipenser sturio). J Exp Mar Biol Ecol. 2012;430–431:68–77.
21. McDougall CA, Blanchfield PJ, Peake SJ, Anderson WG. Movement patterns and size-class influence entrainment susceptibility of lake sturgeon in a small hydroelectric reservoir. Trans Am Fish Soc. 2013;142:1508–21.
22. McDougall CA, Anderson WG, Peake SJ. Downstream passage of lake sturgeon through a hydroelectric generating station: route determination, survival, and fine-scale movements. N Am J Fish Manag. 2014;34:546–58.
23. Johnson GE, Dauble DD. Surface flow outlets to protect juvenile salmonids passing through hydropower dams. Rev Fish Sci. 2006;14:213–44.
24. Adams NS, Plumb JM, Perry RW, Rondorf DW. Performance of a surface bypass structure to enhance juvenile steelhead passage and survival at Lower Granite Dam, Washington. N Am J Fish Manag. 2014;34:576–94.
25. Popper AN, Carlson TJ. Application of sound and other stimuli to control fish behavior. Trans Am Fish Soc. 1998;127:673–707.
26. Marohn L, Prigge E, Reinhold H. Escapement success of silver eels from a German river system is low compared to management-based estimates. Freshw Biol. 2014;59:64–72.
27. Collaboration for Environmental Evidence. Guidelines for systematic review and evidence synthesis in environmental management. Version 4.2. Environ Evid. 2013; http://environmentalevidence.org/wp-content/uploads/2014/06/Review-guidelines-version-4.2-finalPRINT.pdf. Accessed 26 Jan 2017.
28. Bilotta GS, Milner AM, Boyd IL. Quality assessment tools for evidence from environmental science. Environ Evid. 2014;3:14.
29. Higgins JP, Altman DG, Gøtzsche PC, Jüni P, Moher D, Oxman AD, Sterne JA. The Cochran Collaboration's tool for assessing risk of bias in randomized trials. Bmj. 2011;343:d5928.

CHAPTER

5

POPULATION CONNECTIVITY: DAM MIGRATION MITIGATIONS AND CONTEMPORARY SITE FIDELITY IN ARCTIC CHAR

Jens Wollebæk[1,2], Jan Heggenes[2] and Knut H Røed[1]

[1] The Norwegian School of Veterinary Science, Dep. of Basic Sciences and Aquatic Medicine, Box 8146, Dep. 0033 Oslo, Norway

[2] Telemark University College, Dep. of Environmental Sciences, Hallvard Eikas Plass, 3800 Bø i Telemark, Norway.

ABSTRACT

Background

Animal feeding and spawning migrations may be limited by physical barriers and behavioral interactions. Dam constructions (e.g. hydropower) commonly include gateways for fish migrations to sustain ecological connectivity. Relative genetic impacts of fish passage devices versus natural processes (e.g. hybrid inferiority) are, however, rarely studied. We examined genetic (i.e. microsatellite) population connectivity of highly migrating lake-dwelling Arctic char (*Salvelinus alpinus*), introduced 20 generations ago, across and within two subalpine lakes separated by a dam with a subterranean tunnel and spill gates after 7 generations. Due to water flow regime, the time window for fish migration is highly restricted.

Results

Char populations, with similar genetic structuring and diversity observed across and within lakes, were admixed across the dam with fishways during feeding. For spawning, however, statistically significant, but very low population differentiation (θ; 0.002 - 0.013) was found in nine out of ten reproductive site comparisons, reflecting interactions between extensive migration (mean first generation (F_0) = 10.8%) and initial site fidelity. Simulations indicated that genetic drift among relatively small effective populations (mean N_e = 62) may have caused the observed contemporary differentiation. Novel Bayesian analyses indicated mean contributions of 71% F_0 population hybrids in spawning populations, of which 76% had maternal or paternal native origin.

Conclusions

Ecological connectivity between lakes separated by a dam has been retained through construction of fishways for feeding migration. Considerable survival and homing to ancestral spawning sites in hybrid progeny was documented. Population differentiation despite preceding admixture is likely caused by contemporary reduced reproductive fitness of population hybrids. The study documents the beginning stages of population divergence among spatial aggregations with recent common ancestry.

BACKGROUND

Genetic differentiation, fundamental to population genetics, is initiated by restricted gene flow and reproductive isolation mechanisms within gene pools [1, 2]. Habitat and population fragmentation with concomitant disruption of ecological connectivity is threatening biota worldwide, on large scales determined by time since isolation and physical reproductive barriers among demes [3]. On finer scales, genetic differentiation may additionally involve reproductive behaviors and social interactions [4, 5]. Addressing genetic effects of physical barriers and reproductive behaviors in novel populations may forecast evolutionary consequences of human intervention.

Natural ecological connectivity may fluctuate in time and space, but anthropogenic alterations (e.g. water regulation) typically exaggerate this variation. Human impacted water flows and water habitat fragmentation expanded greatly worldwide in the twentieth century [6]. In Europe (excluding Russia) close to all large river systems are fragmented by dams [7], inevitably presenting impediments to migration. Construction of habitat corridors to sustain ecological connectivity may in theory counteract negative fragmentation effects [8], although empirical evidence is limited. A number of often retrospective fish passage improvements have been designed to reduce ecosystem wide effects of (hydropower) dams [9]. Dams commonly allow passive or active downstream drift of fish, but obstruct upstream migration. World Commissions of Dams with contributing papers [10, 11] and recent reviews [12, 13, 14], highlight the need for empirical studies addressing long term effects of fish passage success. Migration barriers may obstruct passage directly or negate otherwise adaptive benefits of homing and assortative mating indirectly [15, 16, 17].

Genetic drift and behavior may induce reproductive isolation mechanisms operating on fine scales. Genetic microstructure even within continuous habitat is reported [18, 19], although the biological relevance of weak differentiation is debated [20, 21, 22]. Within-lake genetic structure for highly mobile fish without obvious migration barriers have similarly been documented [23, 24, 25, 26, 27, 28] for a traditional evolutionary time perspective. Genetic structuring and adaptive microevolution may, however, also occur on contemporary time scales [29, 30, 31, 32, 33]. Recently founded or invasive species provide opportunities to study initial genetic divergence and evolution in the wild [34], but few studies have evaluated contemporary genetic structure in the invasive species [35, 36, 37]. Moreover, we are not aware of assessment of initial genetic structure in recently founded and highly migrating species that are temporally (i.e. seasonally) admixed. Recently founded salmonid species in freshwater lake habitats represent natural small scale experiments that provide opportunities to study the details of early stages of population divergence.

Micro-scale studies under initial admixture introduce challenges concerning sampling design and analytical approaches. For obvious reasons, sample sizes are often restricted both in numbers and temporal replicates. In continuous habitats sampling units may be difficult to delineate. Widely used Bayesian methods to infer cryptic population structure [38, 39, 40] may not, however, detect population substructure of recently established populations with low differentiation. Inclusion of population migrants and hybrids (i.e. admixed ancestry) in presumably distinct populations may lead to underestimates of differentiation and erroneous conclusions regarding reproductive isolation.

Reproductive isolation may be reinforced by homing, as documented within a wide range of species [41, 42]. Successful reproduction of migrants may be common, caused by relaxed selection against non-natives resulting in non-native offspring, or hybrid offspring with admixed ancestry which break down incipient population differentiation [5, 43]. Further differentiation requires reduced reproductive success of non-native and hybrid progeny. Therefore, studies of population admixture and hybrid contribution to population segregation under incomplete divergence are important and may possibly unveil biological relevant differentiation. Traditional measures of population differentiation combined with recently developed Bayesian statistics [44] now permit studies of how inter-population hybrids may affect population differentiation.

Here we study genetic differentiation of Arctic char (*Salvelinus alpinus*), introduced around 1920 (i.e. 20 generations ago), across and within two lakes: Pålsbufjorden (PAL) and the downstream Tunhovdfjorden (TUN) (Figure 1). The two lakes, naturally connected by a short river, were separated by a dam in 1946, creating two hydropower reservoirs. Both lakes have geographically distinct char spawning sites (PAL; P1 - 3, TUN; T1 - 2) for which we assess population connectivity implementing 10 microsatellites. The first objective was to examine genetic structure across lakes to quantify present effects of two migration enhancements at the dam; a subterranean tunnel and spill gates. An additional summer sample (X) of lake feeding and presumably admixed char from PAL was included to assist interpretation of differentiation. We expected to find greater genetic differentiation across lakes than across populations within lakes, and assignment of sample × to PAL populations. We also expected that passive migration downstream would exceed upward migration and result in observed asymmetric migration and associated genetic variation. The second was to examine migration patterns and population hybrid contributions potentially responsible for contemporary genetic substructure. Due to expectations of low population differentiation in the progress towards reproductive isolation, we intend to deliberate over the paradox of population divergence in the face of migration, rather than to state absolute assertions.

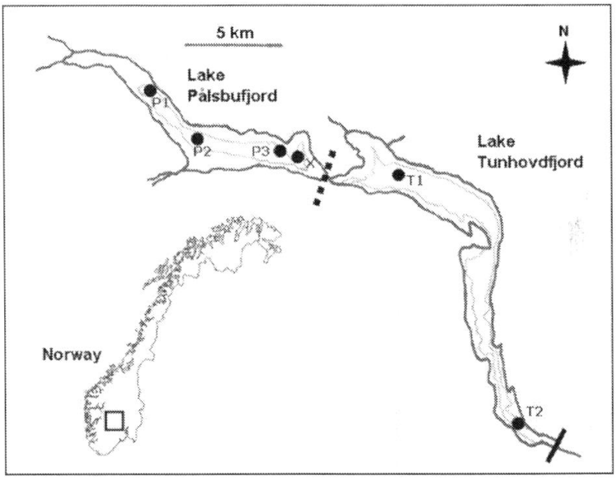

Figure 1: Map of sample localities. Map of the study area with Lake Pålsbufjord (PAL) and the downstream Lake Tunhovdfjord (TUN). Minimum water level (LRV) is indicated by the grey line, and sampling sites are marked with dots and abbreviations (P1 - 3, T1 - 2 and X). Solid bars indicate semi-barrier, while broken bars indicate partial restriction to up- and downstream migration.

RESULTS

Amplification and allele calling of 190 individuals over 6 sites was obtained in 99.8% of the cases, and secondary amplifying and allele calling of a subsample gave consistent results. Number of alleles per locus averaged 15.4 (SD ± 7.2, range 2 - 26). Quality control screening did not reveal indications of scoring error due to stuttering or evidence for large allele dropout. The test for null alleles assuming Hardy-Weinberg equilibrium (HW) suggested that null alleles may be present for locus *Sco204* in sample site T2, by general excess of homozygotes. However, the combined probability for all homozygote class frequencies was not significant ($P > 0.05$). No deviations from HW were found, but nonrandom association of alleles from different loci was indicated in 12 and 9 (*Sco 213/Sfo8, Smm17, Smm24, Str73; Sfo8/Smm17, Smm24, Str73; Smm24/Smm17, Str73*) of 45 tests uncorrected and after false discovery rate (FDR) correction [45] at 5% level, respectively. Genetic variation revealed a mean value of local $F_{ST} = 0.024$ (range 0.013 - 0.036), and inbreeding F = 0.028 (range 0.003 - 0.053) over sites.

Dam Barrier to Migration and Gene Flow

Measured lake water levels and estimated velocities in the fishways indicated possible upstream migration only within very limited windows of time. During 2000 - 2010, migration through the tunnel was feasible in less than 3.6% of the year (mean 13 days per year, SD ± 20, min 0, max 55). Migration through the spill gates was possible in less than 1.7% of the year (mean 6 days per year, SD ± 4, min 1, max 16). Upstream migration was possible in all seasons through gates, but only through the tunnel in spring.

Observed heterozygosity and allelic richness was similar between lakes. Number of private alleles found totaled 22, 6 in TUN and 16 in PAL (data not shown).

The total allele frequency variation consisted of 99.3% ($P < 0.001$) variation within populations, 0.7% ($P = 0.001$) variation among spawning sites within lakes and no variation between lakes ($P = 0.499$). Genetic differentiation of char from the two connected lakes PAL and TUN was, however, highly significant ($P = 0.001$) but very low ($\theta = 0.003$, standardized (θ') = 0.014). Removal of first generation migrants (F_0), amounting to 7 and 9% in PAL and TUN (not significant after correction, 170 tests, threshold $P = 0.0003$), respectively, did not affect this differentiation. STRUCTURE did not detect genetic structure among lakes or sites (K = 1).

Estimated F_{ST} and exact tests revealed that one spawning population (T2) was significantly different from a summer feeding sample (X) in PAL (Table 1). Sample × assigned to both lakes (TUN, 25%), and 30% was most similar to spawning site T1 on a population level (Table 2), indicating upstream migration.

Table 1: Population differentiation

Sample site	**P1**	**P2**	**P3**	**X**	**T1**	**T2**
P1		0.0125*	0.0051*	-0.0046^{NS}	0.0076*	0.0060*
P2	0.0035*		0.0038*	0.0044^{NS}	0.0022*	0.0105*
P3	0.0103*	0.0050*		-0.0045^{NS}	0.0030*	0.0098*
X	0.6572^{NS}	0.1290^{NS}	0.9587^{NS}		< 0.0001^{NS}	0.0026*
T1	0.0085*	0.0473^{NS}	0.0019*	0.5535^{NS}		0.0046^{NS}
T2	0.0004*	< 0.0001*	< 0.0001*	0.0045*	0.0580^{NS}	

Pairwise test of differentiation (θ, upper), and P-values from exact test in TFPGA (lower) between spawning char in PAL (P1 - 3), TUN (T1 - 2) and a mixed summer sample in PAL (X). Indicative adjusted nominal level (5%) in pairwise test for multiple comparisons is 0.00333, initial k = 15. Non-significant (^{NS}) and significant (*) differentiation at 5% level after FDR correction.

Table 2: F_0 migrants and population assignment

					Assigned population			
Sample site	**N**	**F_0**	**P1**	**P2**	**P3**	**T1**	**T2**	**na**
P1	34	0.15	0.32	0.15 [0.03]	0.32 [0.03]	0.18 [0.06]	0.00	0.03 [0.03]
P2	34	0.03	0.12	0.38	0.38 [0.03]	0.12	0.00	0.00
P3	34	0.12	0.03	0.09 [0.03]	0.62	0.15	0.03	0.09 [0.09]
T1	34	0.09	0.18 [0.03]	0.12 [0.03]	0.21 [0.03]	0.38	0.09	0.03
T2	34	0.15	0.21	0.09	0.24 [0.09]	0.35 [0.06]	0.12	0.00
X	20		0.15	0.00	0.55	0.3	0.00	0.00

Frequencies of first generation (F_0, before FDR correction) migrants and char from the spawning sites (PAL; P1 - P3 and TUN; T1 - 2) and a mixed summer sample in PAL (X) assigned (na: non-assigned) to the five spawning sites estimated in GENECLASS. Frequencies of F_0 migrants are given in brackets.

Migration beyond F_0 was assessed with a maximum likelihood approach for assignment and with Bayesian estimates for present and contemporary estimates. Self-assignment of individual char in

GENECLASS2 revealed 93 and 46% correct assignments to Lake PAL and TUN, respectively, suggesting a downstream source-sink structure. Probability estimates of ancestral origin over the last generation estimated in the software program BIMR, also indicated high, but more symmetric migration rates. With the two lakes as populations, mode allele frequencies showed almost equal ancestral native origin for the two lakes (PAL; 56%, highest posterior density intervals (HPDI): 34 - 75, TUN; 59%, HPDI: 25 - 90), with estimated HPDI for non-native origin of 24 - 65% and 9 - 74%, in PAL and TUN respectively.

Mode probability of hybrid ancestral origin for the pooled samples within lakes estimated in BIMR were 50% in PAL (HPDI = 36 - 49) and also 50% (HPDI = 13 - 48) in TUN. Nevertheless, pure origin from the home lake was more common than pure origin from the other lake.

Lentic Population Migration and Admixture

All except one test of genetic differentiation among spawning sites were significant (mean θ = 0.007, SD ± 0.003). Mean standardized differentiation (θ') was 0.030 (SD ± 0.016). Exact tests corroborated the pattern of significantly differentiated spawning sites (Table 1). Using spawning sites as prior populations, 37% of the char assigned to its sampled site, varying from 12 - 62% among sites (Table 2). Mean probability of best assignment, however, was only 51% (SD ± 28), indicating recent admixture, migration, or un-sampled populations. Exclusion of 18 char (11%) found to be F_0 migrants across spawning sites (range 1 - 5, Table 2, not significant after correction, 170 tests, threshold P = 0.0003) increased population segregation and resulted in all spawning populations being significantly different ($\theta > 0.002$, $\theta' > 0.008$, $P < 0.021$). This finding corroborates the estimated migration rates. Analyses in IMMANC verified migrant detection (8%) with a power of 0.968. The Bayesian probability that the sampled alleles within spawning sites originated from the same site last generation (ancestral rate) were 0.47 across sites (SD ± 0.09, min 0.34, max 0.60), and higher than the probability of origin from any other population (mean 0.13, SD ± 0.10, min 0.02 max 0.29).

Mixed ancestry analyses indicated that population hybrids constituted a substantial proportion within spawning sites. The majority of hybrids within all sites were progeny of native and non-native char (mean 54%), whereas hybrids of two non-natives were less common (mean 17%,

Figure 2). Similar to pooled lake estimates, the mode probability of home origin within sites was consistently higher than pure non-native origin within sites, and consequently higher for hybrids with partial native ancestry than for any other mixed ancestry across sites. Standard deviations of mixed ancestry estimates were generally similar to mode estimates.

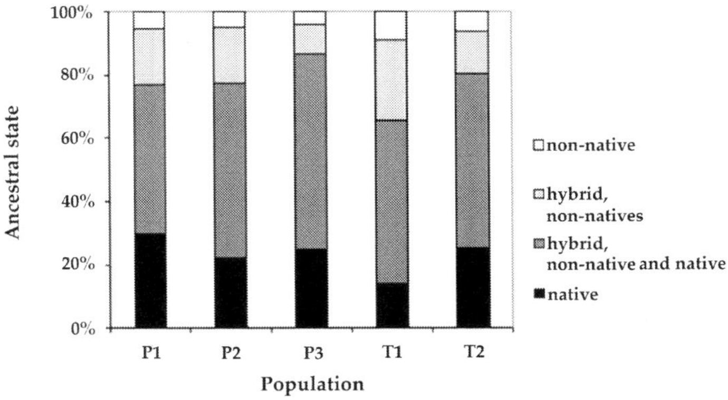

Figure 2: Ancestral origins. Pooled ancestral state proportions of spawning char at site P1 - 3 in Lake Pålsbufjord, and site T1 - 2 in Lake Tunhovdfjord, estimated in BIMR. Hybrid native indicates that one parent is from the local population, while hybrid non-native indicate parental origin of two different non-native populations.

Potential Interpretation Bias

Potential bias and biological relevance of observed data were addressed to assure sound interpretations. F_{ST} tests excluding loci with possible LD (*Sco213* and *Sfo8*) revealed similar differentiations as for all loci combined (data not shown). The F_{ST} estimate per locus averaged 0.0064 (SD ± 0.006, min 0.001, max 0.023). Exclusion of single loci generally weakened spawning site differentiation, but did not change the above pattern, thus indicating neutral behavior of markers.

AMOVA analyses segregated based on year-classes revealed 100% variation within sites and a lack of variation among year-classes (F_{ST} = -0.001, P = 0.624), whereas hierarchical analyses based on spawning sites revealed that 99.45% of genetic variation stemmed from within-population variation and 0.55% (F_{ST} = 0.006, P = 0.002) from among-population variation. Thus, genetic variation was congruent with the high variability of microsatellites used, indicating temporal stability among consecutive year-classes. Pairwise year-classes were not significantly different within any spawning site ($\theta < 0.060$, $P > 0.013$), or pooled across sites ($\theta < 0.004$, $P > 0.184$). Removal of year-classes one-by-one changed the significance in 15.5% of the pairwise tests, but did not affect the pattern of differentiation among spawning sites.

We did not find evidence for family structures affecting population differentiations. Simulations indicated a power of 1.000 and 0.918 to discriminate unrelated from full sib and half sib, respectively. Only 0.1% of pairwise tests had relatedness coefficients ≥ 0.25, indicating half sib relationship, and half of these were across sites. Mean distribution of relatedness (LRM) for the pooled dataset was negative ($\mu = -0.003$, SD ± 0.045, Figure 3), although positively skewed (1.640) and not normally distributed ($P = 0.010$). Sites individually revealed the same pattern (not shown), with lack of bimodal patterns that could have indicated kin-groups.

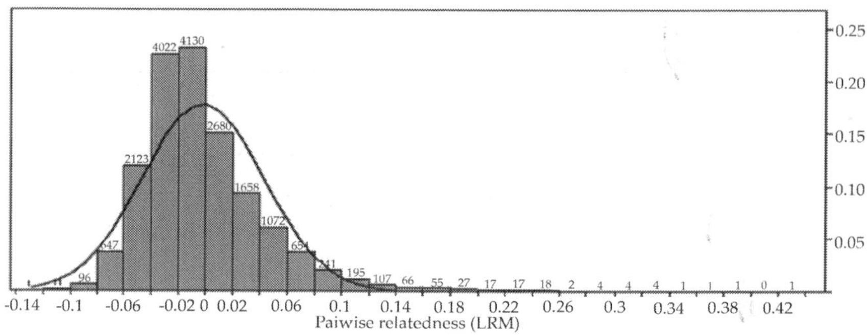

Figure 3: Pairwise relatedness. Ritland and Lynch (1999) pairwise relatedness (LRM;[153]) histogram for the pooled sample, with fitted normal density curve. Pairwise counts are given on top of bars.

Effective population sizes (N_e) were estimated to be 52 - 71 individuals within spawning sites, 133 - 142 within lakes, and 233 in total (Table 3), supporting both across and within lake genetic structure. Estimates were not affected by prior values of N_e. Estimates of effective parental population size (N_b) were generally higher but similar to N_e, except negative and infinite for population P1. This may indicate sampling error, or no evidence for LD caused by genetic drift due to a finite number of parents. Tests for LD did not, however, give significant deviations for any loci combination at this site, after FDR correction. Recent bottlenecks over all loci were not confirmed for any of the tests ($P > 0.080$), assuming a mixed mutation model (TPM). Neither did tests for di-nucleotide and tetra-nucleotide loci separately, under TPM and a single-step mutation model (SMM) respectively, reveal signs of severe loss of allelic diversity (normal L-shaped distribution, $P > 0.060$), except for population P2 and tetra-nucleotide loci, where the Wilcoxon test for heterozygosity excess was significant ($P = 0.030$). Test for isolation by distance (IBD) was positive but not significant ($r = 0.29$, $P_{upper} = 0.250$, $P_{lower} = 0.760$).

Table 3: Effective population sizes

Lake	**Population**	N_e**(ONESAMP)**	N_b**(LDNE)**
PAL	P1	63.6 (47.9 - 106.8)	∞ (229.3 - ∞)
	P2	53.9 (41.3 - 87.5)	58.0 (43.3. - 84.0)
	P3	51.7 (39.9 - 83.0)	56.5 (42.8 - 79.8)
	PAL total	132.6 (111.0 - 173.5)	272.5 (198.7 - 416.5)
TUN	T1	69.8 (51.8 - 124.0)	62.0 (46.2 - 90.3)
	T2	70.6 (48.8 - 135.1)	45.4 (33.3 - 66.8)
	TUN total	142.4 (100.3 - 263.6)	105.6 (79.9 - 149.5)
PAL/TUN	total	232.5 (202.7 - 278.9)	312.6 (244.6 - 422.0)

Estimated mean effective population size (N_e) and parental population size (N_b) with 95% C.I., from ONESAMP and LDNE respectively from PAL (P1 - 3) and TUN (T1 - 2) and pooled data. Negative estimates are reported as infinite (∞).

Forward simulated F_{ST} estimates (Figure 4), indicated that genetic drift alone may have caused the observed population differentiation. For instance, five populations with $N_e = 60$ were significantly differentiated (mean $\theta = 0.005$, SD $\pm$ 0.002, $P < 0.047$) after a mean migration rate of 0.3 in 20 generations, assuming initial admixture after introduction. Ten independent replicates of this scenario indicated similar differentiation (mean $\theta = 0.006$, SD $\pm$ 0.001). IMMANC indicated detection power for migrants (8%) and hybrids (7%) in our sampled dataset to be 0.968 and 0.638, respectively.

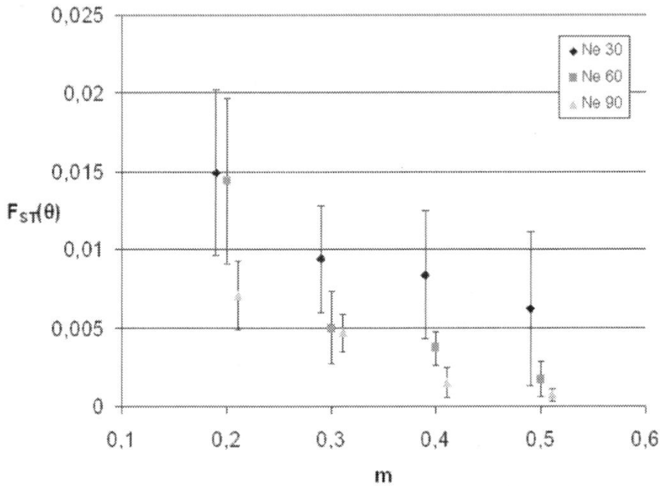

Figure 4: Population differentiation simulation. Simulated effects of relevant migration rates and effective population sizes to mean (SD) pairwise differentiation between 5 populations, after 20 generations.

A forward simulated alternative hypotheses of m = 0.01 revealed high differentiation (mean θ = 0.122, SD ± 0.019, P < 0.001) between reproductive sites, whereas m = 0.99 revealed lack of differentiation (mean θ = 0.001, SD ± 0.001, P > 0.176). A scenario without migration only indicated 5.0% (1.3% after correction) F_0 migrants in GENECLASS, similar to expectations from type 1 errors. Similar analyses in IMMANC indicated 10.7% (1.0% after correction) F_0 migrants and 10.0% (1.3% after correction) F_1 hybrids, with mean power > 0.999. Coalescent simulations supported that drift alone may have caused contemporary site fidelity.

DISCUSSION

Migrations

Our results suggest that constructed gateways for migration ensured alternate habitat utilization and gene flow across the dam separating the two studied lakes. Significant differentiation was found between spawning char from the two interconnected lakes, but differentiation among populations within lakes equaled or exceeded that for lakes. Without assuming any direct relationship between F_{ST} and gene flow (below), these relative measures are low and indicate high connectivity across the dam. Char with genetic assignment to distinct populations in both lakes were caught during feeding migration in the upper lake in summer, indicating upstream migration across the dam. Corroborating this, gene diversity and genetic structure were similar across and within lakes.

Extensive migration reflecting seasonal foraging admixture across lakes and populations is independently corroborated by tagging studies [46, 47]. Barrier type, hydrology and life stages of focal species all interact in determining the impact of barriers [48]. Tunnels associated with turbine outflow at dams are known to attract ascending salmonids [49, 50, 51]. The lack of turbines in the PAL dam tunnel makes migration through the tunnel possible. It is an open question, however, to what extent char in our study lakes utilize the spill gates or the tunnel for (upstream) migration. Ongoing studies in brown trout also reveal high dispersal across the focal dam, with unknown proportional use of the fishways (unpublished). Regardless, results indicate that char, in both cases, take advantage of the highly restricted time window suitable for upstream migration. Disentangling present from previous genetic structure and biological interpretation of low differentiation is, however, challenging (below), and results should therefore be interpreted with caution.

Contemporary Genetic Substructure

The present study documents that genetically differentiated sympatric populations can be established within decades. Pairwise genetic differentiation was found to be significant, but very low, both among lakes and among most spawning sites within lakes, indicating spatial reproductive isolation even within continuous habitat. Estimates of N_e supported this. However, STRUCTURE did not reveal any population structuring. Performance of Bayesian clustering methods decreased for $F_{ST} < 0.02$ [52], and we propose that even the novel algorithm used in STRUCTURE [39] may not detect recently established populations with very low differentiation ($F_{ST} < 0.01$).

Low differentiation is expected due to the short time since species introduction, and the migratory behavior of char. Following the simplistic [53, 54] equation $F_{ST} \approx 1/4Nm + 1)$, our mean local F_{ST} estimates indicate long term Nm = 10.2, and migration rates of 0.17 assuming N_e = 60. Fixed m = 0.5 would reveal the same differentiation if N_e is as low as 20. As an example, 60 individuals and $m = 0.5$ result in the same differentiation as e.g. 1000 individuals with 3.0% gene flow. In contrast, if we assume F_0 analyses (mean $m = 0.11$) are representative for present gene flow, and implement N_e 's of 60, the present Nm is 6.6, likely overestimated due to common reproductive inferiority in migrants. Consequently, mean gene flow between populations may have decreased since introduction. Thus, present genetic structure does not reflect panmixia [55]. Both estimates, however, indicate that gene flow is too small to bring about drift connectivity [53], but too large for Bayesian detection of subdivision [54]. A denotation of the sampled reproductive entities as populations, according to the evolutionary and especially the ecological paradigm [22] is, however, debatable. Maximum F_{ST} estimates will be limited by heterozygosity [21], and estimated differentiation may underestimate true population differentiation [56], especially when gene diversity is high [57]. Similar standardized population differentiation is previously found in char [58]. Direct comparisons of observed θ are complicated, but similarly low and biological relevant differentiations are found in both mammals and fish [59, 60]. Interestingly, similarly low, but significant differentiations are also found among populations of char founded thousands of years ago [61], and population structure of lake-dwelling brook char (*Salvelinus fontinalis*) has been found on a smaller geographical scale under migration-drift equilibrium [62] than in our study. Low but significant pairwise differentiations are also found in invading catadromous crustaceans separated after recent colonization [35]. This study also documented strong genetic drift simultaneously with significant differentiation ($F_{ST} > 0.007$), but among year classes, as opposed to our study.

True substructures in our study were corroborated by N_e estimates, genetic variability, year-class stability and low number of sib-groups within sites. Invasive species, such as char in our lakes, are expected to express low genetic variation and N_e , caused by founder events and population bottlenecks [63, 64, 65]. The high level of genetic variation within the examined loci in our populations does not suggest depauperate populations or bottlenecks on the surveyed scale, which would otherwise certainly support genetic differentiation caused by strong genetic drift in very small founding populations. The concept of founder populations, however, suggests that current structure is novel. We are not aware of evidence from the literature that fish populations bred in admixture may resume historic population structures in novel habitat. Simulations, however, documented that drift alone, even in larger populations, could cause rapid differentiation. Estimation of population size based on single samples relies on a number of simplifying assumptions [66, 67]. Our estimated population sizes should therefore be interpreted with caution. Echosounder and gillnetting experiments [68] suggest a population size of char around 9000 in PAL, returning $N_e/N \approx 0.015$, which is common for marine fish [60] but low compared to most salmonids [69, 70]. Nevertheless, estimated N_e and N_b gave similar results, and suggest population sizes sufficient to confer contemporary genetic stability as well, as opposed to a transitory or strict metapopulation structure [71]. It is possible, however, that estimates are inflated by immigrants and sampling of hybrid zones, even though no deviations from HW were evident. Regardless, the question remains whether one is actually able to sample biological populations correctly, and if hybrid contribution represents true population structure. The presence of immigrants and hybrids, not contributing to the reproductive population gene pool, can easily complicate interpretations of genetic diversity, N_e and genetic differentiation.

Initial Genetic Structure

The low level of interspecies competition and vacant niches fits well with rapid establishment of char in our lakes, and zooplanktivores such as char show a high rate of invasion success [72]. Genetic differentiation by means of the evolution of reproductive isolation can be viewed as non-ecological modes of differentiation such as founder effect or genetic drift in small populations, or by divergent natural selection [73, 74]. Lag time between initial colonization and the onset of rapid population growth is expected in invasive species [75], in particular if evolutionary adaptation is important during colonization. Thus, the rapid expansion in the lakes surveyed [76] corroborates the independent simulations, pointing to genetic drift as an initial diversifying factor for observed population differentiation. However, theory [1, 77, 78], experiments [79] and empirical studies [80, 81, 82, 83, 84, 85] all suggest a potential for rapid evolutionary changes involving adaptation in newly founded populations. Moderate levels of plasticity are also typical for char [86, 87], and may also have been optimal for fast genetic evolution [81, 88, 89, 90].

The combined genetic and ecological data suggest that the initial genetic structure and diversity within our lakes is a result of site fidelity and initial drift. Several evolutionary forces may, however, interact at the same time. Realistic and rapid drivers of cryptic kin selection [91] or discontinuous adaptive variation [32, 92, 93, 94] may add to incipient reproductive isolation. Once sub-structure is established, site fidelity, drift, and low effective migration rates may strengthen differentiation in time and space. Such isolation by adaptation [95] is likely facilitating drift in neutral loci by reduced gene flow as a general barrier mechanism [96], and even neutral markers may detect adaptation in the face of intermediate migration [97].

Differentiation despite Extensive Migration

Substantial migration among reproductive sites was found, without significant IBD, suggesting little present drift. Thus, although common in salmonids, equilibrium between gene flow and genetic drift is not present [98, 99]. Migration analyses may be biased by method assumptions, unsampled populations, low differentiation and convergence problems [44, 100, 101]. However, all methods used to interpret migration in our study, indicated considerable migration among lakes and populations. Extensive migration among populations is often seen in salmonids [43]. Homing and kin discrimination will, however, contribute to genetic structuring and is widely known in fish [102], particularly char [103, 104, 105, 106]. Pairwise genetic differentiation of other neighboring lacustrine char populations has been found to be highly significant, despite long term migration rates of 1.853 - 2.755 individuals per generation [107]. Low levels of effective migration, i.e. gene flow among sympatric populations, contrary indicate breeding site fidelity in established populations [107, 108]. Reproductive divergence despite extensive migration in the study at hand indicates restricted effective gene flow between lakes and populations, even after 20 generations. Thus, at first glance, one may be tempted to conclude that results indicate reproductive selection against immigrants in newly established populations, along the lines of previous studies [77, 83, 109]. Alternatively, one may speculate that breakdown of non-native gene flow is caused by reduced reproductive fitness of population hybrids.

Hybrid Contribution during Naturalization

The hierarchical Bayesian analyses of mixed ancestry confirmed non-native spawning success. The large proportion of hybrids documents reproduction by immigrants, i.e. relaxed selection against non-natives early in differentiation and survival of hybrids. Selection against population hybrids and immigrants strengthen divergence [110, 111]. Hybrids of sympatric and closely related (F_{ST} = 0.070) perch (*Perca fluviatilis*) were found to have reduced fitness in early life stages in laboratory experiments, but the authors did not test for hybrid inferiority in vivo [112]. Novel habitats without competition as in our study may have promoted survival of hybrids, despite potentially reduced fitness. However, hybrids may be less competitive in secondary stages of naturalization, as increased population divergence reduces hybrid fitness [113].

In our study, hybrids without parental origin from their spawning site (Figure 2; non-native hybrids) constituted a minor part of mature char caught at their reproductive site. The considerably larger proportion of individuals with partial native ancestral origin indicates hybrid homing. Reproductive units sustained by kin discrimination are found in a range of species [114, 115, 116, 117], and hybrid juveniles of char will be in close proximity to kin when hatching. Population differentiation, however, could only be established if reproductive success of returning hybrids is low. Admixed individuals would otherwise have caused migration loads obscuring effects of selection and differentiation [109]. Postmating reproductive isolation and hybrid inferiority have been addressed for a century [2, 111]. Both theoretical and empirical studies frequently reveal selection against hybrids during speciation, although its origin seems unclear [74, 96]. Sexual selection (mate choice) against hybrids is found across taxa [e.g. insects; [118], fish; [119]], and may also characterize char. Such mechanisms may evolve faster in small populations [120], as within our sites. Social recognition of relatives in fish using odour cues can induce assortative behavior in fish [121, 122, 123]. Finally, high dispersal rates may increase the absolute number of hybrids, but the increased competition reduces hybrid reproductive fitness, and consequently increases reproductive isolation of the resident population [96].

CONCLUSIONS

Migration corridors between populations separated by dams may be valuable for sustaining evolutionary potential. This study demonstrates that even temporally very limited connectivity through a subterranean tunnel and spill gates between two regulated lakes, likely counteract genetic isolation in char.

Initial stages of genetic divergence of subpopulations, despite high migration rates, are documented in this study. We are not aware of previous studies addressing hybrid migration and reproductive success in recently founded and admixed populations. Our study indicates that population differentiation may be detectable, despite inclusion of hybrid sub-populations, when assessing genetic structure among populations. Combined also with independent mark-recapture data by Aass [46, 47], our genetic approach indicates that hybrids contribute extensively to migration rates in recently founded populations. Population hybrid events may have been important in establishing the diverse population structure in the novel habitat [78]. However, present differentiation implicates nascent non-native and hybrid inferiority, whether biologic or ethologic [32, 91]. Estimates generally gave wide HPDI, an inherent problem expected from low differentiation affecting accuracy [44], and results must be interpreted with caution.

Few population genetic software programs have been evaluated with respect to their performance in detecting low genetic differentiation, making interpretation of differentiation close to admixture difficult [53, 124]. However, restricted power in admixture analyses should not restrict studies of hybridization mechanisms, as they raise consequential questions in micro-evolution and behavioral ecology. Both small scale empirical and larger simulation studies of incomplete reproductive isolation may guide management of invasive and naturalized species, potentially unveiling initial population differentiation mechanisms.

METHODS

The Study Species

The Arctic char is a salmonid fish with Holarctic distribution [125] showing extreme phenotypic and life history variation, exemplified by weight at maturation from 3 g to 12 kg [87]. This reflects the species capability to evolve trophic polymorphism and possible genetic differentiations within drainages [126] and lakes [107, 127, 128]. While most populations of char result from natural postglacial colonization, high altitude populations are introduced by man. When in sympatry, Arctic char typically have a benthopelagic distribution in landlocked habitat. Young char typically feed in the littoral, shifting towards a highly migratory pelagic feeding pattern as adults. Lacustrine char are commonly 20 - 40 cm in length and a few years old when mature, and spawn in the littoral within a few weeks in autumn. Site fidelity is common, but reproductive ecology is poorly understood (see Jonsson *et al.* 2001 [86], Klemetsen *et al.* 2003 [87] and Johnson 1980 [129] for excellent reviews).

The Study Site

Lake Tunhovdfjord (TUN) and Lake Pålsbufjord (PAL) are part of a 35 km long sub alpine hydroelectric reservoir located in south-central Norway (48°E, 67°N, Figure 1), regulated first in 1919. A hydropower dam separating the two lakes was erected in 1946, restricting the previously free migration among the lakes to migration through spill gates in the dam (lower level 734.9 m a.s.l., c 10 m, c 3.1 m^2, neutral gradient) and a subterranean anthropogenic branch (lower level 722.4 m a.s.l., 1300 m, 7.1 m^2, neutral gradient) with outflow 600 m below the dam. Lake PAL now has a surface area of 5.25 - 19.5 km^2 with maximum depth 25 m, 725.5 - 749.1 m a.s.l.. Lake Tunhovdfjord, located immediately downstream, has a surface area of 14 - 25 km^2 and a maximum depth 70 m, 716.4 - 734.4 m a.s.l. Arctic char are sympatric with brown trout (*Salmo trutta*) and invasive European minnow (*Phoxinus phoxinus*) in both lakes. Brown trout have probably been native for > 6000 years [130, 131]. Minnow were introduced coincidentally around 1920.

In 1910, 10.000 fry from a natural char population in Lake Tinnsjø (187.2 - 191.2 m.a.s.l., 51 km^2, mean depth 190 m), situated in a different watershed 80 km south-west of PAL and TUN were stocked in two alpine lakes above PAL [132, 133]. Char were first observed in PAL in 1919. The population expanded rapidly and was present in large numbers in both PAL and TUN after a decade [76]. Densities have thereafter been highly variable, probably because of large water level fluctuations and associated egg mortality [68, 134]. Char in these lakes seldom become more than 8 years old and have a generation time of about 4 years (20 generations since introduction). They are usually < 200 g [68], although cannibalistic individuals of 3 kg have been reported [47]. Extensive mark-recapture studies have revealed substantial char movements within and between the lakes [134], even after the construction of the dam in 1946. Char feeding migration intensity increases in late summer and autumn. The majority of adult char congregate in outlets and narrow parts of the lakes after spawning in autumn when water level drops, followed by c. 10 - 50% passive displacement downstream in the spring [134]. Char emigrating PAL often attempt to re-enter the lake, but successful upstream migration is unknown [47].

Arctic char are low performance swimmers as opposed to brown trout [135]. Maximum relative swimming speed (body length s^{-1}) is about 2.8 [136], thus a water velocity of 150 cm s^{-1} is likely an absolute limit for upstream migration for studied char across the dam. The Mannings formula; $v = M * R_h^{2/3} * \sqrt{I}$ [137], assuming M = 35, was used to estimate elevation heads associated with 150 cm s^{-1} water velocity in the fishways. The water level below the dam in TUN commonly increases (up-arches) to the lower level of the gates. There is also free entrance from TUN to the subterranean tunnel independent of water level. Thus, migration was estimated to be solely restricted by water levels in PAL at 734.9 - 735.2 m a.s.l. in the gates, and 722.4 - 726.4 m a.s.l. in the tunnel. The water level in PAL was registered from year 2000 - 2010.

Sampling

A total of 190 mature char (mean total length 244 mm, SD ± 31, range 170 - 334) were sampled by gillnetting at four sites in PAL and two sites in TUN (Figure 1). Samples represented 54% males (i.e., presence of gametes) and the year classes 1999 - 2003 (N = 31, 37, 48, 37 and 2, respectively, as determined by scales and otoliths of 155 char). All char except those from T2 were aged, but these fish had similar size distribution. Samples from site P1 - 3 and T1 - 2 each included 34 char ready to spawn this season (gonadal stage 4 - 6) caught at traditional and geographically separated spawning sites below minimum water level (LRV) during spawning time, and are therefore treated as discrete populations, although true population boundaries are unknown.

Additional spawning sites within the two lakes are not known. Site × (N = 20) is a randomly chosen mid-lake sample from late summer two months prior to spawning, representing admixtured non-spawning char.

Microsatellite Genotyping and Variability

Tissue samples of approximately 2 mm2 from the adipose fin were preserved in 96% ethanol in the field, and DNA was isolated using DNeasy kit (QIAGEN), following the manufacturer's guidelines. Microsatellite polymorphism was analyzed by means of 10 di- and tetra-nucleotide loci known to be polymorphic in S. alpinus; Mst-85 [138], Sco202, Sco204, Sco213 [139], Sfo-8, Sfo-23 [140], Smm-17, Smm-24 [141], Ssa-85 [142] and Str73 [143]. One primer for each locus was end-labeled with fluorescence (HEX, FAM and NED), and run partly as multiplex PCR reactions; multiplex A; Sfo-8, Smm-17, Smm-24, Str73 and Sco21, multiplex B; Sco202 and Sco204, and multiplex C; Sfo-23 and Ssa-85. Each PCR contained 2 μl genomic template DNA and 8 μl reaction mixtures containing 1 - 2 pmol primer, 50 mM KCl, 1.5 mM MgCl, 10 mM Tris-HCl, 0.2 mM dNTP and 0.25 U Taq polymerase enzymes (Ampliqon). Thermocycling parameters after denaturation at 95°C for 2 min were 24 - 34 cycles of 95°C for 30 sec, annealing temperature of 55°C for 30 sec, followed by an extension at 72°C for 45 sec. The last polymerization step was extended to 10 min. PCR products were added to buffer containing formamide and labeled standard (ROX Std 400, Applied Biosystems), and electrophoresed using an ABI Prism 3100 Genetic Analyzer (Applied Biosystems). The software GENEMAPPER v.3.7 (http://www.applied-biosystems.com) was used to score alleles, and all automated allele calling were controlled for by manual reading. Two positive controls were done in each run, and scoring was repeated twice for several loci, and for all private alleles, to check for consistency.

Quality control screenings were performed by testing for null alleles, large allele drop-outs and scoring errors in MICRO-CHECKER v.2.2.3 [144]. The program TFPGA v.1.3 [145] was used for descriptive statistics (number of alleles, observed and expected heterozygosity). Allelic richness was compared among lakes and spawning sites based on minimum sample size for each comparison separately in FSTAT v.2.9.3.2 [146]. Departure from Hardy-Weinberg (HW) via separate one-tailed tests for heterozygote excess and deficiency for each locus in each site was tested by 60.000 randomizations, and a linkage disequilibrium (LD) test between pairs of loci across populations was performed by 450.000 permutations, both in FSTAT. Local F_{ST}(diversity standardized) and inbreeding coefficient F were calculated in BIMR v.1.1 [44], including 95% C.I., the F-model [147] and default values, except burn-in and sample size of 10^5iterations, to ensure convergence.

Genetic Structure and Temporal Stability

Possible genetic segregation among lakes and sites and their significance were tested with the F_{ST} analogue θ [148] and pairwise test of differentiation in FSTAT, and by exact test (Raymond and Rousset 1995) in TFPGA, all with 15.000 permutations. The estimated θ was chosen since it out-performs other F_{ST} analogues in detecting recently established reproductive isolation [149], and marker neutrality was addressed by jackknifing F_{ST} estimates over loci and populations. Standardized measures of genetic differentiation (θ' [150]) were calculated using the software RECODEDATA v.0.1 (http://www.bentleydrummer.nl/software), to address F_{ST}estimates dependence on the level of genetic variation. Finally, STRUCTURE v.2.3.2 [39] was used to test whether population differentiation was detected in a Bayesian model based algorithm. The admixture locprior model with correlated allele frequencies was run for K = 1 - 10 with a burn in of 200.000 MCMC steps, followed by 500.000 steps in 10 replicates.

Temporal substructure stability was tested for by possible year-class variation, sib-ship analyses, effective size estimates and for isolation by distance (IBD). Allele frequency variation was tested in a hierarchical fashion (AMOVA; among versus within populations), and compared with an AMOVA with segregated year-classes (among versus within year-classes), to test for temporal stability within spawning sites where age was available (P1 - 3 and T1). Analyses were performed in ARLEQUIN v.3.1 [151], under the infinite alleles model (IAM), standard model, unknown gametic phase and 10^4 permutations. Year-class variation was further controlled for by testing pairwise differentiation among year-classes, and excluding one-by-one year-class from site wise θ tests (above). KININFOR v.1.0 [152] simulated the power of applied markers and observed allele frequencies to discriminate between unrelated and full-/half-sib individuals. We ran the simulation with a prior Dirichlet distribution of 1 for Δ_0, Δ_1 and Δ_2, with a 0.05 confidence level, a precision level of 0.01 and 10^6 simulated pairs of genotypes. Family structures within sites, possibly affecting population differentiation estimates, were evaluated in GENALEX v.6.2 [153] by the algorithm of Lynch and Ritland [LRM; 154]. Normality of pairwise relatedness was evaluated in JMP v.8.0 [155].

Contemporary methods were used for estimating effective population size (N_e) and effective parental population size (N_b) within lakes and spawning sites from genetic data. Estimates of N_e using summary statistics and approximate Bayesian statistics were computed in ONESAMP [66], after 50.000 iterations, using prior N_e of 4 - 600 for individual sites and 10 - 2000 for the lake samples. Computations were repeated three times, reporting the median result among tests. Runs with prior values ranging from 2 - 1000 were compared to check for consistency and convergence. Parental N_b was similarly estimated based on linkage disequilibrium using LDNE v.1.31 [67]. The model is assuming closed populations, and is eliminating the possible bias on small sample sizes. We used 0.02 as the lowest allele frequency used in the computation to balance bias and precision [156], including a parametric 95% C.I. in the random mating model. The program BOTTLENECK, v.1.2.02 [157] was used to test for recent bottlenecks using the Wilcoxon test and mode-shift tests with 10^4 replications, with the two phase mutation model (TPM, including 10% infinite alleles model) and the single step mutation model (SMM).

Presence of IBD, indicating equilibrium conditions, was addressed with a Mantel test of correlation between pairwise θ and linear geographic distance among sites in TFPGA, after 1000 iterations.

Migration and Admixture

A combination of individual based maximum likelihood methods and population based Bayesian methods were employed to overcome the potential challenge of detecting gene flow under low differentiation. First generation migrants (F_0) were in GENECLASS v.2.0 [158] identified to evaluate migrants influence on genetic differentiation, using the test statistics L_home/L_max and L_home between lakes and spawning sites, respectively, to account for un-sampled populations and maximizing analyses power. Migrants between spawning sites were assigned to the population with highest (> 5%) self-assignment probability. Estimates were based on the assignment criteria of Rannala and Mountain [159] not assuming genetic equilibrium, and the re-sampling algorithm of Paetkau *et al.* [160] after 10.000 simulations and a threshold score of 5%. We applied IMMANC v.5.0 [159], with 10.000 iterations to verify F_0 estimates and to assess the power of our dataset.

Population patterns of migration over the last generation within lakes and spawning sites were assessed in BIMR. The method implements estimation of inbreeding coefficients (F) to allow for departure from HW, and assumes sampling after reproduction, but before migration. Information used is gametic disequilibrium, and estimates are calculated using a Bayesian approach and MCMC technique, including 95% highest posterior density intervals (HPDI). Burn-in and sample size of 10^5, thinning 50, F-model and default values of pilot runs, priors and incremental values were used, reporting mode estimates from the run with lowest total deviance and acceptance rate of 25 - 45% after 10 replicates to ensure convergence [44, 100, 147]. Bayesian assignment tests were then performed on individual level among lakes and among all sampled spawning sites in GENECLASS (with above settings). The random mid-lake sample was similarly assigned to populations, to test for temporal continuous distribution.

Patterns of recent admixed ancestry were finally evaluated in BIMR to assess hybrid presence within spawning sites with above settings on a population level. Hybrid ancestry was only considered on population level, including HPDI estimates, as hybrid detection under admixture may be hampered by restricted power. IMMANC was used to assess F1 ancestral origin on an individual level. Multiplicity correction procedure of Benjamini and Hochberg was used, balancing the risk of Type 1 and Type 2 errors at $\alpha = 5\%$ [FDR correction; 45].

Forward modeling of population differentiation was performed with EASYPOP v.2.0.1. [161] to evaluate if genetic drift could cause observed differentiation patterns alone, and for power evaluation. Simulations were performed with variation in gene flow (0.01 - 0.99), and N_e (30 - 90), for 5 populations in 20 generations following panmixia ($m = 1.0$) with maximum variation and 15 allelic states, assuming random mating, equal sex ratio and an island migration model. Observed pairwise population differentiation from our sample was compared with matrix scenarios of migration and N_e indicating similar F_{ST} estimates (above) from simulations. Ten independent replicates were obtained for the most likely model to address simulation variation. Similar backward modeling was run in SIMCOAL2 [162] to test the likelihood of the drift model.

ACKNOWLEDGEMENTS

The present study is financed by the Norwegian Ministry of Education and Research. Freshwater Ecology and Inland Fisheries Laboratory, University of Oslo, is acknowledged for assistance in providing samples and age analyses. For help in the laboratory we are in debt to Liv Midthjell.

AUTHORS' CONTRIBUTIONS

All authors; JW, JH and KH conceived the study and participated in the project design. JW conducted the field and laboratory work, and did the genetic analyses. All authors wrote and approved the final manuscript.

REFERENCES

1. Fisher RA: The genetical theory of natural selection. 1930, Oxford: Oxford University Press
2. Dobzhansky T: Genetics and the origin of species. 1937, New York: Colombia University Press
3. Saunders DA, Hobbs RJ, Margules CR: Biological consequences of ecosystem fragmentation: a review. Conserv Biol. 1991, 5: 18-32. 10.1111/j.1523-1739.1991.tb00384.x.
4. Blais J, Plenderleith M, Rico C, Taylor MI, Seehausen O, Van Oosterhout C, Turner GF: Assortative mating among Lake Malawi cichlid fish populations is not simply predictable from male nuptial colour. BMC Evol Biol. 2009, 9:
5. Johnson JR, Fitzpatrick BM, Shaffer HB: Retention of low-fitness genotypes over six decades of admixture between native and introduced tiger salamanders. BMC Evol Biol. 2010, 10:
6. Dynesius M, Nilsson C: Fragmentation and flow regulation of river systems in the northern 3rd of the world. Science. 1994, 266: 753-762. 10.1126/science.266.5186.753.
7. Nilsson C, Reidy CA, Dynesius M, Revenga C: Fragmentation and flow regulation of the world's large river systems. Science. 2005,

308: 405-408. 10.1126/science.1107887.

8. Beier P, Noss RF: Do habitat corridors provide connectivity?. Conserv Biol. 1998, 12: 1241-1252. 10.1046/j.1523-1739.1998.98036.x.
9. Schilt CR: Developing fish passage and protection at hydropower dams. Appl Anim Behav Sci. 2007, 104: 295-325. 10.1016/j.applanim.2006.09.004.
10. World Commissions of Dams: Dams and development A new framework for decisions making. 2000, London: Earthscan Publications Ltd
11. Marmulla G: Dams, fish and fisheries Opportunities, challenges and conflict resolution. 2001, Rome: FAO Fisheries Technical Paper
12. Fullerton AH, Burnett KM, Steel EA, Flitcroft RL, Pess GR, Feist BE, Torgersen CE, Miller DJ, Sanderson BL: Hydrological connectivity for riverine fish: measurement challenges and research opportunities. Freshw Biol. 2010, 55: 2215-2237.
13. Roscoe DW, Hinch SG: Effectiveness monitoring of fish passage facilities: historical trends, geographic patterns and future directions. Fish Fish. 2010, 11: 12-33.
14. Kemp PS, O'Hanley JR: Procedures for evaluating and prioritising the removal of fish passage barriers: a synthesis. Fish Manag Ecol. 2010, 17: 297-322.
15. Taggart JB, McLaren IS, Hay DW, Webb JH, Youngson AF: Spawning success in Atlantic salmon (*Salmo salar* L.): a long-term DNA profiling-based study conducted in a natural stream. Mol Ecol. 2001, 10: 1047-1060. 10.1046/j.1365-294X.2001.01254.x.
16. Gerlier M, Roche P: A radio telemetry study of the migration of Atlantic salmon (*Salmo salar* L.) and sea trout (*Salmo trutta trutta* L.) in the upper Rhine. Hydrobiologia. 1998, 372: 283-293.
17. Thorstad EB, Okland F, Aarestrup K, Heggberget TG: Factors affecting the within-river spawning migration of Atlantic salmon, with emphasis on human impacts. Rev Fish Biol Fish. 2008, 18: 345-371. 10.1007/s11160-007-9076-4.
18. Knutsen H, Olsen EM, Jorde PE, Espeland SH, Andre C, Stenseth NC: Are low but statistically significant levels of genetic differentiation in marine fishes ‹biologically meaningful›? A

case study of coastal Atlantic cod. Mol Ecol. 2011, 20: 768-783. 10.1111/j.1365-294X.2010.04979.x.

19. Nielsen EE, Hemmer-Hansen J, Poulsen NA, Loeschcke V, Moen T, Johansen T, Mittelholzer C, Taranger GL, Ogden R, Carvalho GR: Genomic signatures of local directional selection in a high gene flow marine organism; the Atlantic cod (*Gadus morhua*). BMC Evol Biol. 2009, 9:
20. Waples RS: Separating the wheat from the chaff: Patterns of genetic differentiation in high gene flow species. J Hered. 1998, 89: 438-450. 10.1093/jhered/89.5.438.
21. Hedrick PW: Perspective: Highly variable loci and their interpretation in evolution and conservation. Evolution. 1999, 53: 313-318. 10.2307/2640768.
22. Waples RS, Gaggiotti O: What is a population? An empirical evaluation of some genetic methods for identifying the number of gene pools and their degree of connectivity. Mol Ecol. 2006, 15: 1419-1439. 10.1111/j.1365-294X.2006.02890.x.
23. Strange RM, Stepien CA: Genetic divergence and connectivity among river and reef spawning groups of walleye (*Sander vitreus vitreus*) in Lake Erie. Can J Fish Aquat Sci. 2007, 64: 437-448. 10.1139/f07-022.
24. Massicotte R, Magnan P, Angers B: Intralacustrine site fidelity and nonrandom mating in the littoral-spawning northern redbelly dace (*Phoxinus eos*). Can J Fish Aquat Sci. 2008, 65: 2016-2025. 10.1139/F08-114.
25. Seehausen O, Terai Y, Magalhaes IS, Carleton KL, Mrosso HDJ, Miyagi R, van der Sluijs I, Schneider MV, Maan ME, Tachida H, Imai H, Okada N: Speciation through sensory drive in cichlid fish. Nature. 2008, 455: 620-626. 10.1038/nature07285.
26. Bergek S, Bjorklund M: Cryptic barriers to dispersal within a lake allow genetic differentiation of Eurasian perch. Evolution. 2007, 61: 2035-2041. 10.1111/j.1558-5646.2007.00163.x.
27. Heggenes J, Roed KH, Jorde PE, Brabrand A: Dynamic micro-geographic and temporal genetic diversity in vertebrates: the case of lake-spawning populations of brown trout (*Salmo trutta*). Mol Ecol. 2009, 18: 1100-1111. 10.1111/j.1365-294X.2009.04101.x.

28. Pavey SA, Nielsen JL, Hamon TR: Recent ecological divergence despite migration in sockeye salmon (*Oncorhynchus nerka*). Evolution.2010,64:1773-1783.10.1111/j.1558-5646.2009.00927.x.
29. Hendry AP, Kinnison MT: Perspective: The pace of modern life: Measuring rates of contemporary microevolution. Evolution. 1999, 53: 1637-1653. 10.2307/2640428.
30. Reznick DN, Ghalambor CK: The population ecology of contemporary adaptations: what empirical studies reveal about the conditions that promote adaptive evolution. Genetica. 2001, 112: 183-198.
31. Hendry AP, Nosil P, Rieseberg LH: The speed of ecological speciation. Funct Ecol. 2007, 21: 455-464. 10.1111/j.1365-2435.2007.01240.x.
32. Hendry AP: Ecological speciation! Or the lack thereof?. Can J Fish Aquat Sci. 2009, 66: 1383-1398. 10.1139/F09-074.
33. Rolshausen G, Segelbacher G, Hobson KA, Schaefer HM: Contemporary evolution of reproductive isolation and phenotypic divergence in sympatry along a migratory divide. Current Biology. 2009, 19: 2097-2101. 10.1016/j.cub.2009.10.061.
34. Carroll SP: Facing change: forms and foundations of contemporary adaptation to biotic invasions. Mol Ecol. 2008, 17: 361-372. 10.1111/j.1365-294X.2007.03484.x.
35. Herborg LM, Weetman D, Van Oosterhout C, Hanfling B: Genetic population structure and contemporary dispersal patterns of a recent European invader, the Chinese mitten crab, *Eriocheir sinensis*. Mol Ecol. 2007, 16: 231-242.
36. Dupont L, Jollivet D, Viard F: High genetic diversity and ephemeral drift effects in a successful introduced mollusc (*Crepidula fornicata*: Gastropoda). Mar Ecol Prog Ser. 2003, 253: 183-195.
37. Dupont L, Viard F, Dowell MJ, Wood C, Bishop JDD: Fine- and regional-scale genetic structure of the exotic ascidian *Styela clava* (Tunicata) in southwest England, 50 years after its introduction. Mol Ecol. 2009, 18: 442-453. 10.1111/j.1365-294X.2008.04045.x.
38. Corander J, Siren J, Arjas E: Bayesian spatial modeling of genetic population structure. Comput Stat. 2008, 23: 111-129. 10.1007/s00180-007-0072-x.

39. Hubisz MJ, Falush D, Stephens M, Pritchard JK: Inferring weak population structure with the assistance of sample group information. Mol Ecol Resour. 2009, 9: 1322-1332. 10.1111/j.1755-0998.2009.02591.x.

40. Chen C, Durand E, Forbes F, Francois O: Bayesian clustering algorithms ascertaining spatial population structure: a new computer program and a comparison study. Mol Ecol Notes. 2007, 7: 747-756. 10.1111/j.1471-8286.2007.01769.x.

41. Stabell OB: Homing and olfaction in salmonids: A critical review with special reference to the Atlantic salmon. Biol Rev Camb Philos Soc. 1984, 59: 333-388. 10.1111/j.1469-185X.1984.tb00709.x.

42. Wallraff HG: Navigation by homing pigeons: updated perspective. Ethol Ecol Evol. 2001, 13: 1-48. 10.1080/08927014.2001.9522786.

43. de Leaniz CG, Fleming IA, Einum S, Verspoor E, Jordan WC, Consuegra S, ubin-Horth N, Lajus D, Letcher BH, Youngson AF, Webb JH, Vøllestad LA, Villanueva B, Ferguson A, Quinn TP: A critical review of adaptive genetic variation in Atlantic salmon: implications for conservation. Biol Rev. 2007, 82: 173-211.

44. Faubet P, Gaggiotti OE: A new Bayesian method to identify the environmental factors that influence recent migration. Genetics. 2008, 178: 1491-1504. 10.1534/genetics.107.082560.

45. Benjamini Y, Hochberg Y: Controlling the False Discovery Rate: A practical and powerful approach to multiple testing. J R Stat Soc Series B Methodol. 1995, 57: 289-300.

46. Aass P: Rødingvandringar i Pålsbufjord och Tunhovdfjord. 1965, Drottningholm

47. Aass P: Management and utilization of Arctic charr in Norwegian hydroelectric reservoirs. Biology of the Arctic charr Proceedings of the International Symposium on Arctic charr. Edited by: Johnson L, Burns B. 1984, Winnipeg: Univ. Manitoba Press, 277-291.

48. Northcote TG: Migratory strategies and production in freshwater fishes. Ecology of Freshwater Fish Production. Edited by: Gerking SD. 1978, Oxford: Blackwell Scientific, 326-359.

49. Arnekleiv JV, Ronning L: Migratory patterns and return to the catch site of adult brown trout (*Salmo trutta* L.) in a regulated river. River Research and Applications. 2004, 20: 929-942. 10.1002/rra.799.

50. Rivinoja P, McKinnell S, Lundqvist H: Hindrances to upstream migration of Atlantic salmon (Salmo salar) in a northern Swedish river caused by a hydroelectric power-station. Regulated Rivers-Research & Management. 2001, 17: 101-115. 10.1002/rrr.607.
51. Thorstad EB, Okland F, Kroglund F, Jepsen N: Upstream migration of Atlantic salmon at a power station on the River Nidelva, Southern Norway. Fish Manag Ecol. 2003, 10: 139-146. 10.1046/j.1365-2400.2003.00335.x.
52. Latch EK, Dharmarajan G, Glaubitz JC, Rhodes OE: Relative performance of Bayesian clustering software for inferring population substructure and individual assignment at low levels of population differentiation. Conserv Genet. 2006, 7: 295-302. 10.1007/s10592-005-9098-1.
53. Lowe WH, Allendorf FW: What can genetics tell us about population connectivity?. Mol Ecol. 2010, 19: 3038-3051. 10.1111/j.1365-294X.2010.04688.x.
54. Whitlock MC, McCauley DE: Indirect measures of gene flow and migration: F-ST not equal 1/(4 Nm+1). Heredity. 1999, 82: 117-125. 10.1038/sj.hdy.6884960.
55. Mallet J, Meyer A, Nosil P, Feder JL: Space, sympatry and speciation. J Evol Biol. 2009, 22: 2332-2341. 10.1111/j.1420-9101.2009.01816.x.
56. Hendry AP, Castric V, Kinnison MT, Quinn TP: The evolution of philopatry and dispersal: homing versus straying in salmonids. Evolution illuminated: Salmon and their relatives. Edited by: Hendry AP, Stearns SC. 2004, Oxford: Oxford University Press, 52-91.
57. Jost L: G(ST) and its relatives do not measure differentiation. Mol Ecol. 2008, 17: 4015-4026. 10.1111/j.1365-294X.2008.03887.x.
58. Gomez-Uchida D, Knight TW, Ruzzante DE: Interaction of landscape and life history attributes on genetic diversity, neutral divergence and gene flow in a pristine community of salmonids. Mol Ecol. 2009, 18: 4854-4869. 10.1111/j.1365-294X.2009.04409.x.
59. Petit E, Balloux F, Goudet J: Sex-biased dispersal in a migratory bat: A characterization using sex-specific demographic parameters. Evolution. 2001, 55: 635-640. 10.1554/0014-3820(2001)055[0635:SBDIAM]2.0.CO;2.

60. Hauser L, Carvalho GR: Paradigm shifts in marine fisheries genetics: ugly hypotheses slain by beautiful facts. Fish Fish. 2008, 9: 333-362.

61. Bernatchez L, Rhydderch JG, Kircheis FW: Microsatellite gene diversity analysis in landlocked Arctic char from Maine. Trans Am Fish Soc. 2002, 131: 1106-1118. 10.1577/1548-8659(2002)131<1106:MGDAIL>2.0.CO;2.

62. Dynes J, Magnan P, Bernatchez L, Rodriguez MA: Genetic and morphological variation between two forms of lacustrine brook charr. J Fish Biol. 1999, 54: 955-972. 10.1111/j.1095-8649.1999.tb00850.x.

63. Nei M, Maruyama T, Chakraborty R: Bottleneck effect and genetic variability in populations. Evolution. 1975, 29: 1-10. 10.2307/2407137.

64. Dlugosch KM, Parker IM: Founding events in species invasions: genetic variation, adaptive evolution, and the role of multiple introductions. Mol Ecol. 2008, 17: 431-449. 10.1111/j.1365-294X.2007.03538.x.

65. Roman J, Darling JA: Paradox lost: genetic diversity and the success of aquatic invasions. Trends Ecol Evol. 2007, 22: 454-464. 10.1016/j.tree.2007.07.002.

66. Tallmon DA, Koyuk A, Luikart G, Beaumont MA: ONeSAMP: a program to estimate effective population size using approximate Bayesian computation. Mol Ecol Resour. 2008, 8: 299-301. 10.1111/j.1471-8286.2007.01997.x.

67. Waples RS, Do C: LDNE: a program for estimating effective population size from data on linkage disequilibrium. Mol Ecol Resour. 2008, 8: 753-756. 10.1111/j.1755-0998.2007.02061.x.

68. Brabrand Å, Bremnes T, Saltveit SJ, Aass P, Wollebæk J, Heggenes J, Røed K: Fiskeribiologiske undersøkelser i Pålsbufjorden Hovedrapport. 2008, Oslo

69. Shrimpton JM, Heath DD: Census vs. effective population size in chinook salmon: large- and small-scale environmental perturbation effects. Mol Ecol. 2003, 12: 2571-2583. 10.1046/j.1365-294X.2003.01932.x.

70. Frankham R: Effective population size adult population size ratios

in wildlife - a review. Genet Res. 1995, 66: 95-107. 10.1017/S0016672300034455.

71. Levins R: Some demographic and genetic consequences of environmental heterogeneity for biological control. Bull Entomol Soc. 1969, 15: 237-240.
72. Moyle PB, Light T: Biological invasions of fresh water: Empirical rules and assembly theory. Biol Conserv. 1996, 78: 149-161. 10.1016/0006-3207(96)00024-9.
73. Schluter D: Ecological speciation in postglacial fishes. Philos Trans R Soc Lond, B, Biol Sci. 1996, 351: 807-814. 10.1098/rstb.1996.0075.
74. Rundle HD, Nosil P: Ecological speciation. Ecol Lett. 2005, 8: 336-352. 10.1111/j.1461-0248.2004.00715.x.
75. Sakai AK, Allendorf FW, Holt JS, Lodge DM, Molofsky J, With KA, Baughman S, Cabin RJ, Cohen JE, Ellstrand NC, McCauley DE, O›Neil P, Parker IM, Thompson JN, Weller SG: The population biology of invasive species. Annu Rev Ecol Syst. 2001, 32: 305-332. 10.1146/annurev.ecolsys.32.081501.114037.
76. Huitfeldt-Kaas H: Der Einfluss der Gewasserregelungen auf den Fischbestand in Binnenseen. 1935, Oslo
77. Hendry AP: Selection against migrants contributes to the rapid evolution of ecologically dependent reproductive isolation. Evol Ecol Res. 2004, 6: 1219-1236.
78. Templeton AR: The reality and importance of founder speciation in evolution. Bioessays. 2008, 30: 470-479. 10.1002/bies.20745.
79. Conover DO, Munch SB: Sustaining fisheries yields over evolutionary time scales. Science. 2002, 297: 94-96. 10.1126/science.1074085.
80. Stockwell CA, Hendry AP, Kinnison MT: Contemporary evolution meets conservation biology. Trends Ecol Evol. 2003, 18: 94-101. 10.1016/S0169-5347(02)00044-7.
81. Hendry AP: Adaptive divergence and the evolution of reproductive isolation in the wild: an empirical demonstration using introduced sockeye salmon. Genetica. 2001, 112: 515-534.
82. Kinnison MT, Bentzen P, Unwin MJ, Quinn TP: Reconstructing

recent divergence: evaluating nonequilibrium population structure in New Zealand chinook salmon. Mol Ecol. 2002, 11: 739-754. 10.1046/j.1365-294X.2002.01477.x.

83. Taylor EB: A review of local adaptation in Salmonidae, with particular reference to Pacific and Atlantic salmon. Aquaculture. 1991, 98: 185-207. 10.1016/0044-8486(91)90383-I.

84. Koskinen MT, Haugen TO, Primmer CR: Contemporary fisherian life-history evolution in small salmonid populations. Nature. 2002, 419: 826-830. 10.1038/nature01029.

85. Ayllon F, Davaine P, Beall E, Garcia-Vazquez E: Dispersal and rapid evolution in brown trout colonizing virgin Subantarctic ecosystems. J Evol Biol. 2006, 19: 1352-1358. 10.1111/j.1420-9101.2005.01075.x.

86. Jonsson B, Jonsson N: Polymorphism and speciation in Arctic charr. J Fish Biol. 2001, 58: 605-638. 10.1111/j.1095-8649.2001.tb00518.x.

87. Klemetsen A, Amundsen PA, Dempson JB, Jonsson B, Jonsson N, O›Connell MF, Mortensen E: Atlantic salmon *Salmo salar* L., brown trout *Salmo trutta* L. and Arctic charr *Salvelinus alpinus* L.: a review of aspects of their life histories. Ecol Freshw Fish. 2003, 12: 1-59. 10.1034/j.1600-0633.2003.00010.x.

88. Price TD, Qvarnstrom A, Irwin DE: The role of phenotypic plasticity in driving genetic evolution. Proc R Soc Lond, B, Biol Sci. 2003, 270: 1433-1440. 10.1098/rspb.2003.2372.

89. Gilchrist GW, Lee CE: All stressed out and nowhere to go: does evolvability limit adaptation in invasive species?. Genetica. 2007, 129: 127-132. 10.1007/s10709-006-9009-5.

90. Ghalambor CK, Mckay JK, Carroll SP, Reznick DN: Adaptive versus non-adaptive phenotypic plasticity and the potential for contemporary adaptation in new environments. Funct Ecol. 2007, 21: 394-407. 10.1111/j.1365-2435.2007.01283.x.

91. Hatchwell BJ: Cryptic kin selection: Kin structure in vertebrate populations and opportunities for kin-directed cooperation. Ethology. 2010, 116: 203-216. 10.1111/j.1439-0310.2009.01732.x.

92. Puebla O: Ecological speciation in marine v. freshwater fishes. J Fish Biol. 2009, 75: 960-996. 10.1111/j.1095-8649.2009.02358.x.

93. Berner D, Grandchamp AC, Hendry AP: Variable progress toward ecological speciation in parapatry: Stickleback across eight lake-stream transitions. Evolution. 2009, 63: 1740-1753. 10.1111/j.1558-5646.2009.00665.x.

94. Labonne J, Hendry AP: Natural and sexual selection giveth and taketh away reproductive barriers: Models of population divergence in Guppies. Am Nat. 2010, 176: 26-39. 10.1086/652992.

95. Nosil P, Funk DJ, Ortiz-Barrientos D: Divergent selection and heterogeneous genomic divergence. Mol Ecol. 2009, 18: 375-402. 10.1111/j.1365-294X.2008.03946.x.

96. Thibert-Plante X, Hendry AP: Five questions on ecological speciation addressed with individual-based simulations. J Evol Biol. 2009, 22: 109-123. 10.1111/j.1420-9101.2008.01627.x.

97. Thibert-Plante X, Hendry AP: When can ecological speciation be detected with neutral loci?. Mol Ecol. 2010, 19: 2301-2314. 10.1111/j.1365-294X.2010.04641.x.

98. Hutchison DW, Templeton AR: Correlation of pairwise genetic and geographic distance measures: Inferring the relative influences of gene flow and drift on the distribution of genetic variability. Evolution. 1999, 53: 1898-1914. 10.2307/2640449.

99. Whiteley AR, Spruell P, Rieman BE, Allendorf FW: Fine-scale genetic structure of bull trout at the southern limit of their distribution. Trans Am Fish Soc. 2006, 135: 1238-1253. 10.1577/T05-166.1.

100. Faubet P, Waples RS, Gaggiotti OE: Evaluating the performance of a multilocus Bayesian method for the estimation of migration rates. Mol Ecol. 2007, 16: 1149-1166. 10.1111/j.1365-294X.2007.03218.x.

101. Wilson GA, Rannala B: Bayesian inference of recent migration rates using multilocus genotypes. Genetics. 2003, 163: 1177-1191.

102. Griffiths SW: Learned recognition of conspecifics by fishes. Fish Fish. 2003, 4: 256-268.

103. Olsen KH, Grahn M, Lohm J, Langefors A: MHC and kin discrimination in juvenile Arctic charr, *Salvelinus alpinus* L. Anim Behav. 1998, 56: 319-327.

104. Svenning MA, Grotnes P: Stationarity and homing ability of

landlocked Arctic charr. Nord J Freshw Resour. 1991, 66: 36-43.

105. Fraser DJ, Duchesne P, Bernatchez L: Migratory charr schools exhibit population and kin associations beyond juvenile stages. Mol Ecol. 2005, 14: 3133-3146. 10.1111/j.1365-294X.2005.02657.x.

106. Nordeng H: Char ecology. Natal homing in sympatric populations of anadromous Arctic char *Salvelinus alpinus* L.: Roles of pheromone recognition. Ecol Freshw Fish. 2009, 18: 41-51. 10.1111/j.1600-0633.2008.00320.x.

107. Adams CE, Fraser D, Wilson AJ, Alexander G, Ferguson MM, Skulason S: Patterns of phenotypic and genetic variability show hidden diversity in Scottish Arctic charr. Ecol Freshw Fish. 2007, 16: 78-86. 10.1111/j.1600-0633.2006.00182.x.

108. Adams CE, Hamilton DJ, Mccarthy I, Wilson AJ, Grant A, Alexander G, Waldron S, Snorasson SS, Ferguson MM, Skulason S: Does breeding site fidelity drive phenotypic and genetic sub-structuring of a population of Arctic charr?. Evol Ecol. 2006, 20: 11-26. 10.1007/s10682-005-2489-4.

109. Bolnick DI, Nosil P: Natural selection in populations subject to a migration load. Evolution. 2007, 61: 2229-2243. 10.1111/j.1558-5646.2007.00179.x.

110. Barton NH, Hewitt GM: Analysis of hybrid zones. Annu Rev Ecol Syst. 1985, 16: 113-148. 10.1146/annurev.es.16.110185.000553.

111. Nosil P, Vines TH, Funk DJ: Perspective: Reproductive isolation caused by natural selection against immigrants from divergent habitats. Evolution. 2005, 59: 705-719.

112. Behrmann-Godel J, Gerlach G: First evidence for postzygotic reproductive isolation between two populations of Eurasian perch (*Perca fluviatilis* L.) within Lake Constance. Front Zool. 2008, 5:

113. Edmands S: Between a rock and a hard place: evaluating the relative risks of inbreeding and outbreeding for conservation and management. Mol Ecol. 2007, 16: 463-475.

114. Brown GE, Brown JA: Kin discrimination in salmonids. Rev Fish Biol Fish. 1996, 6: 201-219.

115. Ward AJW, Webster MM, Magurran AE, Currie S, Krause J: Species and population differences in social recognition between fishes: a

role for ecology?. Behav Ecol. 2009, 20: 511-516. 10.1093/beheco/arp025.

116. Gerlach G, Hodgins-Davis A, Avolio C, Schunter C: Kin recognition in zebrafish: a 24-hour window for olfactory imprinting. Proc R Soc, B, Biol Sci. 2008, 275: 2165-2170. 10.1098/rspb.2008.0647.
117. Mehlis M, Bakker TCM, Frommen JG: Smells like sib spirit: kin recognition in three-spined sticklebacks (*Gasterosteus aculeatus*) is mediated by olfactory cues. Animal Cognition. 2008, 11: 643-650. 10.1007/s10071-008-0154-3.
118. Naisbit RE, Jiggins CD, Mallet J: Disruptive sexual selection against hybrids contributes to speciation between *Heliconius cydno* and *Heliconius melpomene*. Proc R Soc Lond, B, Biol Sci. 2001, 268: 1849-1854. 10.1098/rspb.2001.1753.
119. van der Sluijs I, Van Dooren TJM, Hofker KD, van Alphen JJM, Stelkens RB, Seehausen O: Female mating preference functions predict sexual selection against hybrids between sibling species of cichlid fish. Philos Trans R Soc, B, Biol Sci. 2008, 363: 2871-2877. 10.1098/rstb.2008.0045.
120. Nei M, Maruyama T, Wu CI: Models of evolution of reproductive isolation. Genetics. 1983, 103: 557-579.
121. Ward AJW, Webster MM, Hart PJB: Social recognition in wild fish populations. Proc R Soc, B, Biol Sci. 2007, 274: 1071-1077. 10.1098/rspb.2006.0231.
122. Olsen KH, Grahn M, Lohm J: Influence of MHC on sibling discrimination in Arctic char, *Salvelinus alpinus* (L.). J Chem Ecol. 2002, 28: 783-795. 10.1023/A:1015240810676.
123. Courtenay SC, Quinn TP, Dupuis HMC, Groot C, Larkin PA: Factors affecting the recognition of population-specific odours by juvenile coho salmon. J Fish Biol. 1997, 50: 1042-1060.
124. Kalinowski ST: The computer program STRUCTURE does not reliably identify the main genetic clusters within species: simulations and implications for human population structure. Heredity. 2011, 106: 625-632. 10.1038/hdy.2010.95.
125. Behnke RJ: Systematics of salmonid fishes of recently glaciated lakes. J Fish Res Board Can. 1972, 29: 639-671. 10.1139/f72-112.

126. Primmer CR, Aho T, Piironen J, Estoup A, Cornuet JM, Ranta E: Microsatellite analysis of hatchery stocks and natural populations of Arctic charr, *Salvelinus alpinus*, from the Nordic region: implications for conservation. Hereditas. 1999, 130: 277-289.

127. Brunner PC, Douglas MR, Osinov A, Wilson CC, Bernatchez L: Holarctic phylogeography of Arctic charr (*Salvelinus alpinus* L.) inferred from mitochondrial DNA sequences. Evolution. 2001, 55: 573-586. 10.1554/0014-3820(2001)055[0573:HPOACS]2.0.CO;2.

128. Wilson AJ, Gislason D, Skulason S, Snorrason SS, Adams CE, Alexander G, Danzmann RG, Ferguson MM: Population genetic structure of Arctic Charr, *Salvelinus alpinus* from northwest Europe on large and small spatial scales. Mol Ecol. 2004, 13: 1129-1142. 10.1111/j.1365-294X.2004.02149.x.

129. Johnson L: The Arctic charr, *Salvelinus alpinus*. Charrs, Salmonid fishes of the genus Salvelinus. Edited by: Balon EK. 1980, The Hague: Dr. W. Junk Publishers, 15-98.

130. Huitfeldt-Kaas H: Ferskvandsfiskens utbredelse og indvandring i Norge med et tillæg om krebsen. 1918, Kristiania: Centraltrykkeriet

131. Indrelid S: De første bosetterne. Hardangervidda. Edited by: Barth EK. 1985, Oslo: Luther forlag, 97-111.

132. Aagaard B: Utsettelse og gjenfangst av røie i Bredevandet, Hol, Buskeruds amt. Norsk Fiskeritidende. 1915, 34: 225-226.

133. Aass P: The winter migration of charr, Salvelinus alpinus L, in the hydroelectric reservoirs Tunhovdfjord and Pålsbufjord, Norway. 1970, Drottningholm

134. Aass P: Some effects of lake impoundments on salmonids in Norwegian hydroelectric reservoirs. Dissertation. PhD Thesis. 1973, University of Uppsala, Faculty of Science

135. Behlke CE, Kane DL, McLean RF, Travis MD: Fundamentals of culvert design for passage of weak-swimming fish. 1991, Alaska

136. Jones DR, Kiceniuk JW, Bamford OS: Evaluation of swimming performance of several fish species from Mackenzie river. J Fish Res Board Can. 1974, 31: 1641-1647. 10.1139/f74-206.

137. Tritton DJ: Physical fluid dynamics. 1988, Oxford: Clarendon Press

138. Presa P, Guyomard R: Conservation of microsatellites in three

species of salmonids. J Fish Biol. 1996, 49: 1326-1329.

139. Dehaan PW, Ardren WR: Characterization of 20 highly variable tetranucleotide microsatellite loci for bull trout (*Salvelinus confluentus*) and cross-amplification in other *Salvelinus* species. Mol Ecol Notes. 2005, 5: 582-585. 10.1111/j.1471-8286.2005.00997.x.

140. Angers B, Bernatchez L, Angers A, Desgroseillers L: Specific microsatellite loci for brook charr reveal strong population subdivision on a microgeographic scale. J Fish Biol. 1995, 47: 177-185.

141. Crane PA, Lewis CJ, Kretschmer EJ, Miller SJ, Spearman WJ, DeCiccoAL, Lisac MJ, Wenburg JK: Characterization and inheritance of seven microsatellite loci from Dolly Varden, *Salvelinus malma*, and cross-species amplification in Arctic char, *S alpinus*. Conserv Genet. 2004, 5: 737-741. 10.1007/s10592-004-1853-1.

142. OReilly PT, Hamilton LC, McConnell SK, Wright JM: Rapid analysis of genetic variation in Atlantic salmon (*Salmo salar*) by PCR multiplexing of dinucleotide and tetranucleotide microsatellites. Can J Fish Aquat Sci. 1996, 53: 2292-2298.

143. Estoup A, Presa P, Krieg F, Vaiman D, Guyomard R: (Ct)(N) and (Gt)(N) microsatellites: a new class of genetic markers for *Salmo trutta* L. (brown trout). Heredity. 1993, 71: 488-496. 10.1038/hdy.1993.167.

144. Van Oosterhout C, Hutchinson WF, Wills DPM, Shipley P: MICRO-CHECKER: software for identifying and correcting genotyping errors in microsatellite data. Mol Ecol Notes. 2004, 4: 535-538. 10.1111/j.1471-8286.2004.00684.x.

145. Miller MP: Tools for population genetic analyses (TFPGA) ver. 1.3. 1997, Department of Biological Sciences, Nortern Arizona University, Flaggstaff, Arizona

146. Goudet J: FSTAT (Version 1.2): A computer program to calculate F-statistics. J Hered. 1995, 86: 485-486.

147. Gaggiotti OE, Foll M: Quantifying population structure using the F-model. Mol Ecol Resour. 2010, 10: 821-830. 10.1111/j.1755-0998.2010.02873.x.

148. Weir BS, Cockerham CC: Estimating F-statistics for the analysis

of population structure. Evolution. 1984, 38: 1358-1370. 10.2307/2408641.

149. Sefc KM, Payne RB, Sorenson MD: Genetic differentiation after founder events: an evaluation of F-ST estimators with empirical and simulated data. Evol Ecol Res. 2007, 9: 21-39.

150. Meirmans PG: Using the AMOVA framework to estimate a standardized genetic differentiation measure. Evolution. 2006, 60: 2399-2402.

151. Excoffier L, Laval G, Schneider S: Arlequin ver. 3.0: An integrated software package for population genetics data analyses. Evol Bioinform Online. 2005, 1: 47-50.

152. Wang JL: Informativeness of genetic markers for pairwise relationship and relatedness inference. Theor Popul Biol. 2006, 70: 300-321. 10.1016/j.tpb.2005.11.003.

153. Peakall R, Smouse PE: GENALEX 6: genetic analysis in Excel. Population genetic software for teaching and research. Mol Ecol Notes. 2006, 6: 288-295. 10.1111/j.1471-8286.2005.01155.x.

154. Ritland K: Marker-inferred relatedness as a tool for detecting heritability in nature. Mol Ecol. 2000, 9: 1195-1204. 10.1046/j.1365-294x.2000.00971.x.

155. JMP: JMP v.8.0. (8.0). 2008, SAS Institute Inc

156. Waples RS, Do C: Linkage disequilibrium estimates of contemporary N-e using highly variable genetic markers: a largely untapped resource for applied conservation and evolution. Evol Appl. 2010, 3: 244-262. 10.1111/j.1752-4571.2009.00104.x.

157. Cornuet JM, Luikart G: Description and power analysis of two tests for detecting recent population bottlenecks from allele frequency data. Genetics. 1996, 144: 2001-2014.

158. Piry S, Alapetite A, Cornuet JM, Paetkau D, Baudouin L, Estoup A: GENECLASS2: A software for genetic assignment and first-generation migrant detection. J Hered. 2004, 95: 536-539. 10.1093/jhered/esh074.

159. Rannala B, Mountain JL: Detecting immigration by using multilocus genotypes. Proc Natl Acad Sci USA. 1997, 94: 9197-9201. 10.1073/pnas.94.17.9197.

160. Paetkau D, Slade R, Burden M, Estoup A: Genetic assignment methods for the direct, real-time estimation of migration rate: a simulation-based exploration of accuracy and power. Mol Ecol. 2004, 13: 55-65. 10.1046/j.1365-294X.2004.02008.x.

161. Balloux F: EASYPOP (version 1.7): A computer program for population genetics simulations. J Hered. 2001, 92: 301-302. 10.1093/jhered/92.3.301.

162. Laval G, Excoffier L: SIMCOAL 2.0: a program to simulate genomic diversity over large recombining regions in a subdivided population with a complex history. Bioinformatics. 2004, 20: 2485-2487. 10.1093/bioinformatics/bth264.

CHAPTER

6

PERFORMANCE OF AN ACOUSTIC TELEMETRY SYSTEM IN A LARGE FISHWAY

Ki Won Jung[1] , Z. Daniel Deng[1], Jayson J. Martinez[1] , David R. Geist[1] , Geoffrey A. McMichael[1] , John R. Stephenson[1] and Peter J. Graf[2]

[1] Pacific Northwest National Laboratory, Richland, WA 99352, USA

[2] Grant County Public Utility District, Ephrata, WA, USA.

ABSTRACT

Background

The ability to track fish in the vicinity of dams and detect the presence of fish in fishways (also referred to as fish ladders) is critical to understanding the migration biology of upstream migrating fish and their passage success. Acoustic telemetry provides a valuable method within the telemetry toolbox, but has rarely been used in noisy, constrained, swift-flowing, and air-entrained environments such as fishways because of the perceived limitations of acoustic telemetry in such environments relative to other techniques such as radio telemetry. However, there have been no published studies that represent systematic evaluations of both detection efficiency based on number of transmitted signals and detection probabilities based on number of fish detected in fishways for acoustic telemetry. As such, the efficacy of acoustic telemetry in a fishway was evaluated using both controlled field experiments and movements of live fish at the Wanapum Dam fishway on the Columbia River in Washington State. In this study, the Juvenile Salmon Acoustic Telemetry System (JSATS) was selected because of its high carrier frequency (416.7 kHz) and short signal duration yielding good system resilience in shallow, noisy, and constrained environments.

Results

Detection efficiencies from the controlled field experiments were over 80 % (mean ± standard error (SE) = 85 ± 2 %) for most locations within the fishway at Wanapum Dam. Detection efficiency was nearly 100 % (mean ± SE = 97 ± 1 %) in the fishway entrance when the transmitters (also referred to as tags) were within 10 m of the hydrophones. The detection probabilities for the live fish experiment were 100 % at all deployment locations.

Conclusions

This study suggests that a 416.7-kHz acoustic telemetry system with binary phase shift-keyed encoding is capable of reliably detecting fish within a large fishway, thus providing a reliable tool to enable researchers to understand and study fish behavior and their fate near and in fishways.

Keywords: Acoustic telemetry, Fishways, Detection efficiency, Detection Probabilities, Migration

ABBREVIATIONS

JSATS: Juvenile Salmon Acoustic Telemetry System

PIT: passive integrated transponder

GPS: global positioning system

A2LA: American Association for Laboratory Accreditation

BFL: Bio-Acoustics and Flow Laboratory

BACKGROUND

Acoustic telemetry was first applied to the study of fish behavior decades ago [1], but recently, acoustic telemetry has become exceedingly popular and is playing an important role in fundamental and applied fish behavior research in both marine and freshwater systems [2, 3, 4, 5, 6, 7, 8, 9, 10]. In riverine systems with barriers, fishways (i.e., fish ladders) are often installed in an attempt to facilitate passage past the barrier and maintain longitudinal connectivity. Nonetheless, fish passage science is imperfect with not all fish successfully locating or ascending the fishway and those that do may experience sublethal alterations that influence fitness. For example, efficient upstream passage of adult salmonids through fishways at large hydroelectric dams may have implications for energy reserves available for spawning [11, 12, 13]. To know all of these, it may be important to be able to track the detailed movements of the fish from the moment they enter a dam till they leave it (regardless of successfully ascending the fishway or not). However, few studies [14] have been published on the application of acoustic telemetry in tracking fish within fishways or other confined spaces. Also, there have been no published studies that systematically evaluate both detection efficiency, based on the number of transmitted signals, and detection probabilities, based on the number of tagged fish detected, in fishways.

Many studies that involve detecting fish in confined spaces use passive integrated transponder (PIT tag) technology to detect fish at fixed points within juvenile bypass systems [15, 16, 17, 18] or in fishways [19, 20, 21, 22, 23]. Studies have been conducted using a PIT tag with another telemetric technology, such as radio telemetry [13], by dual-tagging fish [15, 16, 19] or only using a PIT tag [17, 18, 20, 21, 22, 23, 24]. Although it has an advantage in terms of longevity of the tag relative to other technologies, PIT tags have short detection ranges that typically are less than 1 m in the tags' long axis, perpendicular to the antenna [24]. Hence, PIT tags have lower detectability compared to other telemetry technologies, such as acoustic and radio telemetry, when a wide coverage area is required. As a result of the low detectability, the sample size of fish tagged may need to be considerably larger when conducting survival studies of migrating fish to achieve the same precision in the survival estimation [16]. Also, it is challenging to install PIT antennas on many passage routes such as through turbines or spillbays. Further, while PIT tags may provide detection data within the confined spaces of fishways, they cannot be detected in dam tailraces or forebays, making them ineffective for studies of overall passage behavior of migratory fish encountering large hydroelectric dams.

Radio telemetry also has been used in fishways by many researchers. For example, Gowans et al. [25] investigated the migratory behavior of adult Atlantic salmon through fish ladders, and Johnson et al. [26] studied the effect of the water velocity at fishway openings on the migratory behavior of adult Pacific lampreys. While radio telemetry is an effective tool for studying fish in shallow freshwater environments, an antenna exterior to the tagged fish is often used to transmit the radio signal [1]. However, an external antenna could become entangled [27, 28] and possibly reduce the swimming performance of fish if the antenna is long [29].

Acoustic telemetry has several advantages over radio telemetry and PIT technology. For example, it typically has long detection ranges, good performance in deep water, and capability of determining three-dimensional positions with high accuracy. Roscoe et al. [14] used an acoustic telemetry system in a fishway as a part of an evaluation of the upstream migration mortality of adult sockeye salmon. However, their study did not focus on evaluating the efficacy of acoustic telemetry in fishways and only reported detection probabilities of individual loggers based on number of fish detected compared to those known to have passed logger locations.

This paper presents the first published study that evaluates the efficacy of acoustic telemetry in a fishway. The study was conducted at Wanapum Dam using the Juvenile Salmon Acoustic Telemetry System (JSATS), which was originally developed by the U.S. Army Corps of Engineers to evaluate the survival and behavior of migrating juvenile salmonids [30, 31, 32]. Since its introduction, it has been used extensively in the Columbia River Basin [16, 30]. In this study, the JSATS was selected in part because it has a relatively short signal duration (744 μs) compared to other acoustic telemetry systems. This decreases the likelihood of signal collision due to multipath signals from highly reflective structures such as fishway walls and the water surface. It was also selected because it has a high operating (carrier) frequency. Use of a higher carrier frequency is critical because background noise in environments close to dams is greater at lower frequency ranges [33]. Controlled field tests using JSATS tags as well as a field experiment using fish implanted with JSATS tags were conducted to measure both detection efficiency and detection probabilities. The objective of this study was to evaluate the detection capability of the JSATS technology in a confined space such as a fishway at a large dam. If this study is successful, the JSATS or other acoustic telemetry systems, with similar frequency and pulse duration, can allow researchers to better understand the behavior and survival of migrating fish in a confined space.

METHODS

Location

Owned and operated by the Public Utility District Number 2 of Grant County, Wanapum Dam is located in southeast Washington State on the Columbia River, 669 river kilometers (rkm) from the mouth of the Columbia River (Figure 1). The dam is 2,633 m long and 56 m tall and consists of a 10-unit powerhouse, a 12-bay spillway, and two fish ladders. The powerhouse has a generating capacity of 1,092 MW. All 10 units are six-blade Kaplan turbines.

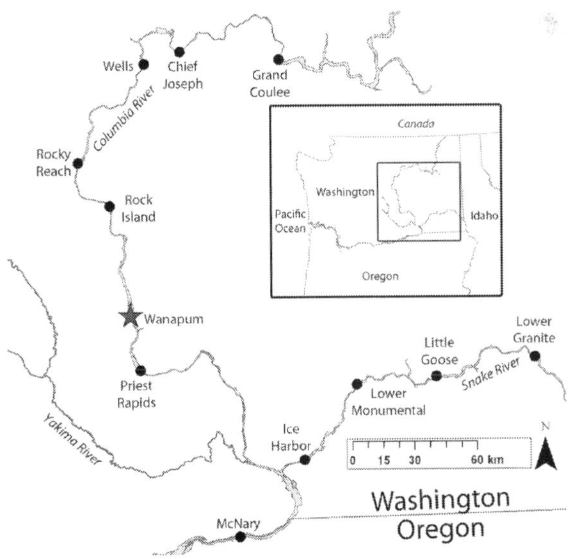

Figure 1: Location of Wanapum Dam on the Columbia River (*red star* on the map).

Acoustic Tags

For this study, two types of JSATS acoustic tags were used. The tags (Advanced Telemetry Systems, tag model SS130; Isanti, MN, USA) used for the controlled field testing were 12.00 mm long, 5.21 mm wide, and 3.77 mm thick; the tags weighed 0.438 g in air. These tags had an average source level of 155 dB re 1 μPa at 1 m, a nominal pulse repetition period of 3.0 s, and a nominal tag life of 33 days. The tags (Advanced Telemetry Systems, tag model SS300) used for the live fish experiment were 10.79 mm long, 5.26 mm wide, and 3.65 mm thick; they weighed 0.346 g in air. These tags had an average source level of 155 dB re 1 μPa at 1 m, a nominal pulse repetition period of 4.2 s, and a nominal tag life of 45 days. The pulse repetition period for each tag did not vary, although there was variability in the pulse repetition period across the tags.

JSATS Cabled Receiver System

A single JSATS cabled receiver system typically consists of one to four cabled hydrophones, a signal conditioning amplifier, a data acquisition computer that contains two 16-bit digital signal processing cards with field-programmable gate arrays each of which has two analog inputs, the detection software, the decoding software, a global positioning system (GPS) card, and a GPS antenna [30, 31, 32]. When an acoustic signal is transmitted from a tag, the hydrophones receive the signal, which is transferred to the amplifier where the signal is band-pass filtered around the carrier frequency of 416.7 kHz and amplified. The filtered and amplified signal then is fed to the digital signal processing card, which contains detection software that checks the incoming data to determine if it contains a potential signal from a tag. If any of the hydrophone signals contain a potential detection from a tag, data are sampled at a frequency of 2.5 MHz, digitized with 16-bit resolution, and written to the hard drive using a filename that includes the high-accuracy GPS time. After the data are written to the hard drive, the decoding software [34] reads in the raw data and determines the presence of a tag code. If a tag code is detected, the decoding software checks the validity of the tag code by performing a cyclic redundancy check.

For this study, three JSATS cabled receiver systems were deployed in the north fish ladder system of Wanapum Dam (Figure 2) at three different locations: (1) near the fish slide at the upper exit (i.e., water outfall structure into the forebay), (2) in the middle portion of the fish ladder, and (3) in the collection channels near the lower entrance. The system near the fish slide had two hydrophones (H1 and H2) installed in the weir pool on 24 April 2014 (Figure 3), just upstream of the fish slide. Also, one hydrophone (H3) was installed in the forebay near the outfall of the fish slide on 29 April 2014. The approximate water depth for both the weir pool and the forebay was 2 to 3 m. The system located in the middle portion of the fish ladder had four hydrophones (H4 to H7), each of which was installed in different pool of the fish ladder on 7 May 2014. The dimension of each pool was 3 × 5 m, and the water depth was approximately 2 m. The system deployed in the collection channels had pairs of hydrophones installed for both the south (H8 and H9) and north (H10 and H11) entrances on 29 April 2014. The width of both collection channels was approximately 5 m, and the water depth was approximately 3 to 4 m. Data were collected by all systems through 15 July 2014.

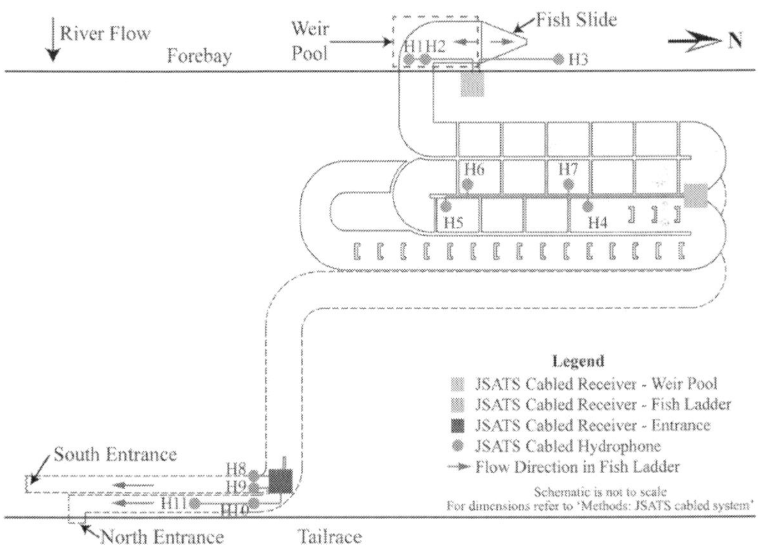

Figure 2: Locations of the three deployed JSATS cabled receiver systems. The locations are in the left (east) bank fish ladder system at Wanapum Dam. Each receiving system monitored three to four hydrophones simultaneously.

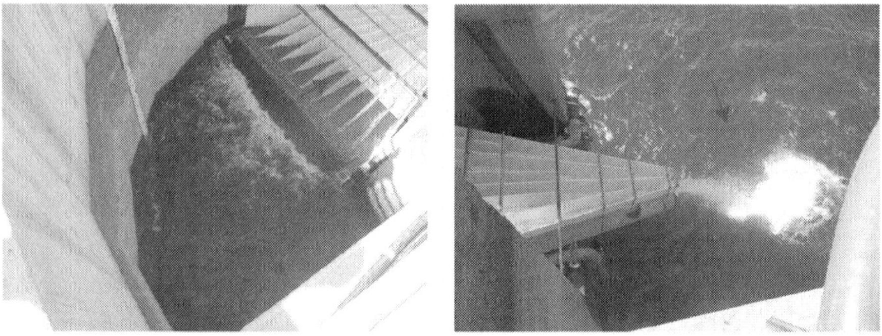

Figure 3: The weir pool (*left panel*) and the forebay (*right panel*) at Wanapum Dam. The weir pool is located adjacent to the fish slide while the forebay is near the outfall of the fish slide. Fish swam out of the weir pool into the top of the fish slide and then passed into the forebay on their upstream migration. The direction of flow is shown with the *red arrow*. Water was pumped up from the forebay to provide flow down the fishway and into the upper end of the fish slide.

Controlled Field Experiments

After each JSATS cabled receiver system was installed, controlled field tests were carried out to evaluate the performance of the system. For the JSATS cabled receiver system near the weir pool of the fish slide (Figure 3), the controlled experiments were conducted separately in the weir pool and the forebay below the weir pool. For these tests, three tags were attached to the line of a fishing rod and were immersed in the water at approximately the middle of the water column. A weight was attached to the terminal end of the line in order for the tags to be fully immersed in the water. The separation between two adjacent tags was approximately 20 cm in the vertical direction, and the tags were approximately 2 to 3 m away from the hydrophones. Each test was performed for 10 min. For the JSATS cabled receiver system inside the middle portion of fish ladder, the three tags were immersed into the water of each pool containing a hydrophone at approximately the middle of the water column, using the same setup as in the weir pool. Their distance from each hydrophone was approximately 2 to 3 m. Each test was performed for 10 min.

System performance data were also collected in the collection channels near the lower entrance of the fishway. Because of the limited access to the water in the collection channels, it was impossible to easily adjust the distance between the tags and the hydrophones by casting the fishing line. Thus, for the JSATS cabled receiver system in the collection channel, the three tags were attached to an approximately 2-m-long piece of rope (with a weight attached to the terminal end of the rope) and were positioned in the mid-depth of water under a 23-cm buoy. Another spool of rope was connected to this 23-cm buoy to control the distances between the tags and the hydrophones by letting the buoy float towards the desired positions. When the buoy was in position, a test was performed for 5 min at each location. At the south entrance, tests were conducted at distances of 1, 10, 30, and 50 m from hydrophones H8 and H9. In the north entrance, tests were conducted at distances of 1, 10, and 20 m from hydrophone H10, which was farther upstream (10, 1, and 9 m, respectively, from the hydrophone H11).

For all tests, the results were processed by calculating the detection efficiency of each tag. The detection efficiency is defined as the ratio of the number of valid detections of the tag to the expected number of tag transmissions within the duration of the test (i.e., 5 or 10 min).

Background Noise Levels

Background noise levels at each location in the confined spaces were estimated from data collected with the deployed JSATS cabled receiver systems. Before being deployed, each hydrophone was calibrated as a function of the frequency in a tank lined with anechoic materials in the Pacific Northwest National Laboratory's Bio-Acoustics and Flow Laboratory (BFL) [35]. The BFL is accredited by the American Association for Laboratory Accreditation (A2LA) to ISO/IEC 17025:2005, which is the international standard for calibration and testing laboratories. The hydrophones (Model SC001, Sonic Concepts Inc., Bothell, WA, USA) have a sensitivity of −180 dB re 1 V/μPa at 416.7 kHz. From data files collected by the JSATS cabled receiver systems, portions that did not contain any transmissions of tag signals were selected and were considered to contain only background noise. Next, from the background noise portions, a frequency range between 375.0 and 458.0 kHz was selected so the bandwidth of the signal becomes 20 % of the carrier frequency of 416.7 kHz. Using the calibration data, the filtered voltage data were used to calculate the pressure spectral density, which was then integrated over the frequency range described above to compute the background noise level. For comparison, the background noise level in the BFL acoustic test tank also was estimated using the same types of instruments as in the field.

Tagged Fish Experiment

As part of a larger effort to evaluate upstream passage through a fish ladder with temporarily reduced flows and a modified exit structure at Wanapum Dam, 50 adult spring Chinook salmon were collected at the Priest Rapids Dam Off-Ladder Adult Fish Trap (rkm 639) from 26 to 29 April 2014, and were surgically implanted with acoustic transmitters (Advanced Telemetry Systems, tag model SS300) while anesthetized [36]. After each tag was implanted, the surgical opening was sutured [37], and the fish were released into the fish ladder at Priest Rapids Dam so they could freely migrate to Wanapum Dam (30 km upstream) on their way to their natal streams upstream of Rock Island Dam (rkm 730). These fish were monitored by the three JSATS cabled receiver systems installed at Wanapum Dam during the migration from Priest Rapids Dam to Rock Island Dam. The results were processed by calculating the detection probability and the median number of detections per tag for each of the three locations where JSATS cabled receiver systems were installed. The detection probability was defined to be the ratio of the detected number of fish to the expected number of fish. The expected number of fish is defined as the number of fish that were supposed to pass this route after the JSATS cabled receiver systems were installed. For example, 43 fish were expected in the weir pool because seven fish were confirmed to have already passed the dam via another route (the fishway on the other bank of the river). In addition, 49 fish were expected in the collection channels due to the fact that one fish passed before the JSATS cabled receiver system was deployed there. Finally, six fish were expected to pass the fish ladder because only six fish were confirmed to remain in the fish ladder when the JSATS cabled receiver system was deployed in this location.

RESULTS AND DISCUSSION

Background Noise

In the BFL acoustic test tank, the estimated background noise level (between 375.0 and 458.0 kHz) was 90 dB re 1 μPa. The mean background noise level near the fish slide was 96 dB re 1 μPa. In the fish ladder, the mean background noise level was 93 dB re 1 μPa, and in the collection channels, the mean background noise levels were 101 dB re 1 μPa and 95 dB re 1 μPa for the south entrance and the north entrance, respectively.

Controlled Field Tests

For the JSATS cabled receiver system deployed near the fish slide, the detection efficiency for each hydrophone is presented in Table 1. The detection efficiencies for hydrophones H1 and H2 individually ranged from 51 to 69 %, and the detection efficiencies for the immediate forebay (hydrophone H3) ranged from 82 to 90 %. Because both hydrophones H1 and H2 were located in the same weir pool, the combined detection efficiency of these two hydrophones is presented to represent the detection efficiency of the transmitted signal in the weir pool. As a result, the combined detection efficiencies for the weir pool were higher than 80 % for all three test tags.

Table 1: Detection efficiency near the weir pool at Wanapum Dam

	Detection efficiency in the weir pool			Detection efficiency in the forebay
Tag number	H1 (%)	H2 (%)	Combined (%)	H3 (%)
1	68	51	85	87
2	59	56	81	90
3	69	54	90	82
Mean	65	54	85	86

Locations of the hydrophones and the JSATS cabled receiver system are indicated in Figure 2

For the system deployed in the middle portion of the fish ladder, downstream of the weir pool, each hydrophone represents the pool in the fish ladder in which it was installed. Among 12 scenarios (i.e., three tags at four pool locations), detection efficiencies in seven scenarios were greater than 80 % (Table 2). Compared to the other locations, the system in the middle portion of the fish ladder had lower detection efficiency when tags were held within 3 m of each hydrophone (H4 to H7). This was the case even though the background noise in the fish ladder at that location was lower than those at the other locations. This lower detection efficiency was likely due to the relatively high water velocity and shallow water inside the fish ladder.

Table 2: Detection efficiency in the middle portion of the fish ladder at Wanapum Dam

	Detection efficiency			
Tag number	**H4 (%)**	**H5 (%)**	**H6 (%)**	**H7 (%)**
1	60	70	65	78
2	81	86	73	80
3	90	93	84	81
Mean	77	83	74	80

Locations of the hydrophones and the JSATS cabled receiver system are indicated in Figure 2

For the system deployed in the collection channels, a pair of two hydrophones was installed in both the south (H8 and H9) and north (H10 and H11) entrances. In the south entrance, the mean detection efficiency was 86 % for H8 and 96 % for H9 when the tags were 1 m away from the hydrophones but decreased to 74 % and 78 % when the distance increased to 10 m (Figure 4). When the tags were 30 to 50 m away from the hydrophones, the mean detection efficiency of an individual hydrophone ranged from 18–53 %. The mean combined detection efficiency was greater than 90 % (98 % and 94 %) up to the distance of 10 m but dropped below 80 % (71 % and 56 %) when the tags were 30 m or more away from the hydrophones. When one-way ANOVA ($p < 0.05$) and Tukey's HSD ($p < 0.05$) were applied to the combined detection efficiency, a statistically significant difference in group means between the groups was found ($p = 7.49 \times 10^{-7}$), but there was no significant difference found between the data at 1 m and at 10 m ($p = 0.42$). On the other hand, the mean detection efficiency of individual hydrophones was greater than 80 % (82 to 99 %), and the mean combined detection efficiency was at least 98 % for all distances in the north entrance (Table 3). The mean detection efficiency of individual hydrophones was consistently higher at the north entrance than at the south entrance when comparing the results from the same distance (1 or 10 m). This observation is consistent with measurements of higher background noise levels at the south entrance (i.e., by approximately 6 dB) than at the north entrance. Overall, the mean combined detection efficiency ranged from 94 to 99 % if the tags were within 10 m of individual hydrophones at each location in the collection channels.

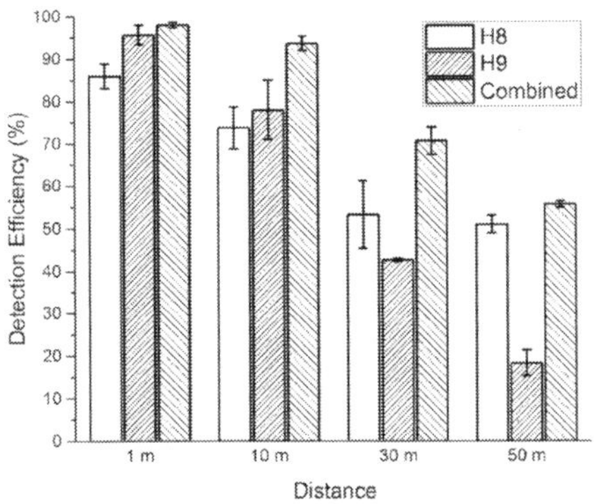

Figure 4: Detection efficiency in the south entrance of the collection channel at Wanapum Dam. Locations of the hydrophones and the JSATS cabled receiver system are indicated in Figure 2. *One half* of the error bar height represents one standard error.

Table 3: Detection efficiency in the north entrance of the collection channel at Wanapum Dam

	Detection efficiency (%)			**Detection efficiency (%)**			**Detection efficiency (%)**		
Tag number	**H10 (1 m)**	**H11 (10 m)**	**Combined**	**H10 (10 m)**	**H11 (1 m)**	**Combined**	**H10 (20 m)**	**H11 (9 m)**	**Combined**
1	86	88	97	88	100	100	88	87	97
2	96	78	99	87	99	99	77	97	100
3	98	93	99	91	99	99	82	97	99
Mean	93	86	98	89	99	99	82	94	99

Locations of the hydrophones and the JSATS cabled receiver system are indicated in Figure 2

Live Fish Experiment

For all three locations, the detection probability was 100 % (Table 4). Here, the expected number of fish is calculated based on the number of fish that were expected to pass the individual route during the time each JSATS cabled receiver system was deployed at each location. For comparison, Roscoe et al. [14] reported detection probabilities of 91 %, 75 %, and 100 % at the bottom, the midpoint, and the top pool of the fishway at Seton Dam in British Columbia, respectively. The difference in detection probabilities between the two studies is likely due to the fact that both the carrier frequency and the pulse duration were different for the systems used in each study. The JSATS uses the carrier frequency of 416.7 kHz, while the system used by Roscoe et al. had a carrier frequency of 69 kHz. Systems that use lower carrier frequencies typically have longer detection ranges in lentic and marine environments than systems that use higher carrier frequencies. However, when working at and near hydroelectric dams, background noise usually affects systems with lower carrier frequencies more because the background noise is typically significantly greater at lower frequency ranges than at higher frequency ranges [33]. In addition, the JSATS utilizes a 744-μs pulse duration compared to the 10-ms pulse duration of the system used by Roscoe et al. [14]. With the narrow width and shallow water depth confined by highly reflective walls in fishways, fewer collisions between the transmitted signals and reflected signals would be expected at the 744-μs pulse duration.

Table 4: Overall detection probabilities and median number of detections per fish at Wanapum Dam

Location	Expected number of fish	Detected number of fish	Detection probability (%)	Median number of detections per fish
Collection channels	49	49	100	4927
Fish ladder	6	6	100	41
Weir pool	43	43	100	86

The number of tagged fish expected in the fish ladder was small because the JSATS cabled receiver system was installed at that location on 7 May 2014, when only six of the tagged fish had not yet been confirmed to have passed the dam. The expected number of fish in the weir pool was 43 because seven fish passed the dam via another route that was not covered by the JSATS cabled receiver systems. Fish spent a median of 3.3 days in the collection channels, although most fish would enter and leave the collection channel multiple times before moving all the way through the ladder. On the other hand, fish spent a median of 10.9 and 16.1 min within the detection range of the JSATS cabled receiver systems in the middle portion of the fish ladder and in the weir pool, respectively. Therefore, the hydrophones in the collection channels had significantly higher numbers of detections than hydrophones installed in the weir pool and the middle portion of the fish ladder.

CONCLUSIONS

Controlled field experiments demonstrated that the JSATS cabled receiver systems can reliably detect the acoustic signal from tags in confined spaces, including the fish ladder entrance (i.e., collection channels), inside the fish ladder, and in the weir pool. The detection efficiency of static tags, based on number of transmitted signals, exceeded 80 % for most locations and ranged from 94 to 99 % in the fish ladder entrance when the tags were within 10 m from all hydrophones. This result may suggest that the JSATS cabled receiver systems can detect the tagged fish in confined spaces such as fishways. A small field study consisting of 50 adult spring Chinook salmon implanted with acoustic tags confirmed the findings of the controlled field experiments. The JSATS cabled receiver systems had a 100 % detection probability of the tagged fish at each deployment location. Even at the locations with the relatively low median number of detections of 41 (fish ladder) and 86 (weir pool), due to the short residence time of the fish, detection probabilities were 100 %, although for the former sample size, there were only six fish. The results of this research will allow scientists to better understand the behavior of fish in confined spaces such as fishways and to design better migration studies involving fishways. The need to implant an additional PIT tag in fish used for a study that concerns one-way migration also might be eliminated if the only purpose of the PIT tag is to confirm the presence of the tagged fish in a fishway. In addition, the results reported in this study are applicable to other acoustic telemetry technologies that have a carrier frequency and pulse duration comparable to JSATS.

ACKNOWLEDGEMENTS

The authors would like to thank the Grant County Public Utility District for providing the opportunity to conduct this study and for funding the work. Without the help of Tao Fu, Yong Yuan, Kyle Larson, Richard Brown, P. Scott Titzler, Kate Deters, Ricardo Walker, Bryan Jones, and Stephanie Liss, this study, including both the controlled experiments and the live fish experiments, would not have been possible. For the live fish experiments, all animal handling was performed under an approval (2014–09) issued by the Institutional Animal Care and Use Committee of Battelle.

AUTHORS' CONTRIBUTIONS

DRG, GAM, PJG, and ZDD conceived, designed, and coordinated this study. KWJ, JRS, and JJM carried out the data acquisition, and KWJ, JJM, and ZDD carried out the data analysis. All authors wrote and approved the manuscript.

REFERENCES

1. Cooke S, Hinch S, Lucas M, Lutcavage M. Biotelemetry and biologging. Fisheries Techniques. 2013;819–860.
2. Michel C, Ammann A, Chapman E, Sandstrom P, Fish H, Thomas M, et al. The effects of environmental factors on the migratory movement patterns of Sacramento River yearling late-fall run Chinook salmon (Oncorhynchus tshawytscha). Environ Biol Fish. 2013;96:257–71.
3. Johnson SL, Power JH, Wilson DR, Ray J. A comparison of the survival and migratory behavior of hatchery-reared and naturally reared steelhead smolts in the Alsea River and Estuary, Oregon, using acoustic telemetry. N Am J Fish Manag. 2010;30:55–71.
4. Perry RW, Skalski JR, Brandes PL, Sandstrom PT, Klimley AP, Ammann A, et al. Estimating survival and migration route probabilities of juvenile Chinook Salmon in the Sacramento–San Joaquin River Delta. N Am J Fish Manag. 2010;30:142–56.
5. Trotter AA, Blewett DA, Taylor RG, Stevens PW. Migrations of common snook from a Tidal River with implications for skipped

spawning. Trans Am Fish Soc. 2012;141:1016–25.

6. McCauley MM, Cerrato RM, Sclafani M, Frisk MG. Diel behavior in white perch revealed using acoustic telemetry. Trans Am Fish Soc. 2014;143:1330–40.
7. Childs A-R, Cowley P, Næsje T, Booth A, Potts W, Thorstad E, et al. Do environmental factors influence the movement of estuarine fish? A case study using acoustic telemetry. Estuar Coast Shelf Sci. 2008;78:227–36.
8. Vinebrooke R, Welch DW, Melnychuk MC, Rechisky ER, Porter AD, Jacobs MC, et al. Freshwater and marine migration and survival of endangered Cultus Lake sockeye salmon (Oncorhynchus nerka) smolts using POST, a large-scale acoustic telemetry array. Can J Fish Aquat Sci. 2009;66:736–50.
9. Heupel MR, Simpfendorfer CA, Fitzpatrick R. Large-scale movement and reef fidelity of grey reef sharks. PLoS ONE. 2010;5, e9650.
10. Abecasis D, Bentes L, Erzini K. Home range, residency and movements of Diplodus sargus and Diplodus vulgaris in a coastal lagoon: connectivity between nursery and adult habitats. Estuar Coast Shelf Sci. 2009;85:525–9.
11. Gowans ARD, Armstrong JD, Priede IG, McKelvey S. Movements of Atlantic salmon migrating upstream through a fish-pass complex in Scotland. Ecol Freshw Fish. 2003;12:177–89.
12. Geist DR, Abernethy CS, Blanton SL, Cullinan VI. The use of electromyogram telemetry to estimate energy expenditure of adult fall Chinook salmon. Trans Am Fish Soc. 2000;129:126–35.
13. Brown RS, Geist DR, Mesa MG. Use of electromyogram telemetry to assess swimming activity of adult spring Chinook salmon migrating past a Columbia River Dam. Trans Am Fish Soc. 2006;135:281–7.
14. Roscoe D, Hinch S, Cooke S, Patterson D. Fishway passage and post-passage mortality of up-river migrating sockeye salmon in the Seton River. British Columbia River Res Appl. 2011;27:693–705.
15. Burke BJ, Jepson MA. Performance of passive integrated transponder tags and radio tags in determining dam passage behavior of adult Chinook salmon and steelhead. N Am J Fish Manag. 2006;26:742–52.

16. Brown R, Oldenburg E, Seaburg A, Cook K, Skalski J, Eppard M, et al. Survival of seaward-migrating PIT and acoustic-tagged juvenile Chinook salmon in the Snake and Columbia Rivers: an evaluation of length-specific tagging effects. Anim Biotelem. 2013;1:8.
17. Muir WD, Smith SG, Williams JG, Sandford BP. Survival of juvenile salmonids passing through bypass systems, turbines, and spillways with and without flow deflectors at Snake River dams. N Am J Fish Manag. 2001;21:135–46.
18. Axel G, Prentice E, Sandford BP. PIT-tag detection system for large-diameter juvenile fish bypass pipes at Columbia River basin hydroelectric dams. N Am J Fish Manag. 2005;25:646–51.
19. Matter AL, Sandford BP. A comparison of migration rates of radio-and PIT-tagged adult Snake River Chinook salmon through the Columbia River hydropower system. N Am J Fish Manag. 2003;23:967–73.
20. McCutcheon CS, Prentice EF, Park DL. Passive monitoring of migrating adult steelhead with PIT tags. N Am J Fish Manag. 1994;14:220–3.
21. Thiem J, Binder T, Dawson J, Dumont P, Hatin D, Katopodis C, et al. Behaviour and passage success of upriver-migrating lake sturgeon Acipenser fulvescens in a vertical slot fishway on the Richelieu River, Quebec. Canada Endanger Species Res. 2011;15:1–11.
22. Foulds W, Lucas MC. Extreme inefficiency of two conventional, technical fishways used by European river lamprey (Lampetra fluviatilis). Ecol Eng. 2013;58:423–33.
23. Franklin AE, Haro A, Castro-Santos T, Noreika J. Evaluation of nature-like and technical fishways for the passage of alewives at two coastal streams in New England. Trans Am Fish Soc. 2012;141:624–37.
24. Castro-Santos T, Haro A, Walk S. A passive integrated transponder (PIT) tag system for monitoring fishways. Fish Res. 1996;28:253–61.
25. Gowans A, Armstrong J, Priede I. Movements of adult Atlantic salmon in relation to a hydroelectric dam and fish ladder. J Fish Biol. 1999;54:713–26.

26. Johnson EL, Caudill CC, Keefer ML, Clabough TS, Peery CA, Jepson MA, et al. Movement of radio-tagged adult Pacific lampreys during a large-scale fishway velocity experiment. Trans Am Fish Soc. 2012;141:571–9.
27. Cooke SJ, Bunt CM. Assessment of internal and external antenna configurations of radio transmitters implanted in smallmouth bass. N Am J Fish Manag. 2001;21:236–41.
28. Adams NS, Rondorf DW, Evans SD, Kelly JE, Perry RW. Effects of surgically and gastrically implanted radio transmitters on swimming performance and predator avoidance of juvenile Chinook salmon (Oncorhynchus tshawytscha). Can J Fish Aquat Sci. 1998;55:781–7.
29. Murchie K, Cooke S, Schreer J. Effects of radio-transmitter antenna length on swimming performance of juvenile rainbow trout. Ecol Freshw Fish. 2004;13:312–6.
30. McMichael GA, Eppard MB, Carlson TJ, Carter JA, Ebberts BD, Brown RS, et al. The juvenile salmon acoustic telemetry system: a new tool. Fisheries. 2010;35:9–22.
31. Deng ZD, Weiland MA, Fu T, Seim TA, LaMarche BL, Choi EY, et al. A cabled acoustic telemetry system for detecting and tracking juvenile salmon: Part 2. Three-dimensional tracking and passage outcomes. Sensors. 2011;11:5661–76.
32. Weiland MA, Deng ZD, Seim TA, LaMarche BL, Choi EY, Fu T, et al. A cabled acoustic telemetry system for detecting and tracking juvenile salmon: Part 1. Engineering design and instrumentation. Sensors. 2011;11:5645–60.
33. Ingraham JM, Deng ZD, Martinez JJ, Trumbo BA, Mueller RP, Weiland MA. Feasibility of tracking fish with acoustic transmitters in the Ice Harbor Dam tailrace. Sci Rep. 2014;4.
34. Ingraham J, Deng Z, Li X, Fu T, McMichael G, Trumbo B. A fast and accurate decoder for underwater acoustic telemetry. Rev Sci Instrum. 2014;85:074903.
35. Deng Z, Weiland M, Carlson T, Eppard MB. Design and instrumentation of a measurement and calibration system for an acoustic telemetry system. Sensors. 2010;10:3090–9.
36. Summerfelt RC, Smith L, Schreck C, Moyle P. Anesthesia, surgery,

and related techniques. Methods for fish biology American Fisheries Society, Bethesda, Maryland. 1990;8:2.

37. Deters KA, Brown RS, Boyd JW, Eppard MB, Seaburg AG. Optimal suturing technique and number of sutures for surgical implantation of acoustic transmitters in juvenile salmonids. Trans Am Fish Soc. 2012;141:1–10.

CHAPTER

7

GENETIC FRAGMENTATION IN INDIA'S THIRD LONGEST RIVER SYSTEM, THE NARMADA

Gulab D Khedkar[1,3], Rahul Jamdade[1] , Amol Kalyankar[1] , Anita Tiknaik[1] , Tetsuzan Benny Ron[2] and David Haymer[3]

[1] Paul Hebert Centre for DNA Barcoding and Biodiversity Studies, Dr. Babasaheb Ambedkar Marathwada University, Aurangabad 431004, India

[2] College of Tropical Agriculture and Human Resources, Department of Human Nutrition, Food and Animal Sciences, University of Hawaii, Honolulu, HI, USA

[3] Department of Cell and Molecular Biology, University of Hawaii, 1960 East-West Rd, Honolulu, HI 96822, USA

ABSTRACT

India's third longest river, the Narmada, is studied here for the potential effects on native fish populations of river fragmentation due to various barriers including dams and a waterfall. The species we studied include a cyprinid fish, *Catla catla,* and a mastacembelid, *Mastacembelus armatus*, both of which are found in the Narmada. Our goal was to use DNA sequence information from the D-loop region of the mitochondrial DNA to explore how this fragmentation could impact the genetic structure of these fish populations. Our results clearly show that these barriers can contribute to the fragmentation of the genetic structure of these fish communities, Furthermore, these barriers enhance the effects of natural isolation by distance and the asymmetry of dispersal flows. This may be a slow process, but it can create significant isolation and result in genetic disparity. In particular, populations furthest upstream having low migration rates could be even more subject to genetic impoverishment. This study serves as a first report of its kind for a river system on the Indian subcontinent. The results of this study also emphasize the need for appropriate attention towards the creation of fish passages across the dams and weirs that could help in maintaining biodiversity.

Keywords: The Narmada river, Genetic fragmentation, Falls, Dams, Fish, Dloop, Migration

INTRODUCTION

India is endowed with diverse aquatic habitats due in part to a unique geological history, highly diverse physiography, a monsoon climate and high biotic diversity. The aquatic habitats include an extensive network of rivers and streams made up of medium and minor river systems (Rao 1975). The majority of these are perennial rivers with large seasonal variation in their flows. In India, over the past 60 years, the landscape of many rivers and streams has also been changed artificially. In response to the growing need of water for agriculture, industrialization and domestic use, many dams and reservoirs are constructed (Khedkar et al. 2014a). Water flow is now highly regulated and is often stored in reservoirs that impound nearly all medium and large rivers. Dams provide benefits in terms of flexibility to use water when it is needed for irrigation, generation of electricity, and other purposes, but have ecological costs as well.

The fragmentation of rivers due to dams or barriers may adversely affect fish populations by diminishing natural habitats required for all life stages and interfering with migration between populations. This in turn may lead to reduced population size, loss of genetic diversity, inbreeding and possible species extinctions. Dams may diminish natural habitats in rivers (Dynesius and Nilsson 1994) and can act as barriers that interfere with migration between populations, even in linked river segments. These barriers may also inhibit recolonization by neighboring populations when local extinctions occur. River segments altered by dams can also lead to environmental disparities in terms of the temperature regimes, nutrient levels, substrate sizes, organic matter transport, the availability of lotic and lentic habitats, and the overall flow regimes (Ward and Stanford 1983). As a result, aquatic communities may experience altered seasonal movements, loss of genetic diversity, reduced population sizes and inbreeding (Ellstrand and Elam 1993; Lynch et al. 1995; Antunes Antunes et al. 2006). These alterations can lead directly to population extinctions (Fagan 2002).

Because the effects of such artificial alterations may have disproportionately greater impacts on the ability of migratory species to complete their entire life cycle, a number of studies have been directed toward recognizing the effects of river fragmentation due to dams on migratory fish populations (Dunham et al. 1997; Neraas and Spruell 2001; Morita and Yamamoto 2002). In addition, understanding the consequences of fragmentation may also be important for non-migratory species, such as cyprinids or mastacembelids, that have low clearing capacities to overcome the obstacle. Specifically it has been shown that although the entire fish community can be affected by weirs (Miranda et al. 2005; Poulet 2007), cyprinid species often make up a major portion of such river communities (Winfield and Townsend 1991).

Moreover, many cyprinid species are known to have to cover considerable distances for reproduction and feeding (McKeown 1984; Northcote 1984). One can hypothesise that such species would be sensitive to obstacles (Bainbridge 1958; Ovidio and Philippart 2002) and would be at great risk of population decline and loss of local genetic diversity (Knaepkens et al. 2004). However, relatively few studies have explored the impacts of river fragmentation due to dams and weirs on cyprinid populations. Of the studies done, many have been limited to small-scale observations such as the effects of single obstacles or have been assessed on individual-scale processes such as survival and / or migration only (Lucas and Frear 1997; Ovidio and Philippart 2002).

In this study we used a broad scale approach to analyze the impact of fragmentation of a river due to a combination of factors including dams and natural barriers on populations of a cyprinid fish, *Catla catla*, and a mastacembelid fish, *Mastacembelus armatus*.

C. catla was selected as a model since it is a cyprinid, and although not considered to be migratory, is known to be able to cover distances of approximately 15-20 km against water current to spawn. *C. catla* is a ubiquitous species, and based on its size (commonly from 5–30 cm), surface dwelling habitat and its swimming ability, it can also be considered to be representative of many other cyprinids inhabiting Indian rivers (Jhingran 1968). Another species, *M. armatus,* was also selected for this study since it is an eel like species with a bottom dwelling habit (Rainboth 1996; Vidthayanon 2002), is very common in Indian rivers and relatively easy to sample. *M. armatus* is not subject to manipulation as a source of material for hatchery based seed production and stocking, factors known to interfere with natural processes (Froese and Binohlan 2000).

We used DNA sequence information from the D-loop region of the mitochondrial DNA to identify markers to differentiate populations (Wilson and Cann 1985; Bremer et al. 1996; Nyakaana et al. 2002; Sato et al. 2004; Khedkar et al. 2013) and to explore how both artificial and natural fragmentation of river habitats could impact the genetic structure of cyprinid and mastacembelid populations in the Narmada river in India. Our specific objectives were to evaluate the extent to which fragmentation could (i) prompt population differentiation at the genetic level, (ii) to identify the gross effects of multiple obstacles on the population's genetic structure in the upstream–downstream riverine regions; and (iii) to assess the effects of dams and a natural water fall on dispersion.

MATERIALS AND METHODS

Study Area

India's third longest river, known as the Narmada, is 1332 km long and covers a drainage basin totaling approximately 98,796 km^2. This study considers only the main stretch of the Narmada river and excludes its tributaries. Most of the study sites selected were dominated by cyprinid and mastacembelid fish. The Narmada is fragmented from upstream to downstream by several dams including the Bargi hydroelectric dam (69 m), the Indira Sagar hydroelectric dam (92 m) and Sardar Sarovar hydroelectric dam (136.5 m). In addition a series of six major natural waterfalls in the Bhedaghat area (~30 m height) occur along this part of the river (Table 1, Figure 1). Both the natural and artificial barriers impact the water flow and can potentially limit both upstream and downstream dispersion of the fish.

Table 1: Details of the sampling stations on the Narmada river

Sampling station	Geographic position			Distance from origin of the river (Km)	Distance from upstream sampling station (Km)	Distance from downstream sampling station (Km)	Presence of barrier upstream and type	Presence of barrier downstream and type
	Latitude	Longitude	Elevation (m)					
Dindori	22°56’32”N	81°04’34”E	2203	62	--	272	--	Bargi Hydroelectric dam (69 m high).
Jabalpur	23°04’55”N	79°54’12”E	1319	334	272	310	Bargi Hydroelectric dam (69 m high).	Series of 6 Natural Waterfalls (30 m high).
Hoshangabad	22°45’52”N	77°44’18”E	996	644	310	274	Series of 6 Natural Waterfalls (30 m high).	Indira Sagar Hydroelectric dam (92 m high).
Mortakka	22°09’03”N	75°29’54”E	545	918	274	280	Indira Sagar Hydroelectric dam (92 m high).	Sardar Sarovar Hydroelectric dam (136.5 m high).
Rajpipla	21°51’48”N	73°30’17”E	113	1198	280	105	Sardar Sarovar Hydroelectric dam (136.5 m high).	--
Bharuch	21°42’05”N	73°00’45”E	28	1303	105	--	--	--

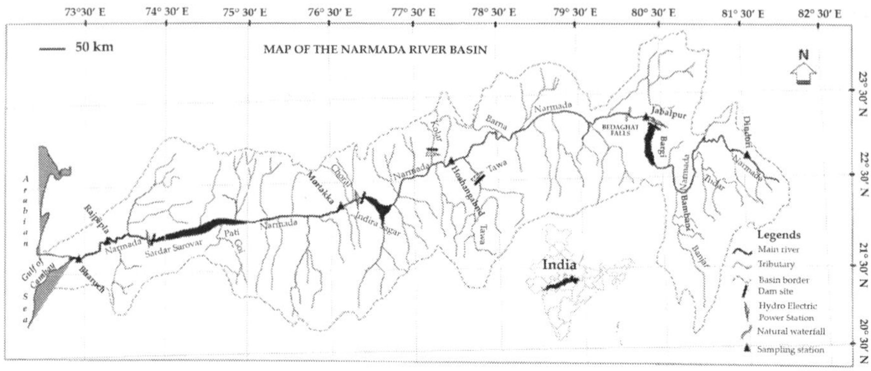

Figure 1: Map of the Narmada river basin showing location of sampling stations and dams.

Sampling

Fish were collected from six sampling stations using various nets and gears during April, 2011 - February, 2013 (Figure 1). Sites were considered for their potential to accommodate cyprinids and mastacembelids (presence of lentic and lotic zones and vegetation shelters) and for their accessibility. Each sampling site was approximately 1500 m long, and all habitats were sampled downstream to upstream until at least 10-12 cyprinid and mastacembelid fish had been caught. We note that *C. catla* fish were not found at Dindori station, and likewise *M. armatus* were not found at Bharuch station. From each specimen a finclip, approximately 1 cm^2, was obtained and stored in absolute ethanol until laboratory analysis as describe below.

DNA Extraction, PCR and DNA Sequencing

Genomic DNA was extracted using the genomic DNA isolation kit (Promega wizard) from a total 43 *C. catla* and 60 *M. armatus* fish. Primers for amplification of the mitchondrial D-loop region of *M. armatus* were designed using the programs PRIMER 3 (Rozen and Skaletsky 2000) and Oligo Calc (Kibbe 2007) by Simgene using a reference sequence obtained from Genbank (EU380216.1). The primers designed for *M. armatus* were: MADF (5′-TTATATGCATTCATTCAGGTACA-3′) and MADR (5′-TAGGGCCCATTTTAACATCT-3′). For *C. catla* the Carp-

Pro and Carp-Phe primers were as described by Thai et al. (2004). Amplifications were carried out using an initial denaturation step at 95°C for 2 min followed by 35 cycles of 94°C for 30 s, 57°C for 1 min and 72°C for 1 min along with a final extension at 72°C for 5 min. The amplified fragments were processed for cycle sequencing using the BigDye®Terminator v.3.1 Cycle Sequencing Kit (Applied Biosystems, Inc.) followed by cleanup and bi-directional sequencing using an ABI 3130 Genetic analyzer (Applied Biosystems, Inc.).

Data Analysis

Sequence Alignments and Topographic Analysis

Sequences were aligned using Codon code aligner v4.0.3 (CodonCode Corporation, Dedham, MA, USA). Within population diversity was estimated by computing haplotype diversity (H) and nucleotide diversity (π) indices using DnaSP v5.10 (Librado and Rozas 2009) and Arlequin v3.5.1.2 (Excoffier and Lischer 2010). Hierarchical relationships among the populations were analyzed using AMOVA and genetic variance was partitioned using Fst. The phylogenetic analysis included finding the best substitution model in Modeltest 2.1.1 (Darriba et al. 2012) using Akaike Information Criterion (AIC) and Baysian Information Criterion (BIC). The haplotype network was computed using Network v4.6.1.1 (Bandelt et al. 1999) where the haplotype pairwise differences were used to determine the number of mutational steps between haplotypes. A statistical test initially developed to analyze selective neutrality of mutations was implemented to test demographic expansion in recent years (de Jong et al. 2011; Ramos-Onsins and Rozas 2002). As shown by de Jong et al. (2011), these tests are designed to distinguish between neutrally evolving sequences in mutation drift equilibrium and sequences evolving under non-neutral processes including directional and balancing selection and demographic expansion or population contraction. These tests were performed in Arlequin v3.1 using 1000 simulations under a selective model of neutrality. For Mantel test, isolation by distance model was analyzed following the method of Jensen et al. (2005).

RESULTS

Diversity Indices

Mitochondrial DNA control region (D loop) sequences of 800 bp were obtained and aligned for the individuals from the Narmada river from to two fish species, *M. armatus* and *C. catla*(NCBI Accession numbers KF468051 to KF468110). The sequences of *M. armatus* (n=60) showed 39 polymorphic sites, accounting for almost 4% of total number of sites examined. Among the polymorphic sites, 7 were singleton variable sites and 2 were parsimony informative sites. Similarly among the individuals representing the *C. catla* population (n=42), there were 15 variable sites (~2%) including 6 singletons and one parsimony informative site.

Genetic Diversity

Eight haplotypes were found in the *M. armatus* populations and thirteen were found in the *C. catla* populations. The distribution and frequency of the haplotypes in each population is shown in Table 2. Haplotypes 2 and 3 are widely distributed in all populations of *M. armatus*.Five unique haplotypes, represented by only single individuals from Jabalpur, Mortakka and Rajpipla populations and three others were found in more than one population (Figure 2). In the *C. catla* populations, haplotype 2 is widely distributed whereas six unique haplotypes represented by only one population and seven others were admixtures of more than one population. Three were found to occur only in the Hoshangabad population (Table 2; Figure 3). Neighbor-joining (NJ) trees showing haplotype relationships were constructed based on the Kimura 2-parameter model, and bootstrap values based on 1,000 replicates shown (Figure 4A and B).

Table 2: Distribution and frequency of different D-loop haplotypes of*M. armatus*and*C. catla*in different populations

Haplotype	Dindori	Jabalpur	Hosangabad	Mortakka	Rajpipla	Bharuch
Mastacembelus armatus						
Hap1	4 *(0.40)*	0	0	1 *(0.076)*	0	--
Hap2	4 *(0.40)*	3 *(0.214)*	12 *(0.923)*	8 *(0.615)*	6 *(0.6)*	--
Hap3	2 *(0.20)*	4 *(0.286)*	1 *(0.076)*	3 (0.231)	3 *(0.3)*	--
Hap4	0	5 *(0.357)*	0	0	0	--

Hap5	0	1 **(0.071)**	0	0	0	--
Hap6	0	1 **(0.071)**	0	0	0	--
Hap7	0	0	0	1 ***(0.076)***	0	--
Hap8	0	0	0	0	1 ***(0.1)***	--
Catla catla						
Hap1	--	0	0	0	0	1 ***(0.10)***
Hap2	--	0	1 ***(0.125)***	2 ***(0.25)***	2 ***(0.20)***	4 ***(0.40)***
Hap3	--	0	0	1 ***(0.125)***	0	1 ***(0.10)***
Hap4	--	0	3 ***(0.375)***	0	0	1 ***(0.10)***
Hap5	--	0	0	2 ***(0.25)***	0	2 ***(0.20)***
Hap6	--	4 ***(0.667)***	0	0	1 ***(0.10)***	1 ***(0.10)***
Hap7	--	1 ***(0.167)***	1 ***(0.125)***	0	0	0
Hap8	--	0	1 ***(0.125)***	0	0	0
Hap9	--	0	1 ***(0.125)***	0	0	0
Hap10	--	0	1 ***(0.125)***	0	0	0
Hap11	--	1 ***(0.167)***	0	0	0	0
Hap12	--	0	0	3 ***(0.375)***	6 ***(0.60)***	0
Hap13	--	0	0	0	1 ***(0.10)***	0

Italicized Values in bracket shows the relative haplotype frequency.

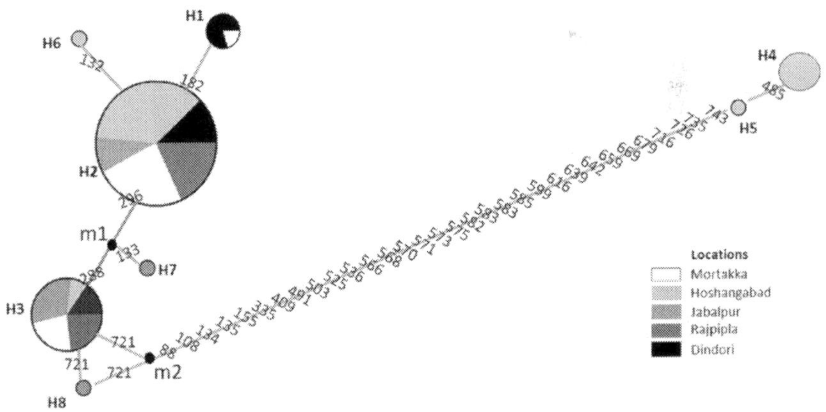

Figure 2: Median joining network map of haplotypes of *M. armatus*.

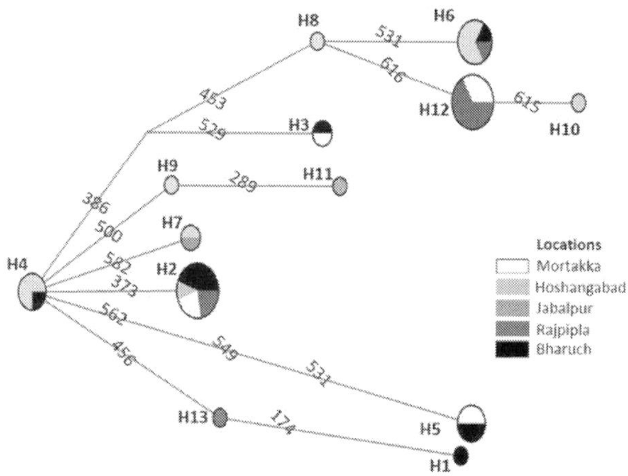

Figure 3: Median joining network map of haplotypes of *C. catla.*

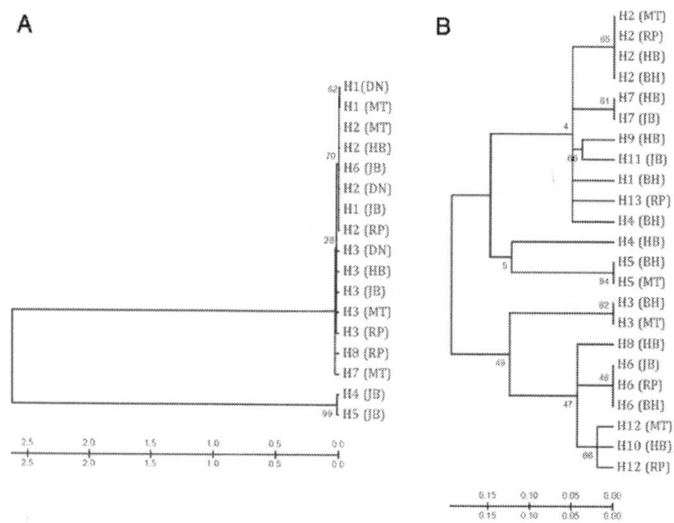

Figure 4: Haplotype based NJ tree (A)*M. armatus*; (B)*C. catla.*

For *M. armatus*, the haplotype diversity and nucleotide diversity values for the Jabalpur population were highest (0.791 and 0.0240 respectively) whereas these values were lowest (0.154 and 0.0003 respectively) in the Hoshangabad population (Table 3). In *C. catla,* for the population from Hoshangabad the haplotype diversity was also highest (0.8929), but for this species the nucleotide diversity was highest (0.2404) in the population from Mortakka (Table 3).

Table 3: D-loop sequence divergence values for*M. armatus*and*C. catla*for populations from the Narmada river

Fish populations	**Parameters**	**Sampling stations**					
		Dindori	**Jabalpur**	**Hoshangabad**	**Mortakka**	**Rajpipla**	**Bharuch**
Mastacembelus armatus	No. of Samples (n)	10	14	13	13	10	-
	No. of Polymorphic sites (PS)	3	39	2	4	3	-
	No. of Haplotypes (k)	3	5	2	4	3	-
	Haplotype diversity (H)	0.711±0.0860	0.791±0.0673	0.154±0.1261	0.603±0.1306	0.600±0.1305	-
	Nucleotide diversity (π)	0.0015±0.0011	0.0240±0.0127	0.0003±0.0004	0.0014±0.0010	0.0015±0.0011	-
Catla catla	No. of Samples (n)	--	6	8	8	10	10
	No. of Polymorphic sites (PS)	--	6	7	8	6	10
	No. of Haplotypes (k)	--	3	6	4	4	6
	Haplotype diversity (H)	--	0.6000±0.2152	0.8929±0.1113	0.8214±0.1007	0.6444±0.1518	0.8444±0.1029
	Nucleotide diversity (π)	--	0.17333±0.1241	0.1404±0.0994	0.2404±0.1551	0.1481±0.1007	0.2044±0.13120

Fst Analysis

Pairwise comparisons of the genetic variation contained in subpopulations relative to the total populations (Fst) of *M. armatus* were significant when the Jabalpur population was compared with all of the others sampled here. In addition, the comparison of the Dindori and Hoshangbad populations was also significant (Table 4). For the C. *catla* populations, these values are significant only for the comparisons of the Rajpipla and Bharuch populations and Jabalpur and Rajpipla populations (Table 5).

Table 4: Population pairwise Fst comparisons for*M. armatus*(below the diagonal)

	Dindori	**Jabalpur**	**Hoshangabad**	**Mortakka**	**Rajpipla**
Dindori	--	*0.00901**	*0.02703**	*0.45946*	*0.22523*
Jabalpur	**0.33467**	--	*0.00000**	*0.00000**	*0.03604**
Hoshangabad	**0.15246**	**0.38583**	--	*0.32432*	*0.08108*
Mortakka	**0.01433**	**0.35912**	**0.04264**	--	*0.62162*
Rajpipla	**0.09018**	**0.31342**	**0.17758**	**-0.04100**	--

*significant p values (<0.05).

Table 5: Population pairwise Fst comparisons for*C. catla (below the diagonal)*

	Jabalpur	**Hoshangabad**	**Mortakka**	**Rajpipla**	**Bharuch**
Jabalpur	--	*0.05405*	*0.18018*	*0.04505**	*0.06306*
Hoshangabad	**0.18719**	--	*0.38739*	*0.07207*	*0.29730*
Mortakka	**0.07238**	**0.01690**	--	*0.38739*	*0.34234*
Rajpipla	**0.17569**	**0.17440**	**0.01864**	--	*0.00901**
Bharuch	**0.18310**	**0.02768**	**-0.00143**	**0.25298**	--

*significant p values (<0.05).

For *M. armatus*, the low pairwise Fst value were noted (-0.0410) between Mortakka and Rajpipla may indicate that these populations undergo genetic exchange events. Among the *C. catla* populations, the low Fst values for the Mortakka and Bharuch comparison (-0.00143) may also suggest high levels of genetic exchange (Tables 4 and 5). Also, based on the AMOVA analysis, the majority of the variation was found to be within populations for both species (Table 6).

Table 6: Analysis of molecular variance (AMOVA) among and within populations

Source of variation	**Among populations of *M. armatus***	**Within *M. armatus* populations**	**Total**	**Fst**	**Among populations of *C. catla***	**Within *C. catla* populations**	**Total**	**Fst**
d.f.	**4**	**55**	**59**	**0.3833***	**4**	**37**	**41**	**0.114****
Sum of squares	**76.321**	**148.212**	**224.533**		**11.295**	**50.300**	**61.595**	
Variance component	**1.372** Va	**2.694** Vb	**4.066**		**0.175** Va	**1.359** Vb	**1.359**	
Percentage of variation	**33.74**	**66.26**	--		**11.45**	**88.55**	--	

Va-Variation among groups; Vb-Variation among populations within groups.

*significant p values (<0.05), **significant p values (<0.01).

To further analyze the potential for gene flow among populations at different locations, we conducted an analysis in ARLEQUIN to obtain M values (Tables 7 and 8). A value greater than two here indicates gene flow between populations (Mallet 2001; Hebert et al. 2004). More than half of our M values are greater than two, and for each of the species, one of the M values is infinite.

Table 7: Matrix of M values for*M. armatus*

	Dindori	Jabalpur	Hoshangabad	Mortakka	Rajpipla
Dindori	--				
Jabalpur	0.99402	--			
Hoshangabad	2.77964	0.79590	--		
Mortakka	34.40182	0.89230	11.22727	--	
Rajpipla	5.04464	1.09529	2.31561	infinite	--

Table 8: Matrix of M values for*C. catla*

	Jabalpur	Hoshangabad	Mortakka	Rajpipla	Bharuch
Jabalpur	--				
Hoshangabad	2.17115	--			
Mortakka	6.40838	29.09091	--		
Rajpipla	2.34597	2.36696	26.32727	--	
Bharuch	2.23077	17.56350	infinite	1.47643	--

Non-significant correlations for isolation by distance were noticed in *M. armatus* populations (Mantel Test; Z=472.20, r=-0.0322, one sided P=0.868 from 10,000 randomizations). RMA regression analysis also revealed a negative isolation-by-distance relationship in case of *M. armatus* (y=0.2607x -2.837, r^2=0.104) (Figure 5A). For the *C. catla* populations there was a positive correlation for isolation by distance (Mantel test; Z=466.05, r=0.408, one sided P=0.1312 from 10000 randomizations) (Figure 5B).

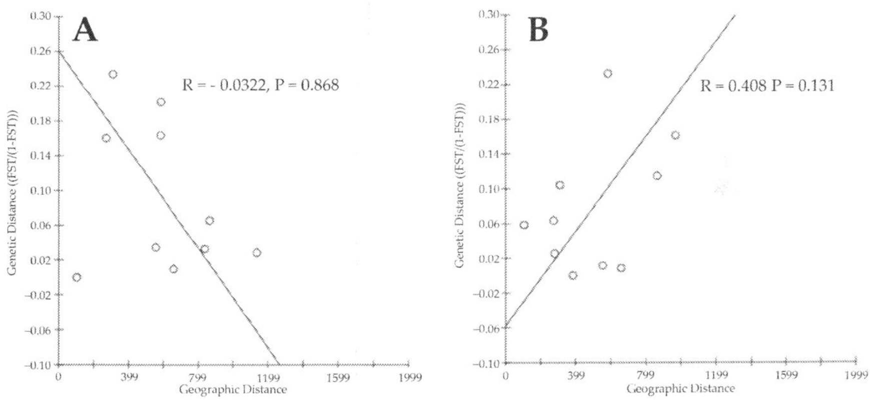

Figure 5: Scatter plot of genetic vs. geographic variation with RMA regression (A)*M. armatus;*(B)*C. catla.*

There was significant support for the model of sudden expansion of population size as well as spatial expansion for the *M. armatus* population at Jabalpur (SSD=0.7924, p=0.000 and SSD=0.129, p=0.0400 respectively). In *C. catla* there is no significant support for either sudden expansion or spatial expansion in any of the populations. The values for Harpending's raggedness index were non-significant in all populations belonging to both species (Table 9).

Table 9: Analysis of populations to deduce demographic and spatial expansion events

Statistics	Dindori	Jabalpur	Hoshangabad	Mortakka	Rajpipla	Bharuch
***M. armatus*Demographic expansion**						
SSD*	0.05702	0.79242	0.03195	0.10070	0.08515	--
p- values	0.08000	0.00000	0.11000	0.10000	0.21000	--
Raggedness index	0.06765	0.37640	0.76331	0.36736	0.31111	--
p- values	0.77000	0.95000	0.66000	0.10000	0.24000	--
Spatial expansion						
SSD*	0.00738	0.12927	0.01297	0.07936	0.05586	--
p- values	0.61000	0.04000	0.20000	0.14000	0.21000	--
***C. catla*Demographic expansion**						
SSD*	--	0.14614	0.01264	0.08586	0.19467	0.08154
p-values	--	0.10000	0.70000	0.04000	0.06000	0.07000
Raggedness index	--	0.27556	0.08291	0.24362	0.62519	0.19605
p- values	--	0.33000	0.73000	0.12000	0.05000	0.08000
Spatial expansion						
SSD*	--	0.05272	0.01264	0.06637	0.13729	0.07079
p- values	--	0.50000	0.67000	0.34000	0.09000	0.27000

(*SSD-sum of squared deviations).

DISCUSSION

Concerns over habitat alterations and the subsequent effects on biodiversity have received much scientific attention during last few decades. Particularly for aquatic diversity, conservationists are concerned that dams may fragment and diminish natural habitats in rivers (Dynesius and Nilsson 1994; Khedkar et al. 2014a) and create environmental disparities (Ward and Stanford 1983) that can result in the loss of genetic diversity, reduced population sizes and inbreeding (Ellstrand and Elam 1993; Lynch et al.1995). Ultimately, this may lead to population extinctions (Fagan 2002). Evaluation of Indian rivers for the possible effects of fragmentation of river habitats has not been adequately addressed, and in part this study is first effort of its kind to evaluate the possible effects of fragmentation of the Narmada river of India using genetic approaches. Prior to this study, no records of genetic studies for the Narmada river fish have been found (Khedkar et al. 2014b).

The results of our study indicate several things. First, using the data from mtDNA control region (D loop), overall levels of genetic variation of *M. armatus* are moderate as compared to other teleostean fishes (Panarari-Antunes et al. 2012; Librado and Rozas 2009; Aboim et al. 2005; Stefanni and Thorley 2003; Salzburger et al. 2003; Lee et al. 1995; Fajen and Breden 1992). In terms of populations, the *M. armatus* population from Jabalpur is naturally isolated due to a waterfall fall on downstream side and a dam on the upstream side. This population had high levels of nucleotide diversity (0.0240), with up to 39 nucleotide substitutions in some cases. This suggests that this population is evolving independently and may become further genetically differentiated. Similar observations were made for populations of *P. squamosissimus* species from the Parniba and Tocantinus river basins in China (Panarari-Antunes et al. 2012).

In contrast, genetic differentiation among the *C. catla* populations in the Narmada river basin appears to be weaker as overall levels of genetic diversity were only moderate to high. Most of the dams constructed across the river basin are almost 30 years old, and the Bedaghat fall is a natural barrier that is much older. The MtDNA data described here uphold the common assumption that analyzed populations of two native fish species existed in this river before the construction of the dams.

From samples across the entire river, we found eight haplotypes among 60 individuals of *M. armatus* vs. thirteen haplotypes among 42 individuals of *C. catla*. The range of haplotype diversity values (H=0.154 to 0.791 and 0.6000 to 0.8929, respectively for these two species) and nucleotide diversity values (π=0.0003 to 0.0240 and 0.1404 to 0.2044 respectively) (Table 3) were similar to values seen in studies by Chen et al. (2006) on Kessle fishes (H=0.9992, π=0.0082), Anthunes et al. (2012) on *P. squamonsissmus* (H=0.690, π=0.0236).

Genetic data presented here suggest the possibility that *M. armatus* populations, which are invasive to the Narmada basin, originated from Jabalpur. This region was naturally isolated from the rest of downstream regions by the Bedaghat falls a few million years ago, and this would naturally limit genetic exchange for this population on the upstream side. The Jabalpur population has relatively moderate genetic and haplotype diversity, and this may also suggest a recent population expansion after a founder event (Shaw et al. 1992; Carvalho et al. 1996). Comparisons of populations at localities serving as drainages, however, could also provide additional resolution for assessing the levels of genetic differentiation of *M. armatus* populations.

The data revealed different levels of genetic differentiation among the different sampling locations across the basin of the Narmada, Haplotypes 2 and 3 may be treated as ancient haplotypes because they are distributed among all populations, whereas the few private haplotypes may be more recently evolved (Figure 2). The genetic diversity within *C. catla* populations documented here strongly suggests that these fishes were likely derived from different sources, possibly through human translocations, since a mixed haplotype distribution was observed (Figure 3). No significant genetic divergences were revealed in *C. catla*. This may be explained by high rates of gene flow between populations of this species within this system.

C. catla is also widely used for aquaculture in India because of its simple culture needs and defined seed production technologies (Jhingran 1968). Many hatcheries are located in the Narmada basin, particularly at the Bharuch region, and artificial seed products may be reentering the river channel further impacting genetic diversity. Also from the median joining networks, a higher degree of haplotype sharing between localities was seen for *M. armatus* as compared to the *C. catla* populations (Figures 2 and 3). Therefore, the patterns of genetic diversity seen within *C. catla* may also reflect both inbreeding and genetic drift type events (Nei et al. 1975; Wishard et al. 1984).

For *M. armatus*, the pairwise Fst values indicate some significant amounts of genetic exchange between the Mortakka and Rajpipla populations in spite of the artificial barrier created by the Sardar Sarowar dam. This suggests that these locations may be connected through a sub tributary network that allows migration. On the other hand, the significantly lower amounts of genetic exchange between the Jabalpur and Dindori populations supports the idea that the presence of the Barghi dam between them promotes fragmentation. Also overall, the pairwise Fst values for the Jabalpur population of *M. armatus* compared to the other populations also indicates significant isolation and supports the role of various barriers in promoting population isolation (Table 4). In the case of *C. catla*, the Fst pairwise comparisons suggest that between some populations, such as the Bharuch and Rajpipla populations, significant amounts of genetic exchange can occur. Here, there is no barrier separating these locations.

The potential for isolation is also considered in the analysis of M values. Here, values of M below 2 imply isolation in the population structure (Mallet 2001). The pairwise comparisons of M values for the *M. armatus* population from Jabalpur with Dindori, Hoshangabad, Mortakka and Rajpipla are all less than this critical value, and this supports the role of dams in promoting genetic fragmentation and population structuring (Table 7). The populations of *C. catla* do not show such structuring, except for the pairwise comparison of M values between the Bharuch and Rajpipla population ($M<2$). This result is consistent with the results based on pairwise Fst values as described in the previous section.

The results of demographic analysis by mismatch distribution show evidence of population expansion. This inference is supported, except perhaps for the Jabalpur population of *M. armatus* (Table 9). We can assume that a recent population expansion might have been affected by genetic drift due to a bottleneck or a founder effect in case of *M. armatus*. The demographic and spatial expansion values for *M. armatus* are significant for Jabalpur population only (Table 9).

Isolation by Distance

Our sampling was restricted to the main channel of the Narmada River, and the sites are lined up in accordance with the stepping stone model of Kimura (1953). According to this model (Kimura and Weiss 1964), in the idealized case the genetic distance between samples increases simply as geographical distance increases. This model also predicts that in the one-dimensional case, the genetic correlation should fall off exponentially with distance. In the case of *M. armatus*, our observations are not in accordance with this predicted pattern because here adjacent populations are significantly differentiated. This is consistent with the notion that other factors, such as the barriers separating these populations, are promoting fragmentation. In contrast, the relationships of the *C. catla* population follow the model pattern in general and do not appear to be so fragmented.

Upstream-Downstream Structure

The variation in genetic diversity along the upstream-downstream gradient is another parameter potentially influenced by the presence of dams. We observed that haplotype and nucleotide diversities are not lined up in a manner consistent with a simple upstream-downstream gradient in populations of both fish species. However, haplotype diversity may be sensitive to sampling bias in that only about 10 individuals per population collected at each sampling site. Additional sampling may reveal the presence of undetected haplotypes.

A number of factors may explain the genetic diversity values seen for the *M. armatus*populations. First, the upper part of the Narmada river studied here is characterized by a natural waterfall and a narrow channel, and it is known that fish population sizes are often correlated to the amount of available habitat (Frankham 1996; Hanfling et al. 2002). Also, *M. armatus* may be subject to a range-edge effect (Arnaud-Haond et al. 2006) since it was more difficult to catch enough specimens in the sites furthest upstream. No individuals were caught at the last locality (Bharuch). Specifically for *M. armatus*, the nature of the dispersion between sampling points, due to the river flow and the obstacles, also suggests that the downstream sites are not receiving new haplotypes from upstream localities.

In the case of *C. catla*, this same dispersion process may maintain a high level of genetic diversity downstream, while genetic drift is less influential in the upstream sampling sites (Lacy 1987). Combining these observations, it can be hypothesized that the *M. armatus* populations within the Narmada river are organized as a single metapopulation composed of local populations exchanging individuals by migration processes (Levins 1970; Hanski and Gilpin 1991). Therefore in this case, the applicability of the stepping-stone model can be rejected.

Dams and or waterfalls, which are designed to prevent upstream movements, seem here in some cases to be sufficient obstacles to induce detectable genetic consequences. In other studies of small species such as bullhead, it has been shown that obstacles only about 20 cm high can prevent upstream movement. Individuals of species such as *Cottus gobio* (Utzinger et al. 1998) and barbel (*Barbus barbus*) individuals are stopped or slowed in their upstream dispersion by a barrier 40 cm high weir (Lucas and Frear 1997). This study confirms that in some cases both dams and water falls can contribute to the asymmetry of dispersion and a decrease in genetic diversity of fish populations.

Data Set Quality and Potential Bias

The Sampling Scheme

Our sampling strategy was to survey as many sites as possible between obstacles that may impact the river habitat structure. The main objective of this study was to assess the extent to which genetic differentiation of populations can be related to fragmentation. The Narmada river system was chosen because there has been extensive fragmentation due to the presence of waterfalls and dams.

In artificially fragmented systems, the number of obstacles is strongly correlated with water way distances (Meldgaard et al. 2003). Dams are generally positioned such that the distance between two consecutive obstacles is about 250 km on average. This holds true for the Narmada system, except that there is also a natural waterfall within 60 kms of the Barghi dam. Due to this spatial configuration, here the effects of waterway distance and the number of obstacles were studied independently.

Lastly, only sampling points of the main Narmada River were considered here. Tributaries are known to have influences on communities inhabiting the main stream of drainage basins, and could also be involved in the demographic and genetic structuring of the Narmada populations (Hitt and Angermeier 2006; Khedkar et al. 2014a). However, for some species, Carlsson et al. (1999) suggested that tributaries may represent ecologically differentiated and locally adapted populations that only rarely disperse in to the main stream, and vice versa. Thus, consideration of only the main Narmada river may be an efficient way to study the genetic structure of these fish populations, although it would be useful to investigate in future studies. the role potentially played by the tributaries.

CONCLUSIONS

The results reported herein for *M. armatus* and *C. catla* populations from the Narmada river clearly show evidence, in some cases, for fragmentation effects by dams and a water fall, and that these barriers contribute to the genetic isolation and differentiation of fish populations. The prevention of movements, either completely from downstream to upstream, or partially from upstream to downstream, by dams or other barriers, enhance the natural effects of isolation by distance and the asymmetry of the dispersal flows. Consequently, populations, especially those furthest upstream, would be expected to have very low immigration rates and be more subject to genetic impoverishment. This study can also provide methodological guidance for future studies of such ecological situations. Sampling a long river segment such as the Narmada with multiple obstacles is certainly a more efficient way to assess the impact on fish dispersal.

Finally, our study does suffer from some biases. Some sample sizes were low, and in some cases it was not possible directly sample upstream and downstream of each barrier. Additional studies in the future may address these biases by including samples of tributaries and other potential sources of input into these populations.

ACKNOWLEDGEMENTS

Funding from NAIP, ICAR, New Delhi, Ministry of Food processing Industries, New Delhi, University Grants commission, New Delhi and CREST Fellowship by Department of Biotechnology, Government of

India to GD Khedkar is deeply acknowledged. We also thank to the authorities of Paul Hebert Centre for DNA Barcoding and Biodiversity Studies for providing access to the molecular genetic analysis facilities.

AUTHOR' CONTRIBUTIONS

Conceived and designed the experiments: GD TBR DH. Performed the experiments: GD RJ AK AT. Analyzed the data: GD DH. Contributed reagents/materials/analysis tools: GD. Wrote the paper: GD TBR DH. All authors read and approved the final manuscript.

REFERENCES

1. Aboim MA, Menezes GM, Schlitt T, Rogers AD: Genetic structure and history of populations of the deep-sea fish *Helicolenus dactylopterus* (Delaroche, 1809) inferred form mtDNA sequence analysis. *Mol Ecol* 2005, 14: 1343-1354.
2. Antunes A, Faria R, Johnson WE, Guyomard RP, Alexandrino P: Life on the edge: the long-term persistence and contrasting spatial genetic structure of distinct brown trout life histories at their ecological limits. *J Heredity* 2006, 97: 193-205.
3. Arnaud-Haond S, Teixeira S, Massa SI, Billot C, Saenger P, Coupland G, Duarte CM, Serrao EA: Genetic structure at range edge: low diversity and high inbreeding in Southeast Asian mangrove (*Avicennia marina*) populations. *Mol Ecol* 2006, 15: 3515-3525.
4. Bainbridge R: The speed of swimming of fish as related to size and to the frequency and amplitude of the tail beat. *J Exper Biol* 1958, 35: 109-133.
5. Bandelt HJ, Forster P, Röhl A: Median-joining networks for inferring intraspecific phylogenies. *Mol Biol Evol* 1999, 16: 37-48.
6. Bremer JRA, Mejuto J, Gomez-Marquez J, Boan F, Carpintero P, Rodrıguez JM, Vinas J, Greig TW, Ely B: Hierarchical analyses of genetic variation of samples from breeding and feeding grounds confirm the genetic partitioning of northwest Atlantic and South Atlantic populations of swordfish (*Xiphias gladius* L.). *J Exp Mar Biol Ecol* 1996, 327: 167-182.
7. Carlsson J, Olsen KH, Nilsson J, Overli O, Stabell OB: Microsatellites

reveal fine-scale genetic structure in stream-living brown trout. *J Fish Biol* 1999, 55: 1290523. 303

8. Carvalho GR, Shaw PW, Hauser L, Seghers BH, Magurran AE: Artificial introductions, evolutionary change and population differentiation in Trinidadian guppies (*Poecilia reticulata*: Poeciliidae). *Biol J Linn Soc* 1996, 57: 219-234.
9. Chen D, Zhang C, Lu C, Zhang X: Polymorphism of D-loop sequence from mitochondrial genomes of different brood-stocks of *Gymnocypris przewalskii* (Kessler). *J Fishery Sci China*2006, 13: 800-805.
10. CodonCode Corporation 101 Victoria Street Centerville, MA 02632;http://www.codoncode.com
11. Darriba D, Taboada GL, Doallo R, Posada D: jModelTest 2: more models, new heuristics and parallel computing. *Nat Methods* 2012, 9(8):772.
12. de Jong MA, Wahlberg N, van Eijk M, Brakefield PM, Zwaan BJ: Mitochondrial DNA signature for range-wide populations of *Bicyclus anynana* suggests a rapid expansion from recent refugia. *PLoS One* 2011, 6(6):e21385. doi:10.1371/journal.pone.0021385
13. Dunham JB, Vinyard GL, Rieman BE: Habitat fragmentation and extinction risk of Lahontan cutthroat trout. *N A J Fish Man* 1997, 17: 1126-1133.
14. Dynesius M, Nilsson C: Fragmentation and flow regulation of river systems in the northern third of the world. *Sci (Washington, DC)* 1994, 266: 753-762.
15. Ellstrand NC, Elam DR: Population genetic consequences of small population size: implications for plant conservation. *A Rev Ecol Syst* 1993, 24: 217-242.
16. Excoffier L, Lischer HEL: Arlequin suite ver 3.5: a new series of programs to perform population genetics analyses under Linux and Windows. *Mol Ecol Reso* 2010, 10: 564-567.
17. Fagan WF: Connectivity, fragmentation, and extinction risk in dendritic meta populations. *Ecology* 2002, 83: 3243-3249.
18. Fajen A, Breden F: Mitochondrial DNA sequence variation among natural populations of the Trinidad guppy, *Poecilia reticulata* .

Evolution 1992, 46: 1457-1465.

19. Frankham R: Relationship of genetic variation to population size in wildlife. *Con Biol* 1996, 10: 1500-1508.
20. Froese R, Binohlan C: Empirical relationships to estimate asymptotic length, length at first maturity and length at maximum yield per recruit in fishes, with a simple method to evaluate length frequency data. *J Fish Biol* 2000, 56(4):758-773.
21. Hanfling B, Hellemans B, Volckaert FAM, Carvalho GR: Late glacial history of the cold-adapted freshwater fish *Cottus gobio* , revealed by microsatellites. *Mol Ecol* 2002, 11: 1717-1729.
22. Hanski I, Gilpin M: Metapopulation dynamics: brief history and conceptual domain. *Biol J Linn Soc* 1991, 42: 3-16.
23. Hebert PDN, Penton EH, Burns JM, Janzen DH, Hallwachs W: Ten species in one: DNA barcoding reveals cryptic species in the neotropical skipper butterfly *Astraptes fulgerator* . *Proc Natl Acad Sci, USA* 2004, 101: 14812-14817.
24. Hitt NP, Angermeier PL: Effects of adjacent streams on local fish assemblage structure in western Virginia: implications for biomonitoring. *A Fish Soc Symp* 2006, 48: 75-86.
25. Jensen JL, Bohonak AJ, Kelley ST: Isolation by distance, web service. *BMC Gen* 2005, 6: 13. v.3.23 http://ibdws.sdsu.edu
26. Jhingran VG: *Synopsis of biological data on ostia, Catla catia (Hamilton, 1822)14-6F290 FAO Fich Synnpc, (32) Rev.1:pag.var.* 1968.
27. Khedkar GD, Chandrashekar Reddy A, Ron TB, Haymer D: High levels of genetic diversity in *Penaeus monodon* populations from the east coast of India. *SpringerPlus* 2013, 2: 671.
28. Khedkar GD, Lutzky S, Rathod S, Kalyankar A, David L: A dual role of dams in fragmentation and support of fish diversity across the Godavari River basin in India. *Ecohydrol* 2014a. doi:10.1002/eco.1470
29. Khedkar GD, Jamdade R, Naik S, David L, Haymer D: DNA Barcodes for the FIshes of the Narmada, One of India's Longest Rivers. *PLoS ONE* 2014b, 9(7):e101460. doi:10.1371/journal.pone.0101460

30. Kibbe WA: OligoCalc: an online oligonucleotide properties calculator. *N Acids Res* 2007, 35(suppl 2):W43-W46.
31. Kimura M: Stepping-stone model of population. *Annu Rep Natl Inst Genet* 1953, 3: 62-63.
32. Kimura M, Weiss GH: The stepping stone model of population structure and the decrease of genetic correlation with distance. *Genetics* 1964, 49: 561-576.
33. Knaepkens G, Verheyen E, Galbusera P, Eens M: The use of genetic tools for the evaluation of a potential migration barrier for the bullhead. *J Fish Biol* 2004, 64: 1737-1744.
34. Lacy RC: Loss of genetic diversity from managed populations: interacting effects of drift, mutation, immigration, selection, and population subdivision. *Conser Biol* 1987, 1: 143-158.
35. Lee WJ, Conroy J, Howell WH, Kocher TD: Structure and evolution of teleost mitochondrial control regions. *J Mol Evol* 1995, 41: 54-66.
36. Levins R: Extinction. In *Some mathematical problems in biology*. Edited by: Gustenhaver M. American Mathematical Society, Providence, Rhode Island; 1970:76-107.
37. Librado P, Rozas J: DnaSP v5: A software for comprehensive analysis of DNA polymorphism data. *Bioinformatics* 2009, 25: 1451-1452.
38. Lucas MC, Frear PA: Effects of a flow-gauging weir on the migratory behavior of adult barbel, a riverine cyprinid. *J Fish Biol* 1997, 50: 382-396.
39. Lynch M, Conery J, Burger R: Mutation accumulation and the extinction of small populations. *Am Nat* 1995, 146: 489-518.
40. Mallet J: Gene Flow. In *Insect Movement: Mechanisms and Consequences*. Edited by: Woiwod IP, Reynolds DR, Thomas CD. CAB International, Wallingford, UK; 2001:337-360.
41. McKeown BA: *Fish Migration*. Croom Helm Ltd., London; 1984.
42. Meldgaard T, Nielsen EE, Loeschcke V: Fragmentation by weirs in a riverine system: a study of genetic variation in time and space among populations of European grayling (Thymallus thymallus) in a Danish river system. *Con Gene* 2003, 4: 735-747.

43. Miranda R, Oscoz J, Leunda PM, Garcia-Fresca C, Escala MC: Effects of weir construction on fish population structure in the River Erro (North of Spain). *Annales De Limnologie-Int J Limn*2005, 41: 7-13.

44. Morita K, Yamamoto S: Effects of habitat fragmentation by damming on the persistence of stream-dwelling charr populations. *Cons Biol* 2002, 16: 1318-1323.

45. Nei M, Maruyama T, Chakraborty R: The bottleneck effect and genetic variability in populations. *Evolution* 1975, 29: 1-10.

46. Neraas LP, Spruell P: Fragmentation of riverine systems: the genetic effects of dams on bull trout (*Salvelinus confluentus*) in the Clark Fork River system. *Mol Ecol* 2001, 10: 1153-1164.

47. Northcote TG: Mechanisms Of Fish Migration In Rivers. In *Mechanisms Of Migration In Fishes*. Edited by: McCleave JD, Arnold GP, Dodson JJ, Neill WH. Plenum, New-York; 1984:317-355.

48. Nyakaana S, Arctander P, Siegismund HR: Population structure of the African savannah elephant inferred from mitochondrial control region sequences and nuclear microsatellite loci. *Heredity* 2002, 89: 90-98.

49. Ovidio M, Philippart JC: The impact of small physical obstacles on upstream movements of six species of fish -Synthesis of a 5-year telemetry study in the River Meuse basin. *Hydrobiologia*2002, 483: 55-69.

50. Panarari-Antunes RS, Prioli AJ, Prioli SMAP, Júlio HF, Agostinho CS, Prioli LM: Molecular variability in Brycon cf. pesu Müller and Troschel, 1845 (Characiformes: Characidae) from the Araguaia-Tocantins Basin. *Gen Mole Res: GMR* 2012, 7(1):95-106. DOI:10.4238/vol7-1gmr365

51. Poulet N: Impact of weirs on fish communities in a piedmont stream. *River Res Appli* 2007, 23: 1038-1047.

52. Rainboth WJ: *Fishes of the Cambodian Mekong. FAO Species Identification Field Guide for Fishery Purposes*. FAO, Rome; 1996:265.

53. Ramos-Onsins SE, Rozas J: Statistical properties of new neutrality tests against population growth. *Mol Biol Evol* 2002, 19: 2092-2100.

doi:10.1093/oxfordjournals.molbev.a004034

54. Rao KL: *India's Water Wealth.* Orient Longman Ltd., New Delhi; 1975:255.
55. Rozen S, Skaletsky H: Primer3 on the WWW for General Users And For Biologist Programmers. In *Bioinformatics Methods And Protocols: Methods In Molecular Biology.* Edited by: Krawetz S, Misener S. Humana Press, Totowa, NJ; 2000:365-386.
56. Salzburger W, Brandstatter A, Gilles A, Parson W, Hempel M, Sturmbauer C, Meyer A: Phylogeography of the vairone (*Leuciscus souffia* , Risso 1826) in Central Europe. *Mol Ecol*2003, 12: 2371-2386.
57. Sato S, Kojima H, Ando J, Ando H, Wilmot RL, Seeb LW, Efremov V, LeClair L, Buchholz W, Jin DH, Urawa S, Kaeriyama M, Urano A, Abe S: Genetic population structure of chum salmon in the Pacific Rim inferred from mitochondrial DNA sequence variation. *Env Biol Fishes* 2004, 69: 37-50.
58. Shaw PW, Carvalho GR, Seghers BH, Magurran AE: Genetic consequences of an artificial introduction of guppies (*Poecilia reticulata*) in N. Trinidad. *Proc R Soc B-Biol Sci* 1992, 248: 111-116.
59. Stefanni S, Thorley JL: Mitochondrial DNA phylogeography reveals the existence of an evolutionary significant unit of the sand goby *Pomatoschistus minutus* in the Adriatic (Eastern Mediterranean). *Mol Phyl Evol* 2003, 28: 601-609.
60. Thai BT, Burridge CP, Pham TA, Austin CM: Using mitochondrial nucleotide sequences to investigate diversity and genealogical relationships within common carp (*Cyprinus carpio* L.). *Ani Gen* 2004, 36: 23-28.
61. Utzinger J, Roth C, Peter A: Effects of environmental parameters on the distribution of bullhead *Cottus gobio* with particular consideration of the effects of obstructions. *J App Ecol*1998, 35: 882-892.
62. Vidthayanon C: *Peat Swamp Fishes of Thailand.* Office of Environmental Policy and Planning, Bangkok, Thailand; 2002:136.
63. Ward JV, Stanford JA: The Serial Discontinuity Concept Of Lotic Ecosystems. In *Dynamics Of Lotic Ecosystems*. Edited by: Fontaine

TD, Bartell SM. Ann Arbor Science, Ann Arbour, MI, USA; 1983:494.

64. Wilson AC, Cann RL: Mitochondrial DNA and two perspectives on evolutionary genetics. *Biol J Linn Soc* 1985, 26: 375-400.

65. Winfield IJ, Townsend CR: The Role Of Cyprinids In Ecosystems. In *Cyprinid Fishes: Systematics, Biology And Exploitation.* Edited by: Winfield IJ, Nelson JS. Chapman and Hall, London; 1991:552-567.

66. Wishard LN, Seeb JE, Utter FM, Stefan D: A genetic investigation of suspected red band trout populations. *Copeia* 1984, 1: 120-132.

SECTION III

FISHERY MANAGEMENT

CHAPTER

8

BOUNDARY SPACES, OBJECTS AND ACTIVITIES IN MIXED-ACTOR KNOWLEDGE PRODUCTION: MAKING FISHERY MANAGEMENT PLANS IN COLLABORATION

Kari Stange, Judith van Leeuwen and Jan van Tatenhove

Environmental Policy Group, Wageningen University & Research, Postbus 81306700, EW, Wageningen, The Netherlands

ABSTRACT

This paper investigates knowledge production in collaborations where the actors have different knowledge and interests. Building on boundary object theory, a conceptual framework is developed to analyze knowledge exchange in two stakeholder-led collaborations to make fishery management plans. The framework introduces *boundary spaces* to conceptualize the dynamic interaction between objects and activities. Within boundary spaces, actors can share, transfer and translate their knowledge, and common understanding can emerge. The collaborations analyzed aimed to produce management plans for the *Nephrops* (*Nephrops norvegicus*) fishery in the North Sea and the boarfish (*Capros aper*) fishery in the Northeast Atlantic. Several boundary spaces were identified in each collaboration. During the production phase, the emerging management plans took on multiple representations as boundary objects that facilitated knowledge exchange. Activities were essential, as these created entry points for different actors to become part of the boundary spaces where they could contribute to knowledge production. Fishing industry representatives in the North Sea Advisory Council and the Pelagic Advisory Council played key roles in initiating and coordinating activities. The case studies demonstrate that Advisory Councils take on pro-active roles in initiatives that aim to expand the knowledge base for European fisheries management. Direct engagement was instrumental to create ownership of the problem addressed in the various collaborative settings that emerged during the management plan initiatives.

Keywords: Knowledge production, Knowledge exchange, Boundaries Boundary objects, Participation, Stakeholders, Fishery management, Fisheries, Advisory Councils, Cooperative research

INTRODUCTION

A key challenge in environmental management is how different kinds of knowledge can be integrated into products that are useful for managers (Roux et al. 2006; Fazey et al. 2013; Raymond et al. 2010). European fishery management under the Common Fisheries Policy (CFP) has a legacy of being informed by science that focuses exclusively on biological aspects of fisheries (Hegland 2006). Critical inquiries have highlighted, however, that knowledge contributions from a broad range of natural and social science disciplines as well as from stakeholders are needed to embrace the complexity of fishery management (Degnbol et al. 2006; Symes and Hoefnagel 2010; Garcia and Charles 2008; Hawkins 2005; Schwach et al. 2007). CFP decision-making should, in accordance with good governance principles, be based on best available scientific advice and broad stakeholder involvement (EU 2013). The two principles can be combined by involvement of stakeholders in research, here referred to as 'the scientific route' to knowledge integration. This scientific route was highlighted in the European Commission's Green Paper, which discussed how the CFP could be improved through reforms. Addressing the knowledge base for the CFP, the Green Paper asked: "How can we better promote stakeholder involvement in research projects, and incorporate stakeholder knowledge in research-based advice?" (European Commission 2009:20–21).

Stakeholders can also contribute to the knowledge base of the CFP by submitting their recommendations to the European Commission through the Advisory Councils, here referred to as 'the stakeholder route' to knowledge integration. Established as Regional Advisory Councils (RACs) as part of the 2002 CFP reform, and as Advisory Councils following the 2013 reform[1](Council 2002, 2004; EU 2013), these are heterogeneous stakeholder forums with 60 % representation from the fishing industry and fishermen's organisations and 40 % representation from other interest groups, including non-governmental organisations (NGOs). The Advisory Councils' role is to provide the European Commission and European Union (EU) member states with advice on issues related to fishery management. Linke et al. (2011) draw attention to challenges associated with incorporating the stakeholder knowledge provided by the Advisory Councils into management and policymaking as these bodies are typically consulted during the final stages of a governance process. Active involvement in earlier stages where knowledge is generated would be better aligned with the original motivation for the establishment of Advisory Councils, i.e. to "enable the Common Fisheries Policy to benefit from the knowledge and experience of the fishermen concerned and of other stakeholders and to take into account the diverse conditions throughout Community waters" (Council 2002:4).

Some Advisory Councils have moved beyond a reactive consultation role by taking initiatives to make long-term management plans. Such plans, also commonly referred to as multiannual plans, are used as tools by the EU bureaucracy to achieve the objectives of the CFP (EU 2013; European Commission 2014). Stakeholder-driven initiatives to make such plans is a recent phenomenon in the EU. The first plan produced by an Advisory Council was a proposal regarding the western stock of Atlantic Horse mackerel (*Trachurus trachurus*) presented by the Pelagic Advisory Council (PELAC) in 2007 (PRAC 2007; Hegland and Wilson 2009). In 2012, the PELAC followed up with a plan for boarfish (*Capros aper*) in the Northeast Atlantic (PRAC 2012a). A third Advisory Council plan was presented by the North Sea Advisory Council (NSAC) in

2015 as their advice concerning management of *Nephrops* (*Nephrops norvegicus*) in the North Sea (NSAC 2015).

Advisory Council initiatives to produce management plans open up for integration of various forms of knowledge into the EU fishery management knowledge base via the stakeholder route. Knowledge exchange processes in such stakeholder-driven settings are largely unexplored. Knowledge exchange research on stakeholder engagement typically focuses on science-driven collaborations and the scientific route to knowledge integration (Fazey et al. 2013, 2014). However, the Advisory Council initiatives are mixed-actor collaborations that are not conducted as formal research projects. Insights about best practices from experiences with science-driven participatory research (Mackinson et al. 2011; Reed et al. 2014; Hegger et al. 2012) are thus not directly applicable.

In this paper, we apply and develop boundary object theory to investigate knowledge exchange processes in stakeholder-led initiatives to make management plans for EU fishery management. Building on previous research on boundary processes (Star and Griesemer 1989; Star 2010; Carlile 2002, 2004; Nicolini et al. 2012) described in the next section, we ask: How do *boundary objects*, supported by *boundary activities*, create *boundary spaces* that facilitate knowledge exchange in stakeholder-led collaborations to make management plans? We address this question by analyzing the initiatives to make long-term management plans for *Nephrops* and for boarfish, mentioned above. The two cases represent unique and recent examples of how stakeholders, who represent the fishing industry's interests, engage in activities with other stakeholders and with scientists in collaborations to produce tools for EU fishery management. The two cases complement each other by illustrating collaborations in different fisheries and stakeholder groups; the *Nephrops* case provide insights into an established demersal fishery and a collaborative process in the NSAC, while the Boarfish case is set within the context of a new pelagic fishery and illustrates collaborations co-ordinated by interests in the PELAC. Our analysis focuses on the interplay between objects and activities, and on activities as entry points for various actors, with the aim of advancing understanding of knowledge production in stakeholder-led settings. The paper contributes to boundary object theory by presenting a theoretical understanding of knowledge exchange in stakeholder-led collaborations. In addition, the findings help fishery stakeholders, scientist and mangers understand the mechanisms and dynamics of developing long-term management plans in collaborative settings.

The next section introduces boundary object theory and the conceptual framework developed to analyze boundary processes and knowledge exchange. This is followed by a description of the methods used to collect and analyze empirical material for the two case studies. The *Nephrops* and Boarfish cases are then presented and discussed, and in the final section conclusions are drawn.

THEORY ON BOUNDARY PROCESSES IN MIXED-ACTOR COLLABORATIONS

The metaphor of boundaries is applicable when analyzing the dynamics between actors in collaborative knowledge production processes. Reflecting on the usefulness of boundary concepts in social science research, Lamont and Molnar (2002:169) comment that "[i]f the notion of boundaries has become one of our most fertile thinking tools, it is in part because it captures a fundamental social process, that of relationality". Seminal contributions are Gieryn's concept of boundary work (Gieryn 1983) and Star and Griesemer's concept of boundary objects (Star and Griesemer 1989). The two concepts represent different interpretations of boundaries as a metaphor. While Gieryn's boundary work addresses demarcations, in particular between science and non-science and between scientific disciplines, Star and Griesemer's boundary objects address convergence, as in creating a common understanding between actors who wish to collaborate. As Riesch (2010:455) puts it: "The groups that Star and Griesemer are concerned with here are not rivals that, as in Gieryn's schema need to protect their interest against outsiders, but rather different groups that may have different values, norms and aims, but nevertheless need to work together". Star and Griesemer were interested in how collaborations between diverse groups of actors involved in scientific work can succeed. They describe boundary objects as

"objects which are both plastic enough to adapt to local needs and the constraints of the several parties employing them, yet robust enough to maintain a common identity across sites. They are weakly structured in

common use, and become strongly structured in individual site use. They may be abstract or concrete. They have different meanings in different social worlds but their structure is common enough to more than one world to make them recognizable, a means of translation" (Star and Griesemer 1989:393).

A clarification of Star and Griesemer's use of the words *boundaries* and *objects* in this context is appropriate. Susan Star explains: "Often, boundary implies something like edge or periphery, as in the boundary of a state or a tumour. Here, however, it is used to mean a shared space, where exactly that sense of here and there are confounded" (Star 2010:602–603). An *object*, in Star's understanding, is "something people … act toward and with. Its materiality derives from action, not from a sense of prefabricated stuff or "thing"-ness" (*ibid*:603).

According to Carlile (2002), objects can perform as boundary objects in mixed-actor knowledge exchanges if they have certain characteristics: 1) they establish a shared syntax or language for individuals to represent their knowledge; 2) they provide concrete means for individuals to specify and learn about their differences and dependencies across a given boundary; and 3) they facilitate a process where individuals can jointly transform their knowledge. Carlile (2004) draws attention to that knowledge-sharing becomes increasingly challenging and complex if the actors in the collaboration have high stakes and if there is novelty involved. Boundary objects with characteristics tailored to the complexity and challenge at hand is therefore needed. Nicolini et al. (2012) highlight that boundary object theory is only one of several possibly useful lenses when seeking to understand the dynamics of mixed-actor collaborations. By using multiple theoretical approaches; theory on boundary objects, epistemic objects, cultural historical activity theory, and objects as infrastructure, they shed light on not only *how*, but also *why* and *when* objects may play a role in cross-disciplinary collaborations. They propose that objects can have several functions: 1) motivate collaboration; 2) facilitate work across different types of boundaries; and 3) provide infrastructure. This clarifies that a boundary object works for a reason; someone makes efforts that triggers activities which enable an object to perform as a boundary object in a particular setting and context. *Activities* thus become important when understanding the role of boundary objects in collaborations. We propose that the interplay between activities and boundary objects create 'boundary spaces' within which actors can share, transfer and translate their knowledge into joint knowledge. Without such activities, here referred to as 'boundary activities', the materiality of objects referred to by Star (2010) will not emerge.

These insights on boundary processes discussed above inspired us to develop a conceptual framework for analysing knowledge exchange in stakeholder-led, mixed-actor collaborations.

The framework (Figure 1) addresses the dynamic interplay between objects, activities and actors. A boundary here refers to an abstract, shared space between collaborating actors with different knowledge. This 'boundary space' is where tacit knowledge becomes explicit and where actors are confronted with, and learn about, each other's interests and perspectives. Actors can be individuals or groups. In collaboration, they want to produce new knowledge, based on their own individual knowledge and interests. Objects become boundary objects when they play a role in connecting these actors and help establish a shared understanding between them. Objects can be both abstract and tangible. We introduce the concept of 'boundary activities' as activities that are instrumental in making objects function as boundary objects. Examples of boundary activities are face-to-face meetings, phone calls, workshops, field work, study visits, presentations, and decisions to initiate projects. These activities create entry points for actors to participate in a collaboration. They create ways to focus efforts around a certain shared idea, concept or product that can potentially be a boundary object. Activities are thus instrumental in making objects function as boundary objects in a specific context. Boundary activities and boundary objects are intrinsically interrelated in the sense that the objects trigger activities while the activities, in return, support the objects.

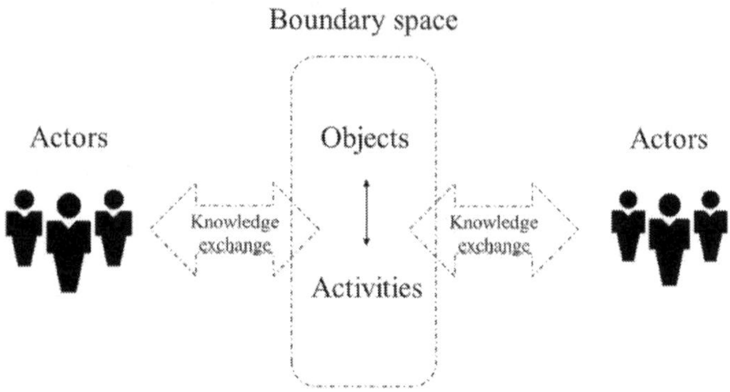

Figure 1: A conceptual framework for understanding knowledge exchange between actors facilitated through an interplay between objects and activities within boundary spaces.

METHODS

Knowledge exchange processes were investigated in two case studies of stakeholder-led collaborations; the *Nephrops* case and the Boarfish case. The cases represent two out of three to us known examples where long-term management plans have been presented by Advisory Councils. A third collaboration to make a plan for western horse mackerel has previously been analyzed from a participatory modelling perspective by Hegland and Wilson (2009). The collaborations studied here were recent or still ongoing, which made it possible to follow the developments in, or close to, real time. Qualitative empirical material was collected from documents, through semi-structured interviews, and from observations. The use of multiple methods gave opportunities for triangulation, i.e. to check and validate findings by combining evidence from multiple sources (Yin 2009). The principal investigator participated as observer in eight NSAC meetings in 2012 and 2013, and in six PELAC meetings in 2014. Documents studied included minutes from NSAC meetings 2006–2015 and PELAC meetings 2006–2015. Other documents consulted were International Council for the Exploration of the Sea (ICES) Advice and Working Group (WG) reports related to *Nephrops* and boarfish, CFP-related documents from the European Commission, and newsletters from fishermen's organisations. The documents were particularly useful to gain an overall understanding of the various actors' involvement with management plans, establish timelines, and identify potential informants for interviews. Semi-structured interviews (n=37) were conducted between May 2012 and November 2014 with fishing industry stakeholders, including Advisory Council members and staff (n=17), scientists who had been involved in *Nephrops* or boarfish research or in the making of management plans (n=14), and civil servants in the European Commission's Directorate-General for Maritime Affairs and Fisheries (DG MARE) (n=6). The interviews were conducted face-to-face, and via telephone or Skype, and were - with a few exceptions - recorded and transcribed.

Documents, interview transcripts and observer notes were assembled, labelled and structured using ATLAS.ti qualitative analysis software. The software aided retrieval and analysis of information from the combined pool of documents. Detailed narratives of the processes of making the *Nephrops* and boarfish management plans were created. The narratives described initiatives taken, actors involved in activities, and tangible output produced by the collaborations. Key informants were asked to review drafts of these narratives to help identify any errors and omissions. The narratives were used to analyse the role of objects, activities and actors with the conceptual framework presented in Figure 1 as a lens. Ideas and interpretations were tested and developed in dialogue between the co-authors in an iterative process through which the conceptual framework evolved.

NEPHROPS CASE

Background

Nephrops are crustaceans that live in burrows in soft sediment. In the North Sea, they are primarily caught by demersal (bottom) trawling within nine areas, called functional units. The functional units are considered to have separate stocks (ICES 2013). A catch quota (Total Allowable Catch, TAC) is currently set for all of the North Sea. ICES advises, however, that management of *Nephrops* should be at the functional unit level "to ensure that catch opportunities and effort are compatible and in line with the scale of the resources in each of the stocks defined by the functional units" (ICES 2013:2). How management at the functional unit level can best be arranged has been subject to debate between stakeholders, managers and scientists for several years. In 2006, the NSAC Demersal Working

Group conceived the idea of producing a long-term management plan for *Nephrops* in the North Sea. The NSAC Executive Committee (ExCom) set work in motion, and in 2015 their proposal could be presented to the European Commission. Here, the dynamic interplay between objects, activities and actors is analyzed to illuminate the knowledge exchange process from idea to proposal. An overview is presented in Table 1.

Table 1: Boundary spaces and knowledge exchange in the *Nephrops* long-term management plan process

Boundary spaces at stages in the collaboration	**Object**	**Actors**	**Activities**	**Knowledge exchange**
Evolution of idea to make a long-term management plan	Long-term management plan as a template	NSAC members	NSAC Demersal Working Group meetings (long term); ExCom decision.	NSAC members develop common interest in taking ownership of a stakeholder-driven, holistic management plan process
Production of draft plans	Long-term management plan as drafts	NSAC members and external experts in the *Nephrops*Focus Group	*Nephrops*Focus Group meetings (ad-hoc, long term)	Focus group members learn about each other's concerns and priorities; invited experts contribute with scientific knowledge
Port visits	Long-term management plan idea and process as a PowerPoint presentation	*Nephrops* Focus Group members, scientists and fishermen	Meetings in fishing ports to discuss management plan ideas and progress (one time)	Presentations and responses; difficult to reach common understanding in "one-off" settings
Functional unit plans	Fishing plan for the Farne Deeps	Farne Deeps Focus Group members	Farne Deeps Focus Group meetings; scientific study. (ad-hoc, short term)	Deliberations to identify functional unit-specific management options; scientific study provides analysis and alternatives
Long-term management plan proposal	Long-term management plan as a NSAC proposal	NSAC ExCom members	NSAC ExCom decision	NSAC-owned plan holds stakeholder produced knowledge

The Idea of a Plan

At the initial stage of the management plan process, the members of the NSAC Demersal Working Group played key roles. Several of them were at the time actively engaged in workshops and conferences where potential benefits of more holistic approaches to fishery management were discussed between stakeholders, scientists and managers (Pope et al. 2006; RACs 2008; ICES 2009). They were inspired by these discussions, and eager to take ownership of a knowledge production process in which stakeholder input would be as important as contributions from scientists and managers. A template for a long-term management plan that would encompass biological, ecological, economic, social and institutional perspectives on fishery management (NSRAC-DWG 2006) became the "scaffolding" around a management plan production process. The NSAC ExCom took action by deciding to establish a *Nephrops* Focus Group as a subset of the Demersal Working Group to bring the hands-on work with the plan forward. With this decision, a new set of actors became involved and another boundary space emerged.

Production of Draft Plans

The *Nephrops* Focus Group consisted of NSAC members with strong interests in the *Nephrops* fishery. Their main activity was meetings; 16 *Nephrops* Focus Group meetings were held between May 2007 and September 2015. Face-to-face interaction over several years allowed the participants to learn from each other about perspectives and priorities within different segments of the fishing fleet. Group members with NGO affiliations brought forward conservation issues including discards, bycatch of vulnerable species, and impact of bottom trawling on the seabed. An activity which further shaped knowledge production within the *Nephrops* Focus Group was to invite scientists. This expansion in terms of actors involved gave opportunities for the *Nephrops* Focus Group members to gain new insights about *Nephrops*biology, recruitment mechanisms, stock assessments, and scientific advice provided by ICES. The scientists contributed with texts on fishery science-specific issues, such as reference points, for inclusion in the long-term management plan. Updated versions of the emerging plan were made available as tangible representations of how the work progressed. As such, the drafts provided a medium for making the participants' tacit knowledge explicit. Furthermore, they served in a boundary object capacity as "containers" for *Nephrops*-related information assembled and ideas discussed by the *Nephrops* Focus Group.

Port Visits

Additional activities were initiated in 2010 to involve more actors in the management plan development process. A press release was sent out to invite fishermen to attend meetings:

"In a ground-breaking initiative that could set a precedent for other stocks, the North Sea Regional Council [sic] is to hold a series of meetings with fishermen and other stakeholders in the main nephrops ports. Chairman of the North Sea RAC ..., said: 'We have been working on a long-term management plan for nephrops in the North Sea for two years. Now it is time to share our work with those in the fishery to test whether we have got things right and to see whether we have missed anything. Long-term management plans are the future and it is therefore extremely important that everyone involved in the nephrops fishery contributes to its development [...]." (NFFO 2010).

The port meetings created entry points for active *Nephrops* fishermen to contribute with their knowledge. In these meetings, PowerPoint slides presented by *Nephrops* Focus Group members set the stage by explaining the rationale behind the NSAC management plan initiative, present progress to date, and share ideas for ways forward. The responses from the fishermen signaled that there were strong and diverging interests and views between different operators and fleet segments. In this boundary space, lack of a shared language between the actors was a barrier to knowledge exchange. While the *Nephrops* Focus Group members and the scientists present were familiar with the scientific and bureaucratic jargon associated with stock assessments and management plans, this was not the case for the other participants. A one-off opportunity for face-to-face interaction, supported by the PowerPoint presentation, was insufficient for common understanding to develop. The feedback from the fishermen did, however, clarify to the NSAC that an overall plan for *Nephrops* management in the North Sea needed to take various local conditions and concerns into account. The next step was therefore to initiate more activities, this time with a local or regional focus.

Functional Unit Management

Based on feedback from the fishermen in the port meetings, and scientific advice from ICES that management of *Nephrops* should be at the functional unit level rather than on the overall North Sea level, the *Nephrops* Focus Group proceeded with the idea to make separate "fishing plans" for functional units. Initiatives to make functional unit fishing plans brought new actors into the overall process. Scientists were engaged to help draft a fishing plan for the Farne Deeps, the functional unit where management measures were most urgent. The scientists interviewed *Nephrops* fishermen and explored which objectives and management options that might be suitable for this particular area (Bailey et al. 2012). The response from fishermen to the scientists' enquiries to participate in the study was low, but for the fishermen who chose to engage, the initiative provided entry points for contributing with their knowledge and concerns. A quote from the Farne Deeps study highlights differences in interests and perspectives between the local fishermen and the more distant operators:

"There was a stark difference of opinion between fishers registered at ports in North East England and those from elsewhere. The main concern of those registered at ports close to the grounds was that the twin rig gears preferred by many visiting skippers were damaging the seabed and taking too much from the fishery. ... There was a strong call from many of the local fleet to ban twin rigging on the Farne Deeps Nephrops grounds. In contrast, those using twin rig gear were of the opinion that claims of seabed damage were unfounded and that bottom contact by heavy weights used by twin rig gear was minimal if it was set up correctly." (Bailey et al. 2012:ii).

The *Nephrops* Focus Group established a Farne Deeps sub-group to discuss possible alternatives to reduce fishing pressure. This initiative reduced the number of actors involved in this boundary space to only a few individuals. After considering effort controls, gear restrictions, spatial measures, and "of which no more than x tonnes" quota-restricting clauses, the Farne Deeps sub-group found the quota-restricting clauses most promising (NSRAC-NFG 2012). However, this conclusion triggered questions from industry stakeholders about who would then be eligible to take the functional unit-specific quota in the Farne Deeps (NSAC-NFG 2014). Exploratory calculations on how such a clause would be materialised in the form of quota allocations to fishermen under the existing management regime highlighted problems related to creating winners and losers. Diverging views on appropriate management measures in the Farne Deeps among fishing industry representatives, who wished to protect the business interests of their constituencies, could not be resolved by arranging yet more Farne Deeps sub-group or *Nephrops* Focus Group meetings. The NSAC finally settled for including a provisional fishing plan for the Farne Deeps as an annex to the overall management plan for the North Sea *Nephrops* fishery. The annex highlighted the "...of which no more than ..." provision as a potentially good solution for the Farne Deeps, while also drawing attention to possible consequences of imposing such quota-restricting measures:

"... all parties accept that in the event of a significant reduction in Nephrops fishing opportunities for operators in the Farne Deeps fishery, quota availability would become a serious issue for locally based vessels dependent on this single fishery. The administrations involved would need to work with the POs (producer organisations) to find the best outcome for those who have a record of fishing in the area and to safeguard the interests of the locally based fleet for the duration of any required quota reductions." (NSAC 2015:38).

The challenges encountered when trying to reach consensus on management measures for the Farne Deeps functional unit clarified how far such a constellation of actors as the Farne Deeps sub-group, the *Nephrops* Focus Group, and ultimately the NSAC, could take a collaborative mixed-actor knowledge production process before encountering politically sensitive issues and handing it over to managers for further work and decision-making (Stange et al. 2015).

A *Nephrops* Long-Term Management Plan Proposal

In February 2015, nine years after the decision to make a management plan was taken by the NSAC ExCom, a 43-page long document *A Long Term Management Plan for North Sea Nephrops* was submitted to DG MARE as NSAC's advice (NSAC 2015). A provisional fishing plan for the Farne Deeps was included in an annex. Another annex, written by scientists, proposed a new reference point for identifying precautionary levels of *Nephrops* stocks. Actors involved at this stage were the NSAC's ExCom as the formal owner and producer of the plan. Introductory statements explained the plan's somewhat unusual length and format:

"Because this is the first plan that has been prepared in this way, with the full involvement of stakeholders, the plan is rather longer [sic] and

more detailed than a conventional Management Plan. The plan includes information on how the management conclusions were reached, and how the plan has progressively evolved. Later versions are expected to be more concise."(NSAC 2015:3).

With this, the NSAC communicated that it had been as important the them to develop a plan with ideas and elements that the fishermen could support as to deliver a product that would fit smoothly into the existing management framework.

Nephrops Case Summary

Producing a long-term management plan for the *Nephrops* fishery became a lengthy learning-by-doing exercise for the NSAC. The process was open and transparent. Stakeholders, scientists and managers with an interest in *Nephrops* science and management were invited to contribute to the hands-on knowledge production process by engaging in the *Nephrops* Focus Group. Several boundary spaces emerged within which actors could exchange knowledge. Draft versions of the long-term management plan served as objects through which this exchange took place. Boundary activities were manifested in the form of numerous group meetings, port visits and an interview study. These activities provided entry points for participation in the knowledge production process; however, direct engagement by active fishermen and managers was limited. It seemed an overly ambitious undertaking to produce a long-term management plan with a holistic approach to management based on broad input from stakeholders with different knowledge and interests. Still, a tangible output was produced and delivered through the stakeholder route. The proposal for a long-term management plan presented to DG MARE represents a milestone for the NSAC as stakeholder contributors to the knowledge base for EU fishery management.

BOARFISH CASE

Background

Boarfish is a small pelagic species that is being caught in increasing amounts on the shelf edge south and west of Ireland. Some consider it a nuisance by-catch species, while others target it as a resource for the fishmeal industry. In 2013, boarfish was added to the list of stocks handled by the PELAC (EU 2013:Annex III). In the PELAC, a majority of the boarfish fishermen's interests is represented through Killybegs Fishermen's Organisation (KFO) and Danish Pelagic Producers Organisation (DPPO). Development of a long-term management plan for the boarfish fishery in the Northeast Atlantic was part of collaborations in 2010–2012 between KFO, DPPO and fishery scientists (Stange 2016). Here, knowledge exchange processes during these collaborations are described with focus on the interplay between objects, activities and actors. An overview of the boundary spaces that emerged is shown in Table 2.

Table 2: Boundary spaces and knowledge exchange in the Boarfish long-term management plan process

Boundary spaces at stages in the collaboration	**Object**	**Actors**	**Activities**	**Knowledge exchange**
Science-industry research initiatives	Problematic knowledge gaps in a rapidly developing fishery	Fishermen's organisation representatives (KFO and DPPO); Scientists (Marine Institute and DTU-Aqua); Boarfish fishermen	Mobilisation of funding for scientific studies; sampling program, acoustic survey, age and maturity studies	New scientific knowledge produced; fishermen develop understanding about the role of scientific data in stock assessment and management
Interim plan proposal development	Interim plan proposal as idea	Fishermen's organization representatives (KFO and DPPO); Scientist (Marine Institute)	Deliberations (one-on-one)	Common interest identified
Long-term management plan proposal development	Long-term management plan as idea	Fishermen's organization representatives (KFO and DPPO); Scientist (Marine Institute); PELAC WG II members	Deliberations (one-on-one); PELAC WG II presentation	Common ground established, facilitated by industry representatives

Long-term management plan proposal	Long-term management plan as proposal	PELAC ExCom members	PELAC ExCom decision	PELAC owned plan holds stakeholder produced knowledge

Science-Industry Collaborations

In 2010, representatives for the fishermen saw a need for the industry to contribute to expand scientific knowledge about boarfish, as explained by the KFO Chief Executive in a newsletter editorial:

"The KFO has embarked on a scientific study of boarfish with the contracting of (name) to carry the necessary scientific work. Very little is known about boarfish and in light of development of the fishery by the RSW (refrigerated seawater) pelagic vessels, the KFO considered it was necessary to have the relevant biological information. Such information is central to devising rational management arrangements that will ensure the long term sustainable future for this fishery. This is a new developing fishery, which has the potential to become a significant economic Irish fishery. Investment in the science at an early stage is paramount to that development." (KFO 2010:4).

Funding for scientific studies on boarfish was made available through contributions from fishermen who had boarfish quotas, as well as from the land-based processing industry. Co-ordination and deliberations within the producer organizations were essential to mobilize the funding. An interviewee explained:

"If the members (of the producer organization) believe that this is the right thing to do, there is never an issue with money. If they don't believe in it, you will never get beyond that first talk." (Interview, Industry representative).

The funding from the industry enabled fishery scientists at the Marine Institute in Ireland and at the Danish Technical University DTU-Aqua to quickly initiate studies on maturity and age verification of boarfish (Farrell et al. 2012; Hussy et al. 2012). Funding also made it possible to arrange annual acoustic surveys to generate abundance data (O'Donnell et al. 2011) and to investigate boarfish-specific acoustic signals (Fässler et al. 2013). The scientific studies provided opportunities for the fishermen to be directly involved in the research undertaken, not only as financers, but also as suppliers of boarfish samples. These activities created entry points for the fishermen to engage in a boundary space where a shared understanding about scientific components to support management of the new boarfish fishery could evolve. Results that emerged from the scientific studies were regularly communicated back to the fishermen in KFO meetings and newsletters, along with explanations of how scientific knowledge on boarfish life history and abundance fit into the contexts of stock assessment, scientific advice, and management decisions. The scientists also published their findings in scientific journals, and brought their new insights on boarfish to the ICES' Working Group for Widely Distributed Stocks (WGWIDE) where it was used to underpin stock assessments and scientific advice (ICES 2012). The scientific studies facilitated knowledge exchange between industry actors and scientists around elements that were relevant for decision-making on boarfish quota and management.

An Interim Management Plan

In parallel with the science-industry collaborations, a first interim management plan for boarfish was developed in 2010 as an Irish-Danish initiative. The initiative was triggered by a mesh size regulation that temporarily closed the boarfish fishery. At the time, industry representatives, scientists and managers were assembled to discuss other pelagic stocks, and they took the opportunity to engage in informal one-on-one discussions around the urgent boarfish situation. A scientist and a representative for a producer organization, who could draw on experiences from previous collaborations around management plans, developed an interim boarfish management plan proposal within a time frame of only a few weeks. An interviewee explained the role of the scientists in this kind of setting:

"My role was to act as a technical advisor to the industry. That doesn't mean that I, or the institute where I work, endorsed those plans. We provided a technical service. We tried to develop something that represented their value system. It was clear to the industry that if they wanted to get a plan accepted, it had to have my value system in it. But ultimately, it is their plan. ... By the time the boarfish plan came along, we had experiences with other plans that had failed because they weren't

precautionary. So that learning curve had already been established and we didn't need to go through that phase again." (Interview, Scientist).

The key actors who were involved in this boundary space had a history of working together. Among previous collaborations was the development of a management plan for horse mackerel (Hegland and Wilson 2009; Clarke et al. 2007). The boarfish interim plan proposal was a 2-page document which emphasized the need for taking a precautionary approach, given the very limited knowledge about this species. A specific TAC-setting rule was proposed, and this element made the plan directly applicable as a tool for managers. The proposal also included measures to avoid by-catch of unwanted species. The interim plan filled a role as a boundary object by facilitating a transformation of thinking for the industry actors from short-term gains to longer term strategies (Stange 2016). To the fishermen, the interim plan implied a dramatic reduction of catch opportunities, at least in the short term. However, it also served as a stimulus to collaborate with scientists. Filling scientific knowledge gaps was urgent to avoid closure of the fishery and could possibly lead to better catch opportunities in the longer term. The interim plan also served as a medium through which the Irish and Danish industry actors could signal their commitment to sustainable management of the rapidly expanding boarfish fishery.

A Long-Term Management Plan Proposal

In 2012, Irish and Danish industry representatives wished to follow up the interim plan initiative with a proper long-term management plan for boarfish that could be endorsed by the PELAC and used as a tool by DG MARE. At this stage, boarfish was not yet among the species formally handled by the PELAC. There was, however, an understanding between the PELAC and the European Commission that boarfish would be added to the PELAC list of species with the 2013 reform of the CFP. With the inclusion of the PELAC as actor, the boundary space widened and the process became more formal and structured. However, the development work was done outside the PELAC by the same key actors who had been involved in the interim plan process. With only a small number of people involved, one-on-one communication co-ordinated by the producer organization representatives was efficient for bringing the process forward. The long-term management plan for boarfish was introduced to the PELAC's WG II in July 2012 in the form of a PowerPoint presentation (Clarke 2012). The meeting minutes reflect how the design of the plan could be interpreted as linking level of knowledge with more and less restrictive quota-setting mechanisms:

"... the more information is available the more generous the TAC can be set whereas the less information is available the higher the uncertainty becomes and therefore the more restricted the TAC would have to be." (PRAC 2012b:12).

The design of the plan thus created incentives for continued collaboration between the scientists and the industry to further expand the scientific knowledge base for boarfish management. The few members of the PELAC WG II who had high stakes in the boarfish fishery had been involved during the drafting phase of the plan, and were already familiar with its content when it was presented to the PELAC. Other PELAC members had little interest in the boarfish fishery. With a few edits, including a change of ownership in the preamble from KFO and DFFO to the PELAC, the plan was brought forward for endorsement by the PELAC ExCom (PRAC 2012b).

The PELAC submitted their *Draft management plan for Boarfish, Caperos aper* as their recommendation to DG MARE in August 2012 (PRAC 2012a). This recommendation was a 5-page long document designed to meet the managers' needs when making decisions about TAC in a new fishery with a small, but growing, scientific knowledge base. Put forward through the formal channel established for delivering stakeholder advice to the EU fisheries management system, i.e. through an AC, the boarfish plan got a "wrapping" as stakeholder produced knowledge. In this boundary space the PELAC ExCom members were the main actors, while the scientists were no longer involved.

Boarfish Case Summary

The analysis highlights that the mixed-actor collaborative process that led towards a long-term management plan proposal for the boarfish fishery in the Northeast Atlantic had two distinct components; one was the production of boarfish-specific scientific knowledge, and another was the making of a long-term management plan that could be proposed through the PELAC as their stakeholder advice to the European Commission. In between, an interim plan helped draw stakeholders', managers' and scientists' attention to the implications of the lack of knowledge on which to build management decisions, and the urgency in building a knowledge base to support this new fishery. Through all stages, representatives for the fishermen co-ordinated initiatives and activities that provided entry points for participation and engagement by various actors. The initial stages were fairly closed and involved only a few, well informed individuals who trusted each other and could work efficiently together. The setting changed from the point when the PELAC took ownership of the long-term management plan. Additional actors then entered the boundary space, and the producer organization representatives followed up on boarfish matters in roles as PELAC members. The widening of context did not trigger any controversies. This indicates that interests were aligned, or that contested issues had been worked through in the preparation phase, during which activities were co-ordinated by PELAC members in roles as representatives of producer organizations.

CONCLUSIONS

This paper has described and analyzed how proposals for long-term management plans for the *Nephrops* fishery in the North Sea and the boarfish fishery in the Northeast Atlantic evolved through collaborations in which fishing industry representatives in the NSAC and PELAC played key roles. The paper contributes to boundary object theory by presenting a theoretical understanding of knowledge exchange in stakeholder-led collaborations. In addition, the findings help fisheries stakeholders, scientist and mangers understand the mechanisms and dynamics of developing long-term management plans in collaborative settings. The *Nephrops* case involved a large number of actors who needed to understand each other and establish ways of working together. The Boarfish case involved only a few actors, and several of them had experiences from previous collaborations. The conceptual framework added dimensions to traditional boundary object theory that were helpful for understanding knowledge exchange in both settings. The analysis focused on the interplay between objects and activities at the boundary between actors, and on how this interplay created spaces for knowledge exchange.

In both cases, management plans were boundary objects in a capacity of motivating the collaborations. Additional boundary objects with different capacities emerged during the course of the collaborations. For example; a management plan template provided "scaffolding" to guide the work with the *Nephrops* plan on from idea to action. In this capacity, the template served as a translational device between actors who needed to develop a common understanding of what a long-term management plan should contain. A more complex boundary process was illustrated in the Boarfish case, where an interim management plan was instrumental in transforming boarfish fishermen's priorities from short-term gains to longer-term precautionary harvesting strategies.

The conceptual framework highlighted the importance of activities as entry point for actors' engagement. In the *Nephrops* case, *Nephrops* Focus Group meetings were key activities. These meetings provided a forum for knowledge-sharing between stakeholders and scientists as well as between different stakeholder interests. Over time, a common understanding evolved around key issues and challenges. Initiatives to address specific problems triggered new activities, which brought different sets of actors into the collaboration. In the Boarfish case, scientific studies created entry points for direct engagement by fishermen in roles as sample providers and financers. These activities created leverage for fruitful knowledge exchange, and a common understanding around the need for management measures for the new boarfish fishery developed.

Analyzing the collaborations as a progression of boundary spaces was useful for pinpointing challenges encountered and strategies chosen to address them. For example; when contested issues are encountered, opening up another boundary space with new actors and objects can be a strategy for bringing the process forward. This was illustrated in the *Nephrops* case when diverging views triggered the need for another boundary space where a tailored functional unit management solution could be elaborated. In the Boarfish case, the creation of a small boundary space with only a few well informed individuals was an efficient way of getting an interim plan produced quickly. These examples demonstrate that multiple boundary spaces evolve during a collaborative knowledge production process, and that inclusion of new actors and exclusion of others are elements in such processes.

The stakeholder-led collaborations analyzed were characterized by *ad hoc* ways of working. Individuals with a strong interest in the issues took initiatives to get collaborative processes started. They used their networks to get others involved, and took on roles as facilitators. These individuals also acted in roles as representatives of fishermen, and were thus not without stakes themselves. In the Boarfish case, the absence of formalities seemed to make the collaborations efficient. As long as there was agreement within and between the producer organizations where the interests of a majority of the boarfish fishermen were represented, decisions could be made quickly and new activities were initiated on short notice. The *Nephrops* process encountered a number of more complex problems which required significant time and efforts to identify ways forward. The approach to problem-solving practiced by the *Nephrops* Focus Group made the development of a long-term management plan for *Nephrops* an inclusive "bottom-up" process, in line with the NSAC's original intensions.

The *Nephrops* and boarfish long-term management plan proposals were submitted as Advisory Council recommendations to the European Commission. This route represents a formally established channel for stakeholder input, and the tangible outcome from the mixed-actor collaborations in the form of management plan proposals could thus be recognized as stakeholder knowledge contributions to the EU fishery management knowledge base. The Boarfish case illustrated that a dual strategy was used by the pelagic industry actors. Their engagement as funders of scientific studies on boarfish resulted in knowledge contributions through the scientific route as well, e.g. in the form of scientific publications and data to underpin ICES stock assessments.

The management plan proposals illustrate Advisory Councils' ability and willingness to take on pro-active roles as producers of knowledge for EU fishery management. The plans produced were different in terms of their form and content. The *Nephrops* plan was long and descriptive, while the Boarfish plan was short with focus on harvest control strategies. It is beyond the scope of this study to evaluate the utility of the two proposals for EU fishery management. It is notable, however, that managers were literally absent from the boundary spaces during the production of the plans. Including managers in the knowledge exchange process would increase possibilities for producing plans that take managers' current priorities and needs into account. This could, however, make the process more complex and time-consuming.

ACKNOWLEDGEMENTS

The authors would like to thank the interviewees for sharing their experiences and insights. Two anonymous reviewers contributed with constructive comments that helped improve the final paper.

ABBREVIATIONS

CFP: Common Fisheries Policy

DG MARE: European Commission's Directorate-General for Maritime Affairs and Fisheries

DPPO: Danish Pelagic Producers Organization

DWG: (NSRAC and NSAC) Demersal Working Group

EU: European Union

ExCom: Executive committee

ICES: International Council for the Exploration of the Sea

KFO: Killybegs Fishermen's Organization

NFFO: National Federation of Fishermen's Organization

NFG: (NSRAC and NSAC) *Nephrops* Focus Group

NGO: Non-governmental organization

NSAC: The North Sea Advisory Council

NSRAC: The North Sea Regional Advisory Council

PELAC: Pelagic Advisory Council

PRAC: Pelagic Regional Advisory Council

RAC: Regional Advisory Council

TAC: Total allowable catch

WG: Working group

WGWIDE: (ICES) Working Group on Widely Distributed Stocks

AUTHORS' CONTRIBUTIONS

KS collected the empirical material, analyzed data, and wrote most of the manuscript under the supervision of JvL and JvT. All authors contributed to development of the conceptual framework and interpretation of results. All authors read, edited and approved the final manuscript.

REFERENCES

1. Bailey, Martin C, Nicholas V. C. Polunin, and Anthony D Hawkins. 2012. A Sustainable fishing plan for the Farne Deeps *Nephrops* fishery. report to the Marine Management Organisation May 2012. Newcastle University. http://www.nsrac.org/wp-content/uploads/2012/05/Agenda-item-4-Nephrops-sust-fishing-plan.pdf. Accessed 7 Oct 2016.
2. Carlile, P.R. 2002. A pragmatic view of knowledge and boundaries: Boundary objects in new product development. *Organization Science* 13(4): 442–455.
3. Carlile, P.R. 2004. Transferring, translating, and transforming: an integrative framework for managing knowledge across boundaries. *Organization Science* 15(5): 555–568.
4. Clarke, M. 2012. Draft interim plan for boarfish. Presentation to the Pelagic Regional Avisory Council, Working Group II, 11 July 2012. http://www.pelagic-ac.org/media/pdf/Presentation%20by%20Maurice%20Clarke.pdf. Accessed 7 Oct 2016.
5. Clarke, M, G van Balsfoort, A Coers, A Campbell, M Dickey-Collas, A Egan, M Ghiglia, I Harkes, C Kelly, S O›Donoghue. 2007. A new scientific initiative with the Pelagic RAC to develop a management plan for western horse mackerel. ICES annual science conference, Helsinki. ICES CM 2007/O:20. International Council for the Exploration of the Sea.
6. Council (Council of the European Union). 2002. Council Regulation (EC) No 2371/2002 of 20 December 2002 on the conservation and sustainable exploitation of fisheries resources under the common fisheries policy. http://eur-lex.europa.eu/legal-content/EN/TXT/PDF/?uri=CELEX:32002R2371&from=EN. Accessed 7 Oct 2016.
7. Council (Council of the European Union). 2004. Council Decision of 19 July 2004, establishing Regional Advisory Councils under

the Common Fisheries Policy (2004/585/EC). Official Journal of the European Union, L256/17, 3.8.2004. http://eur-lex.europa.eu/legal-content/EN/TXT/PDF/?uri=CELEX:32004D0585&from=EN. Accessed 7 Oct 2016.

8. Degnbol, P., H. Gislason, S. Hanna, S. Jentoft, J.R. Nielsen, S. Sverdrup-Jensen, and D.C. Wilson. 2006. Painting the floor with a hammer: Technical fixes in fisheries management. *Marine Policy* 30(5): 534–543.

9. European Commission. 2009. Green paper. Reform of the common fisheries policy. Brussels, COM(2009) 163. http://eur-lex.europa.eu/legal-content/EN/TXT/PDF/?uri=CELEX:52009DC0163&from=EN. Accessed 7 Oct 2016.

10. European Commission. 2014. Communication from the Commission to the European Parliament and the Council - Concerning a consultation on Fishing Opportunities for 2015 under the Common Fisheries Policy. COM(2014) 388 final. http://ec.europa.eu/dgs/maritimeaffairs_fisheries/consultations/fishing-opportunities-2015/doc/com-2014-388_en.pdf. Accessed 7 Oct 2016.

11. EU (European Union). 2013. Regulation (EU) No 1380/2013 of the European Parliament and of the Council of 11 December 2013 on the Common Fisheries Policy, amending Council Regulations (EC) No 1954/2003 and (EC) No 1224/2009 and repealing Council Regulations (EC) No 2371/2002 and (EC) No 639/2004 and Council Decision 2004/585/EC. Official Journal of the European Union. L 354/22, 28.12.2013. http://eur-lex.europa.eu/LexUriServ/LexUriServ.do?uri=OJ:L:2013:354:0022:0061:EN:PDF. Accessed 7 Oct 2016.

12. Farrell, E.D., K. Hussy, J.O. Coad, L.W. Clausen, and M.W. Clarke. 2012. Oocyte development and maturity classification of boarfish (*Capros aper*) in the Northeast Atlantic. *ICES Journal of Marine Science* 69(4): 498–507.

13. Fazey, I., A.C. Evely, M.S. Reed, L.C. Stringer, J. Kruijsen, P.C.L. White, A. Newsham, L.X. Jin, M. Cortazzi, J. Phillipson, K. Blackstock, N. Entwistle, W. Sheate, F. Armstrong, C. Blackmore, J. Fazey, J. Ingram, J. Gregson, P. Lowe, S. Morton, and C. Trevitt. 2013. Knowledge exchange: a review and research agenda for

environmental management. *Environmental Conservation*40(1): 19–36.

14. Fazey, I., L. Bunse, J. Msika, M. Pinke, K. Preedy, A.C. Evely, E. Lambert, E. Hastings, S. Morris, and M.S. Reed. 2014. Evaluating knowledge exchange in interdisciplinary and multi-stakeholder research. *Global Environmental Change-Human and Policy Dimensions* 25: 204–220.
15. Fässler, S.M.M., C. O›Donnell, and J.M. Jech. 2013. Boarfish (*Capros aper*) target strength modelled from magnetic resonance imaging (MRI) scans of its swimbladder. *ICES Journal of Marine Science* 70(7): 1451–1459.
16. Garcia, S.M., and A.T. Charles. 2008. Fishery systems and linkages: Implications for science and governance. *Ocean and Coastal Management* 51: 505–527.
17. Gieryn, T.F. 1983. Boundary-work and the demarcation of science from non-science: strains and interests in professional ideologies of scientists. *American Sociological Review* 48(6): 781–795.
18. Hawkins, T. 2005. The role of partnerships in the governance of fisheries within the European Union. In *Participation in fisheries governance*, ed. T.S. Gray, 65–83. Dordrecht: Springer.
19. Hegger, D., M. Lamers, A. Van Zeijl-Rozema, and C. Dieperink. 2012. Conceptualising joint knowledge production in regional climate change adaptation projects: success conditions and levers for action. *Environmental Science and Policy* 18: 52–65.
20. Hegland, T.J. 2006. Fisheries policy-making: production and use of knowledge. In *The knowledge base for fisheries management*, ed. L. Motos and D.C. Wilson, 219–237. Amsterdam: Elsevier.
21. Hegland, T.J., and D.C. Wilson. 2009. Participatory modelling in EU fisheries management: Western Horse Mackerel and the Pelagic RAC. *Maritime Studies* 8(1): 75–96.
22. Hussy, K., J.O. Coad, E.D. Farrell, L.A.W. Clausen, and M.W. Clarke. 2012. Age verification of boarfish (*Capros aper*) in the Northeast Atlantic. *ICES Journal of Marine Science* 69(1): 34–40.
23. ICES. 2009. *Report of the ICES-STECF Workshop on Fishery Management Plan Development and Evaluation (WKOMSE). ICES CM 2009/ACOM:27. 22 pp.* Copenhagen: International Council

for the Exploration of the Sea. http://www.ices.dk/sites/pub/Publication%20Reports/Expert%20Group%20Report/acom/2009/WKOMSE/WKOMSE09.pdf. Accessed 7 Oct 2016.

24. ICES. 2012. *Report of the Working Group on Widely Distributed Stocks (WGWIDE). ICES CM 2012/ACOM:15.* Copenhagen: International Council for the Exploration of the Sea. http://ices.dk/sites/pub/Publication%20Reports/Expert%20Group%20Report/acom/2012/WGWIDE/WGWIDE%202012.pdf. Accessed 7 Oct 2016.
25. ICES. 2013. ICES Advice 2013, Book 6. *Nephrops* in Subarea IV (North Sea). Copenhagen: International Council for the Exploration of the Sea. http://www.ices.dk/sites/pub/Publication%20Reports/Advice/2013/2013/Neph-IV.pdf. Accessed 7 Oct 2016.
26. KFO (Killybegs Fishermen's Organisation). 2010. KFO news. April 2010, Issue 37. http://www.kfo.ie/assets/newsletter-no.-37_apr2010.pdf. Accessed 7 Oct 2016.
27. Lamont, M., and V. Molnar. 2002. The study of boundaries in the social sciences. *Annual Review of Sociology* 28: 167–195.
28. Linke, S., M. Dreyer, and P. Sellke. 2011. The Regional Advisory Councils: What is their potential to incorporate stakeholder knowledge into fisheries governance? *Ambio* 40(2): 133–143.
29. Mackinson, S., D.C. Wilson, P. Galiay, and B. Deas. 2011. Engaging stakeholders in fisheries and marine research. *Marine Policy* 35(1): 18–24.
30. NFFO (National Federation of Fishermen's Organisation). 2010. Nephrops: Taking it to the Ports. In *FiskerForum 13 June 2010.*. http://www.fiskerforum.dk/en/news/b/Nephrops-Taking-it-to-the-Ports. Accessed 7 Oct 2016.
31. Nicolini, D., J. Mengis, and J. Swan. 2012. Understanding the role of objects in cross-disciplinary collaboration. *Organization Science* 23(3): 612–629.
32. NSAC (North Sea Advisory Council). 2015. A Long term management plan for North Sea *Nephrops*. North Sea Advisory Council. http://www.nsrac.org/wp-content/uploads/2014/12/2-1415-2015-02-16-Nephrops-LTMP.pdf. Accessed 7 Oct 2016.
33. NSAC-NFG (North Sea Advisory Council, Nephrops Focus

Group). 2014. Meeting of the Nephrops Focus Group, 11 November 2014. Brussels, Belgium. http://www.nsrac.org/wp-content/uploads/2014/08/NSAC-NFG-20141111-Brussels-MReport-1.doc. Accessed 7 Oct 2016.

34. NSRAC-DWG (North Sea Regional Advisory Council, Demersal Working Group). 2006. Template for a Long-term Management Plan. Produced from the Edinburgh Workshop. Paper for Den Helder WG. May 14, 2006. http://nsrac.org/wp-content/uploads/2009/09/wp05_wd20060615_Template_LTM_plan.pdf. Accessed 7 October 2016.
35. NSRAC-NFG (North Sea Regional Advisory Council, Nephrops Focus Group). 2012. Meeting of the Nephrops Farne Deeps Focus Group, 13 November 2012. Aberdeen, Scotland. http://nsrac.org/wp-content/uploads/2012/10/NSRAC-NFG-2012-11-13-Aberdeen-MReport-Draft2.doc. Accessed 7 Oct 2016.
36. O'Donnell, C., E.D. Farrell, R.A. Saunders, and A. Campbell. 2011. *Boarfish acoustic survey cruise report, 07 July-28 July, 2011. FSS Survey Series: 2011/03*. Galway: Marine Institute. http://oar.marine.ie/handle/10793/675. Accessed 7 Oct 2016.
37. Pope, John, A.D. Hawkins, Diana Tingley, Simon Mardle, and Nikola Cattermoul. 2006. *Long-term Management of North Sea Fisheries*.. A report to DEFRA and the North Sea Regional Advisory Council, http://www.nsrac.org/wp-content/uploads/2009/09/LTM_REPORT-with-annexes_sc20060302.pdf. Accessed 7 Oct 2016.
38. PRAC (Pelagic Regional Advisory Council). 2007. Management plan for western horse mackerel. http://www.pelagicac.org/media/pdf/20070725%20Management%20Plan%20Horse%20Mackerel%20July%202007.pdf.
39. PRAC (Pelagic Regional Advisory Council). 2012a. Long-term management plan for boarfish. Letter to DG MARE 14 August 2012. http://www.pelagic-ac.org/media/pdf/1112PRAC127%20Boarfish%20management%20plan.pdf. Accessed 7 Oct 2016.
40. PRAC (Pelagic Regional Advisory Council). 2012b. Minutes of Working Group II, 11 July 2012. Amsterdam. http://www.pelagic-ac.org/media/pdf/Minutes%20WG%20II%20meeting%2011%20July%202012.pdf. Accessed 7 Oct 2016.

41. RACs (Regional Advisory Councils). 2008. Joint RACs meeting on Long Term Management Plans 11–12 September 2008. Nantes, France. http://www.nwwac.org/_fileupload/Image/FINAL_REPORT_RAC_LTM%20Seminar_%20Nantes_Ian_Napier_en.pdf. Accessed 7 Oct 2016.
42. Raymond, C.M., I. Fazey, M.S. Reed, L.C. Stringer, G.M. Robinson, and A.C. Evely. 2010. Integrating local and scientific knowledge for environmental management. *Journal of Environmental Management* 91(8): 1766–1777.
43. Reed, M.S., L.C. Stringer, I. Fazey, A.C. Evely, and J.H.J. Kruijsen. 2014. Five principles for the practice of knowledge exchange in environmental management. *Journal of Environmental Management* 146: 337–345.
44. Riesch, H. 2010. Theorizing boundary work as representation and identity. *Journal for the Theory of Social Behaviour* 40(4): 452–473.
45. Roux, D.J., K.H. Rogers, H.C. Biggs, P.J. Ashton, and A. Sergeant. 2006. Bridging the science-management divide: Moving from unidirectional knowledge transfer to knowledge interfacing and sharing. *Ecology and Society* 11(1): 4.
46. Schwach, V., D. Bailly, A.S. Christensen, A.E. Delaney, P. Degnbol, W.L.T. van Densen, P. Holm, H.A. McLay, K.N. Nielsen, M.A. Pastoors, S.A. Reeves, and D.C. Wilson. 2007. Policy and knowledge in fisheries management: a policy brief. *Ices Journal of Marine Science* 64(4): 798–803.
47. Stange, K. 2016. Building a knowledge base for management of a new fishery: boarfish (*Capros aper*) in the Northeast Atlantic. *Fisheries Research* 174: 94–102.
48. Stange, K., J. van Tatenhove, and J. van Leeuwen. 2015. Stakeholder–led knowledge production: Development of a long term management plan for North Sea *Nephrops* fisheries. *Science and Public Policy* 42(4): 501–513.
49. Star, S.L. 2010. This is not a boundary object: Reflections on the origin of a concept. *Science, Technology & Human Values* 35(5): 601–617.
50. Star, S.L., and J.R. Griesemer. 1989. Institutional ecology, 'translations' and boundary objects: Amateurs and professionals in

Berkley's Museum of Vertebrate Zoology, 1907–39. *Social Studies of Science* 19(3): 387–420.

51. Symes, David, and Ellen Hoefnagel. 2010. Fisheries policy, research and the social sciences in Europe: challenges for the 21st century. *Marine Policy* 34(2): 268–275.
52. Yin, R.K. 2009. *Case study research: design and methods*. London: SAGE.

CHAPTER

9

MANAGING SMALL-SCALE FISHERIES IN COLOMBIA

Lina M. Saavedra-Díaz[1], Robert Pomeroy[2] and Andrew A. Rosenberg[3]

[1]Department of Biology, University of Magdalena, Carrera 32 No. 22 – 08, Santa Marta, Magdalena, Colombia

[2] University of Connecticut-Avery Point, Agricultural and Resource Economics, Room 380 - Marine Science Building, 1080 Shennecossett Road, Groton, Connecticut 06340-6048, USA

[3] Center for Science & Democracy, Union of Concerned Scientists, 2 Brattle Square, Cambridge, Massachusetts 02138, USA

ABSTRACT

The small-scale fishermen of Colombia face a wide range of problems and conflicts. While many problems are shared among individuals from both the Atlantic and the Pacific coasts (bi-coastal), others are unique to a subset of the communities, only occur on one of the coasts (uni-coastal) or in an individual locality. To come to grips with the major problems for these fisheries requires establishing a fisheries management strategy that can prioritize solutions at different levels: national, coastal, and local. This study describes the solutions identified by three sets of stakeholders: fishermen, local leaders and fisheries experts, to improve small-scale fisheries management in Colombia. Some cross-cutting solutions were recommended by all three sets of stakeholders. In other cases, only two of the three stakeholder groups agreed on certain proposals, and some isolated solutions were found in only one type of stakeholder. All three sets of stakeholders recommended that the government put in place fundamental regulatory framework for small-scale fisheries including support for alternative employment opportunities to reduce fishing pressure on the resources. Some but not all groups supported specific measures, such as gear restrictions, closed areas and closed seasons. There was also a clear need to distinguish those truly engaged in fishing as their livelihood from opportunists moving in and out of the sector. Specific recommendations are here presented to reform and restructure governance through co-management, and to develop a consensus among the main government and user stakeholders.

Keywords: Colombia, Small scale fisheries, Co-management, Fishery solutions

INTRODUCTION

Small-scale tropical fisheries studies in the last few decades have strongly benefitted from a multidisciplinary approach that includes ecological, economic, sociological, technological and governance information (Christy 1997; Preikshot and Pauly 1998). Many fisheries frameworks and governance models around the world are based on a combination of government and community efforts. These models could be used as points of departure for discussions with important stakeholders in Colombian small-scale fisheries.

A growing global literature describes lessons learned from implementing small-scale fisheries management (Berkes 2004, 2010; Brown and Pomeroy 1999; Chuenpagdaee and Jentoft 2007; Cinner et al. 2012; FAO 2006; Francis et al. 2007; Garcia and Cochrane 2005; Lam and Pauly 2010; Lebel et al. 2006; Olsson et al. 2004, 2006; Pomeroy et al. 2004; Salas et al. 2011; Tyler 2006). Understanding social interactions within the context of fishing, and the dynamic among fishermen within and among fishing communities (St. Martin et al. 2007) are essential in order to develop effective management strategies.

Moreover, the theoretical context for co-management of small-scale fisheries is an ongoing debate among academics and fisheries management practioners, especially in the context of social-ecological systems (Berkes et al. 2001; Ostrom 1990, 2009; Pomeroy et al. 2004, 2011). Weak governance is one of the main causes of the present poor condition of fisheries and is characterized by (but are not limited to) corruption, lack of stakeholder participation, political will and capacity, weak institutional capacity and capabilities, poor enforcement, and inadequate information (CRC et al. 2006). Small-scale fisheries are extremely diverse, complex and dynamic (Berkes et al. 2001). They operate at many different scales and with many different stakeholders. These characteristics make the challenges and concerns that confront small-scale fisheries governance more difficult. Given the range of actors in small-scale fisheries, with their varying perceptions, agendas and power, there is a need to develop a governance system that can include multiple stakeholders (Bavinck et al. 2005). How best to tap their various interests, agendas and capacities remain a challenge.

In the case of Colombia, small-scale fishermen have not been fully recognized as key stakeholders in the fisheries management process (Cuello and Duarte 2010; García 2010; Saavedra-Díaz et al. 2015a). It is critically important to involve communities in co-management practices in order for management to be fully effective and to reduce the propensity for overexploitation. In the Caribbean and Latin America, examples of co-management exist in Mexico, Belize, Brazil, Peru and Chile. Due to the high diversity of the fisheries in these areas, as well as poverty, violence, forced displacement, among other situations, many co-management practices need to be implemented in concert in order to be effective. Also, decision makers need to recognize the importance of supporting management decisions with local knowledge (Begossi and Brown 2003). Recently, fisheries co-management was successfully implemented in Chile by dividing responsibilities for the Management Area System between government agencies and small-scale fishers' organizations (Marin and Berkes 2010). This success raised consciousness among fishermen. It avoided the tragedy of the commons by building rapport between the state and fishermen, raising ecological and management awareness, and promoting teamwork between fishermen and scientists (Schumman 2007). The research presented here explores possible cross-cutting solutions of this kind for the particular realities of the under-researched small-scale fisheries of Colombia.

Government institutions and agencies in Colombia have attempted to organize the fisheries sector and articulate policy through the publication of different sets of national documents. However, a disarticulated institutional framework and a thicket of contradictory policies highlight the fragmented nature of Colombian fisheries management. Unfortunately, national planning documents such as the "Colombian Vision 2019" have set unrealistic goals for increases by 30 % in fisheries captures for 2019 (ECOVERSA 2007). Development goals such as this misunderstand the current status of both fishery resources and the management system and have not directly involved local communities in the development and implementation of policy to achieve sustainable fisheries (Hart 2003; McCay and Jentoft 1996; Schumman 2007). The Colombian legal framework that regulates fishery resources at the national level is the Code of Renewable Natural Resources and Protected Environment, created by the Decree/law 2811 of 1974 and Law 13 that created the General Statute for Fisheries in 1990. The Rural Development and Agriculture Ministry- through the Fishery and Aquaculture National Authority - is in charge at the national level of the Marine Small-Scale Fishery sector since 2011. However, over the last decade five different institutions have been in charge of fisheries. Such institutional instability at the national level has being one of the major challenges for successful management of the small-scale fishery sector.

Another challenge to overcome is the difference of social perceptions among stakeholders –i.e., fishermen, local leaders, and fishing experts and their existing relationships within marine ecoregions. According to Saavedra-Díaz et al. (2015b), fishermen perceive destructive fishing methods as the most significant problem with the highest percentage response among all respondents (70 % in the Caribbean coast and 79 % in the Pacific), followed by the lack of regulation enforcement (60 % in Caribbean and 47 % in Pacific) and pollution and industrial contamination. Local leaders identified the low quality of life of fishermen including the lack of public services, lack of potable water and lack of education focused on environmental awareness. They also noted a weakness of fisher's organizations and leadership. In contrast, the majority of fisheries experts believed that the small-scale fishery sector has been largely ignored by governmental actions. The fisheries experts identified the instability of regulatory authorities and lack of control measures, oversight and monitoring by the agencies in charge of artisanal fishery activity due to lack of personnel, and the lack of infrastructure or equipment to facilitate this work. While many problems occur on both coasts, other problems are coastally unique, and further, some affect only a small portion of fishing communities. In consequence, there is a need to establish a fisheries management strategy that can prioritize solutions at different levels: national, coastal, and local.

Using results from the same survey discussed in Saavedra-Díaz et al. (2015b), the objective of this paper is to discuss the solutions identified by fishermen, local leaders and fisheries experts for improving small-scale fisheries management in Colombia. In addition, we present specific recommendations on a framework for small-scale fisheries governance in Colombia.

METHODS

Study Area

The Colombian coastline is divided between the Caribbean and the Pacific. The Caribbean coastline is 1642 km in length and the Pacific coastline is 2188 km in length. The Caribbean coast has a population of 2,919,348, while the Pacific coast has a population of 543,594 (Posada and Rozo 2005). Colombia has 12 political and administrative coastal states, eight on the Caribbean coast and four on the Pacific coast. Different from the states, coastal and marine environmental divisions separate the Colombian coast into six Coastal and Marine Ecoregions (CME) on the Caribbean and four CMEs on the Pacific coast (Figure 1), plus four Insular Ecoregions on both oceans. CMEs are distinguished by different environmental characteristics such as geo-morphology, hydrography, sedimentology, and coastal and marine ecosystems (Posada and Rozo 2005). Nevertheless, state and CME boundaries are relatively similar, in some cases nearly overlapping. Since the present study focuses on environmental conditions, CMEs provide spatial orientation. Although the Caribbean and Pacific Insular Eco-regions (San Andrés and Providence Archipelago, San Bernardo, Malpelo and Gorgona Islands) are not included this study they should be included in future efforts to improve Colombian fisheries management overall.

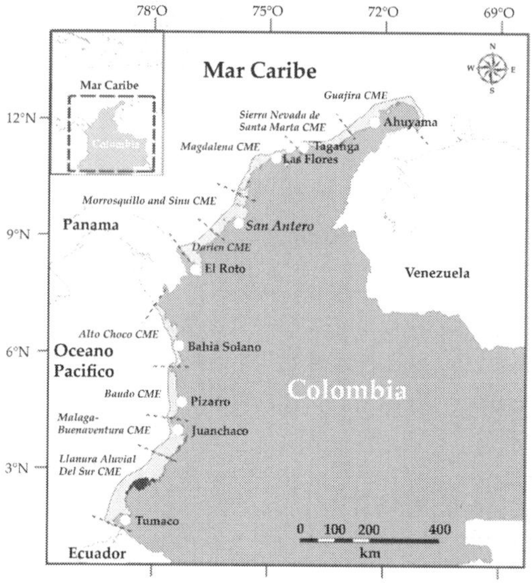

Figure 1: Map of the selected fishing community (*white dots*) in coastal and marine ecoregions (CME) on the Colombian Caribbean coast and Pacific coast (Taken and modified from Posada and Rozo 2005).

The following parameters were employed in identifying one "typical" community in each CME: (1) it was not located in a marine protected area or on an island; (2) it was historically recognized as a fishing community; (3) it relied on fishing as a primary economic or subsistence activity; (4) there existed some level of fisher's organization (formal); (5) the community was involved previously in research projects; (6) there was a low incidence of violence or drug trafficking (for safety); and (7) it had been considered by Fishery experts.

Caribbean coast fishing communities selected in each coastal and marine ecoregion (CME) for this study (Figure 1) include *Ahuyama* in the Guajira CME, *Taganga* in the Sierra Nevada de Santa Marta CME, *Las Flores* in the Magdalena CME, *San Antero* in the Morrosquillo and Sinú CME and *El Roto* in the Darién CME. Pacific coast fishing communities selected in each CME are *Bahía Solano* in the Alto Chocó CME, *Pizarro* in the Baudó CME, *Juanchaco* in the Málaga-Buenaventura CME, and *Tumaco* in the Llanura Aluvial del Sur CME.

Data Sources

The fishermen, local leaders and fisheries experts were interviewed concerning their solutions to the identified problems that are affecting the small-scale fisheries communities and the marine fisheries resources they depend upon for food and livelihood. In addition, fishermen at community meetings were asked questions about their interest in, readiness for and how to move toward fisheries co-management. Fishermen are defined here as coastal marine small-scale fishermen or fisherwomen, including a great variety of racial and ethnic groups and are directly involved in fishing activities. Local leaders include presidents of local fishing associations or persons recognized as influential members of the fishing community including traditional authorities who fishers look to for leadership and who bring a broader perspective than just of a single fishermen. Finally, fisheries experts are scientists and technicians from Colombian fisheries institutions and administration with expertise and experience working with fishing communities, fishermen and marine resources. Semi-structured interviews of 309 people were conducted. Table 1 shows the number of interviews undertaken for the Caribbean and Pacific coasts per stakeholder group per community.

Table 1: Number of semi-structured interviews and number of fishermen performed in each stakeholder group on each community and eco-region on each coastal region

	Caribbean					Pacific				
Methods	*Ahuyama*	*Taganga*	*Las Flores*	*San Antero*	*El Roto*	*Bahía Solano*	*Pizarro*	*Juanchaco*	*Tumaco*	Total number of participants by method
Fishermen interviews	18	23	31	36	17	14	15	18	23	195
Leaders' interviews	2	3	4	6	1	4	2	3	2	27
Fishery experts' interviews	5	8	5	6	7	2	2	18	4	57[a]
Co-management hearings	17	10 and 18	20	27 and 28	20	35	13	13 and 17	17	235

[a]Thirty interviews performed in 2007 (15 from the Caribbean and 15 from the Pacific) in the feasibility study for the present research have been included in this study, for a total of 87 fishing experts interviews in total

Fieldwork performed in the fishing communities took place from August 2008 to August 2009 and lasted approximately one month within each community. To collect the fishermen and local leaders' perceptions about fisheries management solutions, a representative sample of leaders and fishermen in each small- scale fishing community were chosen (Table 1). The great variety of the fishing communities imposed important methodological limitations. For instance, fishing communities vary in population from villages such as *El Roto* with 50 Fishermen, to municipalities such as *Tumaco* with 4000 fishermen. In high population fishing communities such as *San Antero* on the Caribbean and *Tumaco* on the Pacific community leaders identified neighborhoods populated mainly by fishermen, for whom fishing activity and proximity to landing places determine where they live. Some municipal fishing communities also cover a much larger area than others (*Tumaco* extends over 167 counties). Therefore, the number of *Tumaco* fishermen involved in the study was low compared to the total number of fishermen in the community, and most came from the municipal center so that peripheral areas were under-represented.

Small communities with fishing populations of around 200 were simpler to sample. The lead author lived approximately three weeks to one month in each community and developed a greater understanding of the situation those fishermen faced. Recognizing these limitations is important in understanding the results of this research.

Community leaders were interviewed first upon arrival at each community. Local leaders identified the principal fishing gears used locally and explained the general economic, social, and cultural conditions in the community. The interview contained 15 standard questions focused on general subjects pertaining to the local artisanal fishing community. Questions were divided into five main subjects: (1) the role of local leaders in the community; (2) overview of the condition of marine artisanal fisheries; (3) fisheries problems; (4) proposed solutions; and (5) fisheries management.

Given the time and resources available, at least two fishermen from each community were interviewed for every fishing method (i.e., gillnets, mainline, longline, beach seine net with bag, harpoon, surrounding net, bottom trawl, among others). While representativeness of such a survey is always challenging, every effort was made to capture a wide range of perspectives. The interview consisted of 89 questions focused on seven main subjects: (1) personal information; (2) demographics and quality of life (family, education, living conditions, among others); (3) current fishing activity (i.e., time spent fishing, reasons for fishing, technology employed, fishing locations, marketing, among others); (4) long term changes in fishing activity and fisheries problems; (5) proposed solutions; (6) fisheries management; and (7) information regarding the community (i.e., relationships with community members, vulnerability to violence or drug traffic, happiness being a fisherman, among others).

In addition to interviews, we also utilized data from public meetings in each community related to the strategy of co- management. The meetings were held where fishermen brainstormed about top-down and bottom- up fisheries management strategies. Exchanging information and opinions, and discussing the co- management process took place in two main steps:

- Fishermen learned about fisheries management strategies, in particular, about co-management. The basic concepts and features of the major types of top-down and bottom-up fisheries management were explained. Examples of traditional management were analyzed and then contrasted with co-management to show the benefits and drawbacks of this process. This basic information allowed them to understand how fishery management has been applied in other countries, what alternatives exist, the benefits communities might gain from co-management, and the importance of working with other fisheries actors (stakeholder groups) in this process.
- An open discussion focused on the following questions:

What opinions do the fishermen have about management in general and co-management in particular?

What weakness and strengths within the artisanal fisheries community might affect the success of co- management?

What first steps could the community take to start the co-management process?

On the Caribbean coast, 141 community members participating in the interviews, while 81 on the Pacific coast were participated (Table 1). In addition, 140 community members participated in the hearings on the Caribbean coast and 95 on the Pacific coast. Even though it was planned to have one meeting per community, in some communities it was necessary to hold two hearings (*Taganga* and *Juanchaco* communities) because of low fishermen participation.

Data Analysis

Qualitative information from semi-structured interviews and fieldwork activities was coded into quantitative data then organized and content analysis performed using computer-assisted qualitative data analysis software (CAQDAS), N-Vivo/8 software (García-Horta and Guerra-Romos 2009; Ozkan 2004; Saldaña 2009; Thayer et al. 2007; Yin 2003). Based on Miles and Huberman (1994) and Yin (2003), coding was done by the meaning of phrases, following the elemental method and incorporated in N-vivo. Then, structural coding and the elemental method were used to pre-code questions by creating main categories of common subjects that allow different opinions to be combined in a single category. Sixteen main categories were created corresponding to the common subjects under which the codes are aggregated (Miles and Huberman 1994). Chi-Square tests were used to distinguish between the perceptions of fisheries solutions and the three groups of fisheries stakeholders. Differences in the perceptions of the main categories of fisheries solutions among stakeholders were analyzed using a Kruskal-Wallis test. In addition, Mann-Whitney tests were performed in order to measure differences in perceptions of the same main categories between the Pacific and Caribbean regions. Finally, to explore the relationships between stakeholders' perceptions of fisheries solutions and the marine eco-regions, a redundancy analysis (RDA) was performed. Descriptive statistics (percentages) were used to analyze the results of each stakeholder group and among all stakeholders.

RESULTS

Given the range of actors in Colombian small-scale fisheries, with their varying perceptions, agendas and power, it can be expected that there will be differences in the proposed solutions to the challenges and problems facing the fisheries. Fishermen are focused on the fishing activity, leaders on governing, and experts on technical aspects of management. The theoretical debate over fisheries governance is how to develop a consensus among stakeholders for a governance system that can allow for including multiple stakeholders.

The results are grouped into cross-cutting solutions (identified by the fishermen, local leaders and experts at the same time), inter-group solutions (identified by only two of the three stakeholders), and isolated solutions (identified by only one stakeholder). In addition to the interviews, the results from public meetings in each community related to the strategy of co-management are presented.

Cross-Cutting Solutions

Fishermen from both coasts identified 133 solutions, of which 25 were represented by more than 3 % of respondents. Leaders identified 121 solutions, of which 38 were represented by more than 3 % of respondents. Experts identified 397 solutions, of which 135 were represented by more than 3 % of respondents. The responses from the three groups generated 22 cross-cutting solutions in eight categories. The Chi Square Test was used to establish the significance of these cross-cutting solutions to the three participant groups (Table 2).

Table 2: The perception of importance of cross-cutting solutions ranked by the fishermen, local leader and fisheries experts groups and analyzed using the Chi square test

Categories	Cross-cutting solutions	Code	Fisheries experts (%)	Fishermen (%)	Local leaders (%)	X^2	*p*-value
Aquaculture	Explore aquaculture as an option	Aqua	22.50	13.33	37.04	10.66	0.05
Fishermen and communities	Changes in Community attitude	Comm	2.50	5.64	7.40	1.56	0.46
	Changes in Fishermen attitude	Fish	41.25	14.35	3.70	30.28	<0.001
Fishing equipment	Open waters fishing depth sea fish – *pesca de altura*	OpenW	23.75	13.84	14.81	4.09	0.13
Fishing methods	Use friendly gears and recommendations	GearsOk	7.50	9.23	14.81	1.28	0.53
Goverment-Administration	Control over	Control	7.50	4.61	3.70	1.09	0.58
	Find or generate other jobs	Jobs	13.75	14.87	3.70	2.52	0.28

	Government support to real Fishermen	Real-Fisher	1.25	8.20	3.70	5.16	0.08
	Loan program with clear rules	Loan	11.25	1.02	3.70	15.54	<0.001
Industrial fishing activity	Restrictions on Industrial fishing	IndusRest	12.50	7.69	7.40	1.69	0.42
Organization of Fishermen	Strengthen Fishermen associations	F Asso	25.00	6.66	3.70	20.77	<0.001
	Strength Leaders	Leaders	5.00	0.51	3.70	6.31	0.04
Regulations	Create regulations	Regulate	61.25	50.76	62.96	3.34	0.19
	Create Protected Marine Areas	PMA	10.00	0.51	3.70	15.96	<0.001
	Fishing Zoning plan	Zoning	10.00	10.25	18.51	1.75	0.42
	Regulate gears	Gears	11.25	39.48	14.81	24.80	<0.001
	Increase mesh size	Mesh	7.50	6.66	3.70	0.47	0.79
	Prohibit gill nets	Not-Gill	7.50	19.48	3.70	9.36	0.01
	Change unfriendly methods for environmental friendly	C-Unfriend	11.25	40.00	3.70	31.99	<0.001
	Establish minimum size per spp.	Min-size	10.00	2.05	7.40	8.62	0.01
	Veda-Time closed season	Veda	23.75	17.43	37.03	6.07	0.05

Results show that 11 cross-cutting solutions had a *p-value* less than or equal to 0.05, and thus were ranked at different levels of importance to all three groups, while 10 cross-cutting solutions had similar importance. Support for proposed solutions in the category of Government-Administration were not significantly different among the groups with the exception of the need for a loan program for fishermen based on clear rules, which experts felt would be particularly valuable.

All groups perceived with a similar sense of urgency that the government must regulate the fishery sector, that other jobs must be generated to replace fishing and decrease pressure on fragile resources and ecosystems, and that the administration needs to target support for "real fishermen" rather than opportunists (participants used the term "real fishermen" to distinguish between legitimate fishermen like themselves and pretenders who show up when aid is being handed out).

In the category of Regulations, only fishermen placed high importance on the need to regulate the use of gear, to prohibit the use of gillnets, and to exchange damaging gear for methods that are environmentally friendly. Experts placed great importance on creation of marine protected areas and establishing minimum catch size. For local leaders, the creation of close seasons (*veda*) was particularly important. There was significant agreement in the need to create regulations for fishing. In fact, this solution had the highest representation among the three groups. The category, Organization of Fishermen, was assigned a different level of importance by each group, with only experts emphasizing the need to strengthen fishermen associations.

Non-significant differences in importance were found in solutions in the following categories: (1) Changes in community attitude (Fishermen and Communities); (2) The use of vessels that encourage open waters and deep sea fishing (Fishing Equipment); (3) The need to use environmentally friendly gear (Fishing Methods); (4) Restricting industrial fishing (Industrial Fishing Activity); and (5) The need to establish fisheries zoning in each community, and to increase gill net mesh size (Regulations).

To relate the ranked importance of the perceived solutions with different stakeholders and regions, a redundancy analysis (RDA) was carried out (Figure 2). In Figure 2, the first two factors of the RDA explained 63 % of the total variance. The x-axis (F1) represents 44 % of the variation of perceived solutions between fishing communities on the Pacific (positive scores) versus fishing communities on the Caribbean (negative scores). Meanwhile, the y-axis (F2) is showing a gradient of perceived solutions represented 20 % of the variation between the perceived solutions by fishers (negative scores) versus experts (positive scores). Local leaders have low representation, being located almost in the middle of the other two stakeholders.

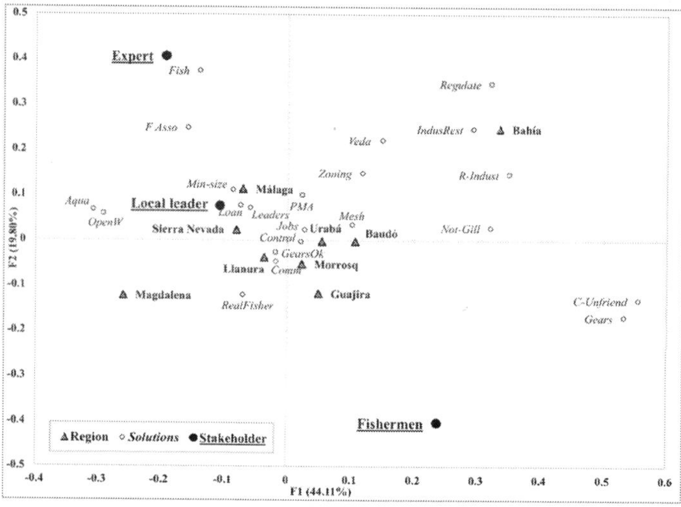

Figure 2: Redundancy analysis of primary solutions (Cross-cutting) obtained through semi-structured interviews related to ecoregions and stakeholders (explaining variables). The x-axis (F1) represents 44 % and the y-axis (F2) 20 % of the variation.

While solutions proposed by Caribbean communities differed from solutions proposed by communities on the Pacific, differences are not as great as was the case with the problems (refer to Saavedra-Díaz et al. 2015b). The eco-regions of *Alto-Chocó* (*Bahía Solano* fishing community) and *Baudó* (*Pizarro*) are representative of the fishery situation on the Pacific, whereas the eco-region of *Magdalena* (*Las Flores*) is representative of the fishery situation on the Caribbean coast. Conversely, the Pacific eco-regions of *Málaga-Buenaventura* (*Juanchaco*) and *Llanura Aluvial del Sur* (*Tumaco*) correspond more closely to the Caribbean situation in terms of perceived solutions. Interestingly the eco-region of *Sierra Nevada de Santa Marta*(*Taganga*) was closer in viewpoint to Pacific communities. On the Pacific coast, the *Bahía Solano* community shared opinions with other communities, but their strong tradition of group deliberation and awareness of the fisheries situation resulted in a clarity of thought that produced more solutions and a wider spectrum of options.

Pacific fishing communities are more aware of the need for regulations (restricting industrial fishing, prohibiting gill nets, establishing closed fishing seasons, fisheries zoning, minimum mesh sizes net, ecologically friendly fishing methods, among others), while Caribbean communities proposed fewer solutions over a narrower range. They focused primarily on the need for oversight and control of fishing activity, change in community attitude, identifying "real fishermen", as well as for strong fishermen's associations, change in fishermen's attitude, exploration of aquaculture and access to open water vessels.

The y-axis (F2) shows how the opinions about solutions of fishers were far different from the opinions of experts. This result supports the impression that the opinions of fishers are not represented by experts. In fact, need for changes in fishermen's attitude (Fish), strengthen fishermen associations (F Asso), and the creation of marine protected areas (MPA) are perceived as important solutions mostly by experts. In contrast, fishers identified solutions that were not considered important by the other two stakeholder groups, such as identifying "real fishermen" (RealFisher), and access to loans (Loan).

All three stakeholder groups identified regulation as the most important solution (fishermen 51 %, experts 61 %, and leaders 63 %). All groups agreed on five additional solutions, but differed significantly on the degree of importance. They are presented in order from the most to the least different. The change from unfriendly (environmentally damaging) to friendly (less damaging) fishing methods (X^2=31.99; *p*- value <0.001) was most important to fishermen (40 %). Changes in fishermen's attitude (X^2=30.28; *p*- value <0.001) was most important to experts (41 %). Gear regulation (X^2=24.80; *p*- value <0.001) was also most important to fishermen (39 %). The fourth and fifth solutions that differ in importance, the need to strengthen fishermen associations (X^2=20.77; *p*- value <0.001), and to create marine protected areas (X^2=15.96; *p*- value <0.001), were most important to the group of experts.

Inter-Group and Isolated Solutions

There are 37 inter-group solutions which were proposed by the stakeholder groups. These solutions are separated into nine categories (coastal uses and infrastructure, fishermen and communities, fishing resources, government-administration, institutions, marine ecosystems threatened, marketing, organization-fishermen association, regulations). The top solutions in each category are presented in Table 3.

Table 3: Main inter-group solutions shared by two stakeholder groups

Main intergroup solutions	% both coasts		
	Fishermen	**Local lead-ers**	**Fisheries experts**
Coastal Uses and Infrastructure			
Tourism		25.90	2.40
Fishermen and Communities			
Invest in Fishermen Education-read and write		11.11	16.67
Assume responsibility as Fishermen and improve attitude	14.36		7.14
Fishing Resources			
Find new target spp.	0.51		10.71
Goverment-Administration			
Recognition of the importance of Fish-ermen role and the job-chain involved		18.52	13.10
Empower community	7.20		17.86
Institutions			
Academia important actor		3.70	13.10
Establish Real Fishery institution	3.60		56.95
Marine Ecosystems Threatened			
Recover mangroves and breeding zones		7.40	3.60
Marketing			
Have or improve their own selling and marketing process		14.81	5.95
Improve product manipulation and quality		7.41	5.95
Organization - Fishery Association			
Promote all Fishermen to be associated and the importance of it		3.70	8.33
Support Fishermen in organizing their F.A.		3.70	8.33
Regulations			
National Fishery policy-regulations		22.22	16.67
Work and control fish buyers or mer-chant		7.41	4.76

In addition to these inter-group solutions, there were 37 isolated solutions separated in 9 categories (aquaculture, coastal activities and infrastructure, fishermen and communities, government- administration, institutions, national situation, marketing, organization-fishermen association, regulations). The largest number of inter-group solutions were proposed by experts. These included improved fishery information, more research, co-management, and learning about successful management from other countries. Two of these 37 solutions contain sub-solutions. Establishing a real fishery institution was described in 10 sub-solutions, and creating a national fishery policy was identified in 22 sub-solutions.

In order to understand differences at the category level, solutions in the first (cross-cutting solutions), second (inter-group) and third (isolated) orders were combined for analysis. The Kruskal–Wallis test (Kruskal and Wallis 1952) was used here to evaluate differences among categories with respect to the three interviewed groups (Table 4). No significant differences were found in the categories of Fishing Equipment, Fishing Methods, Industrial Fishing Activity, and Regulations; all stakeholders viewed these categories with the same level of importance. Leaders and experts contributed to categories such as Marketing and Fishermen Associations. The former gave great feedback about solutions related to the category of Coastal Uses and Infrastructures. The latter brought substantial investigative experience to the national fishery situation through their knowledge in the categories of Institutions and Government Administration Fishermen were not highly represented in any category compared with the other two stakeholders.

Table 4: Kruskal – Wallis test by the categories of solutions through the three stakeholders groups

Categories	Fisheries experts	Fishermen	Local leaders	X^2	*p*-valor
Acuaqulture	0.225	0.133	0.370		
	(0.420)	(0.341)	(0.492)	10.621	0.005
Coastal Uses and Infrastructure	0.100	0.056	0.370		
	(0.302)	(0.231)	(0.492)	26.86	<0.001
Fishermen and Communities	0.438	0.190	0.444		
	(0.499)	(0.393)	(0.438)	21.35	<0.001
Fishing Equipment	0.263	0.159	0.185		
	(0.443)	(0.367)	(0.396)	3.960	0.138
Fishing Methods	0.075	0.092	0.000		
	(0.265)	(0.290)	(0.000)	2.783	0.24
Fishing Resources	0.125	0.000	0.074		
	(0.333)	(0.000)	(0.267)	24.066	<0.001

Goverment-Administration	0.713	0.333	0.593		
	(0.455)	(0.473)	(0.501)	34.948	<0.001
Industrial Fishing Activity	0.125	0.077	0.000		
	(0.333)	(0.267)	(0.000)	4.388	0.111
Institutions	0.625	0.077	0.037		
	(0.487)	0.267	0.192	105.148	<0.001
Threatened Marine Ecosystems	0.063	0.036	0.148		
	(0.244)	(0.187)	(0.362)	6.132	0.047
Marketing	0.238	0.036	0.148		
	(0.428)	(0.187)	(0.362)	26.471	<0.001
National Situation	0.038	0.000	0.000		
	(0.191)	(0.000)	(0.000)	8.381	0.015
Organization of Fishermen	0.300	0.077	0.333		
	(0.461)	(0.267)	(0.480)	27.77	<0.001
Regulations	0.613	0.508	0.630		
	(0.490)	(0.501)	(0.492)	3.333	0.189
Small Scale Fishing	0.050	0.005	0.074		
	(0.219)	(0.072)	(0.267)	8.409	0.015

Mean relative value (and standard deviation between parenthesis) for each category and stakeholder is shown

The Mann-Whitney test was used to evaluate differences among the categories with respect to the Caribbean and Pacific coastal regions (Table 5). These results show that category responses on one coast are not significantly different from the other, as was shown in the Redundancy Analysis (RDA).

Table 5: Mann-Whitney test by categories of solutions, comparing Colombia's Caribbean and Pacific coasts

Mann-whitney test				
Categories	**Caribbean**	**Pacific**	***U***	***p*-valor**
Acuaqulture	0.214	0.122	11743.50	0.043
	(0.411)	(0.328)		
Coastal Uses and Infrastructure	0.112	0.070	11212.0	0.222
	(0.317)	(0.256)		
Fishermen and Communities	0.262	0.304	10297.5	0.427
	(0.441)	(0.462)		

Fishing Equipment	0.166	0.226	10104.0	0.194
	(0.373)	(0.420)		
Fishing Methods	0.086	0.070	10924.5	0.619
	(0.280)	(0.256)		
Fishing Resources	0.032	0.052	10536.5	0.387
	(0.177)	(0.223)		
Goverment-Administration	0.444	0.478	10382.5	0.561
	(0.498)	(0.502)		
Industrial Fishing Activity	0.043	0.148	9623.0	0.001
	(0.203)	(0.356)		
Institutions	0.193	0.261	10017.5	0.164
	(0.395)	(0.441)		
Threatened Marine Ecosystems	0.064	0.035	11068.5	0.270
	(0.246)	(0.184)		
Marketing	0.059	0.165	9608.5	0.003
	(0.236)	(0.373)		
National Situation	0.011	0.009	10774.0	0.868
	(0.103)	(0.093)		
Organization of Fishermen	0.144	0.183	10342.5	0.379
	(0.352)	(0.388)		
Regulations	0.503	0.617	9519.0	0.052
	(0.501)	(0.488)		
Small Scale Fishing	0.037	0.000	11155.0	0.036
	(0.190)	(0.000)		

Mean relative value (and standard deviation between parenthesis) for each category and stakeholder is shown

However, exceptions are seen in categories such as Aquaculture, with higher representation on the Caribbean coast, and Industrial Fishing and Marketing, with greater feedback on the Pacific. This supports the conclusion that solutions on both coasts are basically the same, but that differences in expectations must be part of any management framework.

Fisheries Management

The three stakeholder groups – fishermen, local leaders and experts – were asked a number of questions concerning the concept of fisheries management and co-management (bottom-up management).

Fishermen

Only one percent of the fishermen responded that they knew what fisheries management is. These fishermen used a variety of words or phrases to explain the concept, such as organization and order, closed fishing zones, temporary closures, fishermen's association, marketing, control, changing fishing gears, processing fish products, aquaculture, and managing fishing equipment. There is a lack of understanding among fishermen of what fisheries management is.

When asked if their community was ready for bottom-up management, 37 % answered positively and negatively at the same time (42 % Caribbean vs. 28 % Pacific), only 31 % emphatically answered positively (31 % Caribbean vs. 30 % Pacific), 9 % answered negatively (8 % Caribbean vs. 11 % Pacific), and 23 % did not know or were unsure. Fishermen's answers revealed that they believe fisheries management is necessary because otherwise fishing will get worse (8 %), they need order and rules (7 %), and the community should try and see if it works (4 %). Fishermen reported that some communities have implemented rules by themselves. For example, *Pizarro* established a rule to use gillnets with mesh size larger than 3 in.. *El Roto* established a minimum mesh size of 3 ½ inches for gill nets. The fishermen reported that while most efforts at implementing their own rules have been unsuccessful, they are optimistic that it is possible for fishing communities to come together to establish rules. They felt that it would be possible for the community to implement fisheries management under certain conditions, such as: (1) If all fishermen agree to it; (2) If management is controlled by both community and government; (3) If training in fishery management comes first; and (4) If a subsidy mitigates a fishing restriction. However, fishermen believed that establishing bottom-up management is difficult because: (1) Fishermen's attitude prevents it; (2) It is difficult to get fishers to agree; (3) Some attempts at management had failed in the past; (4) Fishermen cannot stop fishing; (5) It could generate violence among fishers; and (6) Community features complicate the process.

When asked what their community needs to do to manage fisheries, the fishermen believed that: (1) Fishing communities need to get together and agree about implementing management; (2) All fishermen should organize and belong to the fishermen's associations; (3) Fishermen need to face the problems that threaten their livelihood and get specific training to overcome them; (4) Some fishing communities have shown that successful fisheries management is possible, and their experiences can be useful examples; and (5) Fishermen should learn to compromise and explore other job possibilities. When asked what government needs to do to manage fisheries, the fishermen stated that existing regulations needed to be effectively implemented; they should stop foreign fishermen; and training should be provided.

Local Leaders

Similar to the fishermen, local leaders knew nothing about fisheries management. When asked if they believed their community was ready for implementing the bottom-up process, 59 % answered positively (63 % Caribbean vs. 55 % Pacific) and 41 % answered negatively (38 % Caribbean vs. 45 % Pacific). A majority of leaders believed they were ready to implement fishery management by themselves because of:

(1) Rules already established by some communities, (2) Strong fishermen's associations, and (3) Community reliance on fishing for food. Leaders believed that bottom-up management is possible, but only if government and community work together. They reported less confidence in efforts developed solely by communities or imposed by the government.

Leaders who were pessimistic about establishing bottom-up fishery management cited fishermen's attitude - according to which the government must give them everything, but they do not make any effort (7 %). Other difficulties include: (1) Low individual representation; (2) Difficulties in getting many fishermen on the same page; (3) Lack of organization or association among the majority; (4) Prevalence of illegal gear that undercuts participation; (5) Lack of understanding of the need for fisheries management; and (6) No institutional authority is in charge.

When asked what they think that the community needs to do in order to manage its fishing resources; local leaders reported that the entire community must get together and agree to actively participate in implementing fisheries management (the process must include fish sellers). They also stated that fishermen must practice responsible fishing, they must develop a sense of belonging, they must belong to a fishery organization, and they must stop using unfriendly gears. When asked what they thought that the government needs to do in order to manage fisheries resources, local leaders reported that government should: (1) Invest in establishing regulations, including making subsidies available while restrictions apply; (2) Establishing exclusive fishing zones for each community; and (3) Closed seasons. At the same time, leaders believe that it is fundamental that any restriction or regulation be discussed with the community before it is implemented and the government should accord the small-scale fishery sector the importance it deserves.

Experts

Fifty-seven percent of the experts believed that it was possible to establish bottom-up management, but 20 % were skeptical. Experts who believed that bottom-up management should work knew about recent, closely related attempts. These included projects in marine protected areas such as *Guapi* and *Sanquianga* Parks on the Pacific coast, led by WWF; San Andres Island on the Caribbean, led by CORALINA; the Special District in *San Antero* that is zoned and planned to preserve mangrove forests; and a similar effort on the Urabá Gulf. Experts described regional or local committees that could be the basis for coastal zoning or for implementing fisheries management, such as NODOS (Regional Institutional Fishery Councils), INVEMAR (regional Committees for Coastal Management), and local Community Councils (some with their own Natural Resources Code). They also listed fishermen's associations that could be local foundations of bottom-up fisheries management, such as associations in *Las Flores* on the Caribbean, and *Juanchaco* and *Bahía Solano* on the Pacific coast. Experts who were skeptical of bottom- up fisheries management doubted that communities would participate. Other concerns were the poor track records of external institutions going in and out of communities makes implementing fisheries management through external agencies difficult; fishermen's attitude and need for constant supervision and their inability to work alone; and cultural aspects of Indigenous fishing communities.

When asked what they think fishing communities need to do in order to manage their fishing resources, the experts believed that a consensus by the majority of the population in each community was necessary in order to obtain high participation. Additionally, the community should discuss and agree upon their own rules and take responsibility for ensuring compliance; fishermen must belong to a fishermen association, and that the community must have a long term vision.

When asked what they thought that the government needs to do in order to manage fisheries resources, the experts felt that the government needs to be aware that bottom-up management requires a long term vision. Additionally, investing in education in order to increase awareness of fishery management; provide active and consistent inter-institutional support in each community; promote and strengthen fishermen's associations; and adapt to community dynamics.

Public Meetings

A series of public meetings to discuss fisheries management were held in selected communities – five communities on the Caribbean coast and four on the Pacific coast. The meetings introduced fishermen to the basics of fishery management, why it is important, how they can be part of this process and start to work from their own communities. After presentations, open discussions allowed the participants to express their individual opinions about this fisheries management and bottom-up management and whether or not it could be useful and applied in their situation.

Even though all meetings brought up internal weaknesses, most participants believed that bottom-up management was possible if they worked hand in hand with government. The fact that some communities had already established rules resulting from their own deliberations suggested that widespread bottom-up fisheries management may be possible. Most weaknesses identified in the meetings matched those identified in the fishermen survey described above. These include disunity among fishermen, lack of participation, the presence of foreign fishermen, and weak fishermen's associations due to poor leadership. Additional weaknesses include authorities' distrust of fishermen, dislike of authority, distrust among fishermen, lack of education, lack of a sense of belonging, the involvement of fishermen in drug trafficking.

The main strengths each community identified were related to human capital such as strong knowledge and experience in fisheries issues; leaders and traditional authorities who believe that bottom-up management is the only solution; and community experience in establishing their own rules. Some believe that the presence of old, established fishermen associations are key to success. Communities cautioned that government must support local decision making processes and impose agreed-upon rules (*Juanchaco*), otherwise bottom-up management will not work. Communities stated that fisheries management needs to be designed for the long run and applied consistently over time in each community that adopts the process. There was a need to integrate non-traditional fisheries stakeholders, such as fishermen's wives, local schools and universities, and seafood supply chains into fisheries management. This shows the extent to which small- scale fishing is importantly integrated into local and national life.

Although none of the communities expressed that they enjoy optimum conditions for implementing bottom-up management, all realized its importance and expressed an urgent need to start the process. Particular characteristics of each community affected their readiness to put bottom-up fisheries management into practice.

DISCUSSION

The introduction to this paper noted that weak governance has been a major factor contributing to the overexploitation and consequent poor condition of small-scale fisheries and many of the communities that depend upon them around the world. This is no less true and Colombia and the data presented here and in Saavedra-Díaz et al. (2015b) contribute to the global comparison of this important component of the fisheries sector. The latter paper identified a number of different problems facing the Colombia small scale fishery, as perceived by different stakeholders – fishermen, local leaders, and fishing experts. While there were some difference among the stakeholders and between the two regions, several key problems emerged including destructive fishing methods, lack of regulation enforcement, pollution and industrial contamination, lack of public services, weakness of fishermen's organizations and leadership. The majority of fisheries experts believed that the small scale fishery sector has been largely ignored by government actions. That analysis concluded that these problems require establishing a fisheries management strategy that can prioritize solutions at different levels: national, coastal, and local.

Identification of problems in any given fishery is necessary but ultimately insufficient. A path forward to solving those problems must be developed. Our approach here is to draw on expertise of fisheries, community leaders and in-country experts to highlight that path.

More than 50 % of each stakeholder group believed that implementing a bottom-up fisheries management approach, such as co-management, is possible. While it is clear that what the details of "co-management" means is still not fully shared, it is still notable that there is significant interest in attempting to take action at a local level to address the problems each community faces, and they all face in common. This approach to using co-management is not only seen to have potential by Colombian stakeholders, but also in other Latin American countries that face similar issues due to fisheries crises, such as Mexico (Salas et al. 2015), Costa Rica (Solís et al. 2015), Brazil (Futemma and Seixas 2008), Uruguay (Trimble and Berkes 2015), and Chile (Marin and Berkes 2010). Even in Colombia, there are examples of cases that have tried participatory management with varying degrees of success (Beardon 2008; Delgado et al. 2010; Navia et al. 2010).

It was notable that all of the stakeholder groups recognize two overriding needs, 1) that governance of the fishery is urgently needed or conditions will continue to decline, and 2) that whether management is top-down or bottom up, no one group can implement effective management. This is the essence of co-management, shared governance, but shared responsibility. Clearly, fishers perceived the need for the government to help them manage some of the biggest challenges including external challenges (e.g., pollution, criminal elements), while government looked to support from fishers in making real change. Local leaders too, needed a greater willingness of fishers to come together and compromise as needed to make changes, and in dealing with the government.

Surprisingly, Colombian fishermen were the most optimistic about co-management among the three stakeholder groups; however, a majority qualified their answers and listed changes necessary for success. These conditions closely relate to changes mentioned by local leaders and government experts answering similar questions about conditions for success. Optimists believed that existing examples of bottom-up fishery management provide evidence that success is possible. However, pessimists highlighted examples of community or external management efforts that failed and made them concerned about the future of bottom-up management.

From different perspectives, concurrence was also reached on the need to train fishermen in fishery management. Fishermen would share local knowledge, communities would participate in management with greater confidence and the government would provide training that gives everyone a vested interest in the process. Basic changes will be necessary in fishing communities, foremost is increasing membership in fishermen's associations. Experts and leaders agreed on the need to promote participatory research in which fishermen-researchers work closely with teams of biologists.

Each group identified changes in government necessary for insuring successful bottom-up management. Both fishermen and local leaders urgently stressed the need to establish a fisheries agency in order to constantly and consistently oversee the restrictions and regulations agreed to in deliberation. Fishermen and leaders believed that such restrictions would fail without strong subsidies to encourage the compliance of fishermen.

There were some real differences that will need to be resolved in order to make a management system viable. Clearly one of these is the designation of "real" fishermen versus opportunists. In other words, there is a need to develop a shared vision of who is a member of the community and in what capacity, recognized by the community itself, its leaders, and government. Another real difference in perspective, of a different character is the concern fishers have about gear types, particularly gillnets, versus government and leaders support for closed areas. These are not mutually exclusive approaches but a middle ground needs to be developed since it is clearly a major visible issue of concern for many.

Three of the nine study communities were found to have specific features that encourage them to start thinking about fishery management. These are *Ahuyama* and *El Roto* on the Caribbean, and *Bahía Solano* on the Pacific. These are communities with a small fishing population; few fishing methods; small, close- knit fishing neighborhoods; and fishermen's association could easily start the process of fisheries management. These communities were found to have attitudes and activities which support their commitment and awareness of the need for management. It is felt that these three communities could serve as pilot sites for implementing fisheries co-management in the country.

As noted in the introduction, there is a broad theoretical framework for the governance of small-scale fisheries that has been developed by Berkes, Ostrom, Pomeroy and others (Berkes et al. 2001; Ostrom 1990, 2009; Pomeroy et al. 2004, 2011). Colombian small-scale fisheries management does not have all the elements of this framework. However the present paper and previous work in identifying problems and solutions suggests that implementing successful fisheries co-management may now be possible. Long-term direction is needed for fisheries policy in Colombia and it should include a strong governmental framework that enables local bottom-up co-management. Clear a "transition" process, is needed which can be based on these research results, that moves from the current unregulated situation towards sustainable fisheries. Lessons learned as a result of the present study suggest that work during the next five years needs to focus on building strong foundations of social infrastructure during the "transition" stage that will support and maintain viable and resilient fisheries management plans. Particular characteristics of each community affect their readiness to put bottom-up fisheries management into practice. The national framework should be flexible and adaptable to local community needs and priorities. This may be achieved through four activities: (A) small-scale fisheries program of research; (B) social marketing and awareness raising in government and community; (C) restructure governance for co- management; and (D) community organization.

CONCLUSIONS

Understanding small-scale fisheries is key to protecting the health of Colombia's coastal ecosystems and improving the quality of life for coastal fishing communities. All stakeholder groups included in this study generally agreed that both old and new problems could be reduced by implementing co-management strategies at a community level with governmental support. Consistent, long-term policy direction is needed for fisheries in Colombia. If the Colombian administration continues centralized administration of the fishery sector, without adequate support and involvement at the community level, future conditions could well be worse than at present. Socio-ecological systems related to marine and coastal fisheries are fragile, such that fishing communities are vulnerable, food security is at risk, and the health of marine ecosystems is endangered.

All stakeholders engaged in this study believed that bottom-up management is possible in their communities. This concurrence of opinion and the fact that some communities have already established rules resulting from their own deliberations are evidence in favor of co-management. Most communities believed that co-management is possible working hand in hand with government. Communities that successfully established internal rules, which then failed due to the lack of enforcement, prove the need for a partner authority to assist in implementing fisheries management through co-management. This role should fall to national regulatory agencies; however, they often cannot fulfill this role due to frequent changes in policy directions resulting from changes in national administrations.

Effective fisheries management requires a mixture of national and local authority in order to work well, and steps toward a preliminary framework for a two-tiered system are suggested. Hopefully in the future communities will be able to take the lead in local management, and government agencies will grow into trusted partners by coordinating the consistent application of national fisheries policies and protecting national resources against foreign interlopers.

ACKNOWLEDGMENTS

This research was possible because of the participation and collaboration with Colombian marine fishers and local leaders from the communities of Ahuyama, Taganga, Las Flores, San Antero, El Roto, Bahía Solano, Pizarro, Juanchaco, and Tumaco. We also thank the national and regional fisheries experts and fisheries researchers for their open participation and for giving their life's work to make Colombian fisheries better. The main part of the data analysis was done thanks to the help and support of Berta Martín-López. We also thank Karen Alexander for her editorial assistance. This study was supported by the Conservation International Foundation, Universidad del Magdalena, University of New Hampshire, fellowship programs from COLCIENCIAS and Young Women in Science - 2008 (UNESCO – L´Oreal), and Connecticut Sea Grant.

REFERENCES

1. Bavinck, M, R. Chuenpagdee, M. Diallo, P. van der Heijden, J. Kooiman, R. Mahon, and S. Williams. 2005. Interactive fisheries governance: a guide to better practice. Centre for Maritime Research, Amsterdam. Eburon Publishers. http://www.marecentre.nl/fishgovfood/documents/bavinck_interactive.pdf.
2. Beardon, H. 2008. *Del Caos a la Esperanza: cultura, política y protección de los manglares en el Pacífico colombiano*. Cali: WWF – World Wildlife Foundation. WWF-UK. DFID y WWF Reino Unido.
3. Begossi, A, and D Brown. 2003. Experiences with Fisheries co-management in Latin America and the Caribbean. Chapter 8. In: The Fisheries Co-management Experience. Accomplishment, Challenges and Prospects, ed. Douglas Wilson, Jesper Nielsen, and Poul Degnbol, 135 – 152. Fish and Fisheries Series, Volume 6. Dordrecht/Boston/London: Kluwer Academic Publishers.
4. Berkes, F. 2004. Re-thinking community-based conservation. *Conservation Biology* 18(3): 621–630. doi:10.1111/j.1523-1739.2004.00077.x.
5. Berkes, F. 2010. Shifting perspectives on resource management: resilience and the reconceptualization of 'natural resource' and 'management'. *MAST* 9(1): 13–40. http://www.marecentre.nl/mast/documents/Mastvol9no1_Berkes.pdf.
6. Berkes, Fikret, Robin Mahon, Patrick McConney, Richard Pollnac, and Robert Pomeroy. 2001. *Managing Small-Scale Fisheries-Alternative Directions and Methods*. Ottawa: International Development Research Centre.
7. Brown, D.N., and R. Pomeroy. 1999. Co-management of Caribbean Community (CARICOM) fisheries. *Marine Policy* 23(6): 549–570. doi:10.1016/S0308-597X(98)00040-2.
8. Christy, F. 1997. *The development and management of marine fisheries in Latin America and the Caribbean*. Washington, D.C.: Inter-American Development Bank. http://idbdocs.iadb.org/wsdocs/getdocument.aspx?docnum=355303.
9. Chuenpagdaee, R., and S. Jentoft. 2007. Step zero for fisheries

management: what precedes implementation. *Marine Policy* 31(6): 657–668. doi:10.1016/j.marpol.2007.03.013.

10. Cinner, J., T. McClanahan, M. MacNeilc, N. Grahama, T. Dawd, M. Mukmininf, D. Fearyg, A. Rabearisoah, A. Wamukotai, N. Jiddawik, S. Campbellf, A. Bairda, F. Januchowski-Hartleya, S. Hamedk, R. Laharil, T. Morovel, and J. Kuangel. 2012. Co-management of coral reef socio-ecological systems. *PNAS* 109(14): 5219–5222. doi:10.1073/pnas.1121215109.
11. CRC, URI and FIU. 2006. Coastal Resources Center, University of Rhode Island, and Florida International University. Fisheries Opportunities Assessment. http://www.crc.uri.edu/download/Fish_Opp_Assess_Final_012607_508.pdf.
12. Cuello, F., and L.O. Duarte. 2010. El Pescador Artesanal, Fuente de Información Ecológica para la Ordenación Pesquera en el Mar Caribe de Colombia. *Proceedings of the Gulf and Caribbean Fisheries Institute* 62: 463–470. http://nsgl.gso.uri.edu/flsgp/flsgpc09001/data/papers/073.pdf.
13. Delgado, M.F., W. Gualteros, S. Espinosa, C. Lucero, A.M. Roldan, L.A. Zapata, J.R. Cantera, C. Candelo, C. Palacio, O. Muñoz, G. Mayor, and D.L. Gil-Agudelo. 2010. *"Pianguando - Estrategias para el manejo de la piangua" (Cartilla)*, INVEMAR, ASCONAR, WWF Colombia, Universidad del Valle, UAESPNN – PNN Sanquianga; Co-financiado por el Ministerio de Agricultura, 20. Cali: Serie de publicaciones generales INVEMAR No. 45. http://issuu.com/natucreativa/docs/piangua.
14. ECOVERSA. 2007. Justificación sobre la necesidad de una nueva estructura para el manejo y ordenación de la pesca y acuicultura a nivel nacional. Corporación ECOVERSA. 30p. Bogotá, Colombia.
15. FAO. 2006. Food and Agriculture Organization of the United Nations. Report of the Expert Consultation on the Economic, Social and Institutional Considerations of Applying the Ecosystem Approach to Fisheries Management. ftp://ftp.fao.org/docrep/fao/009/a0673e/a0673e00.pdf.
16. Francis, R.C., M.A. Hixon, M.E. Clarke, S.A. Murawski, and S. Ralston. 2007. Ten commandments for ecosystem-based

fishery scientists. *Fisheries* 32(5): 217–233. doi:10.1577/1548-8446(2007)32[217:TCFBFS]2.0.CO;2.

17. Futemma, C.R.T., and C.S. Seixas. 2008. Is there artisanal fishing territoriality in the Ubatumirim Bay (Ubatuba, SP)? Scale issues in community relations. *Biotemas* 21(1): 125–138. http://dx.doi.org/10.5007/2175-7925.2008v21n1p125.
18. García, C. 2010. Conocimiento tradicional: lo que los pescadores artesanales del Caribe colombiano tiene para decirnos. *Pan-American Journal of Aquatic Sciences* 5(1): 77–89. http://www.panamjas.org/pdf_artigos/PANAMJAS_5(1)_78-90.pdf.
19. Garcia, S.M., and K.L. Cochrane. 2005. Ecosystem approach to fisheries: a review of implementing guidelines. *ICES Journal of Marine Science* 62(3): 311–318. doi:10.1016/j.icesjms.2004.12.003.
20. García-Horta, J.B., and M.T. Guerra-Romos. 2009. The use of CAQDAS in educational research: Some advantages, limitations and potential risks. *International Journal of Research & Method in Education* 32: 151–165. doi:10.1080/17437270902946686.
21. Hart, P. 2003. The fisheries co-management experience. Accomplishment, challenges and prospects. *Fish and Fisheries* 5(1): 95–96. doi:10.1111/j.1467-2979.2004.00145_4.x.
22. Kruskal, W., and W. Wallis. 1952. Use of Ranks in One-Criterion Variance Analysis. Journal of the American Statistical Association 47(260): 583–621.
23. Lam, M., and D. Pauly. 2010. Who is right to fish? Evolving a social contract for ethical fisheries. *Ecology and Society* 15(3): 16. http://www.ecologyandsociety.org/vol15/iss3/art16/.
24. Lebel, L., J. Anderies, B. Campbell, C. Folke, S. Hatfield-Dodds, T. Hugues, and J. Wilson. 2006. Governance and the capacity to manage resilience in regional social-ecological systems. *Ecology and Society* 11(1): 19. http://www.ecologyandsociety.org/vol11/iss1/art19/.
25. Marin, A., and F. Berkes. 2010. Network approach for understanding small-scale fisheries governance: the case of the Chilean coastal co-management system. *Marine Policy* 34(5): 851–858. doi:10.1016/j.marpol.2010.01.007.
26. McCay, B.J., and S. Jentoft. 1996. From the bottom up: participatory

issues in fisheries management. *Society & Natural Resources* 9(3): 237–250. doi:10.1080/08941929609380969.

27. Miles, Matthew B., and Michael Huberman. 1994. *Qualitative data analysis: an expanded sourcebook*, 2nd ed. Thousand Oaks: SAGE Publications.
28. Navia, AF, PA Mejía-Falla, J López-García, LA Muñoz y, V Ramírez-Luna. 2010. Pesquería artesanal de la zona norte del Pacífico colombiano: aportando herramientas para su administración, Fase II. Documento técnico Fundación SQUALUS No. FS0110. 100 p. Cali, Colombia.
29. Olsson, P., C. Folke, and F. Berkes. 2004. Adaptive comanagement for building resilience in social- ecological systems. *Environmental Management* 34(1): 75–90. doi:10.1007/s00267-003-0101-7.
30. Olsson, P., L.H. Gudenson, S. Carpenter, P. Ryan, L. Lebel, C. Folke, and C.S. Holling. 2006. Shooting the rapids: navigating transitions to adaptative governance of social-ecological systems. *Ecology and Society* 11(1): 18. http://www.ecologyandsociety.org/vol11/iss1/art18/.
31. Ostrom, Elionor. 1990. *Governing the commons: the evolution of institutions for collective action*. Cambridge: Cambridge University Press.
32. Ostrom, E. 2009. A General framework for analyzing sustainability of social – ecological systems. *Science* 325: 419–422. doi:10.1126/science.1172133.
33. Ozkan, Betul. 2004. Using NVivo to analyze qualitative classroom data on constructivist learning environments. *The Qualitative Report* 9(4): 589–603. http://nsuworks.nova.edu/tqr/vol9/iss4/2/.
34. Pomeroy, R., P. McConney, and R. Mahon. 2004. Comparative analysis of coastal resources co- management in the Caribbean. *Ocean & Coastal Management* 47(9): 429–447. doi:10.1016/j.ocecoaman.2004.09.005.
35. Pomeroy, Robert, Joshua E Cinner, and Jesper Raakjaer Nielsen. 2011. Conditions for Sucessful Co- management: Lessons Learned in Asia, Africa, the Pacific and the Wider Caribbean. In Small-scale Fisheries Management. Frameworks and Approaches for the Developing World, eds. Robert Pomeroy and Neil Andrew. 115–

131. UK: CAB International.

36. Posada, Blanca O., and Daniel M. Rozo. 2005. Marco Geográfico. In *Informe del estado de los ambientes marinos y costeros en Colombia, Series de publicaciones periódicas No.8*, 3–9. Santa Marta: INVEMAR. http://www.invemar.org.co/redcostera1/invemar/docs/3801IER_2005_completo.pdf.

37. Preikshot, Dave B, and Daniel Pauly. 1998. A multivariate interdisciplinary assessment of small-scale tropical fisheries. In Proceedings of the International Symposium on Fishery Stock Assessment Models, eds. T.J. Quinn II, F. Funk. J. Heifetz, J.N. Ianelli, J.E. Powers, J.F. Schweigert, P.J. Sullivan, C.-I. Zhang, 803–814. Alaska, USA: Alaska Sea Grant College Program Report. https://s3-us-west-2.amazonaws.com/legacy.seaaroundus/doc/Researcher+Publications/dpauly/PDF/1998/Books+and+Chapters/MultivariateInterdisciplinaryAssessmentSmallScaleTropical.pdf.

38. Saavedra-Díaz, L, AA Rosenberg, and R Pomeroy. 2015a. Why Colombian marine fisher′s knowledge is a fundamental tool for marine resource management and assessment. In Fishers′ knowledge and the ecosystem approach to fisheries: applications, experiences and lesson in Latin America, eds. Johanne Fishers, John Jogersen, Helga Josupeit, Daniela Kalikoski, and Cristine M. Lucas. 89 – 106. Rome: FAO Fisheries and Aquaculture Technical Paper No. 591. http://www.fao.org/3/a-i4664e.pdf.

39. Saavedra-Díaz, L, AA Rosenberg, and B Martín-López. 2015b. Social perceptions of Colombian small marine fisheries conflicts: insights for management. Marine Policy 56: 61–70. doi:10.1016/j.marpol.2014.11.026.

40. Salas, S., R. Chuenpagdee, A. Charles, and J.C. Seijo (eds.). 2011. *Coastal fisheries of Latin America and the Caribbean*, FAO Fisheries and Aquaculture Technical Paper. No. 544, 430. Rome: FAO. http://www.fao.org/docrep/014/i1926e/i1926e.pdf.

41. Salas, S., R. Regist, C. Zapata, M. Cabrera, and J. Euán. 2015. How much we can learn from fishers about ecology and fisheries management: case studies on spiny lobster fishery in Mexico. In *Fischers′ knowledge and the ecosystem approach to fisheries: applications, experiences and lesson in Latin America*, ed. Fishers

Johanne, Jogersen John, Josupeit Helga, Kalikoski Daniela, and Cristine M. Lucas, 107–124. Rome: FAO Fisheries and Aquaculture Technical Paper No. 591. http://www.fao.org/3/a-i4664e.pdf.

42. Saldaña, Johnny. 2009. *The Coding Manual for Qualitative Researchers*, 1st ed. London: SAGE Publications.

43. Schumman, S. 2007. Co-management and "consciousness": fishers' assimilation of management principles in Chile. *Marine Policy* 31(2): 101–111. doi:10.1016/j.marpol.2006.05.008.

44. Solís, V., A. Muñoz, and M. Fonseca. 2015. Integrating traditional and scientific knowledge for the management of small scale fisheries: an example from Costa Rica. In *Fischers´ knowledge and the ecosystem approach to fisheries: applications, experiences and lesson in Latin America*, ed. Fishers Johanne, Jogersen John, Josupeit Helga, Kalikoski Daniela, and Cristine M. Lucas, 179–190. Rome: FAO Fisheries and Aquaculture Technical Paper No. 591. http://www.fao.org/3/a-i4664e.pdf.

45. St. Martin, K., B.J. McCay, G. Murray, T. Johnson, and B. Oles. 2007. Communities, knowledge and fisheries of the future. *International Journal of Global Environmental Issues* 7(2/3): 221–239. http://umsms.siteturbine.com/faculty/faculty_files/publications/912/St.%20Martin%20et%20al.pdf.

46. Thayer, A., M. Evans, A. McBride, M. Queen, and J. Spyridakis. 2007. Content analysis as a best practice in Technical Communication Research. *Journal Technical Writing and Communication* 37: 267–279. doi:10.2190/TW.37.3.c.

47. Trimble, M., and F. Berkes. 2015. Towards adaptive co-management of small-scale fisheries in Uruguay and Brazil: lessons from using Ostrom´s design principles. *Maritime Studies* 14: 14. doi:10.1186/s40152-015-0032-y.

48. Tyler, Stephen. 2006. *Comanagement of Natural Resources. Local Learning for Poverty Reduction*. Otawa: Centro Internacional de Investigación para el Desarrollo. http://www.idrc.ca/EN/Resources/Publications/openebooks/346-1/index.html.

49. Yin, Robert. 2003. Case study research: design and methods. Applied social research methods series. New York, USA: SAGE Publications.

SECTION IV

ENVIRONMENTAL MONITORING AND MANAGEMENT

CHAPTER

10

DIATOM-BASED WATER QUALITY MONITORING IN SOUTHERN AFRICA: CHALLENGES AND FUTURE PROSPECTS

Tatenda Dalu and P William Froneman

Zoology and Entomology, Rhodes University, Grahamstown 6140, South Africa

ABSTRACT

Diatoms are of significant ecological importance in aquatic ecosystems, which stems from their dynamic position at the base of the trophic web as primary producers. Because diatom communities have specific environmental requirements and respond rapidly to changes in environmental conditions they are often employed as a cost-effective method to assess anthropogenic impacts and health statuses of aquatic ecosystems, particularly in Europe and North America. The purpose of this review is to summarize the challenges and future prospects associated with biological water quality monitoring using diatoms with special focus on southern Africa. Much work still needs to be carried out on diatom tolerances, ecological preferences and ecophysiology. It is recommended that past research pertaining to African diatom taxonomy should be made readily accessible to all through electronic media for use as a reference point. Moreover, following the same approach as for macroinvertebrate biomonitoring, African and other developing countries can resort to intermediate diatom taxonomy (i.e. genus), which is easier, less time-consuming and requires less-skilled personnel. While the lack of capacity and baseline information on diatom community composition and ecological requirements represent significant hurdles, diatom biomonitoring potentially holds much promise for understanding the ecological functioning and management of aquatic ecosystems in southern Africa. The application of diatom-based water quality assessment protocols has direct and immediate value for use as an 'added-value' assessment tool in addition to the use of macroinvertebrates and fish indices as these can indicate anthropogenically impacted and pristine sites.

Keywords: Application, biomonitoring, ecology, taxonomy, rivers, water quality

INTRODUCTION

Aquatic ecosystems form a fundamental component of global biogeochemical cycles, acting as transport pathways, elemental transformations and storage sites (Wehr and Descy, 1998; Bere and Tundisi, 2010). These systems are characterized by a high degree of interaction and complexity among physical, chemical and biological processes. Aquatic ecosystems, e.g., rivers, often present a continuous gradient of physical, chemical and biological conditions that are frequently predictable (Vannote et al., 1980). An important feature of aquatic ecosystems is that a disturbance at one location, through conversion of landscapes into other uses such as agriculture, urban and industrial developments, often affects the functioning, processes and organisms downstream. This has serious implications for ecosystem functioning and for management of these systems (Ndaragu et al., 2004; Ndiritu et al., 2006).

Agricultural activities, e.g., irrigation and livestock, increase inputs of nutrients and sediments (Leland and Porter, 2000; Bere 2007), and urbanization contributes large quantities of organic wastewater and solids (Lobo et al., 1995; Beyene et al., 2009; Bere and Mangadze, 2014, Dalu et al., 2015), while industries are major sources of inorganic pollution (Beyene et al. 2009). Although information on the impacts of anthropogenic activities in river systems is well documented in developed countries (e.g. Dixit and Smol, 1994; Kelly et al., 2008, 2012; Stevenson, 2014), very little is known from most developing countries, yet such data are important for developing appropriate management practices (Fabricius et al., 2003; De la Rey et al., 2004, 2008; Taylor et al., 2007a; Bere and Tundisi, 2010) necessary to meet Millennium Development Goal 7, on ensuring environmental sustainability.

River systems are characterized by longitudinal differences in the physico-chemical and biotic fauna and flora making it difficult to design and implement management strategies due to their heterogeneous state (Bere and Tundisi, 2010). Ecological principles such as networks, flows and nested systems thus play an important role in determining the best management and policy options for these aquatic ecosystems (De la Rey et al., 2004). The characterisation and management of these dynamic physico-chemical and biological conditions in these heterogeneous systems requires innovative approaches, with many management processes attempting to strike a balance between ecological integrity and human needs (Fig. 1) (De la Rey et al., 2004; Stevenson, 2014).

Currently, two basic approaches to water quality assessment within river ecosystem health assessments are the assessment of physico-chemical variables, and biomonitoring (De la Rey et al., 2004; Bere and Tundisi, 2010). The assessment of physicochemical variables in river systems to assess water quality allows only for instantaneous measurements, thereby restricting the knowledge of water conditions to the period when the measurements were taken (De la Rey et al., 2004), giving a 'snapshot' result that ignores temporal variation of physico-chemical variables. The monitoring of water chemistry typically also only involves the determination of macronutrient concentrations and measurement of selected physico-chemical variables including pH (De la Rey et al., 2004; Bere et al., 2013).

The second approach, biomonitoring, involves the use of aquatic organisms to assess their response to environmental change. Biomonitoring aims to provide a direct measure of ecological integrity through integration of various stressors. It allows long-term environmental effects to be detected, providing a broad measure of their synergistic impacts (Harding et al., 2004; Stevenson, 2014). Aquatic communities integrate and reflect the effects of environmental disturbances that occur over extended periods of time, thereby providing a holistic and an integrated measure of health (Chutter, 1998; Ndiritu et al., 2006). The main advantage of using a biological approach is that it examines organisms whose exposure to pollutants is continuous and the organisms present in aquatic ecosystems reflect both the past and present water quality history, allowing detection of environmental changes that might otherwise be overlooked using physico-chemical assessments (Pan et al., 1996; De la Rey et al., 2004).

The key to the use of aquatic organisms as reliable indica-tors of the environmental changes in aquatic ecosystems is untangling the integrated information in species assemblages and the environment (Pan et al., 1996). The use of biomonitor-ing has, therefore, gained momentum in aquatic ecosystem health management programmes due to the shortcomings in the physico-chemical variables based assessments (De la Rey et al., 2004). Biomonitoring is now viewed as an ideal means by which progress towards integrated water resource management can be monitored in that it provides an overview of environmental conditions (Passy, 2007; Stevenson, 2014; Bere, 2016a). As a con-sequence, biomonitoring is now viewed as an important branch of applied ecology where the ecological, scientific and economic interests of our society meet in the management of river systems (De la Rey et al., 2004; Passy, 2007; Stevenson, 2014).

The purpose of this review is to summarise the challenges and prospects for employing diatoms as a biomonitoring tool in water quality monitoring, with specific focus on the sub-Saharan African region.

Diatom Use in Biomonitoring

Diatoms comprise the major component of the microphytoben-thos and form an integral part of primary producers in aquatic ecosystem food webs (Dalu et al., 2014a). Diatoms belong to the class Bacillariophyceae and comprise of a ubiquitous, highly successful and distinctive group of mostly unicellular algae, with the most obvious distinguishing characteristic being the presence of siliceous cell walls (frustules; Round et al., 1990). Because diatoms respond directly to growth stimulants such as nutrients (e.g. Lobo et al., 1995; Bere and Mangadze, 2014; Dalu et al., 2014a; Dalu and Froneman, 2014), physical factors such as habitat and elevation (e.g. Ndiritu et al., 2003, 2006; Passy et al., 2006; Dalu et al., 2014b) and/or stressors such as contaminants and habitat alterations (e.g. Lobo et al., 1995; Passy et al., 2006; Stevenson, 2014), they can be employed to monitor changes (natural and anthropogenic) within aquatic systems.

Diatoms have long been employed in ecological assess-ments for environmental change and pollution around the world (Bere and Tundisi, 2010; Stevenson et al., 2014). Studies by Bate et al. (2002), Bere and Tundisi (2010), Harding et al. (2014) and Stevenson (2014) present a global historical overview of diatom-based monitoring from the time Kolkwitz and Marsson (1908) first used diatoms in water monitoring to the present. Importantly, the global ability to use diatoms to evaluate present (e.g. Ndiritu et al., 2003, 2006; Ndaragu et al., 2004; Bere and Mangadze, 2014; Dalu et al., 2014c) and past conditions (e.g. Haberyn and Hecky, 1987; Stager and Johnson, 2000; Talbot and Lærdal, 2000; Taylor et al., 2005b; Harding and Taylor, 2014) had increased by the turn of the 21st century.

Biomonitoring programmes across sub-Saharan Africa commonly make use of several aquatic biological indicator organisms including fish (Fish Assemblage Integrity Index, e.g., Kleynhans, 1999; Kadye, 2008), macroinvertebrates (e.g. South African Scoring System (SASS); Chutter, 1998; Gratwicke, 1998; Mbaka et al., 2014; M'Erimba et al., 2014; Bere et al., 2016b) and macrophytes (e.g. Riparian Vegetation Index; Kemper, 2001). Until the turn of the century no biomonitoring method based on autotrophic organisms (diatoms) was routinely being used in African waters (De la Rey et al., 2004; Ndiritu et al., 2004, 2006), despite the fact that earlier African scholars had used diatoms as indicators of water quality (Table 1; e.g. Schoeman et al., 1979; Schoeman and Haworth, 1986; Pieterse and Van Zyl, 1988). Béla Jenö Cholnoky (1899–1972) known as 'the father of South African diatomology', placed little faith in performing only physico-chemical analyses for aquatic ecosystem water quality analysis, arguing that the physico-chemical characteristics of a water body could be determined more reliably and cost effectively through the study of associated diatom communi-ties (Harding et al., 2005). The use of diatom biomonitoring, therefore, formed part of the Department of Water Affairs and Forestry (DWAF) of South Africa's National Aquatic Ecosystem

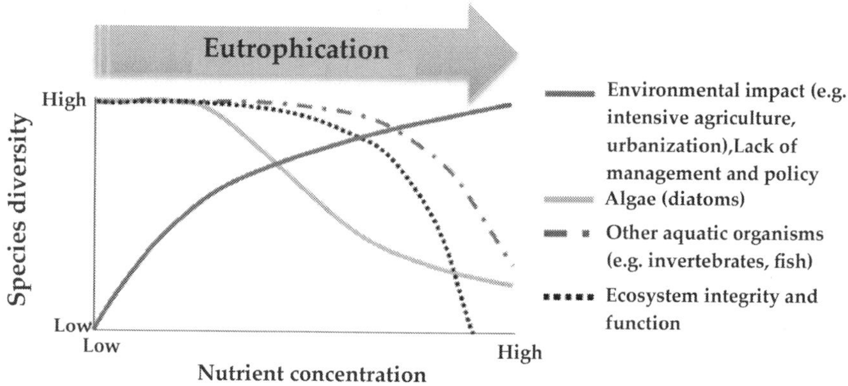

Figure 1: Environmental impact on aquatic biodiversity, ecological integrity and functioning of aquatic ecosystems. Modified from Stevenson (2014).

Biomonitoring Programme or River Health Programme (RHP) as it was considered cost effective and accurate, giving compara-ble data (De la Rey et al., 2004; Harding et al., 2004).

More recently, diatoms have increasingly been used for the assessment of short- and long-term water quality and environ-mental change (Fabricius et al., 2003; De la Rey et al., 2004; Taylor et al., 2007b; De la Rey et al., 2008; Bere and Tundisi, 2010; Bere et al., 2014) across the African continent. Diatom-based information can be gleaned not only from natural surfaces such as sediments, stones-in-current and marginal vegetation (macrophytes), but also from artificial substrates in aquatic ecosystems (De la Rey et al., 2004; Taylor et al., 2007; Bere and Tundisi, 2011; Dalu et al., 2014a, c; Dalu and Froneman, 2014). The use of artificial substrates has been key in the formula-tion of stressor-response models (Hirst et al., 2004; Lacoursière et al., 2011).

Diatom-based Monitoring in Southern Africa

Benthic diatom communities are sensitive indicators of environmental pollution, e.g., nutrient status, in aquatic ecosystems, which explains their widespread utilization across North America and Europe for routine determinations of river bio-integrity (Dixit and Smol, 1994; Lang et al., 2012; Stevenson, 2014). In developed countries, various environmental legislation statutory requirements, such as the Water Framework Directive 2000/60/EC in Europe (Kelly et al., 2008, 2012) and Environmental

Monitoring and Assessment Program Surface Waters EPA/620/R-94/004F (EMAP-SW) in the United States of America (Dixit and Smol, 1994), have been developed to safeguard these aquatic ecosystems as they seek to evaluate the biotic integrity and trophic conditions.

In sub-Saharan Africa, South Africa (e.g. Schoeman et al., 1979; Schoeman and Haworth, 1986; Pieterse and Van Zyl, 1988; De la Rey et al., 2004, 2006; Harding et al., 2004; Taylor et al., 2007a, b; Taylor and Harding, 2014) has been the leader in environmental protection and the use of benthic diatoms in aquatic ecosystems to monitor biotic integrity, and trophic condition assessments are routinely used in both freshwater and estuarine systems (Taylor et al., 2007). In addition, Kenya has also implemented diatom biomonitoring programmes (e.g. Ndiritu et al., 2003; Ndaragu et al., 2004; Ndiritu et al., 2006; Triest et al., 2012), with other countries such as Zambia (Lang et al., 2012) and Zimbabwe (Bere et al., 2014; Bere and Mangadze, 2014) beginning to embrace the benefits of implementing diatom-based water quality assessments of their aquatic ecosystems. Despite the relative cost effectiveness of this approach, however, the practice of using diatoms to monitor anthropogenic impacts on river health in Africa has largely been ignored (Cocquyt et al., 2013).

In South Africa, Water Research Commission (WRC) funded projects have laid the foundation for the implementation and successful use of diatom-based monitoring programmes in both freshwater and estuarine ecosystems (Bate et al., 2002; Harding, 2004; Taylor et al., 2005a, 2007c, d; DH Environmental Consulting, 2015). In Zambia, the Southern African River Assessment Scheme (SAFRASS) study, funded by African, Caribbean and Pacific countries (ACP) Science and Technology Programme, spearheaded the development of a national diatom biomonitoring protocol, guides and procedures (Dallas et al., 2010; Lang et al., 2012). The importance of the provision of standardized methods has been recognized in Europe and America and has taken the form of recommended standard protocols (CEN, 2003; 2004; Kelly et al., 2012). Although the methods and techniques for diatom preparation for viewing under the microscope existed in Africa before the turn of the century (e.g. Schoeman and Archibald, 1977; Schoeman, 1982; Pieterse and Van Zyl, 1988), these methods often made use of harmful chemicals such as nitric and sulphuric acids (Harding et al., 2004; Taylor et al., 2007c). Consequently, techniques and methods for diatom collection, preparation and enumeration were tested and applied based on methods developed in Europe, for the recommendation of a standard set of procedures in South Africa (Harding et al., 2004; Taylor et al., 2005a; Taylor et al., 2007c, d).

Table 1: Published records of use of diatoms in water quality assessments (pre- and post-2000) from different countries of southern Africa

		YEAR	LOCALITY	SOURCE
PRE-2000	SOUTH AFRICA	1956-1957	Jukskei-Crocodile River	Cholnoky (1958)
		1968	South African rivers	Cholnoky (1968)
		1972	South African rivers	Archibald (1972)
		1976	Jukskei-Crocodile River	Schoeman (1976)
		1979	Hennops River	Schoeman (1979)
		1982	Jukskei-Crocodile River	Schoeman (1982)
		1986	Unknown location	Schoeman and Haworth (1986)
	MALAWI	1999	Lake Malawi	Hecky et al. (1999)
		YEAR	LOCALITY	SOURCE
POST-2000	KENYA	2003	Nairobi River	Ndiritu et al. (2003)
		2006	Gombe and Lake Tanganyika Rivers	Bellinger et al. (2006)
		2006	Central Kenya	Ndiritu et al. (2006)
	MALAWI	2003	Lake Malawi	Higgins et al. (2003)
		2012	Blantyre	Kaonga and Monjerezi (2012)
	SOUTH AFRICA	1956, 1957	Jukskei-Crocodile River	Taylor et al. (2005b)
		1960, 2008	Wemmershoek Catchment	Harding and Taylor (2014)
		2002	South African River systems	Bate et al. (2002)
		2002-2003	Vaal and Wilge Rivers	Taylor et al (2007e)
		2004	Swartkops River	Bate et al. (2004)
		2008	Mooi River	De La Rey et al. (2004)
		2005	South Africa	River Health Programme (2005)
		2007	South African River systems	García-Rodriguez et al. 2007
		2007	Berg, Kromme and Mngazi Estuaries	Snow (2007)
		2007	Vaal and Wilge Rivers	Taylor et al. (2007a)
		2007	Crocodile West and Marico Water Management Area	Taylor et al. (2007b)
		2008	St Lucia Estuary	Bate and Smailes (2008)
		2008	Marico-Molopo River catchment	De la Rey et al. (2008)
		2009	Buffelspoort Valley	Taylor et al. (2009)
		2009	Crocodile and Magalies Rivers	Walsh and Wepener (2009)
		2010	Skidaway River Estuary	Verity and Borkman (2010)
		2010-2012	Great Fish River	Holmes and Taylor (2015)
		2012-2013	Kowie River	Dalu et al. (2014d)
		2012-2013	Kowie River	Dalu et al. (2016)
		2013	Sundays River	Janse Van Vuuren and Taylor (2015)
		2007	Sanyati Basin, Lake Kariba	Phiri et al. (2007)

	ZIMBABWE	2007	Nyanga	Bere et al. (2013)
		2012	Chinhoyi	Bere and Mangadze (2014)
		2012	Chinhoyi	Bere et al. (2014)
		2013	Manyame Catchment	Mangadze et al. (2015)
		2007, 2012	Nyanga, Chinhoyi	Bere (2016a)
		2007	Eastern Highlands	Bere (2016b)
		2013	Manyame Catchment	Mangadze et al. (2016)

Present legislation, e.g., the water and environmental acts, governing water resource management in many African countries (e.g. Botswana, Kenya, South Africa, Zambia, Zimbabwe) requires that river health, and aquatic ecosystem health, be assessed in order to determine user strategies and impacts of new developments on the environment (Harding et al., 2004). In support of these requirements, river health assessment protocols have been developed in South Africa (Harding et al., 2004; Taylor et al., 2007c, d; DH Environmental Consulting, 2015) and, more recently, in Zambia methodological recommendations have been developed (Dallas et al., 2010; Lang et al., 2012). The protocols need to be constantly refined as part of the river health policy implementation based on diatom monitoring as new diatoms are identified and ecological tolerances/ preferences determined. Other African countries, e.g., Zimbabwe, have adapted diatom protocols and methods and procedures developed in South Africa for use in river health monitoring (Bere and Mangadze, 2014; Bere et al., 2014).

Recently, diatom- based indices such as the Specific Pollution Index (SPI) and Biological Diatom Index (BDI) have come into the spotlight as potential additions to more established biological indicators such as SASS (De la Rey et al., 2008), with several studies being carried out over the past few years exploring the potential use of diatoms as biological indicators (De la Rey et al., 2004; Harding et al., 2005; Taylor et al., 2007a, b). A standard protocol for assessment using diatoms has also been published (Taylor et al., 2005a) to facilitate comparability of diatom index results. The value of diatoms as biological indicators has been recognized and formed part of the state of the rivers report for the Crocodile West–Marico Water Management Area in South Africa (River Health Programme, 2005; Taylor et al., 2007b; De la Rey et al., 2008).

Challenges and Prospects in Diatom-Based Monitoring in Africa

Round (1993) and De la Rey et al. (2004) have highlighted the benefits of using diatom-based biomonitoring programmes which bear special relevance to South Africa, Kenya and Zimbabwe. These include cost effectiveness, rapidity and accuracy of data collection, and that non-specialists with a biological background can do identifications and counts if they are pro-vided with illustrated guides.

Concerns have, however, been raised by scholars on the confidence in results produced by non-specialists in diatom identification (Kociolek and Stoermer, 2001). The difficulties with diatom taxonomy and nomenclature can now, however, be addressed through recent technology advances enabling comparison of images with electronic keys and rapid e-mail communication with experts (De la Rey et al., 2004; Kociolek and Stoermer, 2001). Indeed, Kociolek and Stoermer (2001) propose that studies on the taxonomy and ecology of diatoms in the 21st Century must be linked through integrated research programmes, facilitated by technological advances that support both accurate taxonomy and improved ecological interpretation, such as through the use of powerful microscopes (i.e. electron microscopes). The authors' personal experience from using the most recent and powerful electron microscopes, supported by good identification guides, along with recently published literature, confirms in the possibility of a marked reduction in the time required for identification (taxonomy) and high confidence in the results obtained.

One major problem that has led to diatom-based biological monitoring not being widely employed is the lack of capacity and the training of diatom taxonomists due to limiting funding and job opportunities. Taxonomists tend to restrict themselves to identifying and naming species and often fail to consider the ecological context (i.e. species tolerances) of the differ-ent species. Such an approach limits funding and employment opportunities. The study of diatoms in southern Africa initially attracted a number of investigators who produced large volumes of published research (Harding et al., 2004). However, relatively few students were trained to extend their work. Moreover, the decision to focus on biomonitoring using macroinvertebrates in the 1980s further contributed to limited job opportunities for diatom specialists. As a consequence, South Africa now only has a handful of diatom experts (< 10). A select group of individuals with limited skills and training are attempting to continue diatom biomonitoring and ecology studies within the region. In South Africa, for example, the bulk of the diatom taxonomic work appears in the grey literature (formerly housed at the Council for Scientific and Industrial Research (CSIR) but now housed at the South African Institute for Aquatic Biodiversity (SAIAB) National Diatom Collection at North-West University, South Africa) and is not readily available to researchers using internet search engines. These data can, however, be accessed on request, but this can often be time consuming (Fig. 2, Table 2; see Harding et al. (2004) for further details). Furthermore, only a handful of papers and books relate to water quality assessment using diatoms. Moreover, most African research institutions and universities do not have the necessary facilities (such as powerful electron microscopes) and expertise to conduct routine diatom analyses. All these challenges have contributed to the difficulty of implementing diatom-based biological monitoring programmes in most sub-Saharan countries. Hence the field of diatoms has gained the reputation of being one of the most technically difficult within the region (Kociolek and Stoermer, 2001; Kelly et al., 2012; Bere, 2016a).

Table 2: A breakdown of the total number of reprints and papers on diatom ecology and taxonomy that are estimated to be housed at the CSIR diatom collection. Bold text indicates data that has been electronically captured (Source: Harding et al., 2004)

Category	No. of papers	No. of boxes
Literature papers by box numbers	**5 453**	**196**
Papers by general author names	360	12
Papers on chemistry	200	8
Papers on ecological information	300	10
Papers on Africa	**1 365**	**39**
Papers by specific authors	**2 000**	**100**
Reprints and general papers	720	36

Figure 2: Selected diatom ecology and taxonomy reprints and publications at the South African Diatom Collection housed at the Council for Scientific and Industrial Research and since moved to South African Institute for Aquatic Biodiversity National Diatom Collection at North-West University, South Africa.

A further challenge that has been highlighted is the use of species data and ecological requirements of diatoms species between the northern and southern hemisphere. Questions remain concerning the range of ecological tolerances of wide-spread/introduced species due to various physico-chemical factors, distance and climate (Round, 1991). This view is sup-ported by Mann and Droop (1996) and Kelly et al. (2012), who argued that a considerable number of diatoms are endemic and/or show a regionally restricted distribution due to specific physico-chemical factors at specific locations that determine this distribution, with geographic location being the determining factor in the distribution of diatom species and the composition of communities.

Besides the abovementioned logistical challenges, there is also a general perception that diatom biomonitoring is onerous and that the sample preparation is time consuming and requires high levels of expertise and taxonomic skills. In South Africa, unlike in other African countries, diatom ecologists are fortunate in that the majority of the taxonomic work had already been undertaken by Cholnoky (1950–1970) in Harding et al. (2004), with more work still to be done in unexplored environments. Benthic diatom taxonomy in the majority of sub-Saharan African countries, excluding countries such as South Africa, Sierra Leone, Gambia, Ghana and Malawi, warrants re-investigation as highlighted by the discovery of several new diatom species in Democratic Republic of Congo (DRC) (Cocquyt et al., 2013; Cocquyt and Taylor, 2015) and

Zambia (Cocquyt et al. 2014; Taylor et al. 2014a, b). There is a strong reliance on the published diatom literature from European and American sources in the identification of species.

The absence of diatom experts and lack of necessary resources (e.g. microscopes, identification guides) in most African countries, represent significant obstacles in ensuring that diatom-based biomonitoring programmes are included as part of national policy, as is the case of South Africa which included diatoms in the River Health Programme and regular biomonitoring assessments. In Europe and the United States, all water systems are protected under one policy, such as the water quality European standards EN 14407:2004 and EN 13946:2003 and the EMAP-SW EPA/620/R-94/004F, which are constantly being amended in line with changing environments and research. In Africa, different regions can develop different policies that ensure that all aquatic ecosystems are properly conserved, managed and protected for future generations.

CONCLUSIONS AND RECOMMENDATIONS

Diatom-based biomonitoring programmes have been implemented with some success in Kenya, South Africa, Zambia and Zimbabwe. This approach has been incorporated into the national river health programme of South Africa (Dallas et al., 2010), now part of the National Aquatic Ecosystem Health Monitoring Programme (DWS, 2016);). It is anticipated that this approach will be adopted by several other African countries in the near future. The standardization of methodology has laid the foundation for the continued and meaningful collection and analysis of diatom samples in South Africa (Taylor et al., 2007a). It is anticipated to yield similar results in Kenya, Zambia, Zimbabwe and many other African countries that are currently in the process of standardizing their diatom methodology, although these protocols must consider endemic diatom taxa (see Bere (2016a), for Zimbabwean experiences).

In terms of diatom tolerances, ecological preferences and ecophysiology, much work still has to be done (Bate et al., 2002). All past research pertaining to African diatom taxonomy should be made readily accessible to all for use as a reference point, such as the work by Giffen (1963), Cholnoky (1968), Schoeman and Archibald (1977) and Taylor et al (2007c) highlighting the diatom flora of South Africa. On this point, the SAIAB National Diatom Collection is a commendable effort which making much of the grey literature readily accessible to other African research-ers. The collection of diatom samples from different African countries, e.g. DRC, Zambia and Zimbabwe, for water quality determination should be complemented with sound taxonomical investigations to update knowledge of Africa's diatom flora (e.g. in Zambia, a new diatom species has been recently discovered (see Cocquyt et al., 2014; Taylor et al., 2014a, b) and also in DRC (see Cocquyt et al., 2013; Cocquyt and Taylor, 2015)). Cholnoky (1950–1970), and Giffen (1960–1980), amongst other authors, only produced diatom line diagrams which are open to misinterpretation, so the digitization of the images of African and South African type specimens needs to be considered (Harding et al., 2004). Additionally, aquatic ecologists should work in conjunction with diatom taxonomists (see Kociolek and Stoermer, 2001). As this information becomes available it may then be resolved into numerical diatom indices. Finally attention should be paid to both the biology and ecology of diatoms occurring in moder-ate to high water quality environments, as several studies (e.g. Taylor et al., 2007; Bere et al., 2013; Mangedze et al., 2014; Bere 2016a) have highlighted that cosmopolitan species are found in a variety of habitats impacted by different types of pollutants.

One of the most important approaches that African and other countries around the world can adopt, which has also been observed with other biomonitoring methods such as macroinvertebrates (Chutter, 1998; Dickens and Graham, 2002; Metzeling et al., 2006), is to resort to intermediate diatom taxonomy, which is easier, less time consuming and requires less-skilled personnel. Development of a diatom biomonitoring method based on intermediate taxonomy, i.e., genus level, will make the process faster and easier. However, care must be taken to ensure that ecological information is not lost when taxonomic resolution is decreased, as this may compromise diatom model inferences. Several studies (e.g. Raunio and Soininen, 2007; Rimet and Bouchez, 2012; Keck et al., 2015) have highlighted that taxonomic resolution has little effect on diatom community structure description, suggesting that little ecological information is lost as a result of a decrease in diatom taxonomic resolution from species to genus level (see Kelly et al., 2016 for detailed information). However, Bowman and Bailey (1997) and Jones (2008) highlighted that information content increases with diatom taxonomic resolution. Taxonomic identifications become less certain at finer resolutions and data noise increases with increasing taxonomic resolution (Rimet and Bouchez, 2012). Therefore, we advocate for the use of genus level identification for water quality assessments and the development of appropriate indices based on genus level identification.

While the lack of capacity and baseline information on diatom community composition and ecological requirements represent significant hurdles, diatom biomonitoring potentially holds much promise for understanding the ecological functioning and informing the management of aquatic ecosystems in sub-Saharan Africa. Interdisciplinary research that will result in the union of environmental management, biogeography, conservation biology, diatom ecology and systematics could result in significantly more funding, if the relevance of diatom studies and biomonitoring is to be further developed. Results from such a multidisciplinary and integrative approach would also likely yield publications with a greater impact. There is a compelling need for diatom systematics research to be undertaken within multi-interdisciplinary projects, rather than focusing on the taxonomic identification, as such an approach will help in addressing challenging ecological problems.

African countries, like many European countries, should consider introducing ring tests to maintain the skills of the analysts, whereby they take part in identification of specific slides and pool the results to look for problem taxa and resolve prevailing issues (Hedden, 2010; Kelly, 2013). These tests can routinely be conducted, similar to the SASS training for macroinvertebrates which is conducted at approximately 6-monthly intervals by GroundTruth (GroundTruth, 2016). In this way, less experienced individuals can familiarize themselves with the various diatom taxa and acquire the necessary skills set to analyze diatom data, while at the same time providing an opportunity to recruit new researchers into the field. This approach might prevent the profession from collapse and enhance collaboration amongst diatom analysts and taxonomists from different regions of Africa. Across the world, identification is moving towards the use of molecular tools such that eventually identifying species using their basic morphology may become redundant. However, it is recognized that in many countries in Africa, the traditional identification techniques will still be required in the foreseeable future, as well as to interpret the research conducted by phylogeneticists and continue describing new species.

In conclusion, the use of diatom biomonitoring has importance in providing information relevant to common concerns about ecological condition and can be employed for short- and long-term monitoring of aquatic ecosystem function and health. Diatom monitoring programmes can provide results that are readily accessible to water managers and non-specialists. Regional examples from Kenya, South Africa, Zambia and Zimbabwe clearly indicate the importance and robustness of this protocol as a scientific biomonitoring tool (e.g. Stevenson, 2014).

ACKNOWLEDGEMENTS

This review article is part of a study into the use of diatoms for biomonitoring in the Eastern Cape, South Africa. The financial support for the study is from Rhodes University and the first author is a recipient of the Rhodes University Postdoctoral Fellowship grant.

REFERENCES

1. ARCHIBALD RM (1972) Diversity in some South African diatom associations and its relation to water quality. *Water Res.* **6** 1229–1238.
2. BATE G, SMAILES P and ADAMS J (2004) A water quality index for use with diatoms in the assessment of rivers. *Water SA* **30** 493–449.
3. BATE GC and SMAILES PA (2008) The response of the diatom flora of St Lucia Lake and estuary, South Africa, to a severe drought. *Afr. J. Aquat. Sci.* **33** 1–16.
4. BATE GC, ADAMS JB and VAN DER MOLEN JS (2002) Diatoms as indicators of water quality in South African river systems. WRC Report No. 814/1/02. Water Research Commission, Pretoria. BELLINGER BJ, COCQUYT C and O'REILLY CM (2006) Benthic diatoms as indicators of eutrophication in tropical streams. *Hydrobiologia* **573** 75–87.
5. BERE T (2007) The assessment of nutrient loading and retention in the upper segment of the Chinyika River, Harare; Implications for eutrophication control. *Water SA.* **33** 279–284.

6. BERE T (2016a) Challenges of diatom-based biological monitoring and assessment of streams in developing countries. *Environ. Sci. Pollut. Res.* **23** 5477–5486.
7. BERE T (2016b) Are diatom-based biotic indices developed in eutrophic, organically enriched waters reliable monitoring metrics in clean waters? *Ecol. Indic.* **62** 312–316.
8. BERE T, DALU T and MWEDZI T (2016) Detecting the impact of heavy metal contaminated sediment on benthic macroinvertebrate commu-nities in tropical streams. *Sci. Total Environ.* **572** 147–156.
9. BERE T and MANGADZE T (2014) Diatom communities in streams draining urban areas: community structure in relation to environ-mental variables. *Trop. Ecol.* **55** 271–281.
10. BERE T and TUNDISI JG (2010) Biological monitoring of lotic ecosys-tems: the role of diatoms. *Braz. J. Biol.* **70** 493–502.
11. BERE T and TUNDISI JG (2011) The effects of substrate type on diatom-based multivariate water quality assessment in a tropical river (Monjolinho), São Carlos, SP, Brazil. *Water Air Soil Pollut.* **216** 391–409.
12. BERE T, MANGADZE T and MWEDZI T (2014) The application and testing of diatom-based indices of stream water quality in Chinhoyi Town, Zimbabwe. *Water SA* **40** 530–512.
13. BERE T, PHIRI P, KADYE WT and UTETE B (2013) Benthic diatom assemblages in mountain streams: community structure in relation to environmental and human pressures. *Afr. J. Ecol.* **51** 625–634.
14. BEYENE A, ADDIS T, KIFLE D, LEGESSE W, KLOOS H and TRIEST L (2009) Comparative study of diatoms and macroinvertebrates as indicators of severe water pollution: Case study of the Kebena and Akaki Rivers in Addis Ababa, Ethiopia. *Ecol. Indic.* **9** 381–392.
15. CHOLNOKY BJ (1958) Hydrobiologische Untersuchungen in Transvaal II. Selbstreinigung im Jukskei-2647Crocodile Flusssystem. *Hydrobiologia* **11** 205-266.
16. CHOLNOKY BJ (1968) *Die* Őkologie *der Diatomeen in Binnengewässern.* J Cramer, Lehre.

17. CHUTTER FM (1998) Research on the rapid biological assessment of water quality impacts in streams and rivers. WRC Report No. 422/1/98. Water Research Commission, Pretoria.

18. COCQUYT C and TAYLOR JC (2015) New and interesting *Surirella* taxa (Surirellaceae, Bacillariophyta) from the Congo Basin (DR Congo). *Euro. J. Taxon.* **133** 1–15.

19. COCQUYT C, DE HAAN M and TAYLOR JC (2013) *Cavinulalilande* (Bacillariophyta) a new diatom species from the Congo Basin, Central Africa. *Diatom Res.* **28** 157–163.

20. COCQUYT C, TAYLOR JC and WETZEL CE (2014) *Stenopterobia cataractarum sp. nov.* (Bacillariophyta), a new benthic diatom from a waterfall in Zambia, Africa. *Phytotaxa* **158** 76–84.

21. COMITÉ EUROPÉEN DE NORMALISATION (CEN) (2003) Water quality – Guidance standard for the routine sampling and pre-treatment of benthic diatoms from rivers. European Standard. EN 13946:2003.

22. COMITÉ EUROPÉEN DE NORMALISATION (CEN) (2004) Water quality – Guidance for the identification and enumeration of ben-thic diatom samples from rivers, and their interpretation. European Standard. EN 14407:2004.

23. DALLAS H, KENNEDY M, TAYLOR J, LOWE S and MURPHY K (2010) WP4: Review of existing biomonitoring methodologies and appropriateness for adaptation to river quality assessment protocols for use in southern tropical Africa. Southern African River Assessment Scheme (SAFRASS). AFS/2009/219013. ACP Science and Technology Programme, Cape Town.

24. DALU T and FRONEMAN PW (2014) Can stable isotopes (δ15N, δ13C) and fatty acid signatures be employed to observe changes in phyto-benthos composition on artificial substrates? *Afr. J. Aquat. Sci.* **39** 425–433.

25. DALU T, BERE T and FRONEMAN PW (2016) Assessment of water quality based on diatom indices in a small temperate river system: a case of the Kowie River, South Africa. *Water SA* **42** 183–193.

26. DALU T, BERE T, RICHOUX NB and FRONEMAN PW (2015) Assessment of the spatial and temporal variations in periphyton

com-munities along a small temperate river system: A multimetric and stable isotope analysis approach. *S. Afr. J. Bot.* **100** 203–212.

27. DALU T, FRONEMAN PW and RICHOUX NB (2014b) Phytoplankton community diversity along a river-estuary continuum. *Trans. R. Soc. S. Afr.* **69** 107–116.
28. DALU T, FRONEMAN PW, CHARI LD and RICHOUX NB (2014a) Colonisation and community structure of benthic diatoms on artificial substrates following a major flood event: A case of the Kowie River (Eastern Cape, South Africa). *Water SA* **40** 471–480.
29. DALU T, RICHOUX NB and FRONEMAN PW (2014c) An assessment, using multivariate analysis and stable isotopes, of the effects of sub-strate type on phytobenthos communities. *Inland Waters* **4** 397–412.
30. DE LA REY PA, ROUX H, VAN RENSBURG L and VOSLOO A (2008) On the use of diatom-based biological monitoring Part 2: A compari-son of the response of SASS 5 and diatom indices to water quality and habitat variation. *Water SA* **34** 61–70.
31. DE LA REY PA, TAYLOR JC, LAAS A, VAN RENSBURG L and VOSLOO A (2004) Determining the possible application value of diatoms as indicators of general water quality: A comparison with SASS 5. *Water SA* **30** 325–332.
32. DH ENVIRONMENTAL CONSULTING 2015). THE USE OF DIA-TOMS FOR WATER QUALITY DETERMINATION AND BIO-MONITORING IN SOUTH AFRICA. URL: WWW.DHEC.CO.ZA/ DIATOMS (Accessed 13 January 2015).
33. DICKENS CWS and GRAHAM PM (2002) The South African Scoring System (SASS) version 5 rapid bioassessment method for rivers. *Afr. J. Aquat. Sci.* **27** 1–10.
34. DIXIT SS and SMOL JR (1994) Diatoms as indicators in the environmen-tal monitoring and assessment program-surface waters (EMAP-SW). *Environ. Monit. Assess.* **31** 275–306.
35. DWS (Department of Water and Sanitation) (2016) National Aquatic Ecosystem Health Monitoring Programme. URL: https://www.dwa.gov.za/iwqs/rhp/naehmp.aspx (Accessed 22 May 2016).
36. FABRICIUS ALM, MAIDANA N, GOMEZ MN and SABATER S (2003) Distribution pattern of benthic diatoms in a Pampean river

exposed to seasonal floods: the Cuarto River (Argentina). *Biodiv. Conserv.* **12** 2443–2454.

37. GARCÍA-RODRÍGUEZ F, BATE GC, SMAILES P, ADAMS JB and METZELTIN D (2007) Multivariate analysis of the dominant and sub-dominant epipelic diatoms and water quality data from South African rivers. *Water SA* **33** 653–658.
38. GIFFEN MH (1963) Contributions to the diatom flora of South Africa. *Hydrobiologia*. **21** 201–265.
39. GRATWICKE B (1998) The effect of season on a biotic water quality index: A case study of the Yellow Jacket and Mazowe Rivers, Zimbabwe. *S. Afr J. Aquat. Sci.* **24** 24–35.
40. GROUNDTRUTH (2016) SASS, miniSASS and Aquatic Ecology Training. URL: http://www.groundtruth.co.za/services/training.html (Accessed 6 September 2016).
41. HABERYAN KA and HECKY RE (1987) The late Pleistocene and Holocene stratigraphy and paleolimnology of Lakes Kivu and Tanganyika. *Palaeogeo. Palaeoclim. Palaeoecol.* **61** 169–197.
42. HARDING WR and TAYLOR JC (2014) Diatoms as indicators of historical water quality: A comparison of samples taken in the Wemmershoek catchment (Western Cape Province, South Africa) in 1960 and 2008. *Water SA* **40** 601–606.
43. HARDING WR, ARCHIBALD CGM and TAYLOR J (2005) Diatomology in South African biomonitoring: the South African Diatom Collection. *Afr. J. Aquat. Sci.* **30** 221–221.
44. HARDING WR, ARCHIBALD CGM, TAYLOR JC and MUNDREE S (2004) The South African diatom collection: An appraisal and over-view of needs and opportunities. WRC Report no.TT 242/04. Water Research Commission, Pretoria.
45. HARRISON TD, COOPER JAG and RAMM AEL (2000) Geomorphology, ichthyofauna, water quality and aesthetics of South African estuaries. ENV-DC 2000-01. Department of Environmental Affairs and Tourism, Pretoria, South Africa.
46. HECKY RE, KLING HJ, JOHNSON TC, BOOTSMA HA and WILKINSON P (1999) Algal and sedimentary evidence for recent changes in the water quality and limnology of Lake Malawi/ Nyasa. *Victoria.* **1** 1–2.

47. HEDDEN H (2010) *The Accidental Taxonomist.* Information Today, Medford, New Jersey
48. HIGGINS SN, KLING HJ, HECKY RE, TAYLOR WD and BOOTSMA HA (2003) The community composition, distribution, and nutrient status of epilithic periphyton at five rocky littoral zone sites in Lake Malawi, Africa. *J. Great Lakes Res.* **29** 181–189.
49. HIRST H, CHAUD F, DELABIE C, JÜTTNER I and ORMEROD SJ (2004) Assessing the short-term response of stream diatoms to acidity using inter-basin transplantations and chemical diffusing substrates. *Freshwat. Biol.* **49** 1072–1088.
50. HOLMES M and TAYLOR JC (2015) Diatoms as water quality indica-tors in the upper reaches of the Great Fish River, Eastern Cape, South Africa. *Afr. J. Aquat. Sci.* **40** 321–338.
51. JANSE VAN VUUREN S and TAYLOR JC (2015) Changes in the algal composition and water quality of the Sundays River, Karoo, South Africa, from source to estuary. *Afr. J. Aquat. Sci.* **40** 339–357.
52. JONES FC (2008) Taxonomic sufficiency: the influence of taxonomic resolution on freshwater bioassessments using benthic macroinverte-brates. *Environ. Rev.* **16** 45–69.
53. KADYE WT (2008) The application of a Fish Assemblage Integrity Index (FAII) in a Southern African river system. *Water SA* **34** 25–31.
54. KAONGA CC and MONJEREZI M (2012) Periphyton and earthworms as biological indicators of metal pollution in streams of Blantyre City, Malawi. In: Balkis N (ed) *Water Pollution.* INTECH Open Access Publisher, DOI: 10.5772/28781. URL: http://www.intechopen.com/ books/water-pollution/periphyton-and-earthworms-as-biological-indicators-of-heavy-metal-pollution-in-streams-of-blantyre-m (Accessed 15 January 2015).
55. KELLY MG, KROKOWSKI J and HARDING JP (2016) RAPPER: A new method for rapid assessment of macroalgae as a complement to diatom-based assessments of ecological status. *Sci. Total Environ.* DOI: 10.1016/j.scitotenv.2015.12.068.
56. KELLY M (2013) Building capacity for ecological assessment using dia-toms in UK rivers. *J. Ecol. Environ.* **36** 89–94.
57. KELLY MG, GÓMEZ-RODRÍGUEZ C, KAHLERT M, ALMEIDA

SFP, BENNETT C, BOTTIN M, DELMAS F, DESCY J-P, DÖRFLINGER G, KENNEDY B, MARVAN P, OPATRILOVA L, PARDO I, PFISTER P, ROSEBURY J, SCHNEIDER S and VILBASTE S (2012) Establishing expectations for pan-European diatom based ecological assessments. *Ecol. Indic.* **20** 177–186.

58. KELLY MG, JUGGINS S, GUTHRIE R, PRITCHARD S, JAMIESON J, RIPPEY B, HIRST H and YALLOP M (2008) Assessment of ecologi-cal status in UK rivers using diatoms. *Freshwater Biol.* **53** 403–422.
59. KEMPER NP (2001) *RVI: Riparian Vegetation Index.* WRC Report No. 850/3/01. Water Research Commission, Pretoria.
60. KLEE R and CASPER SJ (1992) New centric diatoms (Thalassiosirales) of Lake Malawi (formerly Lake Nyassa; Malawi, East Africa). *Arch. Protistenkunde* **142** 179–192.
61. KLEYNHANS CJ (1999) The development of a fish index to assess the biological integrity of South African rivers. *Water SA* **25** 265–270.
62. KOCIOLEK JP and STOERMER EF (2001) Taxonomy and ecology: A marriage of necessity. *Diatom Res.* **16** 433–442.
63. KOLKWITZ R and MARSSON M (1908) Ökologie der pflanzlicheSapro-bien. *Ber. Deut. Bot. Ges.* **26** 505–519.
64. KOTSEDI D, ADAMS JB and SNOW GC (2012) The response of micro-algal biomass and community composition to environmental factors in the Sundays Estuary. *Water SA* **38** 177–190.
65. LACOURSIÈRE S, LAVOIE I, RODRÍGUEZ MA and CAMPEAU S (2011) Modeling the response time of diatom assemblages to simulated water quality improvement and degradation in running waters. *Can. J. Fish. Aquat. Sci.* **68** 487–497.
66. LANG P, TAYLOR JC, BERTOLLI L, LOWE S, DALLAS H, KENNEDY MP, GIBBINS C, SICHINGABULA H, SAILI K, DAY J, WILLEMS F, BRIGGS JA and MURPHY KJ (2012) Proposed procedure for the sampling, preparation and analysis of benthic diatoms from Zambian rivers: a bioassessment and decision support tool applicable to fresh-water ecoregions in tropical southern Africa. Southern African River Assessment Scheme (SAFRASS).

AFS/2009/219013. ACP Science and Technology Programme, Cape Town.

67. LELAND HV and PORTER SD (2000) Distribution of benthic algae in the upper Illinois River basin in relation to geology and land use. *Freshwater. Biol.* **44** 279–301.
68. LOBO EA, KATOH K and ARUGA Y (1995) Response of epilithic diatom assemblages to water pollution in rivers in the Tokyo Metropolitan area, Japan. *Freshwater Biol.* **34** 191–204.
69. MANGADZE T, BERE T and MWEDZI T (2015) Epilithic diatom flora in contrasting land-use settings in tropical streams, Manyame Catchment, Zimbabwe. *Hydrobiologia* **753** 163–173.
70. MANGADZE T, BERE T and MWEDZI T (2016) Choice of biota in stream assessment and monitoring programs in tropical streams: A comparison of diatoms, macroinvertebrates and fish. *Ecol. Indic.* **63** 128–143.
71. MANN DG and DROOP JM (1996) Biodiversity, biogeography and conservation of diatoms. *Hydrobiologia* **336** 19–32.
72. MBAKA JG, M'ERIMBA CM, THIONGO HK and MATHOOKO JM (2014) Water and habitat quality assessment in the Honi and Naro Moru rivers, Kenya, using benthic macroinvertebrate assemblages and qualitative habitat scores. *Afr. J. Aquat. Sci.* **39** 361–368.
73. M'ERIMBA CM, MATHOOKO JM and OUMA KO (2014) Macroinvertebrate distributions in relation to human and animal-induced physical disturbance of the sediment surface in two Kenyan tropical Rift Valley streams. *Afr. J. Aquat. Sci.* **39** 337–346.
74. METZELING L, PERRISS S and ROBINSON D (2006) Can the detection of salinity and habitat simplification gradients using rapid bioassess-ment of benthic macroinvertebrtes be improved through finer taxo-nomic resolution or alternative indices? *Hydrobiologia* **572** 235–252. NDARUGA AM, NDIRITU GG, GICHUKI NN and WAMICHA WN (2004) Impact of water quality on macroinvertebrates assemblages along a tropical stream in Kenya. *Afr. J. Ecol.* **42** 208–216. NDIRITU GG, GICHUKI NN and TRIEST L (2006) Distribution of
75. epilithic diatoms in response to environmental conditions in an urban tropical stream, Central Kenya. *Biodiv. Conserv.* **15** 3267–3293.

76. NDIRITU GG, GICHUKI NN, KAUR P AND TRIEST L (2003) Characterization of environmental gradients using physico-chemical measurements and diatom densities in Nairobi River, Kenya. *Aquat. Ecosyst. Health Manag.* **6** 343–354.
77. PAN Y, STEVENSON RJ, HILL BH, HERLIHY AT and COLLINS GB (1996) Using diatoms as indicators of ecological conditions in lotic systems: a regional assessment. *J. N. Am. Benthol. Soc.* **15** 481–495.
78. PASSY SI (2007) Community analysis in stream biological monitoring: what must we measure and what we don't. *Environ. Monit. Assess.* **127** 409–417.
79. PERISSINOTTO R, IYER K and NOZAIS C (2006) Response of micro-phytobenthos to flow and trophic variation in two South African temporarily open/closed estuaries. *Bot. Mar.* **49** 10–22.
80. PHIRI C, DAY J, CHIMBARI M and DHLOMO E (2007) Epiphytic diatoms associated with a submerged macrophyte *Vallisnera aethiopica*, in the shallow marginal areas of Sanyati Basin (Lake Kariba): a preliminary assessment of their use as biomonitoring tools. *Aquat. Ecol.* **41** 169–181.
81. PIETERSE AJH and VAN ZYL JM (1988) Observations on the relation between phytoplankton diversity and environmental factors in the Vaal River at Balkfontein, South Africa. *Hydrobiologia* **169** 199–207.
82. RIATO L, VAN GINKEL C and TAYLOR JC (2014) Zooplankton and diatoms of temporary and permanent freshwater pans in the Mpumalanga Highveld region, South Africa. *Afr. Zool.* **49** 113–127.
83. RIMET F (2012) Recent views on river pollution and diatoms.
84. *Hydrobiologia* **683** 1–24.
85. RIVER HEALTH PROGRAMME (2005) State-of-Rivers Report: Monitoring and Managing the Ecological State of Rivers in the Crocodile (West) Marico Water Management Area. Department of Environmental Affairs and Tourism, Pretoria.
86. ROUND FE, CRAWFORD RM and MANN DG (1990) *Diatoms: Biology and Morphology of the Genera.* Cambridge University Press, Cambridge.

87. ROUND FE (1991) Diatoms in river water-monitoring studies. *J. Appl.*
88. *Phycol.* **3** 129–145.
89. ROUND FE (1993) *A Review and Methods for the Use of Epilithic Diatoms for Detecting and Monitoring Changes in River Water Quality. Methods for the Examination of Water and Associated Materials.* HMSO Publications, London.
90. SCHOEMAN FR (1973) *A Systematical and Ecological Study of the Diatom Flora of Lesotho with Special Reference to the Water Quality.* V & R Printers, Pretoria
91. SCHOEMAN FR (1976) Diatom indicator groups in the assessment of water quality in the Jukskei-Crocodile river system (Transvaal, Republic of South Africa). *J. Limnol. Soc. S. Afr.* **2** 21–24.
92. SCHOEMAN FR (1979). Diatoms as indicators of water quality in the upper Hennops River. *J. Limnol. Soc. South. Afr.* **5** 73–78.
93. SCHOEMAN FR (1982) The diatoms of the Jukskei-Crocodile River system (Transvaal, Republic of South Africa): a preliminary check-list. *J. S. Afr. Bot.* **48** 295–310.
94. SCHOEMAN FR and ARCHIBALD REM (1977) The diatom flora of Southern Africa. Special report WT50. National Institute for Water Research. Council for Scientific and Industrial Research, Pretoria.
95. SCHOEMAN FR and HAWORTH EY (1986) Diatom as indicator of pol-lution. *Proceedings of the* 8th *International Diatom Symposium.* Koeltz Scientific Books, Koenigstein. 757–766.
96. SNOW GC (2007) Contributions to the use of microalgae in estuarine freshwater reserve determinations. PhD thesis, Nelson Mandela Metropolitan University. South Africa.
97. STAGER JC and JOHNSON TC (2000) A 12,400 14c yr. offshore diatom record from east central Lake Victoria, East Africa. *J. Paleolimnol.* **23** 373–383.
98. STEVENSON J (2014) Ecological assessments with algae: a review and synthesis. *J. Phycol.* **50** 437–461.
99. TALBOT MR and LÆRDAL T (2000) The late Pleistocene–Holocene paleolimnology of Lake Victoria, East Africa, based upon

elemental and isotopic analyses of sedimentary organic matter. *J. Paleolimnol.* **23** 141–164.

100. TAYLOR JC, DE LA REY PA and VAN RENSBURG L (2005a) Recommendations for the collection, preparation and enumeration of diatoms from riverine habitats for water quality monitoring in South Africa. *Afr. J. Aquat. Sci.* **30** 65–75.
101. TAYLOR JC, HARDING WR and ARCHIBALD CGM (2007c) An illustrated guide to some common diatom species from South Africa. Report TT 282/07. Water Research Commission, Pretoria.
102. TAYLOR JC, HARDING WR and ARCHIBALD CGM (2007d). A methods manual for the collection, preparation and analysis of diatom samples. Version 1.0. WRC Report No. TT 281/07. Water Research Commission, Pretoria.
103. TAYLOR JC, HARDING WR, ARCHIBALD CGM and VAN RENSBURG L (2005) Diatoms as indicators of water quality in the Jukskei-Crocodile River system in 1956 and 1957, a re-analysis of diatom count data generated by BJ Cholnoky. *Water SA* **31** 237–246.
104. TAYLOR JC, HARDING WR, ARCHIBALD CGM and VAN RENSBURG L (2005b) Diatoms as indicators of water quality in the Jukskei-Crocodile river system in 1956 and 1957, a re-analysis of diatom count data generated by BJ Cholnoky. *Water SA* **31** 237–246.
105. TAYLOR JC, JANSE VAN VUUREN MS and PIETERSE AJH (2007e) The application and testing of diatom-based indices in the Vaal and Wilge Rivers, South Africa. *Water SA* **33** 51–60.
106. TAYLOR JC, JANSE VAN VUUREN MS and PIETERSE AJH (2007a) The application and testing of diatom-based indices in the Vaal and Wilge rivers, South Africa. *Water SA* **33** 51–60.
107. TAYLOR JC, KARTHICK B, COCQUYT C and LANG P (2014a) *Diploneis fenestrata sp. nov.* (Bacillariophyta), a new aerophilic diatom species from Zambia, Africa. *Phytotaxa* **167** 79–88.
108. TAYLOR JC, KARTHICK B, KOCIOLEK JP, WETZEL CE and COCQUYT C (2014b) *Actinellopsis murphyi gen. et spec. nov.*: A new small celled freshwater diatom (Bacillariophyta, Eunotiales) from Zambia. *Phytotaxa* **178** 128–137.

109. TAYLOR JC, POPE N and VAN RENSBURG L (2009) Community participation in river monitoring using diatoms: a case study from the Buffelspoort Valley. *Afr. J. Aquat. Sci.* **34** 173–182.

110. TAYLOR JC, PRYGIEL J, VOSLOO A, DE LA REY PA and VAN RENSBURG L (2007b) Can diatom-based pollution indices be used for biomonitoring in South Africa? A case study of the Crocodile West and Marico water management area. *Hydrobiologia* **592** 455–464.

111. TRIEST L, LUNG'AYIA H, NDIRITU G and BEYENE A (2012) Epilithic diatoms as indicators in tropical African rivers (Lake Victoria catch-ment). *Hydrobiologia* **695** 343–360.

112. VANNOTE RL, MINSHALL GW, CUMMIS KW, SEDELL JR and CUSHING C (1980) The river continuum concept 1. *Can. J. Fish. Aquat. Sci.* **37** 130–137.

113. VERITY P and BORKMAN D (2010) A decade of change in the Skidaway River Estuary. III. Plankton. *Estuar. Coasts* **33** 513–540.

114. WALSH G and WEPENER V (2009) The influence of land use on water quality and diatom community structures in urban and agriculturally stressed rivers. *Water SA* **35** 579–594.

115. WEHR JD and DESCY J (1998) Use of phytoplankton in large river man-agement. *J. Phycol.* **36** 741–749.

CHAPTER

11

MACROPHYTA AS A VECTOR OF CONTEMPORARY AND HISTORICAL MERCURY FROM THE MARINE ENVIRONMENT TO THE TROPHIC WEB

Magdalena Bełdowska[1], Agnieszka Jędruch[1], Joanna Słupkowska[1], Dominka Saniewska[1] and Michał Saniewski[2]

[1] Institute of Oceanography, University of Gdansk, Pilsudskiego 46, 81-378 Gdynia, Poland

[2] Institute of Meteorology and Water Management - National Research Institute, Maritime Branch, Waszyngtona 42, 81-342 Gdynia, Poland

ABSTRACT

Macrophyta are the initial link introducing toxic mercury to the trophic chain. Research was carried out at 24 stations located within the Polish coastal zone of the Southern Baltic, in the years 2006–2012. Fifteen taxa were collected, belonging to four phyla: green algae (*Chlorophyta*), brown algae (*Phaeophyta*), red algae (*Rhodophyta*) and flowering vascular plants (*Angiospermophyta*), and total mercury concentrations were ascertained. The urbanisation of the coastal zone has influenced the rise in Hg concentrations in macroalgae, and the inflow of contaminants from the river drainage area has contributed to an increase in metal concentration in vascular plants. At the outlets of rivers possessing the largest drainage areas in the Baltic (the Vistula and the Oder), no increases in mercury concentration were observed in macrophyta. Increase in environmental quality and a prolonged vegetative season results in the growing coverage of algae on the seabed and in consequence leads to rapid introduction of contemporary mercury and Hg deposited to sediments over the past decades into the trophic chain. Thriving phytobenthos was found to affect faster integration of Hg into the trophic web.

Keywords: Macrophyta, Mercury Bioaccumulation, Trophic chain, Coastal zone, Baltic Sea

INTRODUCTION

Mercury is considered to be one of the most dangerous contaminants of the environment. The adverse effect of Hg is related to its strong chemical and biological activity, as a result of which it is easily absorbed by organisms and spreads in the environment very rapidly. Mercury compounds become accumulated in tissues and can undergo biomagnification in organisms on higher trophic levels, reaching concentrations many times higher than in the environment itself (Förstner and Wittman 1981; Jackson 1998). Even small amounts of mercury in the system may lead to the disruption of biochemical processes, and it is also known to interfere with enzymatic and hormonal reactions, as well as protein and lipid biosynthesis. Hg is highly neurotoxic and can cause irreversible brain damage which may lead to autism in children and Alzheimer's, Parkinson's, schizophrenia and depression in adults (Zahir et al. 2005; Bose-O'Reilly et al. 2010).

People have used mercury for centuries, but emission of this metal into the environment increased significantly in the twentieth century as a result of industrial development. The main source of mercury pollution of the environment is fossil fuel combustion, but waste and sewage from various branches of industry also have a major impact. Owing to the highly adverse effects of Hg, numerous actions have been undertaken in recent decades aimed at replacing technologies involving mercury with more environmentally friendly alternatives. This has resulted in a much lower Hg inflow into the ecosystem.

Hg is introduced to the marine environment mainly via rivers, which transport pollutants from the drainage area and through atmospheric deposition. For many years, the Baltic Sea was considered to be one of the most polluted basins in the world (Wrembel 1997). However, according to a report by the Baltic Marine Environment Protection Commission, the mercury load introduced to the Baltic has decreased by 44 % since the 1990s (HELCOM 2010). In the case of the Baltic, the volume of Hg inflow is determined by the seven largest rivers that jointly introduce 70 % of the mercury (Wrembel 1993). Two of those large rivers—the Vistula and the Oder—flow through and end in Poland, their basins constituting 12 and 8 % of the Baltic's drainage area, respectively (HELCOM 2004; Galassi et al. 2008). It is currently estimated that the volume of Hg inflow through rivers has fallen by about 16 % since 1990 (HELCOM 2010). In the same period of time, the level of atmospheric Hg deposition has, in turn, dropped by 24 % and the total volume of metals introduced into the Baltic in this way currently amounts to 3.1 t $year^{-1}$ (Bartnicki et al. 2012). The reduction in the volume of Hg introduced into the Baltic Sea has led to decreased concentrations of this metal in these waters.

Organisms have long been used for the evaluation of environmental pollution of the sea. Plants and animals were first used as bioindicators in the 1960s, and benthic plants are considered to be one of the best pollution markers of aquatic environments (Stankovic et al. 2014). These plants absorb substances directly from their surroundings and, being attached to the ground, are permanently exposed to the environmental conditions (including water and sediment pollution) at their sites of growth. Another quality of phytobenthic organisms is their capacity to accumulate elements which do not perform any biological functions in their systems, of which mercury is one (Filipovic-Trajkovic et al. 2012). The accumulation and distribution of mercury in plants depends not only on the concentration of this metal in the environment and the physicochemical conditions but also on the species, the vegetative period, the size of the plant and the build of its root system and leaves (Nagajyoti et al. 2010). As a consequence, metal levels in various plant species occupying the same habitat may vary considerably (Wisłocka et al. 2006).

Macrophytobenthos also plays a key role in the biogeochemical circulation of chemical substances in the marine environment. Organic matter produced by benthic plants is consumed by numerous animals, thereby making these plants an important producer link in the trophic chain. Macrophyta are also one of the main Hg sources for their consumers: invertebrae and fish, both popular foods for humans (Stankovic et al. 2014). Mercury concentration levels in macrophyta and the ability of benthic plants to accumulate it are thus significant issues in studies on the transfer of this toxic metal from abiotic elements of the marine environment to higher levels of the trophic chain, including humans.

In the era of decreasing input of pollutants from anthropogenic sources, an attempt has been made to assess the role of macrophyta as a vector of Hg into marine trophic web, both the one entering the sea at present and that deposited to sediments over the past decades. Studies were conducted in the region of large and small river mouths, as well as away from river mouths (in the gulf waters and in the open sea).

MATERIALS AND METHODS

Study Area

The study was conducted in the Southern Baltic, at 24 stations located in the Polish coastal zone (Figure 1). The study area covered four sections of the Polish coastline: the Pomeranian Bay (Swinoujscie, 53° 54′ 28″ N, 14° 14′ 51″ E; Miedzyzdroje, 53° 55′ 44″ N, 14° 27′ 05″ E), an area of open sea near the coast (Kolobrzeg, 54° 10′ 40″ N, 15° 34′ 37″ E; Ustronie Morskie, 54° 12′ 55″ N, 15° 45′ 17″ E; Ustka, 54° 34′ 43″ N, 16° 52′ 09″ E; Kuznica (open sea), 54° 44′ 01″ N, 18° 34′ 56″ E), the Gulf of Gdansk (Gdynia, 54° 31′ 09″ N, 18° 32′ 22″ E; Orlowo, 54° 28′ 36″ N, 18° 32′ 56″ E; Sopot, 54° 26′ 31″ N, 18° 33′ 35″ E; Sobieszewo, 54° 20′ 11″ N, 18° 52′ 29″ E; Mikoszewo, 54° 20′ 02″ N, 18° 57′ 10″ E; Stegna, 54° 19′ 35″ N, 19° 06′ 44″ E; Katy Rybackie, 54° 20′ 22″ N, 19° 13′ 47″ E; Krynica Morska, 54° 22′ 57″ N, 19° 26′ 40″ E; Piaski, 54° 25′ 54″ N, 19° 36′ 05″ E) and Puck Bay (Hel, 54° 36′ 42″ N, 18° 48′ 29″ E; Jastarnia, 54° 41′ 58″ N, 18° 40′ 36″ E; Kuznica (inner bay), 54° 44′ 01″ N, 18° 34′ 56″ E; Swarzewo, 54° 45′ 31″ N, 18° 23′ 47″ E; Oslonino, 54° 40′ 08″ N, 18° 27′ 24″ E; Rewa, 54° 37′ 58″ N, 18° 30′ 37″ E; Mechelinki, 54° 36′ 32″ N, 18° 30′ 44″ E; Babie Doly, 54° 34′ 52″ N, 18° 32′ 23″ E).

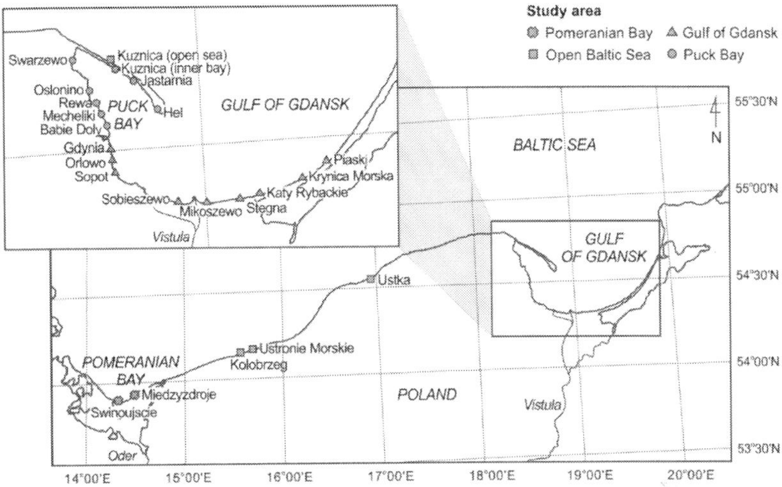

Figure 1: Location of sampling stations in the coastal zone of the Southern Baltic in 2006–2012.

The Pomeranian Bay is located in the southeastern part of the Baltic Sea, formed of a stretch of coast which is part Polish and part German. The environmental quality of this bay is strongly affected by the inflow of highly eutrophicated water from the Szczecin Lagoon, which nevertheless performs the function of a buffer, protecting the water of the Pomeranian Bay from pollutants flowing in from the drainage area of the River Oder. The condition of the Szczecin Lagoon and the coastal waters of the Pomeranian Bay, taking into account the evaluation of both biological and physicochemical parameters, is classified as bad (WIOŚ 2009). This is influenced by the contaminants accumulated in bottom sediments, as well as a constant inflow of pollutants from the drainage area and the vicinity of large urban centers, such as the Szczecin agglomeration (population 750,000). Moreover, Szczecin is a significant economical hub featuring a sea harbor, shipyards, numerous industrial facilities and wastewater treatment plants. An important influence on the water exchange between the Szczecin Lagoon and the Pomeranian Bay is exerted by the constant deepening of the water passage between Swinoujscie and Szczecin.

The Gulf of Gdansk is located in the northeastern part of the Baltic, between Poland and Russia. The main river entering this basin is the Vistula, the largest river flowing through Poland. The Gulf of Gdansk region abounds in anthropogenic pollution sources, and most of these are located within the Tri-city agglomeration, which is situated on the coast with a total population in excess of 1.2 million. The urban centers of this agglomeration are the harbor cities of Gdansk and Gdynia, which together feature numerous facilities related to shipbuilding, petrochemical and energy industry. The western part of the Gulf of Gdansk includes Puck Bay, a small area of water separated from the open sea by the Hel Peninsula. The Puck Bay ecosystem possesses the greatest biodiversity of any area within the Polish coastal zone of the Baltic. In 1978, the Inner Puck Bay became the first sea area to fall under protection as part of the Coastal Landscape Park and, in 1992, it was included in the Baltic Sea Protected Areas (BSPA). It also features on the World Wide Fund for Nature list (WWF).

The stations situated in the open sea area (Kolobrzeg, Ustronie Morskie, Ustka, Kuznica) were located far away from large industrial facilities, in proximity only to small holiday destinations whose populations amount to between 630 (Kuznica) and 47,000 inhabitants (Kolobrzeg).

Sample Collection

Benthic plant samples were collected in the coastal zone in the years 2006–2012. Macrophytobenthos was collected by hand at depths of about 1 m into polyethylene ziploc bags, making sure that each sample contained an entire plant. The collected material was taken to the laboratory, cleansed of epiphytes, organic and mineral particles, and then rinsed in distilled water. Taxonometric analysis was also carried out. The representatives of two species were additionally separated into particular components—roots, stems and leaves in the case of *Potamogeton pectinatus*, and rhizoids and stipes in the case of *Furcellaria lumbricalis*. The plant material was stored at −20 °C until analysis and, immediately prior to testing, was lyophilised and homogenised.

Total mercury concentration (Hg_{TOT}) in macrophytobenthos samples was measured using atomic absorption spectrometry (AAS) on an AMA-245 advanced mercury analyser (Altec Ltd., Czech Republic), following thermal decomposition in pure oxygen. A quality check on the method involved analysing the collected samples in three repeats, together with analysis of certified reference material: BCR-27 (*Ulva lactuca*) and BCR-60 (aquatic plant). The method was characterized by high recovery, amounting to 96–98 %, the relative standard deviation was no higher than 5 %, and the limit of quantification was 0.01 ng g^{-1} dw.

On the basis of the obtained results for total mercury concentrations in macrophytobenthos and its surroundings, it was possible to calculate the bioconcentration factor, which is a parameter reflecting the ability of plant organisms to absorb mercury from water. The bioconcentration factor (BCF) was calculated using the following formula (Szefer and Żbikowski 2010):

$$BCF = \frac{C_X}{C_W}$$

where

C_x: Mercury concentration in wet mass of the organism (ng kg^{-1} ww)

C_w: Mercury concentration in the surrounding water (ng dm^{-3})

In order to determine the mercury BCF for stems, leaves and stems of benthic organisms, near-bottom water was taken into account as the surrounding environment, while in the case of roots and rhizoids, pore water was examined.

Statistical analysis and graphic representation of results was carried out using Statistica 10 software (StatSoft), and the analyzed data was found to be non-parametric (Shapiro-Wilk test, $p<0.05$). In order to determine the relevance of differences, the Mann-Whitney *U* and Kruskal-Wallis tests were used at relevance level of $p=0.05$.

RESULTS AND DISCUSSION

The plant material collected in the Polish coastal zone of the Baltic Sea in the years 2006–2012 consisted of both macroalgae and vascular plants. Among the collected samples, 15 taxa belonging to four phyla were identified—green algae: *Chara* sp., *Cladophora* sp., *Ectocarpus siliculosus*, *Elodea canadensis* and *Enteromorpha* sp.; brown algae: *Chorda filum*, *Fucus vesiculosus*, *Pylaiella littoralis* and *Sphacelaria cirrosa*; red algae: *Furcellaria lumbricalis* and *Polysiphonia* sp.; and flowering vascular plants: *Potamogeton pectinatus*, *Ruppia maritima*, *Zannichellia palustris* and *Zostera marina*. The macrophytobenthic species identified were those typically found in the Southern Baltic coastal region (Pliński and Florczyk 1993; Schubert and Krause 2002).

In all the analysed macrophytobenthos samples, total mercury concentrations (Hg_{TOT}) were above the limit of quantification (Figure 2). The obtained values ranged between 0.8 and 57.2 ng g^{-1} dw (median 7.6 ng g^{-1} dw, mean 10.9 ng g^{-1} dw, $n=248$), while mercury concentrations in vascular plants (median 8.5 ng g^{-1} dw, mean 12.2 ng g^{-1} dw, $n=83$) and macroalgae (median 7.3 ng g^{-1} dw, mean 10.2 ng g^{-1} dw, $n=165$) did not demonstrate differences that were statistically relevant (Mann-Whitney U test, $p=0.21$). In vascular plants, Hg_{TOT} concentrations ranged between 1.3 ng g^{-1} dw for sago pondweed *Potamogeton pectinatus* and 57.2 ng g^{-1} dw for eelgrass *Zostera marina*. In macroalgae, the obtained values ranged from 0.8 ng g^{-1} dw, determined in gut weed *Enteromorpha* sp., to 43.3 ng g^{-1} dw, as determined for black carrageen *Furcellaria lumbricalis* (Table 1, Figure 2).

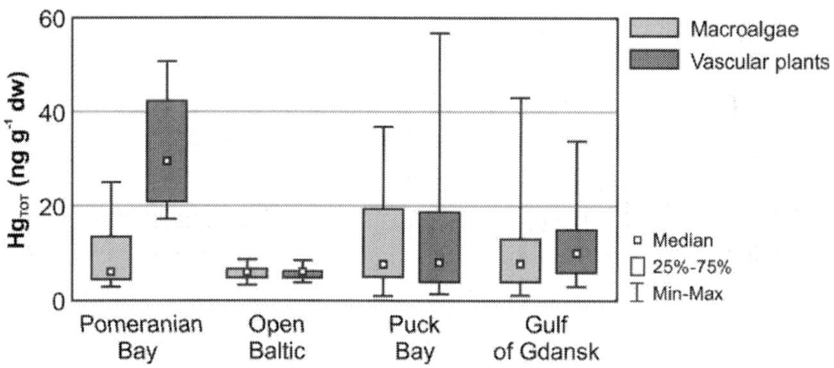

Figure 2: Concentration range of total mercury (Hg_{TOT}) in macrophytobentos: macroalgae and vascular plants, in different regions of the coastal zone of the Southern Baltic in 2006–2012.

Table 1: Median values and concentration ranges of total mercury (Hg_{TOT}, ng g^{-1} dw) in macroalgae (*Chlorophyta*, *Rhodophyta* and *Phaeophyta*) and vascular plants (*Angiospermophyta*) from the Polish coastal zone of the Southern Baltic in 2006–2012

Region	*Chlorophyta*	*Phaeophyta*	*Rhodophyta*	*Angiospermophyta*
Pomeranian Bay	4.7 (3.0–14.3)	6.0 (4.6–7.3)	25.4	29.6 (17.7–50.8)
Open Baltic sea	5.1 (3.8–7.9)	6.3 (4.4–8.4)	7.6 (6.0–8.7)	6.0 (3.7–8.5)
Puck Bay	10.1 (0.8–37.1)	5.2 (4.3–6.8)	11.0	8.3 (1.3–57.2)
Gulf of Gdansk	7.1 (0.9–29.9)	9.1 (4.9–15.6)	7.8 (1.3–43.3)	9.3 (2.8–33.8)

Hg_{TOT} concentrations in the macrophytobenthos of the Southern Baltic coastal zone were found to be much lower in 2006–2012 than those determined in 1995–1998 in the Puck Bay region (Boszke et al. 2003). This is mainly due to the fact that mercury inflow into the Baltic has decreased as the result of a systematic reduction of metal emissions into the environment (HELCOM 2004, 2010). This resulted in a drop in Hg concentration in sea water (Wrembel 1997; Saniewska et al. 2010) and, as a consequence, also in macroalgae. The Hg concentrations in macroalgae presented in this paper were from three to ten times lower than those determined in previous years (Boszke et al. 2003). In the case of vascular plants, which are rooted to the seabed, the decrease in metal concentration was lower with Hg_{TOT} values coming in at 2–2.5 times lower than those from the 1990s. This may be attributed to the fact that mercury concentration in surface sediments is not reduced as rapidly as in surface water (Jędruch et al. 2013b).

Geographic Variation of Hg_{TOT} Concentration

Total mercury concentrations determined in macrophytobenthos collected in the Polish coastal zone were characterized by variability in each study area. Hg_{TOT} concentrations in the organisms from the open sea coastal area were lower in terms of statistical relevance than in the organisms collected in the Pomeranian Bay, the Gulf of Gdansk or Puck Bay (Kruskal-Wallis test, $p=0.03$), and also had the most limited range (Figure 2). In vascular plants, Hg_{TOT} concentrations in the coastal zone of the open sea (median 6.0 ng g^{-1} dw) were lower by 35 % when compared to the other, more sheltered parts of the basin (median 9.2 ng g^{-1} dw). In macroalgae (median 6.1 ng g^{-1} dw), the obtained values were 20 % lower than those from within the bays (median 7.6 ng g^{-1} dw). These differences were mainly due to mercury concentration levels in the surrounding environment (Broekart et al. 1990) and while the concentration of this metal determined in the coastal water of the open sea was similar to values observed in the Baltic Proper (Pempkowiak et al. 1998; Murawiec et al. 2007; Bełdowski et al. 2009), Hg_{TOT} concentrations found within the Gulf of Gdansk and Puck Bay were nearly twice as high (Saniewska et al. 2010). A similar tendency was also observed for the concentration of heavy metals in zooplankton of the Southern Baltic (Pempkowiak et al. 2006). The lower mercury concentrations noted in the coastal waters of the open sea were influenced by, among other factors, the distance from anthropogenic sources of metal emission and outlets of large rivers—total annual run-off from rivers flowing directly into the Baltic Proper is seven times smaller than run-off from the River Vistula and four times smaller than that of the Oder (Bogdanowicz 2009; Niemirycz 2011).

In the Pomeranian Bay, the differences between total mercury concentrations in vascular plants (median 29.6 ng g^{-1} dw) and macroalgae (median 6.0 ng g^{-1} dw) were statistically relevant (Mann-Whitney U test, p=0.03), while concentrations in the higher plants exceeded values found in the other studied regions of the Southern Baltic by over threefold (Figure 2). Higher Hg_{TOT} concentrations in vascular plants, which absorb chemical substances additionally via the root system (Lack and Evans 2005), may be attributed to increased metal concentration in surface sediment compared to other areas of the Polish coastline. Mercury is introduced to the Pomeranian Bay primarily by the River Oder and the waters of the Szczecin Lagoon, one of the most polluted regions of the Polish coast in terms of heavy metals (Glasby et al. 2004; Szefer et al. 2009). Concentrations of these metals in macrophytes have a significant influence on the levels found in organisms occupying higher trophic levels: fish and marine birds. Owing to their placement in the marine food web and the fact that mercury is introduced into their systems mainly through the alimentary canal, birds are held to be good indicators of environmental mercury pollution (Kalisińska et al. 2004). Research carried out in 2004–2005 in the Szczecin Lagoon confirmed the existence of a positive correlation between mercury concentrations in the ingesta and in the tissues of the mallard duck *Anas platyrhynchos*, whose diet is based on phytobenthos (Lisowski 2009). Furthermore, Hg_{TOT} concentrations in the soft tissues of *A. platyrhynchos* (liver 270 ng g^{-1} dw, kidney 250 ng g^{-1} dw, muscle 130 ng g^{-1} dw) were nearly twice as high as those determined for the black-headed gull *Larus ridibundus* from the Gulf of Gdansk, whose diet is composed largely of fish, in 2009–2012 (liver 130 ng g^{-1} dw, kidney 162 ng g^{-1} dw, muscle 60 ng g^{-1} dw) (Kalisińska et al. 2013; Szumiło et al. 2013). High mercury concentrations in *A. platyrhynchos* are indicative of a level of mercury pollution in the Pomeranian Bay which could lead to deterioration of bird's health: disruptions in visual-motor coordination, spatial orientation or a significant reduction in terms of breeding success (Scheuhammer et al. 2008). This is proven by concentrations in the soft tissues of the mallard duck *A. platyrhynchos* exceeding the threshold value of 250 ng g^{-1} dw, above which adverse effects of mercury can be observed in organisms (Kalisińska et al. 2013). This is particularly important in the case of *A. platyrhynchos* as it is the most common gamebird consumed in Poland, the number of specimens shot annually reaching up to 120,000 (Książkiewicz 2006).

Increased mercury concentrations, in comparison with the open sea coastal area, were also observed in vascular plants (median 10.5 ng g^{-1} dw) in the Gulf of Gdansk region (Figure 2). This was mainly as a result of close proximity to the outlet of the Vistula, the largest Polish river and the second largest river flowing into the Baltic (after the Neva) in terms of drainage area size and level of run-off (HELCOM 2004). As noted in previous studies, mercury concentrations in the Vistula (median 6.3 ng dm^{-3}, mean 7.3 ng dm^{-3}) are slightly higher than the global average for unpolluted rivers (mean 5.0 ng dm^{-3}) (Mason et al. 1994; Saniewska et al. 2014). Moreover, owing to its large flow of water, the Vistula contributes a significant mercury load to the Gulf of Gdansk, including the bioavailable dissolved form of the metal which is washed off the drainage area (Saniewska et al. 2014). Additional sources of Hg are storm drain outlets, which discharge the run-off from highly urbanised areas such as the Tri-city agglomeration (Saniewska 2013). Hg_{TOT} concentrations determined in macroalgae (median 7.6 ng g^{-1} dw), absorbing mercury mainly from the near-bottom water, were, as in the Pomeranian Bay, lower than in the more advanced plants. This is probably related to the fact that mercury concentrations determined in pore water were almost twice as high as in surface water (Bełdowska et al. 2013b).

The total mercury concentration in the rooted vascular plants from Puck Bay (median 8.3 ng g^{-1} dw) was 20 % lower than in the Gulf of Gdansk (Figure 2). This is likely to be connected to the physicochemical qualities of Puck Bay, which are conditioned mainly by shallow water depth and the accumulation of organic matter. As a consequence, the near-bottom zone of the bay is characterized by low oxygen levels and periodical occurrence of hydrogen sulfide. In reductive conditions such as these, precipitation of the insoluble mercury sulfide may have occurred in sediments, resulting in a reduction of mercury bioavailability for phytobenthos. Numerous studies confirm that the concentration of mercury dissolved in pore water increases with a rise in Eh potential and decreases as H_2S concentration grows. Moreover, it has been found that the insoluble HgS is one of the main metal forms in the sediment of Puck Bay (Bełdowski and Pempkowiak 2007; Bełdowska et al. 2013c; Jędruch et al. 2013b). In the case of Hg_{TOT} concentration in the macroalgae of Puck Bay (median 7.9 ng g^{-1} dw), no statistically relevant differences were observed in relation to the values determined in the Gulf of Gdansk (Mann-Whitney *U* test, $p=0.16$) (Figure 2). Although Puck Bay experiences little anthropogenic influence and is at a considerable distance from the outlet of the River Vistula, the rivers flowing into it nevertheless turned out to be a significant source of the bioavailable mercury form. These watercourses, transporting mercury washed off urban areas, cultivated fields and forests could have had a significant influence on metal concentrations close to their outlets (Saniewska 2013).

Mercury in *Zostera marina*

The influence of increasing distance from river outlets on Hg concentration levels in macrophytobenthos was particularly noticeable in one vascular plant, *Zostera marina*, collected at most of the test stations. Hg_{TOT} concentrations in this seagrass showed statistically relevant differences at the particular test stations (Kruskal-Wallis test, $p=0.02$), and the highest values were determined close to river outlets (Figure 3a). Studies on mercury speciation in suspended matter, carried out by Bełdowski (2004) in the studied region, showed that the highest proportion of labile Hg forms in suspended matter occurred in the vicinity of river outlets. The maximum Hg_{TOT} concentrations were found in the *Zostera marina* samples collected in Miedzyzdroje (median 50.8 ng g^{-1} dw) and Swinoujscie (median 34.9 ng g^{-1} dw) (Figure 3a), both located close to the mouth of the River Oder, the second largest river in Poland and the third largest river flowing into the Baltic in terms of drainage area (HELCOM 2004). The fact that the water from the Oder flows into the Szczecin Lagoon first causes a large proportion of suspended matter to deposit in the coastal zone, resulting in a rise in pollution. Mercury concentrations which were higher than the Hg median for the study area were also found in the seagrass collected at the Mikoszewo station, located a short distance from the River Vistula outlet. Despite it being the largest river in Poland, however, Hg concentrations in the area of its outlet were not the highest. This was related, among other things, to its strong current transporting the metal into further regions of the sea (Bełdowski and Pempkowiak 2007; Damrat et al. 2013). Even though the mercury load introduced to the Baltic by large rivers is considerable, small rivers can also constitute an ecological problem on a local scale, particularly in the coastal zone (Cyberski et al. 1993; Niemirycz 2011). This is indicated by increased concentrations of the metal in *Zostera marina* in the outlet areas of smaller rivers, at the stations in Swarzewo (River Plutnica), Oslonino (River Gizdepka), Rewa (River Reda, Zagorska Struga, drainage canals) and Orlowo (River Kacza) (Figure 3a). Most mercury of river origin is related to suspended matter and, in the case of small rivers, this tends to undergo sedimentation in the coastal zone close to the outlet (Cossa and Martin 2001; Saniewska 2013; Saniewska et al. 2014). Mercury from coastal surface sediments may then be released into pore water as a result of its oxidation to soluble sulfates (Bełdowski et al. 2009), enabling it to be absorbed by rooted vascular plants.

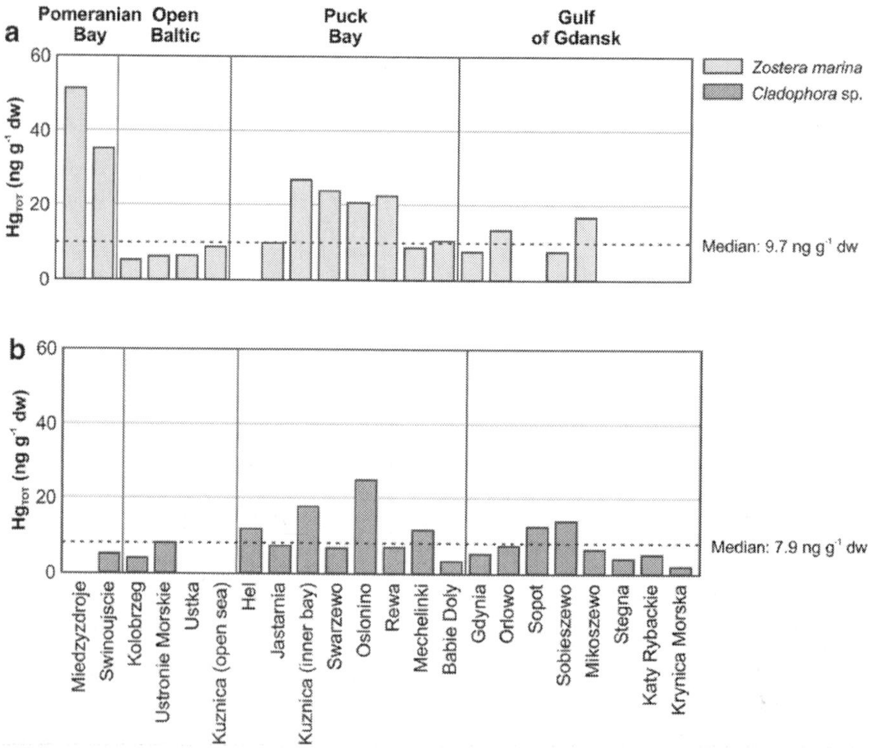

Figure 3: Median values of total mercury (Hg_{TOT}) concentration in *Zostera marina* (**a**) and *Cladophora* sp. (**b**) at different sampling stations in the coastal zone of the Southern Baltic in 2006–2012.

Mercury in *Cladophora* sp.

Proximity to point sources of dissolved mercury had an effect on increased Hg_{TOT}concentrations in macroalgae, exemplified here by *Cladophora* sp., which was collected at most test stations. Concentrations of Hg in this alga displayed statistically relevant spatial variability (Kruskal-Wallis test, $p=0.04$), and the highest concentrations were found in the Oslonino area (median 24.7 ng g^{-1} dw) (Figure 3b). This probably resulted from the presence in the water of mercury in its bioavailable form and its subsequent bioaccumulation in the macroalgae. The mercury concentrations determined in surface water, plankton organisms, surface sediments and pore water in the Oslonino coastal zone were also found to be higher in comparison with the rest of the Gulf of Gdansk coast

(Bełdowska et al. 2013a, d). This may be attributed to the inflow of Hg with the River Gizdepka (Saniewska 2013), and the sediments derived from eroded cliff into the sea. As shown by studies carried out in 2011–2013, a significant load of mercury deposited on land over the course of many years is introduced into the Gulf of Gdansk as a result of coastal erosion (Jędruch et al. 2013c). The level of Hg concentration in the coastal area of Oslonino was also influenced by the particular physicochemical conditions of the test station, which is located in a small bay where intense accumulation of decomposing organic matter takes place as a result of shallow depth and low water dynamics.

Increased Hg_{TOT} concentrations were also found in *Cladophora* sp. samples collected in the Hel region (Figure 3b), and this may have resulted from the proximity of the station to fishing and navy ports. Mercury concentrations in excess of the median value were also detected in Sopot, close to storm drain outlets which introduce pollutants from the Tri-city agglomeration into the basin (Cyberski et al. 1993; Saniewska 2013), and in Mechelinki, where the source of dissolved mercury was probably wastewater discharged from the collector of the Debogorze treatment plant. The influence of this plant on the inflow of bioavailable mercury into the basin has already been confirmed by analyses of metal speciation in surface sediments in the Mechelinki region, which indicate a large proportion of dissolved forms of mercury, including those related to organic matter—fulvic and humic acids (Bełdowski and Pempkowiak 2007).

High concentrations of this metal were also determined in the area of Sobieszewo and were probably related to two bird sanctuaries (Ptasi Raj and Mewia Lacha) located on the island, hosting many different bird species. Mercury can be removed from the systems of birds nesting in these sanctuaries through moulting, excretion with guano or deposition in eggs. These congregations of birds and the consequent presence of guano have become sources introducing toxic substances into the sea water, as confirmed by Falkowska et al. (2007). Additionally, research has shown that a female bird can remove about 5 % of the mercury consumed with food in the laying of eggs (Falkowska et al. 2013a), thus re-introducing Hg back into the trophic chain.

Mercury in the Particular Parts of Macrophytae

Trace metals can be absorbed by benthic plants both through the roots, from pore water, and through leaves and stems, from the surrounding water (Conquery and Welbourn 1994). The diverse ability to absorb mercury from the environment is exemplified here by the vascular plant *Potamogeton pectinatus*, from the Oslonino region, and the red alga *Furcellaria lumbricalis*, collected in the Orlowo region. Both of these species of macrophytobenthos were characterised by statistically relevant variability of mercury BCF (U Mann–Whitney test, $p=0.01$). In *Potamogeton pectinatus*, the BCF from the near-bottom water (median 345.9) was much higher than in *Furcellaria lumbricalis* (median 204.4), indicating a greater ability to accumulate Hg in vascular plants compared to macroalgae. What is more, in vascular plants mercury absorption from the near-bottom water through the parts of the plant that grow above the sediment, occurred seven times more effectively than metal absorption from the pore water through the roots (median 45.2) (Mann-Whitney *U* test, $p=0.01$). This is also confirmed by the Hg_{TOT} levels determined in the particular parts of *Potamogeton pectinatus*, concentrations in the leaves (median 5.1 ng g^{-1} dw, $n=43$) being significantly higher than those found in the roots (median 2.6 ng g^{-1} dw, $n=29$) and stems (median 2.4 ng g^{-1} dw, $n=30$) (Kruskal-Wallis test, $p=0.00$) (Figure 4a). Both the highest mercury concentrations and the highest BCF value (median 395.2) were determined in the leaves of *Potamogeton pectinatus* and resulted from simultaneous absorption of metal from the near-bottom water and transportation into the leaves of Hg absorbed through the roots from pore water (Jackson 1998; Lack and Evans 2005). Increasing concentrations of this metal in the leaves of marine vascular plants, in comparison with the rest of the plant, have also been observed by other researchers (Pergent-Martini 1998; Göthberg et al. 2004; Lafabrie et al. 2011). Higher mercury bioconcentration was noted in the aboveground parts of *Potamogeton pectinatus*, and this was mainly due to the greater absorption surface of the leaves. However, it could also have resulted from the defensive action of the organism, which involves restricting the penetration of metal ions into the roots if mercury concentration in sediments increases (Jędruch et al. 2013b). In the case of macroalgae, which are not rooted to the ground but only attached to it with the use of rhizoids, mercury is absorbed mainly from the surrounding water. This explains the lower

BCF values calculated for *Furcellaria lumbricalis* in comparison with a vascular plant. Furthermore, no statistically relevant differences were found for the BCF values of the rhizoids (median 192.3) and the stipe of the red alga (median 246.4) (Mann-Whitney U test, $p=0.70$). Hg_{TOT} concentration was also evenly distributed in the constituent parts of the plant (Mann-Whitney U test, $p=0.31$), with the level measured in the rhizoids (median 2.9 ng g^{-1} dw; $n=11$) being similar to that of the stipe (median 3.0 ng g^{-1} dw, $n=23$) (Figure 4b). This may be explained by the fact that thallus tissues in macroalgae are not very diversely built, resulting in uniform absorption of mercury through the rhizoids and stipe (Lack and Evans 2005). The differing ability of vascular plants and macroalgae to absorb mercury from the surrounding environment may have an effect on the circulation of this metal in the marine environment. Higher Hg concentrations in vascular plants suggest that, in areas where the seabed is covered in those organisms, mercury will accumulate more effectively than in other areas.

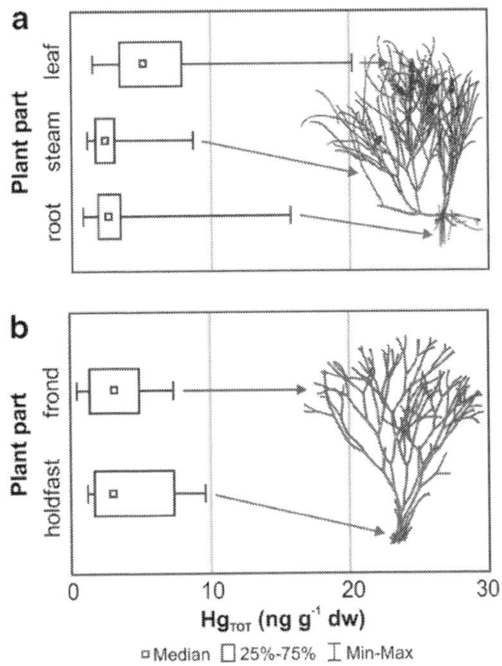

Figure 4: Concentration range of total mercury (Hg_{TOT}) in the different parts of *Potamogeton pectinatus* (**a**) and *Furcellaria lumbricalis* (**b**) from the coastal zone of the Southern Baltic in 2006–2012.

Mercury in Macrophyta Biomass from Puck Bay

The role of macrophytobenthos in the process of introducing pollutants into the trophic chain depends on the state of the marine environment, which can have an influence on conditions limiting the development of seabed flora. In the years 1950–1980, as a result of factors including eutrophication of the basin, there was a deterioration in water quality which caused the macrophytobenthos coverage of the Baltic seabed to be considerably reduced. At the same time, areas covered in benthic plants decreased by as much as 90 % in some parts of the basin (Boström et al. 2014). In recent years, however, some restoration of the seagrass bed of the Baltic has been observed, including within the Polish coastal area (Kruk-Dowgiallo 1991; Gic-Grusza et al. 2009). This has been confirmed by, among others, the results of an ecological assessment study into the condition of seabed flora, conducted by Saniewski (2013) in 2011, in which the rating ranged from moderate inside the bays to good in the shallow area of the central coast. Those results also indicated that some species previously considered as rare (i.e., the brown alga *Polysiphonia* sp. and the red alga *Furcellaria lumbricalis*) had previously covered quite a large part of the seabed in the Polish coastal zone of the Baltic and that their proportion in the total macroalgae biomass had in some areas exceeded 80 % (Osowiecki et al. 2009; Saniewski 2013). This is of particular importance as, in the present study, brown and red algae were characterised by significantly higher Hg_{TOT} concentrations than macroalgae from the Chlorophyta group (Kruskal-Wallis test, $p=0.00$) (Table 1). Increased Hg_{TOT}concentrations in brown and red algae can be explained by the presence of charged polysaccharides in the cell walls (Mamboya 2007). As shown by research, the occurrence of alginates and sulphades fucans with higher affinity for cations, and the presence of physodes containing phenolic compounds, can influence the ability of macroalgae to absorb metals from the surrounding environment (Andrade et al. 2002; Farina et al. 2003; Salgado et al. 2005). Research carried out in the coastal zone of the Baltic also indicated a greater ability of red and brown algae, in comparison with green algae, to bioaccumulate radionuclides, as proven by measurement of the ^{137}Cs and ^{90}Sr isotopes in their tissues (Zalewska and Saniewski 2011; Saniewski 2013). This shows how important these groups are to the circulation of pollutants in the marine environment.

A good quality of water is also reflected by the presence of vascular plants, which are sensitive to eutrophication (Krause-Jensen et al. 2008), and in many areas of the Baltic, the seagrass bed is much more widely distributed than previously reported (Boström et al. 2003, 2014). In the Polish coastal zone, clusters of vascular plants cover the seabed especially in the Inner Puck Bay (Figure 5) and the most valuable species present, in ecological terms, is *Zostera marina*, the clusters of which, according to research carried out in 2007–2009, covered 3 % of the basin's area (Węsławski et al. 2013) (Figure 5). The area of the Puck Bay seabed covered in *Zostera marina* has been increasing in recent years and is now six times larger than it was in the 1980s (Kruk-Dowgiallo 1991; Gic-Grusza et al. 2009; Węsławski et al. 2013). Research carried out in 2010–2011 found the proportion of this species present in the total plant biomass of the seagrass bed (approximately 50 g dw m^{-2}) to be as high as 80 %, while the proportions of particular parts in the biomass of the species were found to be about 60 % for aboveground parts and 40 % for belowground (Jankowska et al. 2014). Taking into account the Hg concentration median for this species, it was estimated that in 1 m^2 of the seabed covered in seagrass, up to 390 ng of total mercury was accumulated, out of which 215 ng m^{-2} was accounted for by the above-sediment parts of the plant (median Hg_{TOT} concentration 8.9 ng g^{-1}dw), and 175 ng m^{-2} by the roots (median Hg_{TOT} concentration 10.9 ng g^{-1} dw) (Figure 5). A sizeable portion of the sea bottom in Puck Bay (53 %) is covered in an assortment of brackish-water flora: *Potamogeton pectinatus*, *Ruppia maritima* and *Zannichellia palustris* (Węsławski et al. 2013). Their combined biomass was twice as high as in the case of seagrass and amounted to about 16 g dw m^{-2} (Brzeska and Saniewski 2012) which, taking into account mercury concentration in plant tissues (median Hg_{TOT} concentration 5.0 ng g^{-1} dw) translates to 81 ng of metal per 1 m^2 of the seabed (Figure 5). Another valuable feature of the Inner Puck Bay are clusters of *Charophyta*, covering 6 % of the sea bottom area of this basin (Węsławski et al. 2013). Despite their low biomass, amounting to about 6 g dw m^{-2} (Brzeska and Saniewski 2012), in 1 m^2 of seabed covered in charophytae, the level of mercury accumulation (82 ng m^{-2}) is similar to that observed in clusters of assorted vascular plants (Figure 5). This is related to the presence of higher mercury concentrations in those organisms (median Hg_{TOT} concentration 13.6 ng g^{-1} dw). The mercury accumulated in marine plants is accessible to organisms from a higher level of the trophic chain. Hg_{TOT} values in macrophytobenthos biomass, ranging from double figures to several hundred nanograms per 1 m^2 of the sea bottom, were an order of magnitude lower than the total mercury concentrations determined in zoobenthos biomass in Puck Bay (Jędruch et al. 2013a). Moreover, Hg_{TOT} values in zoobenthic biomass were found to be the highest in samples containing the largest proportion of *Idotea* sp.—the main consumer of seagrass in the Baltic (Baden et al. 2010).

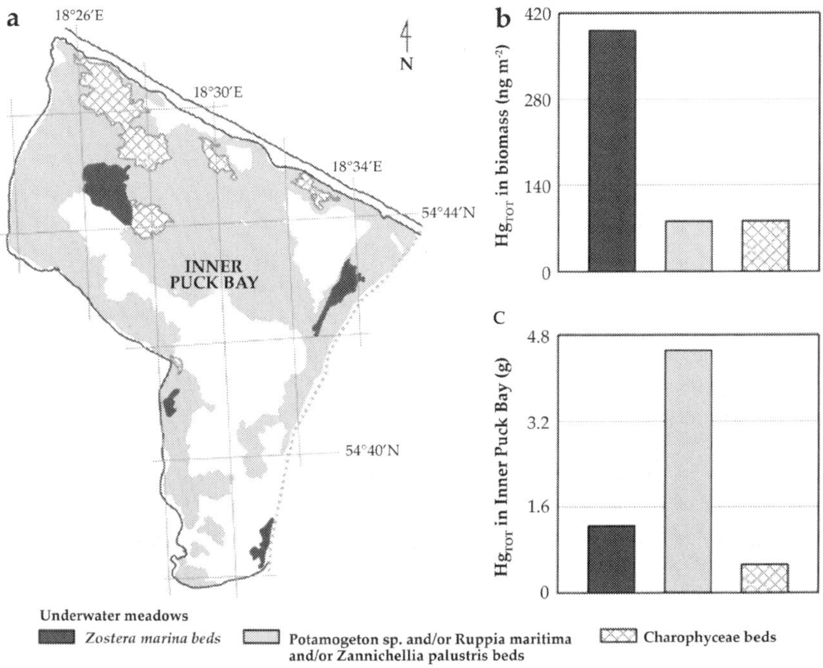

Figure 5: Distribution of macrophyte beds in the Inner Puck Bay, based on Gic-Grusza et al. (2009) (**a**), and median values of total mercury (Hg_{TOT}) in biomass in different types of meadows (**b**) with the amount of Hg_{TOT} accumulated by them (**c**).

Taking into consideration the concentration of mercury in macrophytobenthic biomass and the area of sea bottom covered by it (Gic-Grusza et al. 2009; Węsławski et al. 2013), it was possible to calculate the metal load accumulated in these plants (Figure 5). In the Inner Puck Bay, seabed flora may jointly contain over 6 g of mercury, of which widespread assorted clusters of *Potamogeton pectinatus*, *Ruppia maritima* and *Zannichellia palustris* account for 4.5 g, clusters of *Zostera marina* seagrass account for 1.3 g, and Charophyceae account for 0.5 g (Figure 5). Due to a complete lack of exact data concerning macrophytobenthic biomass in the basin, the calculated load does not take into account other plant clusters, including macroalgae covering the hard seabed, numerous string algae or other vascular plant species. The increased occurrence and subsequent rise in biomass of vascular plants which has been observed in Polish coastal waters in recent years may therefore play a significant role in delivering heightened levels of mercury to the trophic web and increasing concentrations of this metal in more advanced organisms. Another factor influencing this phenomenon is the prolonged vegetative season, which is related to climate changes in the Polish coastal area (Kożuchowski 2009; Bełdowska et al. 2013b). The absence of ice cover on the bay allows Hg absorption from water and sediments to continue for longer, in turn causing the metal to remain in the trophic chain for an extended period of time.

Mercury in Beach-Cast Macophytobenthos Biomass

In many areas of the Baltic, owing to the distribution of currents and the shape of the coastline, large amounts of macrophytobenthos accumulate in the coastal zone or are cast onto beaches. During the summer season in the Gulf of Gdansk, macroalgae accumulation along 1 km of beach can range from several dozen up to as much as 800 t (Filipkowska et al. 2008; Blidberg and Gröndahl 2012). Taking into account the median for Hg_{TOT} concentration 7.6 ng g^{-1} dw, it was estimated that 1 km of beach can accumulate over 6 g of mercury over the course of a few months. Studies into the speed of coast alternations in the Southern Baltic region show that 39 % of the length of the Polish coast is of an accumulative nature (Dubrawski and Zawadzka 2006). This equates to about 200 km of coastline along which processes favourable to the accumulation of phytobenthos may occur. In the summer season, therefore, the total mass of all seabed plants to have been washed up on Polish beaches may contain as much as 1.2 kg of mercury.

In the vicinity of seaside resorts (e.g., Sopot), macrophytobenthos is collected by relevant cleaning services and then taken away to be deposited in designated areas on the outskirts of the city. However, along stretches of coast less frequented by tourists (e.g., Oslonino and

Rzucewo), it is left to accumulate rapidly in the coastal zone, where it serves as an additional source of food for crustaceans, molluscs and birds. What is more, mercury accumulated in beached phytobenthos can be re-emitted into the atmosphere, and this process is at its most intense in the summer season, when the highest temperatures and solar radiation levels are to be observed (Marks and Bełdowska 2001; Lindberg et al. 2005; Falkowska et al. 2013b).

CONCLUSIONS

The urbanization of the coastal zone has contributed to a rise in Hg concentration in macroalgae. Surface run-off entering bays together with the inflow of rainwater from sources such as storm drains, has led to higher mercury concentrations in *Cladophora* sp. in such areas in comparison with areas located further away. Among macroalgae, those which accumulated Hg the most effectively in the Southern Baltic were brown and red algae, a finding related to the chemical composition and construction of the cell wall of the thallus. The inflow of pollutants from river drainage areas also contributed to an increase in the concentrations of this metal in vascular plants, notably *Zostera marina*. Suspended particulate matter introduced to the sea via rivers, particularly in bays, was deposited close to the shore and became a source of Hg entering pore water and, in turn, vascular plants. Additionally, the capacity of vascular plants to accumulate chemical substances from pore water led to the accumulation of historical mercury deposits in the sediments. As a result, *Potamogeton pectinatus* demonstrated a bioconcentration factor 60 % higher than that of *Furcellaria lumbricalis*. Close to the outlets of the two rivers possessing the largest drainage areas in the Baltic (the Vistula and the Oder), no increased mercury concentrations were observed in macrophyta.

In the Polish coastal zone of the Southern Baltic, the reduction of Hg emissions into the environment has led to a decrease in concentrations of this metal in macrophyta. At the same time, however, the growing coverage of red algae on the seabed, stimulated by the increase in environmental quality and a prolonged vegetative season, results in both contemporary mercury from the land and historical deposits of the metal becoming rapidly introduced into the trophic chain. Seagrass beds provide an environment in which animal organisms thrive, often on consumption of phytobenthos (which accumulate Hg directly), and thus rapid growth of these plants may contribute to faster inclusion of Hg in the trophic chain. This is particularly important in areas where seagrass beds of *Zostera marina* are reviving, as their biomass was found to contain the most mercury. Therefore, although these macrophytae purify surrounding water and sediment, they are also responsible for the transfer of Hg, to more advanced organism on higher levels of the trophic chain.

ACKNOWLEDGMENTS

This study has been performed in the frame of the National Science Center project number 2011/01/B/ST10/07697.

REFERENCES

1. Andrade LR, Farina M, Amado Filho GM (2002) Role of Padina gymnospora (Dictyotales, Phaeophyceae) cell walls in cadmium accumulation. Phycologia 41:39–48. doi:10.2216/i0031-8884-41-1-39.1
2. Baden S, Boström C, Tobiasson S, Arponen H, Moksnes P-O (2010) Relative importance of trophic interactions and nutrient enrichment in seagrass ecosystems: a broad-scale field experiment in the Baltic–Skagerrak area. Limnol Oceanogr 55:1435–1448
3. Bartnicki J, Gusev A, Aas W, Valiyaveetil S (2012) Atmospheric supply of nitrogen, lead, cadmium, mercury and dioxins/furans to the Baltic Sea in 2010. EMEP/MSC-W Technical Report 2/2012, EMEP Centres Joint Report for HELCOM, Olso
4. Bełdowska M, Bełdowski J, Jędruch A, Szubska M, Kobos J, Mudrak-Cegiołka S, Zgrundo A, Ziółkowska M, Graca B, Jankowska K, Kotlarska E, Kiełczewska J (2013a) Mercury concentration in the coastal zone of the Gulf of Gdansk as a function of changing climate—preliminary results. In: Vaitkevičienė V (ed) Baltic Sea Science Congress: new horizons for Baltic Sea science. Klaipeda University CORPI, Klaipeda, p 140
5. Bełdowska M, Jędruch A, Bełdowski J, Szubska M (2013b) Mercury concentration in the sediments as a function of changing climate in coastal zone of Southern Baltic Sea—preliminary results. E3S Web

of Conferences 1:06002. doi:10.1051/e3sconf/20130106002

6. Bełdowska M, Jędruch A, Bełdowski J, Szubska M, Kobos J, Mudrak-Cegiołka GB, Zgrundo A, Ziółkowska M, Kiełczewska J, Lewandowska E, Wąsowska K, Falkowska L (2013c) Mercury at coastal zone of Puck Bay (southern Baltic). In: Falkowska L (ed) Mercury in the environment—identification of threats to human health. University of Gdansk, Gdansk, pp 97–103 (in Polish)
7. Bełdowska M, Jędruch A, Kobos J, Mudrak-Cegiołka S, Bełdowski J, Szubska M, Lewandowska A, Falkowska L (2013d) Mercury concentration in phytoplankton and zooplankton in the coastal zone of the Gulf of Gdansk, through the year. In: Vaitkevičienė V (ed) Baltic Sea Science Congress: new horizons for Baltic Sea science. Klaipeda University CORPI, Klaipeda, p 141
8. Bełdowski J (2004) Conditions and the importance of mercury speciation in bottom sediments of the western part of the Gdansk Basin. PhD Thesis, University of Gdansk, Gdansk (in Polish)
9. Bełdowski J, Pempkowiak J (2007) Mercury transformation in marine coastal sediments as derived from mercury concentration and speciation changes along source/sink transport pathway (Southern Baltic). Estuar Coast Shelf Sci 72:370–378. doi:10.1016/j.ecss.2006.10.007
10. Bełdowski J, Miotk M, Pempkowiak J (2009) Mercury fluxes through the sediment water interface and bioavailability of mercury in southern Baltic Sea sediments. Oceanologia 51:263–285
11. Blidberg E, Gröndahl F (2012) Macroalgae harvesting and cultivation. In: Schultz-Zehden A, Matczak M (eds) SUBMARINER compendium. An assessment of innovative and sustainable uses of Baltic marine resources. Maritime Institute of Gdansk, Gdansk, pp 49–76
12. Bogdanowicz R (2009) Water resources of coastal rivers and their variability. In: Bogdanowicz R, Fac-Beneda J (eds) Water resources and water protection. FRUG, Gdansk, pp 47–62
13. Bose-O'Reilly S, McCarty KM, Steckling N, Lettmeier B (2010) Mercury exposure and children's health. Curr Probl Pediatr Adolesc Health Care 40:186–215. doi:10.1016/j.cppeds.2010.07.002

14. Boström C, Baden S, Krause-Jensen D (2003) The seagrasses of Scandinavia and the Baltic Sea. In: Green EP, Short FT (eds) World atlas of seagrasses: present status and future conservation. UNEP, Berkeley, pp 27–37
15. Boström C, Baden S, Bockelmann AC, Dromph K, Fredriksen S, Gustafsson C, Krause-Jensen D, Möller T, Nielsen S, Olesen B, Pihl L, Rinde E (2014) Distribution, structure and function of Nordic eelgrass (Zostera marina) ecosystems: implications for coastal managements and conservation. Aquatic Conserv. doi:10.1002/aqc.2424
16. Boszke L, Siepak J, Falandysz F (2003) Total mercury contamination of selected organisms in Bay of Puck, Baltic Sea, Poland. Pol J Environ Stud 12:275–285
17. Broekart JAC, Gucer S, Adamus F (1990) Metal speciation in the environment. Springer, Berlin
18. Brzeska P, Saniewski M (2012) Phytobentos. In: Zalewska T, Jakusik E, Łysiak-Pastuszak E, Krzymiński W (eds) Southern Baltic Sea in the year 2011. Characteristics of selected environmental compartments. IMGW, Warszawa, pp 120–128
19. Conquery M, Welbourn PM (1994) Mercury uptake from contaminate water and sediment by rooted and submerged aquatic macrophyte Eriocaulon septangulare. Arch Environ Contam Toxicol 26:335–341. doi:10.1007/BF00203560
20. Cossa D, Martin JM (2001) Mercury in Rhône delta and adjacent marine areas. Mar Chem 36:291–302. doi:10.1016/S0304-4203(09)90067-6
21. Cyberski J, Jędrasik J, Kowalewski M (1993) Transport of pollutants. In: Korzeniewski K (ed) The Bay of Puck. University of Gdansk, Gdansk, pp 520–532 (in Polish)
22. Damrat M, Zaborska A, Zajączkowski (2013) Sedimentation from suspension and sedimentaccumulation rate in the River Vistula prodelta, Gulf of Gdańsk (Baltic Sea). Oceanologia 55(4):937–950. doi:10.5697/oc.55-4.937
23. Dubrawski R, Zawadzka E (2006) Characterization and development of sea coasts. In: Dubrawski R, Zawadzka E (eds) Future of polish

sea coast protection. Maritime Institute of Gdansk, Gdansk, pp 12–45 (in Polish)

24. Falkowska L, Bełdowska M, Kwaśniak J, Zawalich K, Miotk M (2007) Mercury in guano from the coast of the Gulf of Gdansk. Oceanol Hydrobiol Stud 36:98–108
25. Falkowska L, Reindl AR, Szumiło E, Kwaśniak J, Staniszewska M, Bełdowska M, Lewandowska A, Krause I (2013a) Mercury and chlorinated pesticides on the highest level of the food web as exemplified by herring from the Southern Baltic Sea and African penguins from the zoo. Water Air Soil Pollut 224:1549. doi:10.1007/s11270-013-1549-6
26. Falkowska L, Witkowska A, Bełdowska M, Lewandowska A (2013b) Waste disposal sites as sources of mercury in the atmosphere in the coastal zone of the Gulf of Gdańsk (Southern Baltic Sea). Oceanol Hydrobiol Stud 42:99–109. doi:10.2478/s13545-013-0064-0
27. Farina M, Keim CN, Corrêa J, Andrade LR (2003) Microanalysis of metal-accumulating biomolecules. Microsc Microanal 9:1512–1513. doi:10.1017/S1431927603447569
28. Filipkowska A, Lubecki L, Szymczak-Żyła M, Kowalewska G, Żbikowski R, Szefer P (2008) Utillisation of macroalgae from Sopot beach (Baltic Sea). Oceanologia 55:255–273
29. Filipovic-Trajkovic R, Ilic SZ, Sunic L, Andjelkovic S (2012) The potential of different plant species for heavy metals accumulation and distribution. J Food Agric Environ 10:959–964
30. Förstner U, Wittman GTW (1981) Metal pollution in the aquatic environment. Springer, Berlin
31. Galassi S, Bettinetti R, Neri MC, Jeannot R, Dagnac T, Bristeau S, Sakkas V, Albanis T, Boti V, Valsamaki T, Falandysz J, Schulte-Oehlmann U (2008) A multispecies approach for monitoring persistent toxic substances in the Gulf of Gdańsk (Baltic Sea). Ecotoxicol Environ Saf 69:39–48. doi:10.1016/j.ecoenv.2006.11.015
32. Gic-Grusza G, Kryla-Straszewska L, Urbański J, Warzocha J, Węsławski JM (2009) Atlas of polish marine area bottom habitats: environmental valorization of marine habitats. Broker-Innowacji, Gdynia

33. Glasby GP, Szefer P, Geldon J, Warzocha J (2004) Heavy-metal pollution of sediments from Szczecin Lagoon and the Gdansk Basin, Poland. Sci Total Environ 330:249–269. doi:10.1016/j.scitotenv.2004.04.004

34. Göthberg A, Greger M, Holm K, Bengtsson BE (2004) Influence of nutrient levels on uptake and effects of mercury, cadmium, and lead in water spinach. J Environ Qual 33:1247–1255. doi:10.2134/jeq2004.1247

35. HELCOM (2004) The fourth Baltic Sea pollution load compilation (PLC-4). Balt Sea Environ Proc 93

36. HELCOM (2010) Towards a tool for quantifying anthropogenic pressures and potential impacts on the Baltic Sea marine environment: a background document on the method, data and testing of the Baltic Sea Pressure and Impact Indices. Balt Sea Environ Proc No. 125

37. Jackson TA (1998) Mercury in aquatic ecosystem. In: Langston J, Bebino MJ (eds) Metal metabolism in aquatic environment. Chapman and Hall, London, pp 178–249

38. Jankowska E, Włodarska-Kowalczuk M, Kotwicki L, Bałazy P, Kuliński K (2014) Seasonality in vegetation biometrics and its effects on sediment characteristics and meiofauna in Baltic seagrass meadows. Estuar Coast Shelf Sci. doi:10.1016/j.ecss.2014.01.003

39. Jędruch A, Bełdowska M, Bełdowski J, Zgrundo A, Ziółkowska M, Szubska M, Graca B, Jankowska K, Kotlarska E, Kiełczewska J (2013a) Variability of mercury concentrations in sediments and benthos of the coastal zone of Puck Bay (Southern Baltic Sea), through the year. In: Vaitkevičienė V (ed) Baltic Sea Science Congress: new horizons for Baltic Sea science. Klaipeda University CORPI, Klaipeda, p 186

40. Jędruch A, Bełdowska M, Falkowska L (2013b) Total mercury in surface sediments of the Gdansk Basin. In: Falkowska L (ed) Mercury in the environment. Identification of threats to human health. University of Gdansk, Gdansk, pp 105–110 (in Polish)

41. Jędruch A, Bełdowska M, Falkowska L, Perec M, Grzesińska K (2013c) The impact of coastal erosion on the mercury input into the Gulf of Gdansk. In: Witak M (ed) Geological processes in the

marine coastal zone. University of Gdansk, Gdynia, p 55 (in Polish)

42. Kalisińska E, Salicki W, Mysłek P, Kavetska KM, Jackowski A (2004) Using the Mallard to biomonitor heavy metal contamination of wetlands in north-western Poland. Sci Total Environ 320:145–161. doi:10.1016/j.scitotenv.2003.08.014
43. Kalisińska E, Kosik-Bogacka DI, Lisowski P, Lanocha N, Jackowski A (2013) Mercury in the body of the most commonly occurring European game duck, the Mallard (*Anas platyrhynchos*L. 1758), from Northwestern Poland. Arch Environ Contam Toxicol 64:583–593. doi:10.1007/s00244-012-9860-6
44. Kożuchowski K (2009) Contemporary climate warming in Poland. Papers on Global Change IGBP 16:41–53
45. Krause-Jensen D, Sagert S, Schuber H, Boström C (2008) Empirical relationships linking distribution and abundance of marine vegetation to eutrophication. Ecol Indic 8:515–529. doi:10.1016/j.ecolind.2007.06.004
46. Kruk-Dowgiallo L (1991) Long-term changes in the structure of underwater meadows of the Puck Lagoon. Acta Ichtiol Pisc 22:135–143
47. Książkiewicz J (2006) A mallard duck—*Anas platyrhynchos* L., means a flat-billed duck. Zootechnical News XLIV:25–30
48. Lack A, Evans D (2005) Instant notes in plant biology. Taylor and Francis, Oxford
49. Lafabrie AJ, Major KM, Major CS, Cebrián J (2011) Arsenic and mercury bioaccumulation in the aquatic plant, *Vallisneria neotropicalis*. Chemosphere 82:1393–1400. doi:10.1016/j.chemosphere.2010.11.070
50. Lindberg SE, Dong W, Chanton J, Qualls RG, Meyers T (2005) A mechanism for bimodal emission of gaseous mercury from aquatic macrophytes. Atmos Environ 39:1289–1301. doi:10.1016/j.atmosenv.2004.11.006
51. Lisowski P (2009) Mercury in tissues and organs of mallard (Anas platyrhynchos L.) and fox (Vulpes vulpes L.) originating from the Szczecin region. PhD Thesis, Siedlce University of Natural Sciences and Humanities, Siedlce (in Polish)

52. Mamboya FA (2007) Heavy metal contamination and toxicity: studies of Macroalgae from Tanzanian Coast. Universitets service, Stockholm
53. Marks R, Bełdowska M (2001) Air-sea exchange of mercury vapour over the Gulf of Gdańsk and southern Baltic Sea J Mar Sys 27:315–324. doi:10.1016/S0924-7963(00)00076-2
54. Mason RP, Fitzgerald WF, Morel FMM (1994) The biogeochemical cycling of elemental mercury: anthropogenic influences. Geochim Cosmochim Acta 58:3191–3198
55. Murawiec D, Gajecka A, Bełdowska M, Falkowska L (2007) Investigation on mercury concentration levels in coastal and offshore waters of the Gulf of Gdańsk. Oceanol Hydrobiol Stud 36:83–97
56. Nagajyoti CP, Lee DK, Sreekanth MVT (2010) Heavy metals, occurrence and toxicity for plants: a review. Environ Chem Lett 8:199–216. doi:10.1007/s10311-010-0297-8
57. Niemirycz E (2011) Inflow of pollutants by rivers. In: Uścinowicz S (ed) Geochemistry of Baltic Sea surface sediments. Polish Geological Institute, Warsaw, pp 93–113
58. Osowiecki A, Krzymiński W, Nowicki W, Kruk-Dowgiałło L, Błeńska M, Brzeska P, Michałek-Pogorzelska M, Dubiński M, Łysiak-Pastuszak E, Góralski J, Chojnacki W, Marcinkow A, Kazała P (2009) Development of the methodology of testing and classification of biological elements in the assessment procedure of the ecological status of marine transitional and coastal waters. Maritime Institute of Gdansk, Gdansk (in Polish)
59. Pempkowiak J, Cossa D, Sikora A, Sanjuan J (1998) Mercury in water and sediments of the southern Baltic Sea. Sci Total Environ 213:185–192. doi:10.1016/S0048-9697(98)00091-6
60. Pempkowiak J, Walkusz-Miotk J, Bełdowski J, Walkusz W (2006) Heavy metals in zooplankton from the Southern Baltic. Chemosphere 62:1697–1708. doi:10.1016/j.chemosphere.2005.06.056
61. Pergent-Martini C (1998) Posidonia oceanica: a biological indicator of past and present mercury contamination in the Mediterranean Sea. Mar Environ Res 45:101–111. doi:10.1016/S0141-1136(97)00023-8
62. Pliński M, Florczyk I (1993) Macrophytobenthos. In: Korzeniewski

K (ed) The Bay of puck. University of Gdansk, Gdansk, pp 281–302 (in Polish)

63. Salgado LT, Andrade LR, Amado Filho GM (2005) Localization of specific monosaccharides in cells of the brown alga *Padina gymnospora* and the relation to heavy-metal accumulation. Protoplasma 225:123–128. doi:10.1007/s00709-004-0066-2
64. Saniewska D (2013) Pathways of mercury migration into the coastal zone of the Gulf of Gdansk. PhD Thesis, University of Gdansk, Gdansk (in Polish)
65. Saniewska D, Bełdowska M, Bełdowski J, Saniewski M, Kwaśniak J, Falkowska L (2010) Distribution of mercury in different environmental compartments in the aquatic ecosystem of the coastal zone of the Southern Baltic Sea. J Environ Sci 22:1144–1150. doi:10.1016/S1001-0742(09)60230-8
66. Saniewska D, Bełdowska M, Bełdowski J, Jędruch A, Saniewski M, Falkowska L (2014) Mercury loads into the sea associated with extreme flood. Environ Pollut 191:93–100. doi:10.1016/j.envpol.2014.04.003
67. Saniewski M (2013) Macrophytobenthos as an indicator of the environmental status of the Baltic Sea. Pol Hyperbaric Res 42:83–102. doi:10.13006/phr.42.4
68. Scheuhammer AM, Basu N, Burges NM, Elliott JE, Campbell GD, Wayland M, Champoux L, Rodrigue J (2008) Relationships among mercury, selenium, and neurochemical parameters in common loons (Gavia immer) and bald eagles (Haliaeetus leucocephalus). Ecotoxicology 17:93–101. doi:10.1007/s10646-007-0170-0
69. Schubert H, Krause JC (2002) Macrophytes in coastal lagoons of the Southern Baltic—recent knowledge and developments. In: Schernewski G, Schiewer U (eds) Baltic coastal ecosystems: structure, function and coastal management. Springer, Berlin, pp 31–40
70. Stankovic S, Kalaba P, Stankovic AP (2014) Biota as toxic metal indicators. Environ Chem Lett 12:63–84. doi:10.1007/s10311-013-0430-6
71. Szefer P, Żbikowski R (2010) Trace metals. In: Bolałek J (ed) Physical, biological and chemical studies of marine bottom

sediments. University of Gdansk, Gdansk, pp 375–383 (in Polish)

72. Szefer P, Glasby GP, Geldon J, Renner RM, Björn E, Snell J, Frech W, Warzocha J (2009) Heavy-metal pollution of sediments from the Polish exclusive economic zone, southern Baltic Sea. Environ Geol 57:847–862. doi:10.1007/s00254-008-1364-3
73. Szumiło E, Falkowska L, Hajdrych J, Grajewska A, Meissner W, Bełdowska M, Bzoma S (2013) Mercury levels in organs and tissues of seabirds of genre Larus living on the Gulf of Gdansk in 2009–2012. In: Falkowska L (ed) Mercury in the environment—identification of threats to human health. University of Gdansk, Gdansk, pp 155–163 (in Polish)
74. Węsławski JM, Kryla-Straszewska L, Piwowarczyk J, Urbański J, Warzocha J, Kotwicki L, Włodarska-Kowalczuk M, Wiktor J (2013) Habitat modelling limitations—Puck Bay, Baltic Sea—a case study. Oceanologia 55:167–183. doi:10.5697/oc.55-1.167
75. WIOŚ (2009) Report on the state of the environment of Western Pomerania in 2006–2007. Regional Inspectorate for Environmental Protection, Szczecin (in Polish)
76. Wisłocka M, Krawczyk J, Klink A, Morrison L (2006) Bioaccumulation of heavy metals by selected plant species from uranium mining dumps in the Sudety Mts., Poland. Pol J Environ Stud 15:811–818
77. Wrembel HZ (1993) Atmosphere and rivers as the major sources of mercury to the Baltic Sea. Slupsk Pomeranian University, Slupsk (in Polish)
78. Wrembel HZ (1997) Mercury in the Baltic Sea. Slupsk Pomeranian University, Slupsk (in Polish)
79. Zahir F, Rizwi SJ, Haq SK, Khan RH (2005) Low dose mercury toxicity and human health. Environ Toxicol Pharmacol 20:351–360. doi:10.1016/j.etap.2005.03.007
80. Zalewska T, Saniewski M (2011) Bioaccumulation of gamma emitting radionuclides in red algae from the Baltic Sea under laboratory conditions. Oceanologia 53:631–650. doi:10.5697/oc.53-2.631

CHAPTER

12

THE USE OF BIOMARKERS TO ASSESS THE HEALTH OF AQUATIC ECOSYSTEMS IN BRAZIL: A REVIEW

Thaıs Dalzochio[1], Gabriela Zimmermann Prado Rodrigues[1], Ismael Evandro Petry[2], Gunther Gehlen[1], Luciano Basso da Silva[1]

[1] Graduate Program in Environmental Quality, Feevale University, RS 239, 2755, Novo Hamburgo, RS CEP 93352-000, Brazil

[2] Laboratory of Comparative Histology, Feevale University, RS 239, 2755, Novo Hamburgo, RS CEP 93352-000, Brazil

ABSTRACT

Organisms in polluted environments are typically exposed to a complex mixture of chemical contaminants. The great concern about the health of aquatic ecosystems has led to the increased use of biomarkers over the past years. The aim of this work was to review the papers published from 2000 to 2015, which used biomarkers to assess the health of aquatic ecosystems in Brazil. A research resulted in 99 eligible papers. More than 80% of studies were conducted in the states of São Paulo and Rio Grande do Sul. Approximately 63% of studies used fish as bioindicator, whereas the micronucleus test and biochemical analyses were the most used biomarkers. A multibiomarker approach was used by 60.6% of studies, while 39.4% used one single biomarker. Furthermore, 68% were field studies and more than 75% of these used control animals sampled at reference sites. A relationship between the biomarker responses and pollution was reported by 87% of studies; however, 43.4% of studies analyzed only one sampling period, limiting comparisons and comprehension about possible seasonal variations. This review evidenced some weak points in studies using biomarkers in Brazil, especially related to the lack of studies in two important biomes (the Pantanal and the Amazon Rainforest) and experimental designs (small sample size, sampling in one single period, use of one single biomarker). Thus, future studies should consider mainly the use of multiple biomarkers, greater sample size, seasonal sampling and water physicochemical parameters to better diagnose the health of aquatic ecosystems.

Keywords: Biomarkers, Environmental pollution, Toxicology assessment, Water quality

INTRODUCTION

Pollution of the aquatic ecosystems is considered a serious and growing problem. Increasing amount of industrial, agricultural and urban pollutants discharged into the aquatic environment have led to various deleterious effects on aquatic organisms and also on human health (McGlashan and Hughies 2001). In Brazil, surface water quality is monitored only by the means of physicochemical and microbiological (coliforms) parameters, according to limits established by the Brazilian National Environment Council (CONAMA) (Brasil 2005). The CONAMA Resolution 357/2005 provides maximum values for certain substances in surface waters, but does not provide specific information on the use of ecotoxicological tests for assessing water quality. Furthermore, physicochemical analyses provide only information about the nature of the contaminants and their concentrations in the environment, and they cannot predict bioavailability or potential effects on biota (Seriani et al. 2015). On the other hand, ecotoxicological approaches represent a useful indicator of water quality (Gonzalez et al. 1993; Araújo et al. 2014), because they reflect the real conditions of interaction by synergy and/or antagonism among the contaminants and the effects on the organisms (Azevedo et al. 2013; Fuzinatto et al. 2013).

A biomarker can be defined as a quantitative measure of changes in molecular or cellular components, processes, structures and functions related to exposure to environmental chemicals (Depledge et al. 1995; He et al. 2012; Fasulo et al. 2013). Classes of biomarkers have been proposed according to the extent that they reflect exposure to environmental stressors, or adverse health effects from contaminant exposures (WHO 1993; van der Oost et al. 2003; Viarengo et al. 2007; Hook et al. 2014). Biliary fluorescent aromatic compounds, vitellogenin, cytochrome P4501A mRNA or protein, hepatic ethoxyresorufin-*O*-deethylase (EROD) and metallothioneins (MT) are examples of biomarkers of exposure. They show an early response to contaminants and are typically specific to a particular class of contaminants (Broeg et al. 2005). Biomarkers of effect are related to measurable biochemical, physiological or other alterations within tissues or body fluids of an organism that can be recognized as associated with an established or possible health impairment or disease. Heat shock proteins (HSP70 or HSP90), markers of oxidative stress [superoxide dismutase (SOD), glutathione, catalase (CAT), lipid peroxidation (LPO)], condition indices (condition factor, hepatosomatic index, gonadal index), histopathology evaluation, DNA damage and acetylcholinesterase (AChE) (which indicates both exposure and effects) are examples of biomarkers of biological effects (Hook et al. 2014). Yet, another class of biomarkers is described as "biomarkers that integrate chemical exposure and biological effects". They include AChE and also genomic approaches (Hook 2010). In general, some biomarkers allow the specific identification of exposure to a class of xenobiotics or alterations of physiological function, but the majority of biomarker applications monitor a general response to disturbance (Trapp et al. 2014). Nevertheless, it is important to note that many non-pollution factors may interfere with biomarker responses. These "confounding" factors include the organisms' health, sex, age, nutritional status, metabolic activity, migratory behavior, reproductive and development status, and population density, as well as factors like season, ambient temperature, heterogeneity of the environmental pollution and so forth (van der Oost et al. 2003).

The use of biomarkers with the purpose of biomonitoring natural aquatic systems by the use of bioindicator species is necessary to efficiently measure the degree of exposure in aquatic organisms to chemical contaminants (Sureda et al. 2011). Biomonitoring or biological monitoring can be defined as the systematic use of biological responses to evaluate changes in the environment (Cairns and van der Schalie 1980). In this context, biomarkers are increasingly worldwide-recognized tools for the assessment of pollution impacts, and some are already incorporated in environmental monitoring programs in other countries (Viarengo et al. 2007), although their systematic and large scale application is rare (Trapp et al. 2014). In this context, considering the increased contamination of water resources and the potential risk to biodiversity conservation and human health, as well as the importance of the use of biomarkers to complement the water physicochemical analysis, this study aims to review the use of biomarkers in the assessment of aquatic ecosystems health in the last 16 years in Brazil.

DATA SURVEY

The papers were searched using the databases Science Direct and Scientific Electronic Library Online (SCIELO). The keywords used on the search were biomarker, biomonitoring, water pollution and Brazil. The criteria for the selection of papers included original articles and short communications published between January 2000 and July 2015. Studies should be performed in field and/or laboratory (with native or commercially acquired organisms) and use at least one type of biomarker in any organism exclusively aquatic (including larvae) to assess the water quality of natural environments, such as rivers, lakes, wetlands, streams and sea in Brazil. The exclusion criteria were studies related to the assessment of water quality from environments altered by constructions of reservoirs and dams, as well as studies using semiaquatic organisms (biphasic life cycle).

A qualitative analysis was conducted considering the bioindicator organism, type of study (field, in laboratory or with caged organisms), type of biomarkers, use of control or reference site, number of sampling sites, number of collections, sample size, combination of biomarkers, physicochemical parameters analyses and the response of the bioindicator to the environmental contamination.

PUBLISHED STUDIES

Considering the criteria aforementioned, 99 papers published in national and international scientific journals were selected (Table 1).

Table 1: Summary of studies using biomarkers to assess the health of aquatic ecosystems in Brazil, in chronological order

	Study	**Bioindicator**	**Species**
1	Bainy et al. (2000)	Bivalve	*Perna perna*
2	Schulz and Martins-Junior (2001)	Fish	*Astyanax fasciatus*
3	Torres et al. (2002)	Bivalve	*Mytella guyanensis*
4	Ventura et al. (2002)	Fish	*Orthopristis ruber*
5	Andrade et al. (2004)	Fish	*Mugil* sp. and *Netuma* sp.
6	Geracitano et al. (2004)	Worm	*Laeonereis acuta*
7	Parente et al. (2004)	Fish	*Oreochromis niloticus*
8	Ranzani-Paiva and Silva-Souza (2004)	Fish	*Mugil platanus*
9	Alberto et al. (2005)	Fish	*Astyanax fasciatus*
10	Prá et al. (2005)	Planaria	*Girardia schubarti*
11	Amado et al. (2006a)	Fish	*Micropogonias furnieri*
12	Amado et al. (2006b)	Fish	*Paralichthys orbignyanus*
13	Silva et al. (2006)	Fish	Multispecies
14	Souza and Fontanetti (2006)	Fish	*Oreochromis niloticus*
15	Tortelli et al. (2006)	Fish	*Micropogonias furnieri* and *Cathorops spixii*
16	Villela et al. (2006)	Bivalve	*Limnoperna fortunei*
17	Zanette et al. (2006)	Bivalve	*Crassostrea rhizophorae*
18	Camargo and Martinez (2007)	Fish	*Prochilodus lineatus*
19	Domingos et al. (2007)	Bivalve	*Crassostrea rhizophorae*
20	Fernandez et al. (2007)	Gastropod	*Stramonita haemastoma* and *Thais rústica*
21	Ferreira-Cravo et al. (2007)	Worms	*Laeonereis acuta*
22	Francioni et al. (2007)	Bivalve	*Perna perna*

23	Junior et al. (2007)	Plant	*Allium cepa*
24	Lemos et al. (2007)	Fish	*Pimephales promelas*
25	Lüchmann et al. (2007)	Shrimp	*Farfantepenaeus brasiliensis*
26	Lupi et al. (2007)	Fish	*Oreochromis niloticus*
27	Oliveira et al. (2007)	Fish	Multispecies
28	Pereira et al. (2007)	Bivalve	*Perna perna*
29	Silva and Martinez (2007)	Fish	*Astyanax altiparanae*
30	Villela et al. (2007)	Bivalve	*Limnoperna fortunei*
31	David et al. (2008)	Bivalve	*Mytella falcata*
32	Lemos et al. (2008)	Fish	*Astyanax jacuhiensis*
33	Linde-Arias et al. (2008a)	Fish	*Oreochromis niloticus*
34	Linde-Arias et al. (2008b)	Fish	*Oreochromis niloticus*
35	Medeiros et al. (2008a)	Bivalve	*Crassostrea gigas*
36	Medeiros et al. (2008b)	Bivalve	*Crassostrea gigas*
37	Miranda et al. (2008)	Fish	*Hoplias malabaricus*
38	Parente et al. (2008)	Fish	*Oreochromis niloticus* and *Geophagus brasiliensis*
39	Ruas et al. (2008)	Fish	*Oreochromis niloticus, Tilapia rendalli* and *Geophagus brasiliensis*
40	Zanette et al. (2008)	Bivalve	*Crassostrea rhizophorae* and *Crassostrea gigas*
41	Barberio et al. (2009)	Plant	*Allium cepa*
42	Cardoso et al. (2009)	Fish	*Trichiurus lepturus*
43	Galindo and Moreira (2009)	Fish	*Bathygobius soporator*
44	Katsumiti et al. (2009)	Fish	*Cathorops spixii*
45	Kirschbaum et al. (2009)	Fish	*Centropomus parallelus*
46	Barbosa et al. (2010)	Fish and plant	*Oreochromis niloticus* and *Allium cepa*
47	Carvalho-Neta and Abreu-Silva (2010)	Fish	*Sciades herzbergii*
48	Egito et al. (2010)	Fish	*Crenicichla menezesi*
49	Franco et al. (2010)	Fish	*Oreochromis niloticus*
50	Montes et al. (2010)	Fish	*Brachyplatystoma rousseauxii*
51	Rechenmacher et al. (2010)	Rat	Wistar rats

52	Rocha et al. (2010)	Fish	*Brachyplatystoma rousseauxii*
53	Sáenz et al. (2010)	Bivalve	*Perna perna*
54	Santos et al. (2010)	Fish	*Mugil curema*
55	Scalon et al. (2010)	Fish	*Hyphessobrycon luetkenii*
56	Seriani et al. (2010)	Fish	*Micropogonias furnieri*
57	Flores-Lopes and Thomaz (2011)	Fish	*Astyanax fasciatus* and *Cyanocharax alburnus*
58	Lemos et al. (2011)	Cell culture	Human linphocytes
59	Nunes et al. (2011)	Cell culture and plant	Chinese hamster lung fibroblasts and *Allium cepa*
60	Pereira et al. (2011)	Bivalve	*Perna perna*
61	Souza-Bastos and Freire (2011)	Fish	*Atherinella brasiliensis*
62	Toste et al. (2011)	Gastropod	*Stramonita haemastoma*
63	Anzolin et al. (2012)	Fish	*Trichechus manatus*
64	Azevedo et al. (2012a)	Fish	*Cathorops spixii*
65	Azevedo et al. (2012b)	Fish	*Cathorops spixii*
66	Carvalho et al. (2012)	Fish	*Oreochromis niloticus*
67	Hauser-Davis et al. (2012a)	Fish	*Mugil Liza*
68	Hauser-Davis et al. (2012b)	Fish	*Oreochromis niloticus, Mugil liza and Gephafus brasiliensis*
69	Nascimento et al. (2012)	Fish	*Oligosarcus hepsetus, Hypostomus auroguttatus* and *Geophagus brasiliensis*
70	Oliveira et al. (2012)	Plant	*Allium cepa* and *Eichhornia crassipes*
71	Rola et al. (2012)	Bivalve	*Mytilus edulis*
72	Seriani et al. (2012)	Fish	*Oreochromis niloticus*
73	Souza et al. (2012)	Bivalve	*Crassostrea gigas*
74	Azevedo et al. (2013)	Fish	*Cathorops spixii*
75	Bastos et al. (2013)	Fish	*Mugil* sp.
76	Davanso et al. (2013)	Crab	*Goniopsis cruentata*
77	Fuzinatto et al. (2013)	Fish	*Oreochromis niloticus*
78	Melo et al. (2013)	Fish	Multispecies
79	Ribeiro et al. (2013)	Fish	*Atherinella brasiliensis*
80	Seriani et al. (2013)	Fish	*Centropomus parallelus*

81	Sousa et al. (2013)	Fish	*Sciades herzbergii* and *Bagre bagre*
82	Souza et al. (2013)	Fish	*Centropomus parallelus*
83	Venancio et al. (2013)	Reptile	*Phrynops geoffroanus*
84	Batista et al. (2014)	Fish	*Astyanax bimaculatus*
85	Castro et al. (2014)	Fish	*Hoplias malabaricus*
86	Costa et al. (2014)	Plant	*Tradescantia pallida* var. *purpúrea*
87	Factori et al. (2014)	Plant	*Landoltia punctata*
88	Osório et al. (2014)	Fish	*Geophagus brasiliensis*
89	Pereira et al. (2014)	Bivalve	*Crassostrea rhizophorae* and *Perna perna*
90	Procópio et al. (2014)	Fish	*Prochilodus argenteus*
91	Vieira et al. (2014)	Fish	*Astyanax altiparanae*
92	Barrilli et al. (2015)	Fish	*Astyanax paranae, Phalloceros harpagos*and *Poecilia reticulata*
93	Bueno-Krawczyk et al. (2015)	Fish	*Astyanax bifasciatus*
94	Cruz et al. (2015)	Fish	*Oreochromis niloticus*
95	Gomes et al. (2015)	Plant	*Allium cepa*
96	Maceda et al. (2015)	Fish and plant	*Astyanax altiparanae* and *Allim cepa*
97	Prado et al. (2015)	Fish	*Achirus lineatus*
98	Seriani et al. (2015)	Fish	*Oreochromis niloticus*
99	Zanette et al. (2015)	Barnacles	*Balanus improvisus*

Regarding the number of publications along the years, less than five papers were published per year from 2000 to 2005. A peak of publications was reached in 2007 (13 papers) and 6–11 papers were published per year from 2008 to 2015 (Figure 1). These data show that scientific publication in the field of aquatic ecotoxicology is stable and without tendency of increase in Brazil.

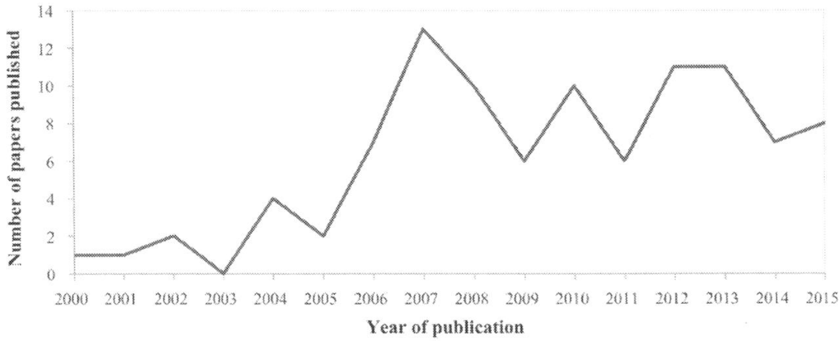

Figure 1: Number of papers published in journals between January 2000 and July 2015 regarding the use of biomarkers to assess the water quality in Brazil.

A higher number of studies carried out in Southeastern and Southern regions were observed, corresponding to more than 80% of total studies (Table 2). Most studies were carried out in the states of São Paulo, in Southeastern region (26%) and Rio Grande do Sul, in Southern region (21%), probably because some of the most polluted rivers in Brazil are located in these states, such as Sinos River, Gravataí River and Caí River, in Rio Grande do Sul; and Tietê River and Paraíba do Sul River, in São Paulo (Hupffer et al. 2013). In addition, these states present well-structured universities and research centers, and also receive more funding for this type of research. It is relevant to note the lack of studies published about the Midwestern region (one single study) and Northern region (three studies) despite the existence of two important biomes of Brazil in these regions—the Pantanal and Amazon Rainforest, respectively. According to Carvalho-Neta and Abreu-Silva (2010) and Montes et al. (2010), the lack of studies using biomarkers as predictors of aquatic health in these regions, and also in the Northeast, indicates the need of biomonitoring studies that might estimate the potential effects suffered by native species.

Table 2: Studies using biomarkers to assess the health of aquatic ecosystems in Brazilian regions

Region	Number of papers	%
Southeast	42	42.4
South	42	42.4

Northeast	9	9.1
North	3	3.0
More than one region[a]	2	2.0
Midwest	1	1.0
Total	99	99.9

[a]Studies with sampling sites located in different Brazilian regions

BIOINDICATOR ORGANISMS

Various organisms have been considered as bioindicators of environmental quality. In Brazil, 62.6% were carried out exclusively with fish, 35.4% used other organisms, such as bivalves, plants and gastropods, and 3% used two types of bioindicators simultaneously (Table 3). Studies using biomarkers in larvae were not found.

Table 3: Bioindicator organisms of studies using biomarkers to assess the health of aquatic ecosystems in Brazil

Organism	**Number of studies**	**%**
Fish	62	62.6
Bivalves	17	17.2
Plants	6	6.1
Others	11	11.1
More than one type of bioindicator	3	3.0
Total	99	100

Fish have been considered suitable organisms for biomonitoring studies as they are sensible to changes in the aquatic environment. Their biological responses change, even at low levels of pollution (Linde-Arias et al. 2008b; Souza et al. 2013). Fish can be found virtually everywhere in the aquatic environment and they play a major ecological role in aquatic food webs as carriers of energy from lower to higher trophic levels (van der Oost et al. 2003). Thus, genetic, biochemical, behavioral and morphological responses represent useful biomarkers in environmental biomonitoring (Pesce et al. 2008; Ballesteros et al. 2009). In addition, the high percentage of studies carried out using fish as bioindicator can also be explained by the simple fish sampling techniques.

Bivalves were the second group most common (17.2%). Bivalves present a wide geographic distribution, availability in different types of aquatic environments, possibility of breeding in aquacultures and are adequate for studies with caged organisms (Viarengo et al. 2007). Other bioindicator organisms, such as crabs, shrimps, worms and planarians were occasionally used. Additionally, two studies applied cell cultures of mammals' fibroblasts (V79 line) and human lymphocytes to assess water quality. Cell cultures are a useful tool in environmental evaluation, being an alternative methodology due to its easy manipulation and sensibility when exposed to physical and chemical agents; in addition, it presents good reproducibility (Rogero et al. 2003; Zegura et al. 2009). Despite the wide use in other countries (Leme and Marin-Morales 2009), the biomarkers analysis in plants to assess water quality is still unusual in Brazil since only six studies were found in the present review.

TYPES OF STUDIES

The studies using biomarkers for the evaluation of aquatic ecosystems can present different approaches. They can be field studies (organisms are sampled in situ), laboratory studies (water samples are collected in an area of interest and transported to the laboratory, where bioassays of exposure are conducted) and with caged organisms (animals are exposed in cages in the study area for a period of time). Considering studies carried out in Brazil, 68% were field studies, 19% were laboratory studies, 10% were with caged organisms and 3% combined laboratory and caged animals.

Complex exposure dynamics to pollutants and resulting biological responses found in the field are seldom replicated in laboratory studies (Crane et al. 2007), thus effects of pollution might be under or overestimated. Field studies comparing impacted and reference areas enable an evaluation of the health condition of organisms in their own environment, although organisms can move and potentially avoid contaminant hot spots in the field (Ward et al. 2013). In addition, it is not always possible to determine with precision the causal agent of any given alteration (Alberto et al. 2005). In this context, laboratory studies are extremely necessary to investigate the potential of an organism to be used as a bioindicator in field studies, and also enable a better understanding of chemical modes of toxicity. However, limitations of laboratory studies include the difficulty in incorporating native species into laboratory tests, problems with size and number of organisms that can be held in the laboratory and the inability of reproducing complex behaviors, such as spawning aggregation or migration (Hook et al. 2014). Moreover, less than 17% of laboratory studies were conducted with replicate. On the other hand, studies with caged organisms are more realistic than experiments conducted in laboratory in environmental assessment and present the advantage of using organisms with a known life history (Crane et al. 2007).

REFERENCE AREAS AND LABORATORY CONTROLS

Usually, tap and mineral water (for fish and bivalves assays) or distilled water (for plant assays) is used for the control group in laboratory experiments, while samplings in reference areas (areas under minor anthropogenic impact) are conducted in the field studies. However, finding clean areas can be difficult, and then, researchers choose to use controls in laboratory with animals acquired at breeding facilities and maintained in tap water, or do not use any type of control. For biomarker analysis, the use of negative control is important, although it is also possible to monitor temporal variations of a biomarker response in only one sampling site. In Brazil, taking into account only field studies and studies with caged animals, more than 75% used controls animals sampled at reference areas, whereas others used animals kept under laboratory conditions or did not use any type of control. Considering studies which reported sampling at reference sites, 27.1% found altered biomarker responses in organisms from these areas, evidencing the difficulty in finding clean areas under minor anthropic influence, and thus complicating the comparison of biomarker results.

BIOMARKERS USED IN BRAZIL

Various types of biomarkers have been used to assess the effects of exposure to pollutants in water, including morphometric indexes, histopathological alterations and molecular analyses. In Brazil, the biomarkers most frequently used were the micronucleus test, the biotransformation enzyme GST, the antioxidant enzyme CAT and histopathological analyses (Table 4). Other biomarkers not listed in Table 4 include the evaluation of physiological and hematological parameters,

HSP70, vitellogenin, EROD, glutathione peroxidase (GPx), hepatic CYP1A and gene expression.

Table 4: Main biomarkers used in the assessment of health of aquatic ecosystems in Brazil

	Number of studies	**%**
Micronucleus test	31	31.3
Glutathione-S-transferase (GST)	28	28.3
Catalase (CAT)	26	26.3
Histopathological analyses	23	23.2
Comet assay	18	18.2
Morphometric index	18	18.2
Acetylcholinesterase/cholinesterase activity (AChE)	16	16.1
Lipoperoxidation analysis (LPO)	14	14.1
Metallothionein (MT)	9	9.1
Superoxide dismutase (SOD)	9	9.1
Others[a]	52	52.5

[a]Sum of studies which used other biomarkers

The DNA damage was assessed mainly by the comet assay (or single cell gel electrophoresis) and the micronucleus test. These techniques are sensitive, rapid and extensively used as genotoxic biomarkers (Zapata et al. 2016). The micronucleus test is one of the biomarkers most widely used for in situ monitoring of genotoxic pollution (Al-Sabti and Metcalfe 1995; Bolognesi et al. 2006; Udroiu 2006). This technique is based on the quantification of whole or fragmented chromosomes that are not incorporated into the main nucleus during mitosis (Al-Sabti and Metcalfe 1995). The comet assay is also an indicator of genotoxicity and an effective biomarker for detecting DNA strand breaks, cross-links and alkali labile sites in aquatic organisms (Tice et al. 2000; Frenzilli et al. 2009).

Xenobiotic metabolism is the central detoxification process that occurs in all the organisms. Phase I enzymes are involved in the first stage of detoxification of xenobiotics compounds and implicates in enzymatic transformation of a chemically modifying lipid soluble toxin into water-soluble toxin. Most of the transformation reactions in this phase include a broad family of enzymes, cytochrome P450s (Lardone et al. 2010). Cytochrome P450s are monooxygenases responsible by a set of functions for controlling homeostasis, including the metabolism of drugs and other xenobiotics (McDonnell and Dang 2013). Phase II enzymes are involved in the second stage of the detoxification process related to enzymatic conjugation. The enzymes of this phase (as GST) modify phase I products into more water-soluble and less toxic forms (Hassan et al. 2015). Antioxidant enzymes, such as CAT and SOD, are considered biomarkers of oxidative damage. Contaminant-stimulated "reactive oxygen species" (ROS) production and resulting oxidative damage may be a mechanism of toxicity in aquatic organisms exposed to a variety of pollutants (Livingstone 2001; Azevedo et al. 2013).

Histopathological analyses represent useful tools for environmental diagnosis and monitoring. This type of analysis provides a method for the detection of morphological alterations in multiple organs (Johnson et al. 1993), as gills and liver. The analysis of gills of fish and bivalves has been widely used because changes in this organ may lead to the impairment of several functions, including gas exchange, ion regulation and excretion of metabolites (Cruz et al. 2015). In addition to these characteristics, the low cost justifies the use of histopathological analyses in ecotoxicological studies.

The morphometric indexes were used in 18.2% of studies, mostly in fish species. The condition factor of the whole body (calculated using the weight and length) provides information on potential pollution impacts. Although this parameter is not very sensitive and may be affected by non-pollutant factors, such as season, disease and nutritional level, it is used as an initial screening biomarker to indicate exposure and effects or to provide information on energy reserves (Mayer et al. 1992; Linde-Arias et al. 2008b). Its low cost, ease and rapidity makes it a valuable tool to assess preliminary effects of pollutants in fish (van der Oost et al. 2003).

The AChE enzyme occurs in cholinergic synapses and motor end plates, and is responsible for the hydrolysis of the neurotransmitter acetylcholine into choline and acetic acid. Inhibition of AChE has been associated with the mechanism of toxic action of organophosphates and carbamates insecticides (Galgani and Bocquene 1990; Payne et al. 1996; Valbonesi et al. 2003; Andreescu and Marty 2006). However, some studies assessing metal and polycyclic aromatic hydrocarbons (PAHs) exposure have also evidenced the inhibition of this enzyme (Zinkl et al. 1991; Akaishi et al. 2004; Richetti et al. 2011).

Other biomarkers analyzed include MTs and LPO. MTs are low molecular weight proteins, high cysteine content, and good heat stability that can be used as biomarkers (Langston et al. 1998). They consist of thiol groups (sulfur–hydrogen) that bind to metals, preventing oxidative stress to the organism. MT induction as a response to metal exposure is well documented in many species and is known to play a role in the detoxification to toxic metals (Amiard et al. 2006). LPO is a consequence of the decomposition of polyunsaturated fatty acid peroxides of membrane lipids, producing a complex mixture of hydroperoxides and secondary products of oxidation, as malondialdehyde (MDA) (Banerjee et al. 1999; Akhgari et al. 2003). LPO can be enhanced by exposure to xenobiotics and some trace metal in ionic form, leading to cellular damage (Viarengo et al. 1990; Montine et al. 2004; Filipak Neto et al. 2008).

Genomics is an emerging approach in biomarkers assessment. Genomics deals with the analysis of the complete genome to understand the function of single genes. On the other hand, functional genomics is based on the analysis of gene expression (transcriptomics) and comprehensive proteins/metalloproteins analysis (proteomics/metallomics) (González-Fernández et al. 2008). New approaches in functional genomics and bioinformatics can help discriminate individual chemicals, or group of chemicals among complex mixtures that may contribute to adverse biological effects (Hook et al. 2014). Furthermore, environmental metabolomics is an emerging field referred to the application of metabolomics to characterize the interactions of living organisms with their environment (García-Sevillano et al. 2015). In Brazil, these approaches are still rare since gene expression was assessed only in 3% of studies.

Multibiomarker approach (combination of two or more biomarkers) allows a better understanding of stress responses due to pollutants exposure (Domingos et al. 2007). This approach may provide results that can be complementary and help in cases when a single biomarker response is affected by non-pollutant factors. Most Brazilian studies used this approach, while 39.4% were carried out using a single biomarker (Table 5).

Table 5: Number of biomarkers used by studies to assess the health of aquatic ecosystems in Brazil

Number of biomarkers	Number of studies	%
One	39	39.4
Two	21	21.2
Three	11	11.1
Four or more	15	27.8
Total	99	100

In 87.8% of studies, the authors reported a relationship between the biomarker responses and pollution in the sampling areas; however, such relation is subjective in most studies given the lack of well-designed experiments or statistical support. Furthermore, the level of environmental contamination was superficially diagnosed, since 51.5% of studies reported only the analysis of physicochemical parameters that can be obtained with portable apparatus (water temperature, salinity, pH or dissolved oxygen). The concentrations of at least one metal were assessed in 17.2% of studies, and one single study assessed the presence of PAHs in water. Pesticides, hormones and other emergent contaminants were not assessed. Most studies provided data on the possible type of pollution (oil spills, domestic sewage discharges, agricultural runoffs and industrial effluents) only based on previous studies, and/or local observations.

EXPERIMENTAL DESIGNS

In addition to the use of a multibiomarker approach, the assessment of more than one sampling site in different sampling periods (seasonal variation) may help in the interpretation of biomarker responses. The number of sampling sites that were assessed in Brazil is shown in Table 6. Most studies were carried out in two or three sampling sites; however, 9.1% of studies analyzed one single sampling site. This is problematic because it precludes comparisons of the biomarker responses between organisms from different areas and does not provide significant information on the impacted area. Moreover, ~43% of studies were carried out in one single sampling period. Therefore, comparisons between periods were not performed, and consequently, these studies do not provide data on possible seasonal variations in biomarker responses and the real contamination scenario.

Table 6: Number of sampling sites of studies using biomarkers to assess the health of aquatic ecosystems in Brazil

Number of sampling sites	Number of studies	%
One	9	9.1
Two	31	31.3
Three	21	21.2
Four	18	18.2
Five	9	9.1
Six or more	11	11.1
Total	99	100

An important aspect in studies related to biomarker responses is the sample size—number of individuals analyzed per site in each sampling period or exposure experiment. Small sample size may lead to inconclusive results. As sample sizes increase, their variability tends to decrease, leading to a better hypothesis testing, a higher statistical power and smaller confidence intervals (Cochran 1977). However, sometimes it is not possible to maintain a certain sample size along the experimental design, especially in field studies. Table 7 shows the sample sizes in studies with biomarkers in the assessment of aquatic ecosystems in Brazil. Studies regarding biomarker responses with less than five organisms in at least one site or sampling period corresponded to 19.1%. Information on sample size was not provided or was not clear in 7.1% of studies, since only the total sample size (sum of organisms from all sampling sites studied and/or sampling periods) was reported.

Table 7: Sample size (number of organisms analyzed per sampling site or exposure experiment) of studies using biomarkers to assess the health of aquatic ecosystems in Brazil

Sample size	Number of studies	%
Less than 5	19	19.1
6–10	41	41.4
11–20	16	16.2
21–49	9	9.1

50 or more	6	6.1
Not applied	1	1.0
Not informed/not clear	7	7.1
Total	99	100

CONCLUSIONS

In this review, we provide information on the use of biomarkers to assess the health of aquatic ecosystems in Brazil, in the last 16 years. In general, the approaches used in Brazilian studies did not differ from other countries. However, some shortcomings were observed. The data analyses points towards a limited use of this approach in the country (basically restricted to two regions), a great variety of organisms used as bioindicators (regional biodiversity) and different sampling patterns. A great number of studies were conducted using fish as bioindicator organisms, therefore, the analysis of others organisms should be stimulated. Furthermore, most studies used biomarkers which are easy, fast and cheap to assess, whereas biomarkers which require more funding and/or more sophisticated equipments were rare. Differently from other countries, water physicochemical analyses are still poor and fail in providing information to establish relations between biomarker responses and contaminants. In general, a good relationship between biomarker responses and environmental pollution has been observed by the authors. However, experimental designs with multiple biomarkers, greater sample size, long-term biomonitoring and knowledge about organisms' ecological aspects may enable a better data interpretation on the environmental quality, as well as the interference of non-polluting factors in the biomarker responses. Additionally, new approaches in the genomics field may be a promising tool to better understand the impacts of sublethal concentrations of pollutants on living organisms, as well as to provide information on pollution-induced genetic changes in organisms' tolerance.

ACKNOWLEDGEMENTS

This work was supported by Fundação de Amparo à Pesquisa do Estado do Rio Grande do Sul (FAPERGS), Conselho Nacional de Desenvolvimento Científico e Tecnológico (CNPq) and Universidade Feevale.

REFERENCES

1. Akaishi FM, Silva De Assis HC, Jakobi SCG, Stjean S, Couternay SC, Lima E, Wagner ALR, Scofield A, Oliveira Ribeiro CA (2004) Morphological and neurotoxicological findings in tropical freshwater fish (*Astyanax* sp.) after waterborne and acute exposure to water soluble fraction (WSF) of crude oil. Arch Environ Contam Toxicol 46:244–253
2. Akhgari M, Abdollahi M, Kebryaeezadeh A, Hosseini R, Sabzevari O (2003) Biochemical evidence for free radical-induced lipid peroxidation as a mechanism for subchronic toxicity of malathion in blood and liver of rats. Human Exp Toxicol 22:205–211
3. Alberto A, Camargo AFM, Verani JR, Costa OFT, Fernandes MN (2005) Health variables and gill morphology in the tropical fish *Astyanax fasciatus* from a sewage-contaminated river. Ecotoxicol Environ Safe 61:247–255
4. Al-Sabti K, Metcalfe CD (1995) Fish micronuclei for assessing genotoxicity in water. Mutat Res 343(2–3):121–135
5. Amado LL, Rosa CE, Leite AM, Moraes L, Pires WV, Pinho GLL, Martins CMG, Robaldo RB, Nery LEM, Monserrat JM, Bianchini A, Martinez PE, Geracitano LA (2006a) Biomarkers in croakers *Micropogonias furnieri* (Teleostei: Sciaenidae) from polluted and non-polluted areas from the Patos Lagoon estuary (Southern Brazil): evidences of genotoxic and immunological effects. Mar Pollut Bull 52:199–206
6. Amado LL, Robaldo RB, Geracitano L, Monserrat JM, Bianchini A (2006b) Biomarkers of exposure and effect in the Brazilian flounder *Paralichthys orbignyanus* (Teleostei: Paralichthyidae) from the Patos Lagoon estuary (Southern Brazil). Mar Pollut Bull 52:207–213
7. Amiard JC, Amiard-Triquet C, Barka S, Pellerin J, Rainbow PS (2006) Metallothioneins in aquatic invertebrates: their role in metal detoxification and their use as biomarkers. Aquat Toxicol 76:160–202
8. Andrade VM, Freitas TRO, Silva J (2004) Comet assay using mullet (*Mugil* sp.) and sea catfish (*Netuma* sp.) erythrocytes for the detection of genotoxic pollutants in aquatic environment. Mutat Res

Gen Toxicol Environ Muyagen 560(1):57–67

9. Andreescu S, Marty JL (2006) Twenty years research in cholinesterase biosensors: from basic research to practical applications. Biomol Eng 23:1–15
10. Anzolin DG, Sarkis JES, Diaz E, Soares DG, Serrano IL, Borges JCG, Souto AS, Taniguchi S, Montone RC, Bainy ACD, Carvalho PSM (2012) Contaminant concentrations, biochemical and hematological biomarkers in blood of West Indian manatees *Trichechus manatus* from Brazil. Mar Pollut Bull 64:1402–1408
11. Araújo ALS, Guimarães ET, Seriani R (2014) Mutagenesis in *Tradescantia pallida* as a biomarker of the effects of water polluted with urban effluent. Holos Environ 14(1):97–102
12. Azevedo SA, Lopes B, Katsumiti A, Braga ES, Roche H, Ribeiro CAO, Bebianno MJ (2012a) Evidence of contamination by oil and oil products in the Santos-São Vicente Estuary, São Paulo, Brazil. Braz J Oceanogr 60(2):117–126
13. Azevedo SA, Souza J, Braga ES, Ribeiro C, Oliveira A (2012b) Nuclear abnormalities and morphometric indexes in the catfish *Cathorops spixii* (Ariidae) from different sites on the Southeastern Brazilian coast. Braz J Oceanogr 60(3):323–330
14. Azevedo JS, Braga ES, Assis HCE, Ribeiro CAO (2013) Biochemical changes in the liver and gill of *Cathorops spixii* collected seasonally in two Brazilian estuaries under varying influences of anthropogenic activities. Ecotoxicol Environ Safe 96:220–230
15. Bainy ACD, Almeida EA, Müller IC, Ventura EC, Medeiros ID (2000) Biochemical responses in farmed mussel *Perna perna* transplanted to contaminated sites on Santa Catarina Island, SC, Brazil. Mar Environ Res 50:411–416
16. Ballesteros ML, Durando PE, Nores ML, Díaz MP, Bistoni MA, Wunderlin DA (2009) Endosulfan induces changes in swimming motility and Acetylcholinesterase activities of *Jenynsia multidentata* (Anablepidae, Cyprinodontiformes). Environ Pollut 157:1573–1580
17. Banerjee BD, Seth V, Bhattacharya A, Pasha ST, Chakraborty AK (1999) Biochemical effects of some pesticides on lipid peroxidation and free-radical scavengers. Toxicol Lett 107:33–47
18. Barberio A, Barros L, Voltolini JC, Mello MLS (2009) Evaluation

of the cytotoxic and genotoxic potential of water from the River Paraíba do Sul, in Brazil, with the *Allium cepa* L. test. Braz. J Biol 69(3):837–842

19. Barbosa JS, Cabral TM, Ferreira DN, Agnez-Lima LF, Medeiros SR (2010) Genotoxicity assessment in aquatic environment impacted by the presence of heavy metals. Ecotoxicol Environ Safe 73(3):320–325
20. Barrilli GHC, Rocha O, Negreiros NF, Verani JR (2015) Influence of environmental quality of tributaries of the Monjolinho River on the relative condition factor (Kn) of the local ichthyofauna. Biota Neotrop 15(1):1–9
21. Bastos FF, Hauser-Davis RA, Tobar SAL, Campos RC, Ziolli RL, Cunha Bastos VLF, Cunha Bastos J (2013) Enzymatic GST levels and overall health of mullets from contaminated Brazilian Lagoons. Aquat Toxicol 126:414–423
22. Batista MTO, Junior ER, Feijó-Oliveira M, Ribeiro AC, Rodrigues E, Suda CNK, Vani GS (2014) Tissue levels of the antioxidant enzymes superoxide dismutase and catalase in fish *Astyanax bimaculatus* from the Una River Basin. Rev Ambient Água 9(4):621–631
23. Bolognesi C, Perrone E, Roggieri P, Pampanin DM, Sciutto A (2006) Assessment of micronuclei induction in peripheral erythrocytes of fish exposed to xenobiotics under controlled conditions. Aquat Toxicol 78(Supp 1):S93–S98
24. Brasil (2005) National Environment Council—CONAMA Resolução CONAMA 357, de 2005. http://www.mma.gov.br/port/conama/res/res05/res35705.pdf. Accessed 14 July 2016
25. Broeg K, Westernhagen HV, Zander S, Korting W, Koehler A (2005) The “bioeffect assessment index” (BAI): a concept for the quantification of effects of marine pollution by an integrated biomarker approach. Mar Pollut Bull 50:495–503
26. Bueno-Krawczyk AC, Guiloski IC, Piancini LD, Azevedo JC, Ramsdorf WA, Ide AH, Guimarães AT, Cestari MM, Silva De Assis HC (2015) Multibiomarker in fish to evaluate a river used to water public supply. Chemosphere 135:257–264
27. Cairns J Jr, van der Schalie WH (1980) Biological monitoring, Part

I—Early warning systems. Water Res 14:1179–1196

28. Camargo MMP, Martinez CBR (2007) Histopathology of gills, kidney and liver of a Neotropical fish caged in an urban stream. Neotrop Ichthyol 5(3):327–336
29. Cardoso TP, Mársico ET, Medeiros RJ, Tortelly R, Sobreiro LG (2009) Concentração de mercúrio e análise histopatológica em músculo, rim e cérebro de peixe-espada (*Trichiurus lepturus*) coletados na praia de Itaipu, Niterói, Rio de Janeiro, Brasil. Ciência Rural 39(2):540–546
30. Carvalho CS, Bernusso V, Araújo HSS, Espíndola ELG, Fernandes MN (2012) Biomarker responses as indication of contaminant effects in *Oreochromis niloticus*. Chemosphere 89:60–99
31. Carvalho-Neta RNF, Abreu-Silva AL (2010) Sciades herzbergii oxidative stress biomarkers: an in situ study of estuarine ecosystem (São Marcos's Bay, Maranhão, Brazil). Braz J Oceanogr 58:11–17
32. Castro JS, Silva JS, Freitas LC, Carvalho-Neta RNF (2014) Biomarcadores histopatológicos na espécie *Hoplias malabaricus* (Pisces, Osteichthyes, Erythrinidae) em uma Unidade de Conservação de São Luís (MA). Arq Bras Med Vet Zootec 66(6):1687–1694
33. Cochran WG (1977) Sampling techniques, 3rd edn. Wiley, New York
34. Costa GM, Cassanego MBB, Petry CT, Benvenuti T, Rubio MAK, Rodrigues MAS, Droste A (2014) Monitoramento químico e do potencial genotóxico para o diagnóstico da qualidade de corpos hídricos. Rev Bras Ciênc Amb 32:65–74
35. Crane M, Burton GA, Culp JM, Greenberg MS, Munkittrick KR, Ribeiro R, Salazar MH, St-Jean SD (2007) Review of aquatic in situ approaches for stressor and effect diagnosis. Integr Environ Assess Manag 3(2):234–245
36. Cruz AL, Prado TM, Maciel LAS, Couto RD (2015) Environmental effects on the gills and blood of *Oreochromis niloticus* exposed to rivers of Bahia, Brazil. Ecotoxicol Environ Safe 111:23–31
37. Davanso MB, Moreira LB, Pimentel MF, Costa-Lotufo LV, Moledo DSA (2013) Biomarkers in mangrove root crab *Goniopsis cruentata* for evaluating quality of tropical estuaries. Mar Environ Res 91:80–

88

38. David JAO, Salaroli RB, Fontanetti CS (2008) The significance of changes in *Mytella falcata*(Orbigny, 1842) gill filaments chronically exposed to polluted environments. Micron 39:1293–1299

39. Depledge MH, Aagaard A, Gyorkost P (1995) Assessment of trace metal toxicity using molecular, physiological and behavioral biomarkers. Mar Pollut Bull 31:19–27

40. Domingos FXV, Azevedo M, Silva MD, Randi MAF, Freire CA, Silva De Assis HC, Oliveira Ribeiro CA (2007) Multibiomarker assessment of three Brazilian estuaries using oysters as bioindicators. Environ Res 105:350–363

41. Egito LCM, Santos PE, Amaral VS, Medeiros SRB, Agnez-Lima LF (2010) Use of native species *Crenicichla menezesi* (Ariidae) as a model for in situ evaluation of genotoxicity in surface water. Sci Total Environ 408:6042–6046

42. Factori R, Leles SM, Novakowski GC, Rocha CLSC, Thomaz SM (2014) Toxicity and genotoxicity of water and sediment from streams on dotted duckweed (*Landoltia punctata*). Braz J Biol 74(4):769–778

43. Fasulo S, Mauceri A, Maisano M, Giannetto A, Parrino V, Gennuso F, D'Agata A (2013) Immunohistochemical and molecular biomarkers in *Coris julis* exposed to environmental pollutants. Ecotoxicol Environ Safe 73(5):873–882

44. Fernandez MA, Pinheiro FM, Quadros JP, Camillo E Jr (2007) An easy, non-destructive, probabilistic method to evaluate the imposex response of gastropod populations. Mar Environ Res 63:41–54

45. Ferreira-Cravo M, Piedras FR, Moraes TB, Ferreira JL, Freitas DP, Machado MD, Geracitano LA, Monserrat JM (2007) Antioxidant responses and reactive oxygen species generation in different body regions of the estuarine polychaeta *Laeonereis acuta* (Nereididae). Chemosphere 66(7):1367–1374

46. Filipak Neto F, Zanata SM, Silva de Assis HC, Nakao LS, Randi MAF, Ribeiro CAO (2008) Toxic effects of DDT and methyl mercury on the hepatocytes from *Hoplias malabaricus*. Toxicol Vitro 22:1705–1713

47. Flores-Lopes F, Thomaz AT (2011) Alterações histopatológicas

observadas nas brânquias de peixes como instrumento no monitoramento ambiental. Braz J Biol 71(1):179–188

48. Francioni E, Wagener ALR, Scofield AL, Depledge MH, Cavalier B (2007) Evaluation of the mussel *Perna perna* as a biomonitor of polycyclic aromatic hydrocarbon (PAH) exposure and effects. Mar Pollut Bull 54:329–338
49. Franco JL, Trevisan R, Posser T, Trivella DB, Hoppe R, Martins-Rosa J, Fernandes-Dinslaken D, Decker H, Inês-Tasca C, Bainy-Leal R, Freire-Marques MR, Dias-Bainy AC, Luiz-Dafre A (2010) Biochemical alterations in caged Nile tilapia *Oreochromis niloticus*. Ecotoxicol Environ Safe 73(5):864–872
50. Frenzilli G, Nigro M, Lyons BP (2009) The Comet assay for the evaluation of genotoxic impact in aquatic environments. Mutat Res 681(1):80–92
51. Fuzinatto CF, Flohr L, Melegari SP, Matias WG (2013) Induction of micronucleus of *Oreochromis niloticus* exposed to waters from the Cubatão do Sul River, southern Brazil. Ecotoxicol Environ Safe 98:103–109
52. Galgani F, Bocquene G (1990) In vitro inhibition of acetylcholinesterase from four marine species by organophosphates and carbamates. Bull Environ Contam Toxicol 45:243–249
53. Galindo TP, Moreira LM (2009) Evaluation of genotoxicity using the micronucleus assay and nuclear abnormalities in the tropical sea fish *Bathygobius soporator* (Valenciennes, 1837) (Teleostei, Gobiidae). Genet Mol Biol 32(2):394–398
54. García-Sevillano MA, García-Barrera T, Gómez-Ariza JL (2015) Environmental metabolomics: biological markers for metal toxicity. Electrophoresis 36:2348–2365
55. Geracitano LA, Monserrat JM, Bianchini A (2004) Oxidative stress in *Laeonereis acuta*(Polychaeta, Nereididae): environmental and seasonal effects. Mar Environ Res 58:625–630
56. Gomes JV, Teixeira JTS, Lima VM, Borba HR (2015) Induction of cytotoxic and genotoxic effects of Gandu River waters in the *Allium cepa* system. Rev Ambient Água 10(1):48–58
57. Gonzalez G, Crespo S, Brusle J (1993) Histo-cytological study

of the liver of the cabrilla sea bass, *Serranus cabrilla* (Teleostei, Serranidae), an available model for marine fish experimental studies. J Fish Biol 43(3):363–373

58. González-Fernández M, García-Barrera T, Jurado J, Prieto-Álamo MJ, Pueyo C, López-Barea J, Gómez-Ariza JL (2008) Integrated application of transcriptomics, proteomics, and metallomics in environmental studies. Pure Appl Chem 80(2):2609–2626
59. Hassan I, Jabir NR, Ahmad S, Shah A, Tabrez S (2015) Certain phase I and II enzymes as toxicity biomarker: an overview. Water Air Soil Pollut 226:153
60. Hauser-Davis RA, Bastos FF, Oliveira TF, Ziolli RL, Campos RC (2012a) Fish bile as a biomarker for metal exposure. Mar Pollut Bull 64:1589–1595
61. Hauser-Davis RA, Gonçalves RA, Ziolli RL, Campos RC (2012b) A novel report of metalothioneins in fish-bile: SDS-PAGE analysis, spectrophotometry quantification and metal speciation characterization by liquid chromatography coupled to ICP-MS. Aquat Toxicol 116–117:54–60
62. He X, Nie X, Yang Y, Liu X, Pan D, Cheng Z, Liang X (2012) Multi-biomarker responses in fishes from two tropical marine aquaculture regions of South China. Mar Pollut Bull 64:2317–2324
63. Hook SE (2010) Promise and progress in environmental genomics: a status report on the applications of microarray studies in ecologically relevant fish species. J Fish Biol 77:1999–2022
64. Hook SE, Gallagher EP, Batley GE (2014) The role of biomarkers in the assessment of aquatic ecosystem health. Integr Environ Assess Manag 10(3):327–341
65. Hupffer H, Figueiredo JAS, Tundisi JG (2013) Pagamento por serviços ambientais: incentivos para a proteção dos recursos hídricos e restauração da mata ciliar. Entremeios, Porto Alegre
66. Johnson LL, Stehr CM, Olson OP, Myers MS, Wigren CA, McCain BB, Varnasi U (1993) Chemical contaminants and hepatic lesions in winter flounder (*Pleuronectes americanus*) from the Northeast coast of the United States. Environ Sci Technol 27:2759–2777
67. Junior HM, Silva J, Arenzon A, Portela CS, Ferreira IC, Henriques

JA (2007) Evaluation of genotoxicity and toxicity of water and sediment samples from a Brazilian stream influenced by tannery industries. Chemosphere 67(6):1211–1217

68. Katsumiti A, Domingos FXV, Azevedo M, Silva MD, Damian RC, Almeida MIM, Silva De Assis HC, Cestari MM, Rando MAF, Oliveira Ribeiro CA, Freire CA (2009) An assessment of acute biomarker responses in the demersal catfish *Cathorops spixii* after the Vicuña Oil Spill in a harbour estuarine area in Southern Brazil. Environ Monit Assess 152:209–222
69. Kirschbaum AA, Seriani R, Pereira CDS, Assunção A, Abessa DMS, Rotundo MM, Ranzani-Paiva MJT (2009) Cytogenotoxicity biomarkers in fat snook *Centropomus parallelus* from Cananéia and São Vicente estuaries, SP, Brazil. Genet Mol Biol 32(1):151–154
70. Langston WJ, Bebianno MJ, Burt GR (1998) Metal handling strategies in mollusks. In: Langston WJ, Bebianno MJ (eds) Metal metabolism in aquatic environments. Kluwer Academic Publishers, London, pp 219–283
71. Lardone MC, Castillo P, Valdevenito R, Ebensperger M, Ronco AM, Pommer R, Piottante A, Castro A (2010) P450-aromatase activity and expression in human testicular tissues with severe spermatogenic failure. Int J Androl 33(4):650–660
72. Leme DM, Marin-Morales MA (2009) *Allium cepa* test in environmental monitoring: a review on its application. Mutat Res 682:71–81
73. Lemos CT, Rödel PM, Terra NR, Oliveira NCD, Erdtmann B (2007) River water genotoxicity evaluation using micronucleus assay in fish erythrocytes. Ecotoxicol Environ Safe 66:391–401
74. Lemos CT, Iranço FA, Oliveira NCD, Souza GD, Fachel JMG (2008) Biomonitoring of genotoxicity using micronuclei assay in native population of *Astyanax jacuhiensis*(Characiformes: Characidae) at sites under petrochemical influence. Sci Total Environ 406:337–343
75. Lemos AO, Oliveira NCD, Lemos CT (2011) In vitro micronuclei tests to evaluate the genotoxicity of surface water under the influence of tanneries. Toxicol In Vitro 25:761–766
76. Linde-Arias AR, Inácio AF, Alburquerque C, Freire MM, Moreira

JC (2008a) Biomarkers in an invasive fish species, *Oreochromis niloticus*, to assess the effects of pollution in a highly degraded Brazilian River. Sci Total Environ 399:186–192

77. Linde-Arias AR, Inácio AF, Novo LA, Alburquerque C, Moreira JC (2008b) Multibiomarker approach in fish to assess the impact of pollution in a large Brazilian river, Paraíba do Sul. Environ Pollut 156:974–979
78. Livingstone DR (2001) Contaminant-stimulated reactive oxygen species production and oxidative damage in aquatic organisms. Mar Pollut Bull 42:656–666
79. Lüchmann KH, Toledo-Silva G, Bainy ACD, Marques MRF (2007) Glutathione S-transferase cytosolic isoform in the pink-shrimp, *Farfantepenaeus brasiliensis*, from Conceição Lagoon, Santa Catarina Island, SC, Brazil. Environ Int 33:546–549
80. Lupi C, Nhacarini NI, Mazon AF, Sá OR (2007) Avaliação da poluição ambiental através das alterações morfológicas nas brânquias de *Oreochromis niloticus* (tilapia) nos córregos Retiro, Consulta e Bebedouro, município de Bebedouro-SP. Rev Fafibe On Line 3:1–6
81. Maceda EB, Grisolia AB, Vaini JO, Candido LS (2015) Uso de biomarcadores para monitoramento das águas do Córrego Arara no município de Rio Brilhante, MS, Brasil. Rev Ambient Água 10(1):117–129
82. Mayer FL, Versteeg DJ, Mckee MJ, Folmar LC, Graney RL, McCume DC, Rattner BA (1992) Metabolic products as biomarkers, 1992. In: Huggett RJ, Kimerly RA, Mehrle PM Jr, Bergman HL (eds) Biomarkers: biochemical, physiological and histological markers of anthropogenic stress. Lewis Publishers, Chelsea, pp 5–86
83. McDonnell AM, Dang CH (2013) Basic review of the cytochrome p450 system. J Adv Pract Oncol 4(4):263–268
84. McGlashan DJ, Hughies JM (2001) Genetic evidence for historical continuity between populations of the Australian freshwater fish *Craterocephalus stercusmuscarum* (Atherinidae) east and west of the Great Diving Range. J Fish Biol 59:55–67
85. Medeiros ID, Siebert MN, Toledo-Silva G, Moraes MO, Marques

MRF, Bainy ACD (2008a) Differential gene expression in oyster exposed to sewage. Mar Environ Res 66:156–157

86. Medeiros ID, Siebert MN, Toledo-Silva G, Rodrigues TB, Marques MRF, Bainy ACD (2008b) Induced gene expression in oyster *Crassostrea gigas* exposed to sewage. Environ Toxicol Pharmacol 26(3):362–365
87. Melo KM, Alves IR, Pieczarka JC, David JAO, Nagamachi CY, Grisolia CK (2013) Profile of micronucleus frequencies and nuclear abnormalities in different species of electric fishes (Gymnotiformes) from the Eastern Amazon. Genet Mol Biol 36(3):425–429
88. Miranda AL, Roche H, Randi MAF, Menezes ML, Oliveira-Ribeiro CA (2008) Bioaccumulation of chlorinated pesticides and PCBs in the tropical freshwater fish *Hoplias malabaricus*: histopathological, physiological, and immunological findings. Environ Int 34:939–949
89. Montes CS, Ferreira MAP, Santos SSD, Von Ledebur EICF, Rocha RM (2010) Branchial histopathology study of *Brachyplatystoma rousseauxii* (Castelnau, 1855) in the Guajará bay, Belém, Pará State, Brazil. Acta Sci Biol Sci 32:93–99
90. Montine KS, Quinn JF, Zhang J, Fessel JP, Roberts LJ, Morrow JD (2004) Isoprostanes and relates products of lipid peroxidation in neurodegenerative diseases. Chem Phys Lipids 128:117–124
91. Nascimento AA, Araújo FG, Gomes ID, Mendes RM, Sales A (2012) Fish gills alterations as potential biomarkers of environmental quality in a eutrophized tropical river in south-eastern Bazil. Anat Histol Embryol 41(3):209–216
92. Nunes EA, Lemos CT, Gavronski L, Moreira TN, Oliveira NCD, Silva J (2011) Genotoxic assessment on river water using different biological systems. Chemosphere 84:47–53
93. Oliveira MM, Silva-Filho MV, Cunha-Bastos VLF, Fernandes FC, Bastos JC (2007) Brain acetylcholinesterase as a marine pesticide biomarker using Brazilian fishes. Mar Environ Res 63:303–312
94. Oliveira JP, Santos RN, Pibernat CC, Boeira JM (2012) Genotoxicity and physical chemistry analysis of waters from Sinos River (RS) using *Allium cepa* and *Eichhornia crassipes* as bioindicators. Biochem Biotechnol Rep 1(1):15–22
95. Osório FHT, Silva LFO, Piancini LDS, Azevedo ACB, Liebel S,

Yamamoto FY, Philippi VP, Oliveira MLS, Ortolani-Machado CF, Neto FF, Cestari MM, Assis HCS, Ribeiro CAO (2014) Water quality assessment of the Tubarão River through chemical analysis and biomarkers in the Neotropical fish *Geophagus brasiliensis*. Environ Sci Pollut Res Int 21(15):9145–9160

96. Parente TEM, De Oliveira ACAX, Silva IB, Araujo FG, Paumgarten FJR (2004) Induced alkoxyresorufin-*O*-dealkylases in tilapias (*Oreochromis niloticus*) from Gandu River, Rio de Janeiro, Brazil. Chemosphere 54:1613–1618
97. Parente TEM, De Oliveira ACAX, Paumgartten FJR (2008) Induced cytochrome P450 1A activity in cichlid fishes from Guandu River and Jacarepaguá Lake, Rio de Janeiro, Brazil. Environ Pollut 152:233–238
98. Payne JF, Mathieu A, Melvin W, Fancey LL (1996) Acetylcholinesterase, an old biomarker with a new future? Field trials in association with two urban rivers and a paper mill in Newfoundland. Mar Poll Bull 32:225–231
99. Pereira CDS, Abessa DMS, Bainy ACD, Zaroni LP, Gasparro MR, Bícego MC, Taniguchi S, Furley TH, Sousa ECPM (2007) Integrated assessment of multilevel biomarker responses and chemical analysis in mussels from São Sebastião, São Paulo, Brazil. Environ Toxicol Chem 26(3):462–469
100. Pereira CDS, Martin-Díaz ML, Zanette J, Cesar A, Choueri RB, Abessa DMS, Catharino MGM, Vasconcellos MBA, Bainy ACD, Sousa ECPM, Valls TAD (2011) Integrated biomarker responses as environmental status descriptors of a coastal zone (São Paulo, Brazil). Ecotoxicol Environ Safe 74:1257–1264
101. Pereira CDS, Abessa DMS, Choueri RB, Almagro-Pastor V, Cesar A, Maranho LA, Martin-Diaz LM, Torres RJ, Gusso-Choueri PK, Almeida JE (2014) Ecological relevance of sentinel's biomarker responses: a multi-level approach. Mar Environ Res 96:118–126
102. Pesce SF, Cazenave J, Monferrán MV, Frede S, Wunderlin DA (2008) Integrated survey on toxic effects of lindane on neotropical fish: *Corydoras paleatus* and *Jenynsia multidentata*. Environ Pollut 156:775–783
103. Prá D, Lau AH, Knakievicz T, Carneiro FR, Erdtmann B (2005)

Environmental genotoxicity assessment of an urban stream using freshwater planarians. Mutat Res 585:79–85

104. Prado LRGB, Felix C, Abessa DMS, Buruaem LM, Abujamara LD, Kirschbaum AA, Turatti GCR, Ranzani-Paiva MJT, Correia AT, Seriani R (2015) Hematological parameters and nuclear abnormalities in peripheral erythrocytes of *Achirus lineatus* (*Pleuronectiformes: Achiridae*). Comp Clin Pathol 24:169–175
105. Procópio MS, Ribeiro HJ, Pereira LA, Oliveira Lopes GA, Castro AC, Rizzo E, Sato Y, Russo RC, Corrêa JD (2014) Sex-response differences of immunological and histopathological biomarkers in gill of *Prochilodus argenteus* from a polluted river in southeast Brazil. Fish Shellfish Immunol 39(1):108–117
106. Ranzani-Paiva MJT, Silva-Souza AT (2004) Co-infestation of gills by different parasite groups in the mullet, *Mugil platanus* Günther, 1880 (Osteichthyes, Mugilidae): effects on relative condition factor. Braz J Biol 64:677–682
107. Rechenmacher C, Siebel AM, Goldoni A, Klauck CR, Sartori T, Rodrigues MT, Gehlen G, Ardenghi P, Silva LB (2010) A multibiomarker approach in rats to assess the impact of pollution on Sinos River, southern Brazil. Braz J Biol 70(4):1223–1230
108. Ribeiro CAO, Katsumiti A, França P, Maschio J, Zandoná E, Cestari MM, Vicari T, Roche H, Assis HCS, Neto FF (2013) Biomarkers responses in fish (*Atherinella brasiliensis*) of Paranaguá Bay, Southern Brazil, for assessment of pollutant effects. Braz J Oceanogr 61(1):1–11
109. Richetti SK, Rosemberg DB, Ventura-Lima J, Monserrat JM, Bogo MR, Bonan CD (2011) Acetylcholinesterase activity and antioxidant capacity of zebrafish brain is altered by heavy metal exposure. Neurotoxicology 32:116–122
110. Rocha RM, Coelho RP, Montes CS, Santos SSD, Ferreira MAP (2010) Avaliação histopatológica de *Brachyplatystoma rousseauxii* (Castelnau, 1855) da Baía do Guajará, Belém, Pará. Ci Anim Bras 11(1):101–109
111. Rogero SO, Lugão AB, Ikeda TI, Cruz AS (2003) Teste in vitro de citotoxicidade: estudo comparativo entre duas metodologias. Mater Res 6:317–320

112. Rola RC, Monteiro MC, Reis SRS, Sandrini JZ (2012) Molecular and biochemical biomarkers responses in the mussel *Mytilus edulis* collected from Southern Brazil coast. Mar Pollut Bull 64:766–771

113. Ruas CBG, Carvalho CS, Araújo HSS, Espíndola ELG, Fernandes MN (2008) Oxidative stress biomarkers of exposure in the blood of cichlid species from a metal-contaminated river. Ecotoxicol Environ Safe 71:86–93

114. Sáenz LA, Seibert EL, Zanette J, Fiedler HD, Curtius AJ, Ferreira JF, Almeida EA, Marques MRF, Bainy ACD (2010) Biochemical biomarkers and metals in *Perna perna* mussels from mariculture zones of Santa Catarina, Brazil. Ecotoxicol Environ Safe 73:796–804

115. Santos PE, Kummrow F, Albergaria-Barbosa ACR, Bícego MC, Umbuzeiro GA (2010) Mutagenicity of blue rayon extracts of fish bile as a biomarker in a field study. Environ Mol Mutagen 51:173–179

116. Scalon MCS, Rechenmacher C, Siebel AM, Kayser ML, Rodrigues MT, Maluf SW, Rodrigues MAS, Silva LB (2010) Evaluation of Sinos River water genotoxicity using the comet assay in fish. Braz J Biol 70(4):1217–1222

117. Schulz UH, Martins-Junior H (2001) *Astyanax fasciatus* as bioindicator of water pollution of Rio dos Sinos, RS, Brazil. Braz J Biol 61(4):615–622

118. Seriani R, Moreira LB, Abessa DMS, Abujamara LD, Carvalho NSB, Maranho LA, Kirschbaum AA, Ranzani-Paiva MJT (2010) Hematological analysis of *Micropogonias furnieri*, Desmarest, 1823, Scianidae, from two estuaries of Baixada Santista, São Paulo, Brazil. Braz J Oceanogr 58:87–92

119. Seriani R, Abessa DMS, Kirschbaum AA, Pereira MJT, Ranzani-Paiva MJT, Assunção A, Silveira FL, Romano P, Mucci JLN (2012) Water toxicity and cyto-genotoxicity biomarkers in the fish *Oreochromis niloticus* (Cichlidae). J Braz Soc Ecotoxicol 7(2):67–72

120. Seriani R, Abessa DMS, Pereira CDS, Kirschbaum AA, Assunção A, Ranzani-Paiva MJT (2013) Influence of seasonality and pollution on the hematological parameters of the estuarine fish *Centropomus*

parallelus. Braz J Oceanogr 61(2):105–111

121. Seriani R, Abessa DMS, Moreira LB, Cabrera JPG, Sanches JQ, Silva CLS, Amorim FA, Rivero DHRF, Silva FL, Fitorra LS, Carvalho-Oliveira R, Macchione M, Ranzani-Paiva MJT (2015) In vitro mucus transportability, cytogenotoxicity, and hematological changes as non-destructive physiological biomarkers in fish chronically exposed to metals. Ecotoxicol Environ Safe 112:162–168
122. Silva AG, Martinez CBR (2007) Morphological changes in the kidney of a fish living in an urban stream. Environ Toxicol Pharmacol 23:185–192
123. Silva DAM, Buzitis J, Krahn MM, Bícego MC, Pires-Vanin AMS (2006) Metabolites in bile of fish from São Sebastião Channel, São Paulo, Brazil as biomarkers of exposure to petrogenic polycyclic aromatic compounds. Mar Pollut Bull 52:175–183
124. Sousa DBP, Almeida ZS, Carvalho-Neta RNF (2013) Biomarcadores histológicos em duas espécies de bagres estuarinos da Costa Maranhense, Brasil. Arq Bras Med Vet Zootec 65(2):369–376
125. Souza TS, Fontanetti CS (2006) Micronucleus test and observation of nuclear alterations in erythrocytes of Nile tilapia exposed to waters affected by refinery effluent. Mutat Res 605:87–93
126. Souza DSM, Ramos APD, Nunes FF, Moresco V, Taniguchi S, Leal DAG, Sasaki ST, Bícego MC, Montone RC, Durigan M, Teixeira AL, Politto MR, Delfino N, Franco RMB, Melo CMR, Bainy ACD, Barardi CRM (2012) Evaluation of tropical water sources and mollusks in southern Brazil using microbiological, biochemical, and chemical parameters. Ecotoxicol Environ Safe 76:153–161
127. Souza IC, Duarte ID, Pimentel NQ, Rocha LD, Morozesk M, Bonomo MM, Azevedo VC, Pereira CDS, Monferrán MV, Milanez CRD, Matsumoto ST, Wunderlin DA, Fernandes MN (2013) Matching metal pollution with bioavailability, bioaccumulation and biomarkers response in fish (*Centropomus parallelus*) resident in neotropical estuaries. Environ Pollut 180:136–144
128. Souza-Bastos LR, Freire CA (2011) Osmoregulation of the resident estuarine fish *Atherinella brasiliensis* was still affected by an oil

spill (Vicuña tanker, Paranaguá Bay, Brazil), 7 months after the accident. Sci Total Environ 409:1229–1234

129. Sureda A, Box A, Tejada S, Blanco A, Caixach J, Deudero S (2011) Biochemical responses of *Mytilus galloprovincialis* as biomarkers of acute environmental pollution caused by the Don Pedro oil spill (Eivissa Island, Spain). Aquat Toxicol 101:540–549

130. Tice RR, Agurell E, Anderson D, Burlinson B, Hartmann A, Kobayashi H, Miyamae Y, Rojas E, Ryu JC, Sasaki YF (2000) Single cell gel/comet assay: guidelines for in vitro and in vivo genetic toxicology testing. Environ Mol Mutagen 35(3):206–221

131. Torres MA, Testa CP, Gáspari C, Masutti MB, Panitz CMN, Curi-Pedrosa R, Almeida EA, Mascio P, Filho DW (2002) Oxidative stress in the mussel *Mytella guyanensis* from polluted mangroves on Santa Catarina Island, Brazil. Mar Pollut Bull 44(9):923–932

132. Tortelli V, Colares EP, Robaldo RB, Nery LEM, Pinho GLL, Bianchini A, Monserrat JM (2006) Importance of cholinesterase kinetic parameters in environmental monitoring using estuarine fish. Chemosphere 65:560–566

133. Toste R, Fernandez MA, Pessoa IA, Parahyb MA, Dore MP (2011) Organotin pollution at Arraial do Cabo, Rio de Janeiro State, Brazil: increasing levels after the TBT ban. Braz J Oceanogr 59(1):111–117

134. Trapp J, Armengaud J, Salvador A, Chaumot A, Geffard O (2014) Next-generation proteomics: toward customized biomarkers for environmental biomonitoring. Environ Sci Technol 48:13560–13572

135. Udroiu I (2006) The micronucleus test in piscine erythrocytes. Aquat Toxicol 79(2):201–204

136. Valbonesi P, Sartor G, Fabbri E (2003) Characterization of cholinesterase activity in three bivalves inhabiting the North Adriatic sea and their possible use as sentinel organisms for biosurveillance programmes. Sci Total Environ 312(1–3):79–88

137. Van der Oost R, Beyer J, Vermeulen NPE (2003) Fish bioaccumulation and biomarkers in environmental risk assessment: a review. Environ Toxicol Pharmacol 13:149–157

138. Venancio LP, Silva MI, Silva TL, Moschetta VA, Campos-Zuccari DA, Almeida EA, Bonini-Domingos CR (2013) Pollution-induced

metabolic responses in hypoxia-tolerant freshwater turtles. Ecotoxicol Environ Safe 97:1–9

139. Ventura EC, Gaelzer LR, Zanette J, Marques MRF, Bainy ACD (2002) Biochemical indicators of contaminant exposure in spotted pigfish (*Orthopristis ruber*) caught at three bays of Rio de Janeiro coast. Mar Environ Res 54:775–779
140. Viarengo A, Canesi L, Pertica M, Poli G, Moore MN, Orunesu M (1990) Heavy metal effects on lipid peroxidation in the tissues of *Mytilus galloprovincialis* lam. Comp Biochem Physiol Part C Toxicol Pharmacol 97(1):37–42
141. Viarengo A, Lowe D, Bolognesi C, Fabbri E, Koehler A (2007) The use of biomarkers in biomonitoring: a 2-tier approach assessing the level of pollutant induced stress syndrome in sentinel organisms. Comp Biochem Physiol C Toxicol Pharmacol 146:281–300
142. Vieira CED, Almeida MS, Galindo BA, Pereira L, Martinez CBR (2014) Integrated biomarker response index using a neotropical fish to assess the water quality in agricultural areas. Neotrop Ichthyol 12(1):153–164
143. Villela IV, Oliveira IM, Silva J, Henriques JAP (2006) DNA damage and repair in haemolymph cells of golden mussel (*Limnoperna fortunei*) exposed to environmental contaminants. Mutat Res 605:78–86
144. Villela IV, Oliveira IM, Silveira JC, Dias JF, Henriques JAP, Silva J (2007) Assessment of environmental stress by the micronucleus and comet assays on *Limnoperna fortunei* exposed to Guaíba hydrographic region samples (Brazil) under laboratory conditions. Mutat Res 628:76–86
145. Ward D, Simpson S, Jolley D (2013) Slow avoidance response to contaminated sediments elicits sublethal toxicity to benthic invertebrates. Environ Sci Technol 47:5947–5953
146. WHO International Programme on Chemical Safety (IPCS) (1993) Biomarkers and risk assessment: concepts and principles. Environmental Health Criteria 155. World Health Organization, Geneva
147. Zanette J, Monserrat JM, Bianchini A (2006) Biochemical biomarkers in gills of mangrove oyster *Crassostrea rhizophorae*

from three Brazilian estuaries. Comp Biochem Physiol C Toxicol Pharmacol 143(2):187–195

148. Zanette J, Nunes FF, Medeiros ID, Siebert MN, Mattos JJ, Lüchmann KH, Melo CMR, Bainy ACD (2008) Comparison of the antioxidant defense system in *Crassostrea rhizophorae* and *Crassostrea gigas* exposed to domestic sewage discharges. Mar Environ Res 66:196–198
149. Zanette J, Monserrat JM, Bianchini A (2015) Biochemical biomarkers in barnacles *Balanus improvisus*: pollution and seasonal effects. Mar Environ Res 103:74–79
150. Zapata LM, Bock BC, Orozco LY, Palacio JA (2016) Application of the micronucleus test and comet assay in *Trachemys callirostris* erythrocytes as a model for in situ genotoxic monitoring. Ecotoxicol Environ Safe 127:108–116
151. Zegura B, Heath E, Cernosa A, Filipic M (2009) Combination of in vitro bioassays for the determination of cytotoxic and genotoxic potential of wastewater, surface water and drinking water samples. Chemosphere 75:1453–1460
152. Zinkl JG, Lockhart WL, Kenny SA, Ward FJ (1991) The effects of cholinesterase inhibiting insecticides on fish. Cholinesterase-inhibiting insecticides chemicals in agriculture, vol 2. Elsevier, Amsterdam, pp 109–125

CHAPTER

13

TRACKING ANIMALS IN FRESHWATER WITH ELECTRONIC TAGS: PAST, PRESENT AND FUTURE

Steven J Cooke[1], Jonathan D Midwood[1] , Jason D Thiem[1] , Peter Klimley[2] , Martyn C Lucas[3] , Eva B Thorstad[4] , John Eiler[5] , Chris Holbrook[6] and Brendan C Ebner[7]

[1] Fish Ecology and Conservation Physiology Laboratory, Department of Biology, Carleton University, 1125 Colonel By Dr., Ottawa, ON, Canada

[2] Biotelemetry Laboratory, Department of Fish, Wildlife and Conservation Biology, University of California at Davis, Davis, CA, USA

[3] Aquatic Animal Ecology Research Group, School of Biological and Biomedical Sciences, Durham University, Durham, UK

[4] Norwegian Institute for Nature Research, Trondheim, Norway

[5] Auke Bay Laboratories, Alaska Fisheries Science Center, National Marine Fisheries Service, National Ocean and Atmospheric Administration, Juneau, AK, USA

[6] Hammond Bay Biological Station, United States Geological Survey, Millersburg, MI, USA

[7] Tropical Landscapes Joint Venture, CSIRO Ecosystem Sciences & TropWATER, James Cook University, Atherton, QLD, Australia

ABSTRACT

Considerable technical developments over the past half century have enabled widespread application of electronic tags to the study of

animals in the wild, including in freshwater environments. We review the constraints associated with freshwater telemetry and biologging and the technical developments relevant to their use. Technical constraints for tracking animals are often influenced by the characteristics of the animals being studied and the environment they inhabit. Collectively, they influence which and how technologies can be used and their relative effectiveness. Although radio telemetry has historically been the most commonly used technology in freshwater, passive integrated transponder (PIT) technology, acoustic telemetry and biologgers are becoming more popular. Most telemetry studies have focused on fish, although an increasing number have focused on other taxa, such as turtles, crustaceans and molluscs. Key technical developments for freshwater systems include: miniaturization of tags for tracking small-size life stages and species, fixed stations and coded tags for tracking large samples of animals over long distances and large temporal scales, inexpensive PIT systems that enable mass tagging to yield population- and community-level relevant sample sizes, incorporation of sensors into electronic tags, validation of tag attachment procedures with a focus on maintaining animal welfare, incorporation of different techniques (for example, genetics, stable isotopes) and peripheral technologies (for example, geographic information systems, hydroacoustics), development of novel analytical techniques, and extensive international collaboration. Innovations are still needed in tag miniaturization, data analysis and visualization, and in tracking animals over larger spatial scales (for example, pelagic areas of lakes) and in challenging environments (for example, large dynamic floodplain systems, under ice). There seems to be a particular need for adapting various global positioning system and satellite tagging approaches to freshwater. Electronic tagging provides a mechanism to collect detailed information from imperilled animals and species that have no direct economic value. Current and future advances will continue to improve our knowledge of the natural history of aquatic animals and ecological processes in freshwater ecosystems while facilitating evidence-based resource management and conservation.

Keywords: Acoustic telemetry, Biotelemetry, Coded tags, Passive integrated transponder, Radio telemetry

BACKGROUND

Freshwater ecosystems around the globe face many threats and stressors [1, 2, 3, 4], which have collectively led to dramatic reductions in freshwater biodiversity for a range of taxa [5, 6]. Human activities are the primary source of these multiple stressors [7], such that Strayer and Dudgeon [8] eloquently describe what they call, 'the collision between humans and biodiversity in freshwater ecosystems.' Freshwater ecosystems provide vital resources for human beings (including water for drinking and irrigation, food, electricity and transport) and demands for freshwater have risen dramatically over the last century. The 125,000 species of freshwater animals that have been described represent 9.5% of all known animal species on the planet (including one-third of all vertebrate species), even though fresh waters cover just 0.8% of the Earth's surface [6, 9], emphasizing that freshwater ecosystems are hotspots for biodiversity [8]. Unfortunately, owing to the proximity of human population centers to sources of freshwater and the associated demands on this resource, these systems are also hotspots for endangerment [6, 8]. Given the immense ecosystem services provided by freshwater ecosystems [10], the importance of the animals that occupy these waters [11], and the livelihoods and well-being of the human beings that depend upon freshwater ecosystems [12, 13], there is a need to understand the biology of freshwater animals.

Basic life history information (including habitat requirements, migratory routes, foraging and reproductive sites, dispersal and home-range characteristics) is fundamental to understanding animals in relation to their environment, and when studied in the context of responses to human activities, provides resource managers and conservationists with practical knowledge to make informed decisions. However, studying macroscopic animals in freshwater ecosystems comes with its own unique set of challenges. Extreme depths, and high turbidity, discharge or current speed, ice cover, habitat complexity, and macrophyte density constrain the ability to view and collect freshwater animals. Additional constraints are imposed by the fact that some animals are small, cryptic or highly mobile. Although certain types of behavior and physiology can be easily studied in captivity, relying solely on this type of study can be misleading since such studies fail to mimic natural scenarios involving complex interactions among multiple species and large-scale processes.

Today's field biologists have a variety of tools for studying animals in the natural environment, but given the challenges with visual observation in aquatic habitats, electronic tagging technology is regarded as one of the most effective means of studying the behavior and ecophysiology of animals in their natural environment [14]. A variety of electronic tagging tools have been used in freshwater for the study of vertebrates and some of the larger invertebrates (summarized in [15, 16, 17, 18]; See Table 1). Active telemetry tools that remotely transmit information between a transmitter and receiver, including radio and acoustic telemetry, and passive tools, such as passive integrated transponders (PIT tags), have been applied to a variety of freshwater taxa (Figure 1). Archival devices that record information from sensors (such as temperature, pressure, activity) that can be downloaded onto retrieving devices (also known as biologging) are also increasingly common in freshwater systems. These types of sensor have also been used extensively with active telemetry tools to enhance information collected regarding animal locations.

Table 1: Summary of the various technologies used for tracking animals in freshwater with a brief summary of strengths, limitations, and common applications

Technology	**Summary of technology and techniques**	**Strengths**	**Limitations**	**Applications**
Acoustic telemetry; manual tracking	Uses a transducer to convert electrical energy to acoustic energy that is detected by an underwater hydrophone	Deep water (>20 m)	Hydrophone must be submerged in water	Some applications in FW
	Animals usually tracked by boat using bearings and triangulation	Can provide detailed movement data depending on tracking method and conditions	Ineffective in shallow or turbulent water	Mostly for fish, some use with alligator, FW mammals and elasmobranch
	Not suitable for transmitters with long pulse interval (time between pulses)		Interference from macrophytes and noise (for example, boats, entrained air)	

Acoustic telemetry; fixed stations	As above	Can be deployed as gates, grids, or arrays to monitor animal movements for long periods, including under ice	Generates large datasets	Widespread use in FW
	Autonomous or cabled hydrophones and associated loggers store time-stamped data when tagged animals enter reception zone	Multiple stations can provide precise two or three-dimensional tracks of animals	Requires significant post-processing and analytical efforts (can be challenging to recover data with some systems)	Mostly on fish, some use with mammals that move between marine and FW environments
		Some systems provide real-time data transmission	Interference from macrophytes and noise (for example, boats, entrained air)	
Radio telemetry; manual tracking	Emit electromagnetic energy in the radio frequency range (usually in the VHF band between 30 and 300 MHz)	Shallow water (<10 m)	Deep water (>15 m)	Widespread in FW
	Signals detected by antennas (aerial or underwater) and a receiver (some have logging capability)	Low-conductivity (<500 µS/cm)	High conductivity Sensitive to localized interference	Fish and other taxa, especially amphibious species (for example, basking turtles, amphibians, some mammals)
	Tracking can occur from boat, vehicle, air, foot	Relatively inexpensive		
		Functions in moving water and through ice as well as on land and in air		

Radio telemetry; fixed stations	As above	As above	Antennas visible, thus can attract vandals	Widespread use in FW
	Fixed stations with multiple antennas detect and log tags when in the vicinity of an antenna	Suitable for long-term deployments	Not possible to obtain precise two-dimensional positions (mostly presence or absence in a given location)	Mostly used with fish
	Most often deployed in riverine systems to detect migration		Sensitive to interference	
			Effectiveness dependent on local geomorphology and station placement [19]	
Passive integrated transponder telemetry; manual tracking	Integrated circuit chip and coil antenna that transmits a unique identity code when energized by a low-frequency radio signal (generally 125 to 400 kHz)	No battery (therefore long life)	Largely limited to shallow and restricted lotic systems owing to small detection range (usually <1 m)	Widespread use in FW
	Detected using hand-held (usually in a backpack with wand waved in or above water surface while wading or walking on ice) or boat-mounted readers	Small	Multiple tags in the same location can prevent signal transmission	Mostly in small fishes, reptiles, amphibians and invertebrates
		Inexpensive	Detection range depends on tag orientation relative to the reader	

Passive integrated transponder telemetry; fixed stations	As above	No battery (therefore long life)	As above	Widespread use in FW
	Antennas deployed in systems that record the time and unique ID of a given tag detection	Small	Remote stations require power source to function	Mostly for fish
	A variety of antenna designs, mostly for small lotic systems, fishways or other areas where animals congregate	Inexpensive		
Low-frequency electromagnetic telemetry	Similar principle as PIT tags, but tags are continually powered	Effective in areas with high noise levels where acoustic and radio may be problematic	Site-specific detection only	Limited use in FW
	Greater detection range on the order of several tens of meters (for example, [20, 21])		Limited tag life	The few examples use fish in lowland rivers [20]
Combined acoustic radio transmitters	Both acoustic and radio output modules can either transmit simultaneously or, if dynamic, can switch between the two	Works in marine and FW	Large devices	Some use in large diadromous fish species and marine mammals that move between marine and fresh waters
	Switching driven by environmental cues (depth or conductivity) [22]	Works for animals that use a wide range of depths or move into higher conductivity waters	More expensive than either radio or acoustic tags	

Archival tags (no transmitting capability)	Biologging tags (for example, data storage tags, time-depth recorders, archival geolocation tags)	Continuous monitoring of desired parameter (for example, temperature, depth)	Require tag recovery to download data from onboard memory	Some use in FW, mostly for logging temperature, depth, geolocation, and other environmental parameters (especially for ectothermic animals)
	Do not transmit data remotely			Occasionally log physiological information (for example, acceleration, heart rate)
	Always equipped with some form of sensor			
Communicating histogram acoustic transponders and business card tags	Hybrid loggers and transmitters, first logging data to memory and later transmitting data when interrogated by a receiver	Useful for wide-ranging fishes that have low recapture probabilities but return to download sites [23]	Reasonably large	Not applied in FW to our knowledge (mostly marine applications)
	Business card tags are a variant where the logging function actually serves as a mini acoustic receiver and can log other tagged animals that it encounters [24]		Expensive	
			Some versions require that tags are retrieved to download data	

Smart position-only tag with real-time GPS	Satellite platform transmitter which identifies tag location by real-time uplinks to orbiting Argos satellites whenever tag breaches the water surface (some variants may include archival capability rather than only transmitting in real time)	Real-time positioning	Relatively expensive	Mostly marine
		Provides information on broad-scale movement of animals (for example, ocean basin scale)	Individual must breach water	Some use in large FW animals that surface (for example, alligators, manatees)
		Detailed records of environmental conditions experienced	Limited to large animals	
Archival pop-up satellite archival transmitter tags	Tags programmed to jettison at a given time or when tags experience a specific event (for example, fail to change depth for a specified period of time) at which time they float to surface and transmit summary data or packets to satellites	Provides information on broad-scale movement of animals (for example, ocean basin scale)	Most release links based on corrosive link, switched on via applied voltage and require seawater to function	Mostly marine
	Geopositioning commonly based on light levels	Detailed records of environmental conditions experienced	Limited to large animals	Most FW applications involve diadromous animals tagged during FW phase with pop-off scheduled for when they are in marine systems
	Can also log depth and temperature		Relatively expensive	Exception: studies on FW porpoises [25]
			Tags and downloading technology not always reliable, needs further refinement	

FW, freshwater; GPS; global positioning system; VHF, very high frequency.

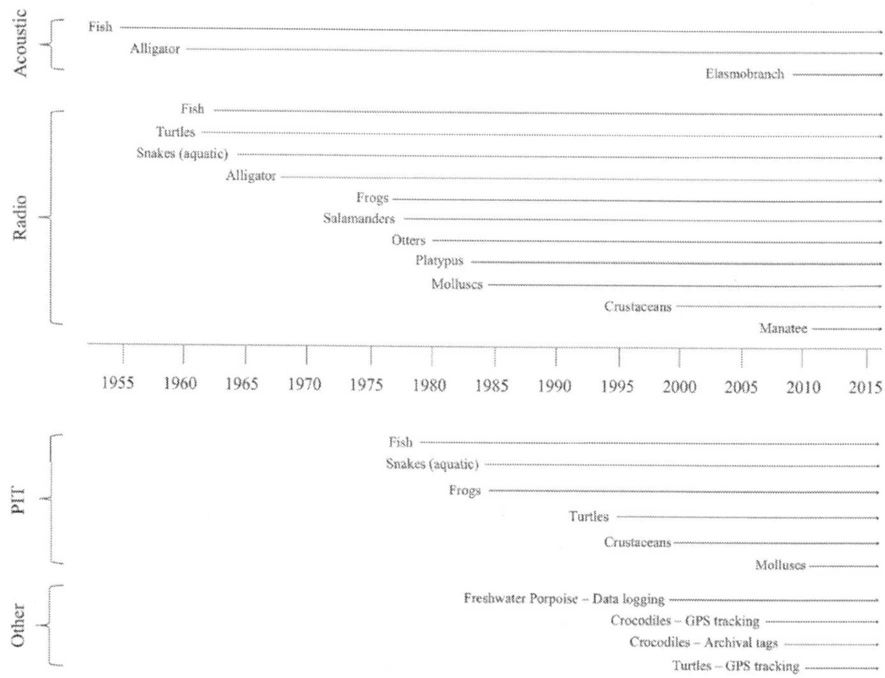

Figure 1: Technology timeline, showing applications for various freshwater species. Early adoption tended to be for larger species of economic importance (that is, salmonids). With miniaturization of tagging technology in the late 1980s, biotelemetry can now be applied to a greater diversity of species.

Despite the breadth of current telemetry technology, there are still many challenges associated with using electronic tags to study freshwater species. These range from the problem of attaching devices to animals, to the technical performance of the electronic technology, and methods of analyzing data. By considering historical developments that have led to our contemporary electronic tagging toolbox, it is possible to identify specific constraints and technical gaps that still exist and provide the foundation needed to foster future advances. To that end, the objective of this synthesis is to consider the past, present and future of electronic tagging and tracking techniques and technology. In order to narrow the scope of the review and focus on environments that share similar challenges, we focus on animals found in freshwater systems, including fish, mammals, reptiles, amphibians and invertebrates (mostly molluscs and arthropods) that are sufficiently large to be studied with electronic tags. We exclude all birds, since most associated tracking studies for birds focus on large-scale migrations. We include diadromous fish (but only their freshwater phase) as well as mammals that spend significant time in freshwater (for example, muskrats, manatees). This is not intended to be an exhaustive review. Rather the information presented represents the collective knowledge and perspectives of the authors with experience in using a variety of electronic tagging technologies, in different regions, and on different taxa.

Key Developments in Freshwater Tracking

The creation and use of new technologies, methods and approaches is a gradual process. As such, many of the key developments are a culmination of innovation and discovery rather than a single discovery (or paper). Table 2 presents a list of key telemetry technologies used to study freshwater animals in the wild, and their associated strengths, weaknesses and typical applications.

Table 2: Summary of key constraints (categorized by abiotic, anthropogenic, environmental/habitat, and biotic) associated with using electronic tags to study animals in freshwater systems, existing technical solutions, and additional technical needs that need to be addressed to improve equipment performance and science outcomes

Category	Condition	Constraint	Existing technical solutions	Additional technical needs
Abiotic	Ice cover in lentic systems	In some regions ice cover for part of the year impedes access the water	When ice conditions are safe animals can be tracked through the ice with radio receivers on foot (skiing, snowshoeing) or driving (for example, snowmobile)	Little information on acoustic performance under ice but some evidence of noise from cracking or shifting reducing detections (Mike Parsley, personal communication)
		Modifies signal properties (for example, via reflections, noise)	Acoustic receivers can be deployed below ice but must be moored so that they are not dragged during ice breakup (for example, do not use surface buoys [26])	Understand the performance of acoustic telemetry (especially for logging and automated positioning) under the ice [27]
		May create problems during breakup by moving or damaging equipment		

Abiotic	Ice dynamics in lotic systems	Winter ice dynamics pose problems for gear deployed in water due to potential damage (for example, PIT antennas, hydrophones [28])	Radio telemetry is favored for tracking [29, 30]	For PIT and acoustic systems there is need for antenna and receiver designs that are not subject to ice damage
		Loss of reception associated with noise and physical obstruction (especially acoustic telemetry [31])	Flatbed PIT antennas reasonably robust to ice processes [32]	
			In small streams, use mobile PIT antennas through the ice [33]	
Abiotic	Shallow water	Exceptionally shallow water (for example, 0 to 30 cm or even ephemeral systems, like billabongs) can expose underwater hydrophones and make manual acoustic tracking nearly impossible [34]	Radio telemetry and mobile PIT antennas can be used in shallow water	Limitations due to physics and physical constraints unlikely to be overcome with technology
			Care must be taken not to disturb or chase animals being tracked	
Abiotic	Deep water	Deep water attenuates radio signals [35, 36] and makes use of radio transmitters and PIT tags ineffective for positioning (in the latter case by virtue of the current difficulty of deploying PIT antennae at depth, though technically feasible if linked *in situ* with waterproofed logger and power supply)	Acoustic telemetry performs well in deep water but positioning accuracy will depend on depth	Unlikely that technical developments will improve performance of radio and PIT systems in deep water
			Can use CART tags [22]	
			A downrigger device has been developed that enables tracking of radio-tagged fish at depth, but has limited reception zone [37]	
			Low-frequency radio tags can improve performance in deep water [35] as can use of more powerful batteries or stronger transmitter output [36]	
Abiotic	Thermocline	Thermoclines can impair the performance of acoustic transmitters [38, 39], but radio transmitters are unaffected (aside from the fact that thermoclines often occur at depth)	Hydrophones can be placed both above and below thermoclines to improve reception, or at the same level as the fish being tracked if they do not frequently cross the thermocline	Unlikely that technical developments will entirely address this issue with acoustic tags
			Rarely is performance so degraded that acoustic detections are not possible, but may impede ability to achieve fine-scale positioning	

Abiotic	High flows, currents and turbulence	High water flows in riverine systems or from lentic currents (creating noise on hydrophones) and associated turbulence and entrained air (even in low flows) can impede the performance of acoustic telemetry systems [40]	Increasing the density of acoustic receivers can improve detection probability [40]	There is need for technical developments further refining code detection and noise filtering in acoustic telemetry systems deployed in high flows, currents and turbulence
			Fixed acoustic receiver station deployment innovations enable their placement in large rivers [41]	Shielding of hydrophones may help to reduce noise (Dale Webber, unpublished data)
			Radio telemetry is unimpeded by such conditions	
			Fixed PIT systems may be difficult to maintain in high flows (get washed out)	
Abiotic	Turbidity	High turbidity (especially if associated with suspended matter) can impair performance of acoustic systems [42]	Acoustic systems usually perform well except in extreme turbidity	Need to develop alternative ways of positioning animals with sensors that do not require light levels (that is, PSATs)
		Makes use of light sensors to position animals ineffective (for example, loggers or PSATs)	Radio and PIT are much less affected	
Abiotic	Conductivity	High conductivity can severely attenuate VHF and HF radio signals in particular but can also reduce performance of acoustic telemetry systems [34]	The extent of radio signal attenuation varies by radio frequency such that choice of lower frequencies (for example, 30 MHz) can improve performance [43, 44]	Innovations in the marine realm for devices that function well in high salinity (and thus high conductivity) potentially benefit freshwater research
			Conductivity presumably also influences PIT system performance [45]	
			Compared with radio, acoustic performance is relatively unaffected by high conductivity [33, 46]	
			Can use CARTs [22]	

Abiotic	Salinity	Saline environments have very high conductivity and make VHF and HF telemetry unusable (as above)	In saline environments, PSATs based on corrosive releases that enable tags to be jettisoned can be used	Development of PSAT releases that work in freshwater would provide new tracking opportunities
		High salinity has an effect on how much sound energy water absorbs [47], which is relevant for acoustic tracking	Corrosive releases do not work in low-salinity environments; limits use of PSATs in inland systems	
			Most inland systems would not have problems with radio attenuation or acoustic energy absorption due to salinity	
Abiotic	Air-water interface	Some animals move in and out of water frequently (for example, turtles) despite being primarily aquatic, which leads to varied signal strength for radio tags and no signal for acoustic tags when in the air	Most studies of animals at the interface use radio telemetry although satellite and GPS tags can also be effective	Unlikely that technical solutions will address these issues
			Acoustic tags can still be used to quantify timing and duration of out of water activities (for example, basking in turtles)	
			CART technologies can be used	
Anthropogenic	Human activity and vandalism	In areas with human activity, some telemetry infrastructure is conspicuous and can attract vandals [48]	Acoustic receivers, if deployed at depths and using gear that the public cannot easily reach, are usually safe from vandals	Development of less conspicuous deployments for radio and PIT gear, such as autonomous underwater antennas and receiving systems (as is common in acoustic telemetry)
		Swim-through PIT loops, with the top edge above the water surface should not be deployed where boats occur	Most fixed-radio and PIT tag antenna arrays (unless in controlled access areas) are subject to vandalism even when using 'vandal-resistant' gear (for example, fixed and locked boxes [48])	
			Miniaturization of power systems and autonomous receivers has helped to deploy receivers out of sight	
			Labeling the equipment with information on content, purpose and value of equipment may reduce likelihood of vandalism in some areas	

Anthropogenic	Human infrastructure and construction materials	Human infrastructure (for example, fishways, dams) are often studied using electronic tags but the construction materials (for example, concrete, metal), operations (for example, entrained air, high flows), and electromagnetic activity (for example, from hydro wires) can interfere with radio, PIT and acoustic systems	Underwater radio antennas can be used in areas near electrical infrastructure to reduce interference [49]	Continued development of equipment that is robust and shielded from electromagnetic activity
		Background radio noise over a wide range of frequencies (for example, communication systems and electrical devices)	Many efforts have been devoted to developing acoustic systems that can work around infrastructure	Development of improved radio telemetry coding systems
			PIT arrays work well in such areas but need to be cautious around metal surfaces	
			Use existing coding systems to avoid background interference	
			Prior to starting a study, identify frequencies that are prone to high levels of interference in your area	
Anthropogenic	Noise	Boat traffic, waves, rain and biological noise can all make detection of acoustic signals difficult	Radio and PIT unaffected unless electrical interference from vessels (including electromagnetic noise interference from tracking boat [22])	Improvements in coding to improve performance when signal to noise ratio is not optimal for acoustic systems
		For example, rainfall can produce or increase noise levels of 15 to 25 dB in the water column [47]	Much less biological noise in freshwater than marine systems, so not a problem in most situations	Conducting detailed system-performance tests under different conditions needed to determine acoustic system efficiency (for example, in a shipping canal)
Environment or habitat	Remote areas	Many inland regions are remote, creating challenges for tending and operating electronic equipment [48, 50]	Solar panels and special battery banks have provided power in remote locations for fixed stations (mostly radio and PIT systems [50])	Need for technological improvements to address challenges with maintaining systems in winter or damp environments (for example, better weatherproofing of equipment [48])

			Satellite connections enable data to be retrieved and receivers to be configured from afar (already exists for radio [51] and recently developed for acoustic systems with a surface cable and GPS or cell interface)	More remote communication options to enable servicing from afar
			Aircraft-mounted antennas can be used for radio tracking animals over great distances in remote locales [52, 53]	Batteries that last longer or systems that require less power
Environment or habitat	Habitat configuration	Geomorphological configuration of habitats (for example, type and configuration of banks, fjords, roughness of substrate, type of substrate) can influence performance of both acoustic [40, 54] and radio telemetry [55]	Increasing density of acoustic receivers can improve detection probability in some areas (for example, such as near banks in rivers or areas with high levels of substrate roughness [56])	Need for additional studies to better understand how habitat configuration influences detection of radio and acoustic telemetry signals
			Manual radio tracking can be confusing around some habitats (for example, rock walls) such that practice is needed to become efficient at locating tags	
Environment or Habitat	Cryptic habitats	Some animals make use of cryptic habitats that make tracking difficult (for example, crayfish and mammals can use burrows [57]; some fish hide under rocks [37]) due to signal attenuation and deflection; acoustic telemetry ineffective in such situations	Use tags that are as powerful as possible (for example, high-output radio tags)	An intrinsic challenge that cannot be easily overcome with technology
		Radio telemetry and PIT tracking possible if animals are not too deep into cover		Efforts should focus on training of team members to track in such environments

Biotic	Macrophytes	Macrophytes can impede sound when using acoustic telemetry [34] and can foul external telemetry transmitters [58]	Unclear if specific frequencies or other characteristics of acoustic telemetry systems would improve performance in dense macrophyte beds	Need for rigorous testing of acoustic telemetry gear in and around macrophytes to identify ideal frequencies for local conditions
			Radio telemetry performs well in dense macrophytes	
			Use internal tag implantation in animals that are in regions with lots of macrophytes	
Biotic	Animal damage	Vermin of various types (mostly mammals such as muskrats or insects such as ants) can chew through wires for cabled systems (including cabled acoustic arrays [59]) or damage-sensitive electronics [48]	Wires can be protected with sheaths and various pest control strategies can be used (for example, pesticides in land-based fixed stations)	Further developments such that cabled systems are not needed
			External tags can be designed to minimize chewing by mammals (for example, was a problem for box turtles [60])	Development of strategies for camouflaging and protecting wires and making gear sealed to prevent entry of insects
Biotic	Biofouling	Algae and molluscs (for example, *Dreissena* spp.) in particular can build up on underwater equipment, such as fixed hydrophones or underwater antennas, reducing performance [61]	Underwater equipment such as hydrophones and antennas can be cleaned (for example, by divers) or various biofouling materials (for example, special paints [61]) can be used	Need for biofouling paints and materials that are minimally toxic
		Fouling of external tags can reduce performance and burden tagged animals [62]	Tags (and antennas) can be placed internally [62]	Need for methods of geopositioning animals without light
		Fouling of light sensors makes use of geopositioning tags in freshwater difficult		

CART, combined acoustic radio transmitter; GPS, global positioning system; HF, high frequency; PIT, passive integrated transponder; PSAT, Pop-up satellite tag; VHF, very high frequency.

Reduction in Tag Size

The progression of smaller and more efficient electronic components has been a major advance for aquatic telemetry, making it possible to reduce tag size substantially, and tag and track a wider range of species and life stages. Miniaturization has also made it possible to increase the capabilities of transmitters used on larger animals in terms of operating life, reception range and environmental and physiological sensors. Like many technologies, this has been a gradual process, and many of the advances in the 1970s set the stage for future work. Specific details in journal articles were often relegated to a few lines in the methods section, but the overall trend can be inferred from the steady increase in telemetry papers focused on habitat selection and movement patterns in smaller species and juvenile life stages of fish [63, 64], and increasingly smaller nonfish taxa, such as some amphibians (black-bellied salamanders [65]) and invertebrates (crayfish [66]). Radio tags as small as 0.25 g, acoustic tags as small as 0.30 g, and PIT tags as small as 10 mg are commercially available at the time of publication.

Coded Signals

The advent of coded transmitter signals is a technical advance that made a major impact on the study of freshwater animals. Prior to the mid-1980s, most telemetry studies were limited to small numbers of individuals (usually <40) in relatively confined areas. The need for more quantitative data on large aggregates of fish, such as the timing and number of returning adult salmon and their migration through entire watersheds [67], made it necessary to develop new approaches for dealing with large numbers of wide-ranging individuals. Because of the commercial importance of Pacific salmon returns in north-western North America, funding was often available to support these developments, which subsequently had transferable benefits for studying other species. In addition to increasing the number of frequencies being monitored, transmitters encoded with simple, audibly distinguishable signal patterns (for example, single, double and triple-pulsed transmissions) or varying pulse rates (for example, 60 vs. 90 pulses per minute) were used to increase the number of unique individuals that could be identified. By the late 1980s, more sophisticated signal patterns were adopted, which together with receivers able to distinguish the encoded signal, dramatically increased the number of individuals that could be identified [42]. For acoustic applications, the advent and application of code division multiple access technology enabled multiple animals to be detected simultaneously [51]. Coded tags are now taken for granted, but represented a significant innovation for freshwater systems.

Stationary Tracking

Passive telemetry systems (where receivers are deployed at stationary locations to record tagged animals that come within range of an antenna or hydrophone) in freshwater were based on coded transmitter technology. Receivers deployed throughout watersheds, and linked via satellite to relay data back to a central processing facility, were first deployed for Pacific salmon to study their upstream spawning migration [67]. Receivers could be programmed to scan a variety of frequencies and antennas, either sequentially or simultaneously (for example, using digital spectrum processing [68, 69]), sometimes in an intelligent manner such that when fish were present they would prioritize 'listening' on a given antenna or frequency [19]. Stationary acoustic receiver systems were first used in marine systems [70, 71] and are now routinely applied in inland waters, especially large rivers, lakes and reservoirs to assess migration patterns [72]. Hydrophones, coding and signal-processing advances mean that freshwater animals can now be tracked in river systems that were too turbulent and noisy for previous acoustic systems [71, 73, 74, 75]. Autonomous omnidirectional acoustic receiver-loggers can be deployed in arrays to efficiently record diadromous fishes migrating between freshwater and marine systems, by enabling the same individuals to be employed in both habitats. This is even possible for small fish, thanks to the availability of relatively small transmitters compared with the combined acoustic and radio tags (CARTs) previously available). This technology, when used in combination with archival tags, has enabled researchers to describe detailed behavior of steelhead trout during the riverine and oceanic migration [76]. Many of the passive acoustic systems can be deployed subsurface and operate autonomously for extended periods (over a year) without requiring additional maintenance and battery changes, which has allowed for continuous tracking of the behavior of freshwater species [23, 38]. Recent developments include links to communication systems (for example, satellites, cellular networks, underwater acoustic modems) to enable data collection without retrieval of equipment. Large-scale acoustic telemetry arrays have now been installed in the Laurentian Great Lakes and many large river systems around the globe, and are based on international collaborations in transboundary waters focusing on wide-ranging animals. If configured with a minimum of three time-synchronized hydrophones, hyperbolic navigation using time difference in signal arrival can be used to generate highly accurate and precise animal positions [42, 70]. To date, there are only a few published examples using such technology in freshwater, which have focused on questions of spatial distribution and variation in temporal activity [26, 27, 77], but many projects using that approach are now underway, including some that use relatively inexpensive technology to look at simultaneous fine-scale movements of multiple individuals [78]. Passive PIT telemetry arrays (discussed below) have also provided similar benefits. Development in antenna technology robust to dynamic river and stream conditions [32] has enabled longer-term field deployments.

The technical contributions noted above have led to a number of important and unique ecological advances. Unlike many marine and terrestrial ecosystems, rivers are by nature linear. The work of Fausch *et al.* [32] was instrumental in discussing some of the problems associated with trying to conduct ecological research broadly within the 'riverscape'. Telemetry applications provide a rare exception; a research method that actually capitalizes on the physical structure of dendritic freshwater ecosystems in capturing movement processes. For this reason, stationary receiver-loggers (often termed automatic listening stations) of radio, acoustic and inductive coupling forms have been deployed along many reaches of rivers to determine rates of movement and survival [20, 22, 75, 79, 80] and along diverging paths to the sea to determine route selection [20, 81]. While the use of acoustic telemetry arrays in marine environments has provided important spatial data for large-bodied marine fauna, there has been an even greater rate of return for investment in telemetry infrastructure in rivers. This is largely a function of the contained and branching nature of rivers, when compared with the unconstrained shape of most marine areas in relation to the relatively limited range of acoustic receivers.

PIT Telemetry

Passive integrated transponder telemetry has provided an affordable tool that enables researchers to study smaller individuals or juvenile life stages than is possible with radio and acoustic telemetry using both active and passive tracking. In addition, tags remain viable for long periods (years), owing to the lack of onboard batteries, allowing long-term tracking of individual fitness proxies, such as growth and survival [16, 82, 83, 84]. Owing to the much shorter range (less than 2 m perpendicular to the antenna plane) than radio and acoustic transmitters, PIT telemetry has mainly been used in smaller systems. It is especially valuable in assessing fish movement through narrow passageways, such as fish passes [85] and streams [86, 87], but is increasingly used in wider swim-through channels (30+ m × 0.8 m deep) using an array of tag readers. Passive systems yield high-quality long-term recording of large samples of animals at discrete locations [88]. Mobile PIT detectors (backpack and boat or trawl-mounted models for example [89]) have been developed as a less obtrusive method for studying the ecology of refuge-seeking or cryptic species of fish, invertebrates (for example crustaceans [90, 91], molluscs) and amphibians [92]. One prominent innovation with PIT telemetry has been the ability to visit bird colonies to estimate predation on small fish (mostly salmonid smolts) by locating PIT tags on land [93]. Further, the cost of PIT tags remains about one hundredth of that of battery-powered radio and acoustic transmitters, facilitating large sample sizes with PIT telemetry.

Sensor Tags

The incorporation of sensors into radio and acoustic transmitters and archival tags has provided novel insight into animal-environment associations and the energetic costs of different behaviors (for example, basking) and stressors (for example, fisheries interactions, fishway use). Some of the earliest sensors incorporated into electronic tags were simple activity or motion sensors that provided the status of a fish (that is, whether it is moving or stationary [94]). The success of these tags encouraged the development of other sensor types. Physiologically oriented sensors (for example, electrocardiography; electromyography, acceleration) have been used primarily on fish [95, 96]. Electromyography sensors incorporated in radio transmitters were developed in the 1970s to record opercular muscle activity or overall fish locomotor activity [97, 98, 99], and are today the most commonly used physiological sensor tool for studying energetics and migration in freshwater fish [17]. Electrocardiography or other cardiac metrics, such as heart rate (see [100] for review) have been used in radio (for alligators [101] and for fish [102]), acoustic (for fish [103]) and archival (for fish [104, 105]) tags. Of late, acceleration loggers [105, 106] and acoustic acceleration transmitters (SJC, unpublished data) have been used to assess temporal changes in energy expenditure and movement rate and are expected to grow in popularity for application in freshwater, akin to the uptake of this technology in terrestrial and marine applications [107, 108]. Environmental sensors, particularly temperature and pressure (that is, depth [109]), have been used on a wide range of reptiles (for example, turtles [110]) and fish [111] and have been used in radio and acoustic transmitters and archival tags to link location with the environmental parameters measured by the sensors. Other environmental sensors linked to transmitters have also been designed to assess the range of key parameters, such as conductivity [112] and oxygen [113], that fish are exposed to during normal activity. Relatively cheap archival temperature tags now available on the market enable collection of long-term thermal information and large sample sizes, provided that the tags can be retrieved (for example, alligators [114], turtles [115], salmon [116]). Early sensor tags were developed by individual researchers, who would build the tags either themselves or in collaboration with engineers [112, 117, 118, 119]. However, many of these sensors now represent 'standard' options for commercially available tags, and their use is greatly simplified. Collectively, these tools have provided novel insights in animal behavior, energetics and physiology that would not be possible by simply measuring the location of an animal.

Integration of Peripheral Technologies

In addition to improvements specific to telemetry and logging equipment, advances in other, often unrelated, technologies have dramatically increased the options available to researchers and made them easier to adopt. For example, availability of low-cost GPS, some even integrated with receiving systems, has improved the positional accuracy and usability of the telemetry data collected (a vast improvement over drawing circles on a map [52]) allowing fine-scale assessment of movement. Uplink technology (for example, satellite, Wi-Fi, cellular, Bluetooth) and nonvolatile data storage has made it faster and easier to recover telemetry data from stationary tracking systems, particularly those in remote or inaccessible areas. Computerized databases have increased the efficiency of data analyses, and links to the internet make the information easier to disseminate and widely available.

Tagging Techniques

More so than in marine environments, significant effort has been devoted to the development and validation of various attachment and implantation procedures (including pre- and post-tagging care), particularly for fish (reviewed in [58, 120, 121]) but also for reptiles [122]. If the tags or associated tagging procedures result in altered behavior, energetics, condition or survival relative to untagged conspecifics, then findings emanating from tagging studies may be biased or incorrect [28]. Most research efforts have focused on identifying the maximum tag burdens for a range of taxa (including turtles, fish and amphibians), testing and optimizing tagging techniques, assessing different tag types, [123, 124], testing anaesthetics (for example, for turtles [125] and fish [126]), and identifying species- or taxon-specific challenges [120]. These challenges often relate to the composition, rigidity or permanence of an animal's outer surface in regard to externally tagging (for example, reptiles [122, 127], an elasmobranch [128], a slippery fish [129], crayfish [91]) or physiological acceptance of foreign bodies following internal tagging [130, 131]. Particularly in freshwater, fouling of external tags (for example, with macrophytes) and associated impacts on drag have led to innovations in internal surgical procedures [132] that have advanced tagging for marine fish. Indeed, many of the refinements in animal handling associated with the development of tagging procedures for freshwater fish have led to broader improvements in and attention to animal welfare for aquatic animals used in research [133].

Combining Electronic Tagging With Other Research Methods

Although early studies focused largely on using electronic tags to identify habitat use and movements (that is, descriptive; reviewed in [15]), electronic tagging is increasingly being combined with other techniques and tools. Such advances have increased the complexity of questions that can be addressed and provide more mechanistic explanations of animal behavior through hypothesis testing (reviewed in [134]). This can be accomplished with integrative studies that combine techniques in parallel (for example, a study that has a laboratory and field component), by obtaining nonlethal biological samples from tagged animals (for example, stable isotope analysis [135], genetic tests [136], physiological status studies [27, 137]), or by experimentally manipulating tagged animals (for example, hormone manipulations [138], controlling feeding [139], cortisol injections [59]) or their environment (for example, temporary holding studies followed by release [140]). Some of the earliest of such studies occurred on Pacific salmon in Japan and involved evaluating homing mechanisms [141] and the endocrinology of migration (reviewed in [142]). Recently, the first example of combining genomic studies with field telemetry of Pacific salmon identified the factors that are associated with migration failure in freshwater [143]. Integration of fish telemetry with detailed hydrodynamic studies has been used to optimize fish passage facilities through improved understanding of fish cognition and behavior [144]. The number of such integrative studies appears to be increasing.

Data Analysis Tools and Techniques

Following their development for terrestrial applications, many of the analytical techniques now being applied in marine systems were first applied in freshwater [145]. Of particular note is the use of mark-recapture models (and software for fitting models) for estimating detection, survival and state transitions during migration through rivers [146, 147]. These papers adapted Cormack-Jolly-Seber and similar models (developed in the 1960s for survival through time) to salmon migration. More recently, survival analyses have been adapted for telemetry data from fishway assessments [148]. Voluminous amounts of data can be collected using telemetry but the advent of easy-to-use, computerized databases [149] has made it possible to organize and process data efficiently and present information in a highly useable form. Records of tag detections from many freshwater species have been entered into large databases managed by software such as ACCESS and HYDRA. The development of geographic information system (GIS) methods (for example, [150]), space-use analyses [151] and such computer programs as ARCVIEW, ARCINFO, RANGES and R has simplified and opened new possibilities for analyzing movement data in freshwater habitats. For example, fish telemetry data analyzed with home-range software led to important findings related to home-range establishment and shifts

[152, 153]. Nevertheless, most space-use analyses have been developed for terrestrial or marine applications and careful assessment of their applicability and bias is needed for the restricted and sometimes near-linear habitats of freshwater systems [154].

International Collaborations

Although not unique to this field, there has been a notable increase in the number of freshwater telemetry studies in remote areas around the world, and also in many less economically developed countries. These efforts have contributed significantly to solving basic and applied research questions in regionally important freshwater animals, and have broadened the habitat types and animal taxa covered by such studies. As noted previously, given the threatened status of freshwater ecosystems and their animal inhabitants, there is a great need for studies on fishes that support commercial and subsistence fisheries in less economically developed regions (for example, central Africa [62], the Mekong basin [155], the Amazon basin [48]). While biotelemetry has long been recognized as an important tool for freshwater research in these regions [156], its adoption has been driven by an apparent increase in available funding for international studies and collaborations, and for studies in developing countries, together with facilitation of such collaborations with the development of email communication and the internet, and greater affordability of equipment. Workshops have also been targeted towards building capacity for doing freshwater telemetry studies in less economically developed countries [157].

Future Needs

The technical advances discussed in this paper have provided a strong foundation for the development of new technologies and technical innovations. Many of these advances will undoubtedly have an impact on the methods used to study aquatic animals and their surroundings. Although several emerging trends are evident in the literature (for example, advances in integrating and summarizing large data sets and the adaption of telemetry tools to novel species and environments), most can only be imagined. Here we provide an overview of what we regard as desirable future technical innovations and research needs related to the study of animals in freshwater systems (not ranked in any particular order). Most of these advances relate to transmitter design, tagging methods, tracking capabilities and tools for analyzing the large amounts of data collected. As in the past, many of the new technologies needed to overcome the constraints identified (Table 1) will probably come from advances in other fields that are then applied to fish and wildlife applications.

A major limitation with current telemetry systems is the size and functionality of the transmitters. Larger tags can provide increased reception range, operating life and sensor options, but limit the size of the individual and life stage that can be studied. Battery size is currently a limiting factor that dictates the ultimate size of the transmitter, the number and type of sensors that it can contain, and the duration of its longevity. Developing smaller, more efficient batteries to reduce transmitter size and increase operating life and reception range would help, but only to a limited degree. These technologies need to be combined with either hardware or software solutions that allow for more efficient power usage. Ultimately, more advanced tags will hopefully allow for near-lifetime tracking of large, long-lived animals from juvenile to adult in large freshwater systems (for example, large tropical catfishes).

While great strides have been made towards miniaturizing tags, there are limits to how far we can go with current technologies. In short, a technological wall has been reached that might significantly limit the extent to which we can study aquatic species. To that end, we preface this section by noting that in many ways we need a paradigm shift in the type of equipment we use to tag and track freshwater species. A component of this shift will be the development of tags that conserve battery power (to increase their lifespan and power peripheral sensors) until they are either required to collect data, or are transmitting data for collection. While not necessary for all applications, this transition will open up new avenues of research and innovation aimed at lifetime tracking of individual behavior and environmental interactions. While the development of very small, long-lasting, inexpensive tags with a long range under a variety of environmental conditions remains an obvious but unlikely desire (dare to dream…), here we focus primarily on technical needs.

Environmental Conditions

Given the extreme range of environmental conditions (for example, subarctic areas or tropical systems) that are home to freshwater species, adapting receiver technology for these conditions is essential. The production of field-ready (that is, waterproof, shock-proof) receiver housings would greatly expand the application of expensive telemetry equipment.

Sensors

While many sensors have been developed for a wide variety of applications, there are still areas where improvements to sensors or entirely new sensors would benefit the study of freshwater organisms. General improvements to both the reliability and ease of using complex sensors would make them more widely applicable. For example, while geolocation tags that use light or a magnetic field have produced valuable results, there is room to improve the algorithms used to derive their position. Reduced battery size has allowed for the development of PIT tags with simple sensors (for example, temperature). Further development of sensor and data storage PIT tags, as well as systems capable of transferring data from the tags, has great potential for large-scale studies. Work on this is ongoing in non-wildlife fields ([158]), but freshwater researchers should look to incorporate these technologies into their research.

While the concept has already been used in marine systems, freshwater animals may represent potentially useful bioprobes that can aid in environmental monitoring. A necessary component is the development of multisensor devices that provide spatial and environmental data that can be tracked over large spatial and temporal scales. At the individual level, mortality tags have been developed that broadcast when an individual stops moving for a set amount of time. As tag sizes have decreased, prey species have increasingly been the focus of telemetry studies. With motion-sensing mortality tags, prey that has been consumed by a predator may still appear to be alive; therefore, development of a tag that senses when an individual has been eaten would provide a more accurate indication of mortality.

The combination of real-time, simultaneous, hydroacoustic echo-sounding (horizontal or vertical) with interrogation of individuals with transmitters or transponders is a concept that dates back to Mitson and Storeton-West [159]. Given the numerous technological improvements since the 1970s, this approach should now be explored using state-of-the-art hydroacoustic equipment to simultaneously combine local-scale habitat and animal (especially fish) distribution and abundance to telemetered behavior and ecophysiology of tagged individuals.

Archival Tags

The major limitation associated with archival tags is the need to recapture tagged individuals to acquire stored data. A potential solution would be the development of archival tags with rapid download capability, probably downloaded as an individual moves through a remote gate. Since not all fish will pass near enough to such a structure, this design could be further improved through the addition of a 'chat' function that allows data to be exchanged between individuals and stored. This technology would be perfectly suited for species that are known to aggregate at some point in their lifecycle but may have low individual recapture rates (for example, adult sturgeon, which aggregate to spawn).

Although some large diadromous fish can be tagged with pop-up satellite archival transmitters (PSATs) in freshwater and pop-off transmitters in marine waters (for example, Atlantic sturgeon tagged in Hudson River [160]), there is a need for PSAT tags that function well in terms of both generating positions and 'popping-off' in freshwater. Presently, most pop-off assemblies are based on corrodible links that require a saline environment to degrade the attachment material, thereby limiting their application in freshwaters.

Tag Transmission

Reception range, which should ideally be maximized due to the sizable areas typically associated with *in-situ* studies, is often marginal or inadequate. To expand the use of these studies to more complex environments, efforts should focus on improving monitoring systems that are capable of tracking animal movements in large, powerful river systems. Solutions will need to include overcoming underwater noise and improved deployment techniques for acoustic telemetry equipment. To combat background noise that interferes with detections (a common problem in urban areas or near infrastructure, such as power plants), unique radiotelemetry coding systems could be developed or site-specific frequencies could be selected. Similarly, novel coding-schemes may help to improve the effectiveness of acoustic telemetry when species are located among macrophytes.

Passive integrated transponder (PIT) tags have been used successfully to study a variety of aquatic species under controlled settings, but the reception range (typically less than a meter) is insufficient for many studies in large, unconstrained environments. The development of more robust and powerful PIT systems and antennas that work in larger environments and under a broader range of environmental conditions will extend the application of PIT tags. Possible solutions include the further development of PIT trawls and meshless static PIT 'gillnets' that detect tagged fish as they swim through them. In addition, when multiple PIT-tagged individuals pass through a stationary array, one tag may block the signal from another. The development of a method that prevents or circumvents this blockage is essential, to ensure that all individuals are detected.

For many telemetry studies, collecting large amounts of data is not a challenge. However, some variables (that is, continuous monitoring of depth or acceleration) are currently limited to archival storage tags. Advances in telemetric transfer and decoding of large or complicated datasets could expand the type of telemetry tags used for collecting continuous or long-term data. Similarly, enhanced telemetric transmission (in terms of rate and volume of data) would improve the transmission of video images (for example, 'critter cams') either in real time or using archival tags.

Tagging Techniques

While a great deal of work has been directed towards improving tagging techniques (especially in fishes), further refinement of tagging procedures and validation of tagging techniques using the species being tagged (rather than surrogates [120]) are essential. This is particularly true for invertebrates, especially those that moult (for example, crustaceans). In addition to refining current techniques, there is also a niche for developing a rapid, internal tagging technique that can be used with aquatic vertebrates and large invertebrates (for example, development of injectable tags that are minimally invasive and require no anaesthetic).

Arrays

Two and three-dimensional arrays are one of the most promising new methods for tracking tagged individuals. Unfortunately, high costs and the complicated nature of both the system themselves and the data they produce limit their broad application. A cheaper and more user friendly two- and three-dimensional system would open new possibilities and improve studies of detailed individual movements in areas above and below dams and through fishways or other possible migration barriers. One of the major challenges associated with two- and three-dimensional arrays is the overwhelming amount of data that can be acquired. To encourage the application of these arrays, data filtering and spatial location calculation methods need to be developed that can be easily applied, ideally by the scientist rather than the vendor. While passive telemetry systems have been installed in small lakes and both large and small rivers, there is a need for the development of large-scale communal telemetry systems at the scale of watersheds and entire lakes [27]. This would allow for ecosystem telemetry studies and manipulations of individuals in their entire natural environment.

Databases and Analysis

A broad theme for all of the aforementioned future developments is to improve user interfaces and make them more readily accessible to nonexperts. This concept of user-friendliness also holds true for computer programs designed to organize and analyze large and cumbersome databases. Software improvements that allow users to integrate and summarize large databases easily would be a great asset, and built-in functions, such as analysis of home range, would be useful (especially in spatially complex environments). Also, from a statistical perspective, models are needed that estimate spatial movement processes and survival probabilities while accounting for observation errors. State-space models seem to have some promise, but a general model framework and software are lacking.

Collaboration

An important part of realizing the seven future needs that we have identified is the fostering of a strong relationship between users of telemetry equipment and those that design and produce it. These relationships have been and will be central to technological advances in freshwater telemetry. Early on, there were relatively few commercial vendors so most researchers either built their own equipment or worked with technical staff within their own agency or institution. Indeed, many of the first commercial suppliers were former researchers themselves, and the spirit of cooperation continued as they developed their product lines and became established. Although the quality of telemetry gear has vastly improved in recent decades, the importance of collaboration between suppliers and users remains paramount. Most equipment purchases are based on predesigned items that may or may not be ideal. Many researchers assume that this is what is available and so make do, and vendors are often reluctant to make changes unless there is a clear upside for sales. Although this is understandable from an economic standpoint, it is not conducive to advancing the technology in a field where funding is often limited. Collaboration between field researchers and product designers and manufacturers is likely to be a key factor in relation to the utility and effectiveness of future advances.

CONCLUSION

Over the years, advances in electronic tags have had a significant impact on the ability to study aquatic animals under a variety of challenging conditions. Here, we reflected on the history of these advances and, in doing so, we were able to characterize the key technical developments that have made electronic tags such a useful tool in research. Most research activities and innovation have arisen from work on fish, but there is certainly an opportunity for greater application of such technology to other taxa and to smaller animals including invertebrates and amphibians.

Many constraints make the use of electronic tags challenging, but for most applications there is at least one form of technology that will work. The reality is that technological limitations require researchers to make trade-offs, which necessitates preparation of clear research questions and research priorities during the study design process. Different research questions may require the application of very different technologies. Compared to where the field was 50 years ago, we have come a long way. Innovations continue, as summarized here, but there are still a variety of future needs that, if achieved, could open more doors for studying freshwater animals. Electronic tag practitioners and equipment manufacturers need to work closely if such advances are to be meaningful and rapid. However, not all of those needs are purely technological, as continued work is needed on analytical and statistical procedures to handle the large volumes of data generated. Further, efforts to refine and validate tagging procedures to ensure that the data are representative of untagged conspecifics and that the welfare status of tagged animals is maintained, are still needed. Many meaningful gains in conservation and management will likely be ascribed to electronic tagging innovations in freshwater in the next 10 to 20 years, and we expect novel discoveries relating to fundamental animal and environmental biology. Indeed, with the growing threats facing freshwater ecosystems and their fauna, the need for further technical and technological improvements is urgent if the scientific community is to provide the data needed to inform conservation efforts.

ACKNOWLEDGEMENTS

SJC is supported by the Canada Research Chairs program and the Natural Sciences and Engineering Research Council of Canada (OTN Canada). We thank Greg Bulté for directing us towards relevant reptilian literature. The findings and conclusions in the paper are those of the authors and do not necessarily represent the views of the National Marine Fisheries Service). Reference to trade names or commercial products does not imply endorsement or recommendation by the National Marine Fisheries Service, NOAA, the USGS or the US Government. This manuscript is contribution number 1745 of the USGS Great Lakes Science Center.

ABBREVIATIONS

CART: Combined acoustic radio transmitter

CDMA: Code division multiple access

CHAT: Communicating histogram acoustic transponder

GIS: Geographic information system

GPS: Global positioning system

HF: High frequency

PIT: Passive integrated transponder

PSAT: Pop-up satellite tag

SPOT: Smart position-only tag

USGS: United States Geological Survey

VHF: Very high frequency.

AUTHORS' CONTRIBUTIONS

All authors helped to identify important historical developments, current constraints and future needs for freshwater biotelemetry in their specific areas of expertise. In addition, JDM helped to synthesize these contributions and SJC wrote the initial draft of the manuscript. All authors were also involved in subsequent revisions of the manuscript. All authors read and approved the final manuscript.

REFERENCES

1. Brinson MM, Malvárez AI: Temperate freshwater wetlands: types, status and threats. *Environ Conserv* 2002, 29: 115–133.
2. Brönmark C, Hansson LA: Environmental issues in lakes and ponds: current state and perspectives. *Environ Conserv* 2002, 29: 290–306.
3. Malmqvist B, Rundle S: Threats to the running water ecosystems of the world.*Environ Conserv* 2002, 29: 134–153.
4. Vörösmarty CJ, McIntyre PB, Gessner MO, Dudgeon D, Prusevich A, Green P, Glidden S: Global threats to human water security and river biodiversity. *Nature* 2010, 467: 555–561. 10.1038/nature09440
5. Moyle PB, Leidy RA: Loss of biodiversity in aquatic ecosystems: evidence from fish faunas. In *Conservation Biology: The Theory and Practice of Nature Conservation, Preservation, and Management.* Edited by: Fiedler PL, Jain SK. New York: Chapman and Hall; 1992:127–169.
6. Dudgeon D, Arthington AH, Gessner MO, Kawabata ZI, Knowler DJ, Leveque C, Naiman RJ, Prieur-Richard AH, Soto D, Stiassny MLJ, Sullivan CA: Freshwater biodiversity: importance, threats, status and conservation challenges. *Biol Rev* 2006, 81:163–182.
7. Ormerod SJ, Dobson M, Hildrew AG, Townsend CR: Multiple stressors in freshwater ecosystems. *Freshwater Biol* 2010,55(Suppl):1–4.

8. Strayer DL, Dudgeon D: Freshwater biodiversity conservation: recent progress and future challenges. *J N Am Benthol Soc* 2010, 29: 344–358. 10.1899/08-171.1
9. Balian EV, Segers H, Lévêque C, Martens K: An introduction to the freshwater animal diversity assessment (FADA) project. *Hydrobiologia* 2008, 595: 3–8. 10.1007/s10750-007-9235-6
10. Costanza R, d'Arge R, de Groot R, Farber S, Grasso M, Hannon B, Limburg K, Naeem S, O'Neill RV, Paruelo J, Raskin RG, Sutton P, van den Belt M: The value of the world's ecosystem services and natural capital. *Ecol Econ* 2010, 25: 3–15.
11. Holmlund CM, Hammer M: Ecosystem services generated by fish populations.*Ecol Econ* 1999, 29: 253–268. 10.1016/S0921-8009(99)00015-4
12. Millennium Ecosystem Assessment: *Ecosystems and Human Well-Being: Current State and Trends*. Washington, DC: Island Press, Volume 1; 2005.
13. Welcomme RL, Cowx IG, Coates D, Béné C, Funge-Smith S, Halls A, Lorenzen K: Inland capture fisheries. *Philos Trans R Soc Lond B Biol Sci* 2010, 365: 2881–2896. 10.1098/rstb.2010.0168Central
14. Lucas MC, Baras E: *Migrations of Freshwater Fishes*. Oxford, UK: Blackwell Scientific Publications; 2001.
15. Lucas MC, Baras E: Methods for studying spatial behaviour of freshwater fishes in the natural environment. *Fish Fish* 2000, 1: 283–316. 10.1046/j.1467-2979.2000.00028.x
16. Gibbons JW, Andrews KM: PIT tagging: simple technology at its best. *Bioscience*2004, 54: 447. 10.1641/0006-3568(2004)054[0447:PTSTAI]2.0.CO;2
17. Cooke SJ, Thorstad E, Hinch SG: Activity and energetics of free-swimming fish: insights from electromyogram telemetry. *Fish Fish* 2004, 5: 21–52. 10.1111/j.1467-2960.2004.00136.x
18. Cooke SJ, Hinch SG, Lucas MC, Lutcavage M: Biotelemetry and biologging. In *Fisheries Techniques*. 3rd edition. Edited by: Zale AV, Parrish DL, Sutton TM. Bethesda, MD: American Fisheries Society; 2013:819–860.
19. Eiler JH: Tracking aquatic animals with radio telemetry. In

Telemetry Techniques: A User Guide for Fisheries Research. Edited by: Adams NS, Beeman JW, Eiler JH. Bethesda, MD: American Fisheries Society; 2012:163–204.

20. Breukelaar AW, bij de Vaate A, Fockens KTW: Inland migration study of sea trout (*Salmo trutta*) into the rivers Rhine and Meuse (The Netherlands), based on inductive coupling radio telemetry. *Hydrobiologia* 1998, 372: 29–33.
21. Ramm DC: Electromagnetic tracking of rock lobsters (*Jasus novachollandiae*). *Mar Freshwater Res* 1980, 31: 263–269. 10.1071/MF9800263
22. Solomon DJ, Potter E: First results with a new estuarine fish tracking system. *J Fish Biol* 1988, 33: 127–132.
23. Klimley AP, Voegeli F, Beavers SC, Le Boeuf BJ: Automated listening stations for tagged marine fishes. *Mar Technol Soc J* 1998, 32: 94–101.
24. Holland KN, Meyer CG, Dagorn LC: Inter-animal telemetry: results from first deployment of acoustic 'business card' tags. *Endang Species Res* 2010, 10: 287–293.
25. Akamatsu T: Diving behaviour of freshwater finless porpoises (*Neophocaena phocaenoides*) in an oxbow of the Yangtze River, China. *ICES J Mar Sci* 2002, 59: 438–443. 10.1006/jmsc.2001.1159
26. Hasler CT, Suski CD, Hanson KC, Cooke SJ, Philipp DP, Tufts BL: Effect of water temperature on laboratory swimming performance and natural activity levels of adult largemouth bass. *Can J Zool* 2009, 87: 589–596. 10.1139/Z09-044
27. Cooke SJ, Niezgoda GH, Hanson KC, Suski CD, Phelan FJS, Tinline R, Philipp DP: Use of CDMA acoustic telemetry to document 3-D positions of fish: relevance to the design and monitoring of aquatic protected areas. *Mar Technol Soc J* 2005, 39:31–41. 10.4031/002533205787521659
28. Brown RS, Hubert WA, Daly SF: A primer on winter, ice, and fish: what fisheries biologists should know about winter ice processes and stream-dwelling fish.*Fisheries* 2011, 36: 8–26. 10.1577/03632415.2011.10389052
29. Brown RS, Mackay WC: Spawning ecology of cut-throat trout (*Oncorhynchus clarkii*) in the Ram River, Alberta. *Can J Fish Aquat*

Sci 1995, 52: 983–992. 10.1139/f95-097

30. Cooke SJ, Thorstad EB: Is radio telemetry getting washed downstream? The changing role of radio telemetry in studies of freshwater fish relative to other tagging and telemetry technology. *Am Fish S S* 2012, 76: 349–369.
31. Brown RS, Power G, Beltaoa S: Winter movements and habitat use of riverine brown trout, white sucker and common carp in relation to flooding and ice break-up. *J Fish Biol* 2001, 59: 1126–1141. 10.1111/j.1095-8649.2001.tb00180.x
32. Johnston P, Bérubé F, Bergeron NE: Development of a flatbed passive integrated transponder antenna grid for continuous monitoring of fishes in natural streams. *J Fish Biol* 2009, 74: 1651–1661. 10.1111/j.1095-8649.2009.02211.x
33. Roussel JM, Cunjak RA, Newbury R, Caissie D, Haro A: Movements and habitat use by PIT-tagged Atlantic salmon parr in early winter: the influence of anchor ice. *Freshwater Biol* 2004, 49: 1026–1035. 10.1111/j.1365-2427.2004.01246.x
34. Stasko AB, Pincock DG: Review of underwater biotelemetry, with emphasis on ultrasonic techniques. *Can J Fish Aquat Sci* 1977, 34: 1261–1285.
35. FreundJG,HartmanKJ:Influenceofdepthondetectiondistanceoflow-frequency radio transmitters in the Ohio River. *N Am J Fish Manage* 2002, 22:1301–1305. 10.1577/1548-8675(2002)022<1301:IODOD D>2.0.CO;2
36. Peters LM, Reinhardt UG, Pegg MA: Factors influencing radio wave transmission and reception: use of radiotelemetry in large river systems. *N Am J Fish Manage*2008, 28: 301–307. 10.1577/M06-146.1
37. Bunt CM, Cooke SJ, Philipp DP: Mobility of riverine smallmouth bass related to tournament displacement and seasonal movements. In *Black Bass: Ecology Conservation and Management*. Edited by: Philipp DP, Ridgway MS. Bethesda, MD: American Fisheries Society; 2003:356–363.
38. Heupel MR, Semmens JM, Hobday AJ: Automated acoustic tracking of aquatic animals: scales, design and deployment of listening station arrays. *Mar Freshw Res* 2006, 57: 1–13. 10.1071/MF05091

39. Zeibell CD: Ultrasonic transmitters for tracking channel catfish. *The Progressive Fish-Culturist* 1973, 335: 28–32.

40. Bergé J, Capra H, Pella H, Steig T, Ovidio M, Bultel E, Lamouroux N: Probability of detection and positioning error of a hydro acoustic telemetry system in a fast-flowing river: intrinsic and environmental determinants. *Fish Res* 2012, 125/126: 1–13.

41. Titzler PS, McMichael GA, Carter JA: Autonomous acoustic receiver deployment and mooring techniques for use in large rivers and estuaries. *N Am J Fish Manage* 2010, 30: 853–859. 10.1577/M09-143.1

42. Stuehrenberg L, Giorgi A, Bartlett C: Pulse-coded radio tags for fish identification.*Am Fish S S* 1990, 7: 370–374.

43. Velle JJ, Weeks LR, Long F: An investigation of the loss mechanism encountered in propagation from a submerged fish telemetry transmitter. *2nd International Conference of Wildlife Biotelemetry* 1979, 228–237.

44. Shroyer SM, Logsdon DE: Detection distances of selected radio and acoustic tags in Minnesota lakes and rivers. *N Am J Fish Manage* 2009, 29: 876–884. 10.1577/M08-189.1

45. Aymes JC, Rives J: Detection efficiency of multiplexed passive integrated transponder antenna is influenced by environmental conditions and fish swimming behaviour. *Ecol Freshw Fish* 2009, 18: 507–513. 10.1111/j.1600-0633.2009.00373.x

46. Fisher WL, Wilkerson ML: Assessment of radio and ultrasonic telemetry system in a polyhaline reservoir. *Air and Water Management Association 90 – Annual Meeting* 1997, 231–238.

47. Voegeli FA, Pincock DG: Overview of underwater acoustics as it applies to telemetry. In *Underwater Biotelemetry*. Edited by: Baras E, Philippart JC. Liege: University of Liege; 1996:22–30.

48. Hahn L, Agostinho AA, English KK, Carosfeld J, da Câmara LF, Cooke SJ: Use of radiotelemetry to track threatened dorados *Salminus brasiliensis* in the upper Uruguay River, Brazil. *Endang Species Res* 2011, 15: 103–114. 10.3354/esr00363

49. Cooke SJ, Bunt CM, McKinley RS: Winter residency of smallmouth bass in a thermal discharge canal: implications for tempering pump operation. *N Am J Fish Manage* 2000, 20: 288–295.

10.1577/1548-8675(2000)020<0288:WROSBI>2.0.CO;2

50. Beddow TA, Deary C, McKinley RS: Migratory and reproductive activity of radio-tagged Arctic char (*Salvelinus alpinus L.*) in northern Labrador.*Hydrobiologia* 1998, 371: 249–262.
51. Niezgoda G, Benfield MC, Sisak M, Anson P: Tracking acoustic transmitters by code division multiple access (CDMA) based telemetry. *Hydrobiologia* 2002, 483:275–286. 10.1023/A:1021368720967
52. Hockersmith EE, Peterson BW: Use of the global positioning system for locating radio-tagged fish from aircraft. *N Am J Fish Manage* 1997, 17: 457–460. 10.1577/1548-8675(1997)017<0457:UOTGPS >2.3.CO;2
53. McCleave JD, Power JH, Rommel SA Jr: Use of radio telemetry for studying upriver migration of adult Atlantic salmon (*Salmo salar*). *J Fish Biol* 1978, 12:549–558. 10.1111/j.1095-8649.1978.tb04201.x
54. Cato DH, Bell MJ: *Ultrasonic ambient noise in Australian shallow waters at frequencies up to 200 kHz. Report no MRL-TR-91–23.* Ascot Vale, Australia: Materials Research Lab; 1992.
55. Garrott RA, White GC, Bartmann RM, Weybright DL: Reflected signal bias in biotelemetry triangulation systems. *J Wildlife Manage* 1986, 50: 747–752. 10.2307/3800995
56. Melnychuk MC, Christensen V: Methods for estimating detection efficiency and tracking acoustic tags with mobile transect surveys. *J Fish Biol* 2009, 75: 1773–1794. 10.1111/j.1095-8649.2009.02428.x
57. Ahlers AA, Heske EJ, Schooley RL, Mitchell MA: Home ranges and space use of muskrats *Ondatra zibethicus* in restricted linear habitats. *Wildlife Biol* 2010, 16:400–408. 10.2981/10-044
58. Bridger C, Booth R: The effects of biotelemetry transmitter presence and attachment procedures on fish physiology and behavior. *Rev Fish Sci* 2003, 11:13–34. 10.1080/16226510390856510
59. O'Connor CM, Gilmour KM, Arlinghaus R, Hasler CT, Philipp DP, Cooke SJ: Seasonal carryover effects following the administration of cortisol to a wild teleost fish.*Physiol Biochem Zool* 2010, 83: 950–957. 10.1086/656286
60. Belzer W, Seibert S: A transmitter attachment method for terrestrial turtles, designed to protect the radio module from mammalian

chewing. *Turtle and Tortoise Newsletter* 2009, 14: 18–21.

61. Heupel MR, Reiss KL, Yeiser BG, Simpfendorfer CA: Effects of biofouling on performance of moored data logging acoustic receivers. *Limnol Oceanogr: Methods* 2008, 6: 327–335.
62. Thorstad EB, Hay CJ, Næsje TF, Økland F: Movements and habitat utilisation of three cichlid species in the Zambezi River, Namibia. *Ecol Freshw Fish* 2001, 10:238–246. 10.1034/j.1600-0633.2001.100406.x
63. Knight AE, Marancik G, Layzer JB: Monitoring movements of juvenile anadromous fish by radiotelemetry. *The Progressive Fish-Culturist* 1977, 39: 148–150. 10.1577/1548-8659(1977)39[148:MMOJAF]2.0.CO;2
64. Jepsen N, Berg S: Use of winter refuges by roach tagged with miniature radio transmitters. *Hydrobiologia* 2002, 483: 167–173. 10.1023/A:1021379528719
65. Peterman WE, Crawford JA, Semlitsch RD: Productivity and significance of headwater streams: population structure and biomass of the black-bellied salamander (*Desmognathus quadramaculatus*). *Freshwater Biol* 2008, 53:347–357.
66. Schütze S, Stein H, Born O: Radio telemetry observations on migration and activity patterns of restocked noble crayfish *Astacus astacus (L.)* in the small river Sempt, north-east of Munich, Germany. *Freshwater Crayfish* 1999, 12: 688–695.
67. Eiler JH: A remote satellite-linked tracking system for studying Pacific salmon with radio telemetry. *Trans Amer Fish Soc* 1995, 124: 184–193. 10.1577/1548-8659(1995)124<0184:ARSLTS>2.3.CO;2
68. Bjornn TC, Tolotti KR, Hunt JP, Keniry PJ, Ringe RR, Peery CA: *Passage of Chinook Salmon Through the Lower Snake River and Distribution Into the Tributaries, 1991–1993: Part 1 of Final Report for Migration of Adult Chinook Salmon and Steelhead Past Dams and Through Reservoirs In the Lower Snake River and Into Tributaries*. Walla Walla District, Washington: US Army Corps of Engineers; 1998.
69. Colavecchia M, Katopodis C, Goosney R, Scruton DA, McKinley RS: Measurement of burst swimming performance in wild

Atlantic salmon (*Salmo salar L.*) using digital telemetry. *Regul Rivers Res Mgmt* 1998, 14: 41–51. 10.1002/(SICI)1099-1646(199801/02)14:1<41::AID-RRR475>3.0.CO;2-8

70. Hawkins AD, MacLennan DN, Urquhart GG, Rob C: Tracking cod *Gadus morhua L.*in a Scottish loch. *J Fish Biol* 1974, 6: 225–236. 10.1111/j.1095-8649.1974.tb04541.x

71. Voegeli FA, Lacroix GL, Anderson JM: Development of miniature pingers for tracking Atlantic salmon smolts at sea. *Hydrobiologia* 1998, 371/372: 35–46.

72. Mathes MT, Hinch SG, Cooke SJ, Crossin GT, Patterson DA, Lotto AG, Farrell AP: Effect of water temperature, timing, physiological condition, and lake thermal refugia on migrating adult Weaver Creek sockeye salmon (*Oncorhynchus nerka*). *Can J Fish Aquat Sci* 2009, 67: 70–84.

73. Lacroix GL, Voegeli FA: Development of automated monitoring systems for ultrasonic transmitters. In *Advances in Fish Telemetry*. Edited by: Moore A, Russell I. Lowestoft, UK: CEFAS; 2000:37–50.

74. Steig TW, Skalski JR, Ransom BH: Comparison of acoustic and PIT-tagged juvenile Chinook, steelhead, and sockeye salmon (*Oncorhynchus* spp.) passing dams on the Columbia River, USA. Aquatic telemetry: advances and applications. *Proceedings of the 5th Conference on Fish Telemetry: 9–13 June 2003; Ustica, Italy* http://www.fao.org/docrep/008/y5999e/y5999e00.htm

75. Melnychuk MC, Welch DW, Walters CJ, Christensen V: Riverine and early ocean migration and mortality patterns of juvenile steelhead trout (*Oncorhynchus mykiss*) from the Cheakamus River, British Columbia. *Hydrobiologia* 2007, 582: 55–65. 10.1007/s10750-006-0541-1

76. Teo SLH, Sandstrom PT, Chapman ED, Null RE, Brown K, Klimley AP, Block BA: Archival and acoustic tags reveal the post-spawning migrations, diving behavior, and thermal habitat of hatchery-origin Sacramento River steelhead kelts (*Oncorhynchus mykiss*). *Environ Biol Fishes* 2013, 96: 175–187. 10.1007/s10641-011-9938-4

77. Hanson KC, Cooke SJ, Suski CD, Niezgoda G, Phelan FJS, Tinline R, Philipp DP: Assessment of largemouth bass (*Micropterus*

salmoides) behavior and activity at multiple spatial and temporal scales utilizing a 3-D whole-lake ecological telemetry observatory. *Hydrobiologia* 2007, 582: 243–256. 10.1007/s10750-006-0549-6

78. Espinoza M, Farrugia TJ, Webber DM, Smith F, Lowe CG: Testing a new acoustic telemetry technique to quantify longterm, fine-scale movements of aquatic animals. *Fish Res* 2011, 18: 364–371.
79. Fausch KD, Torgersen CE, Baxter CV, Li HW: Landscapes to riverscapes: bridging the gap between research and conservation of stream fishes. *Bioscience* 2002, 52: 483–498. 10.1641/0006-3568(2002)052[0483:LTRBTG]2.0.CO;2
80. Laughton R, Smith GW: The relationship between the date of river entry and the estimated spawning position of adult Atlantic salmon (*Salmo salar L* .) in two Scottish major east coast rivers. In *Wildlife Telemetry: Remote Monitoring and Tracking Of Animals*. Edited by: Priede IG, Swift SM. New York: Ellis Horwood; 1992:423–433.
81. Perry GL, Bond NR: Spatially explicit modeling of habitat dynamics and fish population persistence in an intermittent lowland stream. *Ecol Appl* 2009, 19:731–746. 10.1890/08-0651.1
82. Prentice EF, Flagg TA, McCutcheon CA: Feasibility of using implantable passive integrated (PIT) tags in salmonids. *Am Fish S S* 1990, 7: 317–322.
83. Prentice EF, Flagg TA, McCutcheon S: PIT tag monitoring systems for hydroelectric dams and fish hatcheries. *Am Fish S S* 1990, 7: 323–334.
84. Prentice EF, Flagg TA, McCutcheon CA, Brastow DF, Cross DC: Equipment, methods and automated data-entry station for PIT tagging. *Am Fish S S* 1990, 7: 335–340.
85. Castro-Santos T, Haro A, Walk S: A passive integrated transponder (PIT) tag system for monitoring fishways. *Fish Res* 1996, 28: 253–261. 10.1016/0165-7836(96)00514-0
86. Barbin Zydlewski G, Haro A, Whalen KG, McCormick SD: Performance of stationary and portable passive transponder detection systems for monitoring of fish movements. *J Fish Biol* 2001, 58: 1471–1475. 10.1111/j.1095-8649.2001.tb02302.x
87. Bolland JD, Cowx IG, Lucas MC: Dispersal and survival of stocked cyprinids in a small English river: comparison with wild fishes

using a multi□method approach. *J Fish Biol* 2009, 74: 2313–2328. 10.1111/j.1095-8649.2009.02244.x

88. Barbin Zydlewski G, Horton G, Dubreuil T, Letcher B, Casey S, Zydlewski J: Remote monitoring of fish in small streams a unified approach using PIT tags.*Fisheries* 2006, 31: 492–502. 10.1577/1548-8446(2006)31[492:RMOFIS]2.0.CO;2
89. Jorgensen T, Ovredal JT, Huse I, Spedicato MT, Lembo G, Marmulla G: A new pass-through PIT tag detection system for marine use. In *Aquatic Telemetry: Advances and Applications. Proceedings of the Fifth Conference on Fish Telemetry*. Edited by: COISPA Tecnologia and RicercaUstica. Italy: Food and Agriculture Organization of the United Nations; 2005:9–13.
90. Bubb DH, Lucas MC, Thom TJ, Rycroft P: The potential use of PIT telemetry for identifying and tracking crayfish in their natural environment. *Hydrobiologia*2002, 483: 225–230. 10.1023/A:1021352217332
91. Bubb DH, Thom TJ, Lucas MC: Movement patterns of the invasive signal crayfish determined by PIT telemetry. *Can J Zoolog* 2006, 84: 1202–1209. 10.1139/z06-100
92. Cucherousset J, Marty P, Pelozuelo L, Roussel JM: Portable PIT detector as a new tool for non-disruptively locating individually tagged amphibians in the field: a case study with Pyrenean brook salamanders (*Calotriton asper*). *Wildl Res*2009, 35: 780–787.
93. Good TP, McClure MM, Sandford BP, Barnas KA, Marsh DM, Ryan BA, Casillas E: Quantifying the effect of Caspian tern predation on threatened and endangered Pacific salmon in the Columbia River estuary. *Endang Species Res*2007, 3: 11–21.
94. Eiler JH: Radio transmitters used to study salmon in glacial rivers. *Am Fish S Sm*1990, 7: 364–369.
95. Lucas MC, Johnstone ADF, Priede IG: Use of physiological telemetry as a method of estimating metabolism in fish in the natural environment. *Trans Am Fish Soc*1993, 122: 822–833. 10.1577/1548-8659(1993)122<0822:UOPTAA>2.3.CO;2
96. Cooke SJ, Hinch SG, Wikelski M, Andrews RD, Wolcott TG, Butler PJ: Biotelemetry: a mechanistic approach to ecology. *Trends Ecol Evol* 2004, 19: 334–343. 10.1016/j.tree.2004.04.003

97. Oswald RL: The use of telemetry to study light synchronization with feeding and gill ventilation rates in *Salmo trutta* . *J Fish Biol* 1978, 13: 729–739. 10.1111/j.1095-8649.1978.tb03487.x

98. Sayre PD: *A Radiotelemetry System for Monitoring the Activity Levels of Free Swimming Fish*. Canada: MSc Thesis. University of New Brunswick; 1978.

99. Ross LG, Watts W, Young AH: An ultrasonic biotelemetry system for the continuous monitoring of tail-beat rate from free-swimming fish. *J Fish Biol*1981, 18: 479–490. 10.1111/j.1095-8649.1981.tb03789.x

100. Priede IG: Heart rate telemetry from fish in the natural environment. *Comp Biochem Physiol* 1983, 76A: 515–524.

101. Smith EN: Multichannel temperature and heart rate radio-telemetry transmitter. *J Appl Physiol* 1974, 36: 252–255.

102. Nomura NA, Ibaraki T: Electrocardiogram of the rainbow trout and its radio transmission. *Jap J Vet Sci* 1969, 31: 135–137. 10.1292/jvms1939.31.135

103. Armstrong JD, Lucas MC, Priede IG, De Vera L: An acoustic telemetry system for monitoring the heart rate of pike, *Esox lucius L.* , and other fish in their natural environment. *J Exp Biol* 1989, 143: 549–552.

104. Clark TD, Hinch SG, Taylor BD, Frappell PB, Farrell AP: Sex differences in circulatory oxygen transport parameters of sockeye salmon (*Oncorhynchus nerka*) on the spawning ground. *J Comp Physiol B* 2009, 179: 663–671. 10.1007/s00360-009-0349-1

105. Clark TD, Sandblom E, Hinch SG, Patterson DA, Frappell PB, Farrell AP: Simultaneous biologging of heart rate and acceleration, and their relationships with energy expenditure in free-swimming sockeye salmon (*Oncorhynchus nerka*). *J Comp Physiol B* 2010, 180: 673–684. 10.1007/s00360-009-0442-5

106. Kawabe R, Nashimoto K, Hiraishi T, Naito Y, Sato K: A new device for monitoring the activity of freely swimming flatfish, Japenese flounder *Paralichthys olivaceus* . *Fisheries Sci* 2003, 69: 3–10. 10.1046/j.1444-2906.2003.00581.x

107. Wilson RP, Shepard ELC, Liebsch N: Prying into the intimate details of animal lives: use of a daily diary on animals. *Endang Species Res*

2008, 4: 123–137.

108. Shepard ELC, Wilson RP, Quintana F, Laich AG, Liebsch N, Albareda DA, Halsey LG, Gleiss A, Morgan DT, Myers AE, Newman C, Macdonald DW: Identification of animal movement patterns using tri-axial accelerometry. *Endang Species Res* 2008, 10:47–60.
109. Warner EJ, Quinn TP: Horizontal and vertical movements of telemetered rainbow trout (*Oncorhynchus mykiss*) in Lake Washington. *Can J Zool* 1995, 73: 146–153. 10.1139/z95-017
110. Manning B, Grigg GC: Basking behaviour is not of thermoregulatory significance to the 'basking' freshwater turtle *Emydura signata* . *Copeia* 1997, 579–584.
111. Beeman JW, Haner PV, Maule AG: Evaluation of a new miniature pressure-sensitive radio transmitter. *N Am J Fish Manage* 1988, 18: 458–464.
112. Priede IG: An ultrasonic salinity telemetry transmitter for use on fish in estuaries. *Biotelem Patient Monit* 1982, 9: 1–9.
113. Priede IG, Solbe JFLG, Nott JE: Short communication an acoustic oxygen telemetry transmitter for the study of exposure of fish to variations in environmental dissolved oxygen. *J Exp Biol* 1988, 140: 563–567.
114. Seebacher F, Elsey RM, Trosclair PL III: Body temperature null distributions in reptiles with nonzero heat capacity: seasonal thermoregulation in the American alligator (*Alligator mississippiensis*). *Physiol Biochem Zool* 2003, 76:348–359. 10.1086/375426
115. Grayson KL, Dorcas ME: Seasonal temperature variation in the painted turtle (*Chrysemys picta*). *Herpetologica* 2004, 60: 325–336. 10.1655/03-43
116. Donaldson MR, Cooke SJ, Patterson DA, Hinch SG, Robichaud D, Hanson KC, Olsson I, Crossin GT, English KK, Farrell AP: Limited behavioural thermoregulation by adult up-river migrating sockeye salmon (*Oncorhynchus nerka*) in the Lower Fraser River mainstem, British Columbia. *Can J Zool* 2009, 87: 480–490. 10.1139/Z09-032
117. Frank TH: Telemetering the electrocardiogram of free swimming *Salmo irideus* . *IEEE Trans Biomed Eng* 1968, 2: 111–114.

118. Priede IG, Young AH: The ultrasonic telemetry of cardiac rhythms of wild brown trout *(Salmo truffa L.*) as an indicator of bio-energetics and behaviour. *J Fish Biol* 1977, 10: 299–318. 10.1111/j.1095-8649.1977.tb04064.x

119. Klimley AP: Highly directional swimming by scalloped hammerhead sharks,*Sphyrna lewini* , and subsurface irradiance, temperature, bathymetry, and geomagnetic field. *Mar Biol* 1993, 117: 1–22. 10.1007/BF00346421

120. Ebner BC: Preface. Tagging for telemetry of freshwater fauna. *Mar Freshw Res*2009, 60: 281–283. 10.1071/MF09006

121. Cooke SJ, Woodley C, Eppard MB, Brown RS, Nielsen JL: Advancing the surgical implantation of electronic tags in fish: a gap analysis and research agenda based on a review of trends in intracoelomic tagging effects studies. *Rev Fish Biol Fish.* 2011, 21: 127–151. 10.1007/s11160-010-9193-3

122. Doody JS, Roe J, Mayes P, Ishiyama L: Telemetry tagging methods for some freshwater reptiles. *Mar Freshw Res* 2009, 60: 293–298. 10.1071/MF08158

123. HartLG,SummerfeltRC:Surgicalproceduresforimplantingultrasonic transmitters into flathead catfish (*Pylodictis olivaris*). *Trans Amer Fish Soc*1975, 104: 56–59. 10.1577/1548-8659(1975)104<56:SPFIUT>2.0.CO;2

124. Jepsen N, Koed A, Thorstad EB, Baras E: Surgical implantation of telemetry transmitters in fish: how much have we learned? *Hydrobiologia* 2002, 483: 239–248. 10.1023/A:1021356302311

125. Mosley CAE: Anesthesia and analgesia in reptiles. *Semin Avian Exot Pet Med* 2005, 14: 243–262. 10.1053/j.saep.2005.09.005

126. Ross LG, Ross B: *Anaesthetic and Sedative Techniques for Aquatic Animals*. Oxford: Blackwell Scientific Publications; 1999.

127. Franklin CE, Read MA, Kraft PG, Liebsch N, Irwin SR, Campbell HA: Remote monitoring of crocodilians: implantation, attachment and release methods for transmitters and data-loggers. *Mar Freshw Res* 2009, 60: 284–292. 10.1071/MF08153

128. Whitty JM, Morgan DL, Peverell SC, Thorburn DC, Beatty SJ: Ontogenetic depth partitioning by juvenile freshwater sawfish (*Pristis microdon:* Pristidae) in a riverine environment. *Mar Freshw*

Res 2009, 60: 306–316. 10.1071/MF08169

129. Broadhurst BT, Ebner BC, Clear RC: Radio-tagging flexible-bodied fish: temporary confinement enhances radio-tag retention. *Mar Freshw Res* 2009, 60: 356–360. 10.1071/MF08141

130. Summerfelt RC, Mosier D: Transintestinal expulsion of surgically implanted dummy transmitters by channel catfish. *Trans Am Fish Soc* 1984, 113: 760–766. 10.1577/1548-8659(1984)113<760:TEOSID>2.0.CO;2

131. Broadhurst BT, Ebner BC, Clear RC: Effects of radio-tagging on two-year-old, endangered Macquarie perch (*Macquaria australasica:* Percichthyidae). *Mar Freshw Res* 2009, 60: 341–345. 10.1071/MF08142

132. Baras E, Jeandrain D: Evaluation of surgery procedures for tagging eel *Anguilla (L.)* with biotelemetry transmitters. *Hydrobiologia* 1998, 371/372: 107–111.

133. Mulcahy DM: Surgical implantation of transmitters into fish. *ILAR J* 2003, 44:295–306. 10.1093/ilar.44.4.295

134. Cooke SJ, Hinch SG, Farrell AP, Patterson DA, Miller-Saunders K, Welch DW, Donaldson MR, Hanson KC, Crossin GT, Mathes MT, Lotto AG, Hruska KA, Olsson IC, Wagner GN, Thomson R, Hourston R, English KK, Larsson S, Shrimpton JM, Van der Kraak G: Developing a mechanistic understanding of fish migrations by linking telemetry with physiology, behavior, genomics and experimental biology: an interdisciplinary case study on adult Fraser River sockeye salmon. *Fisheries*2008, 33: 321–338. 10.1577/1548-8446-33.7.321

135. Cunjak RA, Roussel JM, Gray MA, Dietrich JP, Cartwright DF, Munkittrick KR, Jardine TD: Using stable isotope analysis with telemetry or mark-recapture data to identify fish movement and foraging. *Oecologia* 2005,2005(144):636–646.

136. Östergren J, Nilsson J, Lundqvist H: Linking genetic assignment tests with telemetry enhances understanding of spawning migration and homing in sea trout *Salmo trutta L* . *Hydrobiologia* 2012, 1: 123–134.

137. Nielsen C, Aarestrup K, Nørum U, Madsen SS: Future migratory behaviour predicted from premigratory levels of gill Na^+ /K^+

-ATPase activity in individual wild brown trout (*Salmo trutta*). *J Exp Biol* 2004, 207: 527–533. 10.1242/jeb.00783

138. Sato A, Ueda H, Fukaya M, Kaeriyama M, Zohar Y, Urano A, Yamauchi K: Sexual differences in homing profiles and shortening of homing duration by gonadotropin-releasing hormone analog implantation in lacustrine sockeye salmon (*Oncorhynchus nerka*) in Lake Shikotsu. *Zoolog Sci* 1997, 14: 1009–1014. 10.2108/zsj.14.1009
139. Brodersen J, Nilsson PA, Hansson LA, Skov C, Brönmark C: Condition-dependent individual decision-making determines cyprinid partial migration. *Ecology*2008, 89: 1195–1200. 10.1890/07-1318.1
140. Crossin GT, Hinch SG, Cooke SJ, Welch DW, Batten SD, Patterson DA, Van Der Kraak G, Shrimpton JM, Farrell AP: Behaviour and physiology of sockeye salmon homing through coastal waters to a natal river. *Mar Biol* 2007, 152: 905–918. 10.1007/s00227-007-0741-x
141. Ueda H, Kaeriyama M, Mukasa K, Urano A, Kudo H, Shoji T, Tokumitsu Y, Yamauchi K, Kurihara K: Lacustrine sockeye salmon return straight to the natal area from open water using both visual and olfactory cues. *Chem Senses* 1998, 23: 207–212. 10.1093/chemse/23.2.207
142. Ueda H: Physiological mechanisms of imprinting and homing migration in Pacific salmon *Oncorhynchus* spp. *J Fish Biol* 2012, 81: 543–558. 10.1111/j.1095-8649.2012.03354.x
143. Miller KM, Li S, Kaukinen KH, Ginther N, Hammill E, Curtis JMR, Patterson DA, Sierocinski T, Donnison L, Pavlidis P, Hinch SG, Hruska KA, Cooke SJ, English KK, Farrell AP: Genomic signatures predict migration and spawning failure in wild Canadian salmon. *Science* 2011, 331: 214–217. 10.1126/science.1196901
144. Nestler JM, Goodwin RA, Smith DL, Anderson JJ, Li S: Optimum fish passage and guidance designs are based in the hydrogeomorphology of natural rivers.*River Res Appl* 2008, 24: 148–168. 10.1002/rra.1056
145. Rogers KB, White GG: Analysis of movement and habitat use from telemetry data. In *Analysis and Interpretation of Freshwater*

Fisheries Data. Edited by: Guy C, Brown M. Bethesda, Maryland: American Fisheries Society; 2007:625–676.

146. Burnham KP, Anderson DR, White GC, Brownie C, Pollock KH American Fisheries Society Monographs. In *Design and analysis methods for survival experiments based on release-recapture*. Bethesda MD: American Fisheries Society, Volunme 5; 1987.
147. Skalski JR, Smith S, Iwamoto RN, Williams JG, Hoffman A: Use of passive integrated transponder tags to estimate survival of migrant juvenile salmonids in the Snake and Columbia rivers. *Can J Fish Aquat Sci* 1998, 55: 1484–1493. 10.1139/f97-323
148. Castro-Santos T, Haro A: Quantifying migratory delay: a new application of survival analysis methods. *Can J Fish Aquat Sci* 2003, 60: 986–996. 10.1139/f03-086
149. Eiler JH, Masters MA: A database-GIS mapping program for summarizing salmon telemetry data for the Yukon River basin. In *Biotelemetry 15: Proceedings of the 15th International Symposium on Biotelemetry. Juneau, Alaska USA*. Edited by: Eiler JH, Alcorn DJ, Neuman MR. Wageningen: International Society on Biotelemetry; 2000:138.
150. Flavelle LS, Ridgway MS, Middel TA, McKinley RS: Integration of acoustic telemetry and GIS to identify potential spawning areas for lake trout (*Salvelinus namaycush*). *Hydrobiologia* 2002, 483: 137–146. 10.1023/A:1021371327810
151. Hodder KH, Masters JEG, Beaumont WRC, Gozlan RE, Pinder AC, Knight CM, Kenward RE: Techniques for evaluating the spatial behaviour of river fish. *Hydrobiologia*2007, 582: 257–269. 10.1007/s10750-006-0560-y
152. Crook D: Is the home range concept compatible with the movements of two species of lowland river? *J Anim Ecol* 2004, 73: 353–366. 10.1111/j.0021-8790.2004.00802.x
153. Crook DA: Movements associated with home-range establishment by two species of lowland river fish. *Can J Fish Aquat Sci* 2004, 61: 2183–2193. 10.1139/f04-151
154. Knight CM, Kenward RE, Gozlan RE, Hodder KH, Walls SS, Lucas MC: Home-range estimation within complex restricted environments: importance of method selection in detecting seasonal

change. *Wildl Res* 2009, 36: 213–224. 10.1071/WR08032

155. Hogan Z, Heng K, Zalinge N: Underwater biotelemetry to study fish migrations in the Mekong River. *Mekong Fish Catch Culture* 2002, 7: 9–13.
156. Hocutt CH, Seibold SE, Jesien RV: Potential use of biotelemetry in tropical continental waters. *Revue d'hydrobiologie tropicale* 1994, 27: 77–95.
157. Baras E, Benech V, Marmulla G: Outcomes of a pilot fish telemetry workshop for developing countries. *Hydrobiologia* 2002, 483: 9–11. 10.1023/A:1021382201886
158. Pais S, Symonds J: Data storage on a RFID tag for a distributed system.*International Journal of UbiComp* 2011, 2: 26–39. 10.5121/ iju.2011.2203
159. Mitson RB, Storeton-West TJ: A transponding acoustic fish tag. *Radio Electronic Engineering* 1971, 41: 483–489. 10.1049/ ree.1971.0147
160. Erickson DL, Kahnle A, Millard MJ, Mora EA, Bryja M, Higgs A, Mohler J, DuFour M, Kennery G, Sweka J, Pikitch EK: Use of pop - up satellite archival tags to identify oceanic–migratory patterns for adult Atlantic sturgeon, *Acipenser oxyrinchus oxyrinchus* Mitchell, 1815 . *J Appl Ichthyol* 2011, 27: 356–365. 10.1111/j.1439-0426.2011.01690.x

CHAPTER

14

MANGROVE SYSTEM DYNAMICS IN SOUTHEAST ASIA: LINKING LIVELIHOODS AND ECOSYSTEM SERVICES IN VIETNAM

Steven Emmerson Orchard, Lindsay Carman Stringer and Claire Helen Quinn

Sustainability Research Institute, School of Earth and Environment, University of Leeds, Leeds, West Yorkshire LS2 9JT, UK

ABSTRACT

Across Southeast Asia, human activity has caused rapid mangrove system degradation and loss. In Vietnam, a country undergoing economic transition, mangrove systems are vital to the livelihoods of coastal rural communities. This paper studies three mangrove system-dependent communities on Vietnam's northern coast. Guided by the sustainable livelihood framework, the paper adopts a mixed methods approach. It presents current uses of mangrove system goods and the factors shaping past livelihood responses to mangrove system change, using livelihood trajectory analysis. Findings demonstrate that communities depend on mangrove systems to different degrees for income, subsistence and to respond to change. However, the rapid development of aquaculture is associated with a significantly reduced and degraded mangrove system commons necessary to support the livelihoods of low-income households. Three distinct livelihood trajectories are identified: consolidator groups able to use their access to a wide range of resources, locked into resilient trajectories; accumulator groups able to use their access to limited resources to move from vulnerable to more resilient trajectories; and marginalized groups facing increasingly reduced access to resources locked into vulnerable trajectories. Vietnam faces challenges in reconciling a more market-orientated economy with the maintenance of mangrove system functions and processes that shape the vulnerability and resilience of livelihood trajectories. Policies and projects promoting the sustainable management of mangrove systems should acknowledge the substantial contribution and multiple uses of mangrove systems in livelihoods, particularly of the poor, and the impact of aquaculture on income equality and livelihood diversity that shapes household resilience and vulnerability.

Keywords: Resilience, Vulnerability, Coastal zone management, Sustainable development, Natural resource management

INTRODUCTION

Southeast Asian mangrove systems are the most biodiverse in the world (Friess et al. 2012), and the goods and services they provide are important components of coastal livelihoods (Van Hue and Scott 2008). Nevertheless, rapid development throughout the region has significantly altered mangrove systems, causing widespread degradation and loss (Gopal 2013). In Vietnam, political and economic reform has facilitated rapid development and vast areas of mangroves have been converted to large-scale, intensive aquaculture (Tri et al. 1998). This trend is particularly alarming in rural areas, where many livelihoods depend on mangrove system commons for collection of fish, crustaceans and other sea life for food and income. Additional goods and services from mangrove systems include: wood for fuel and energy; protection of shorelines from erosion, flooding and storm damage; and to filter pollutants and maintain water quality (Spalding et al. 2010).

Mangrove system degradation and loss occurs through deliberate and inadvertent actions resulting from undervaluation of ecosystem functions and processes (Vilardy et al. 2011). Political, socio-economic and environmental disturbances can negatively impact the structure, function and flow of the goods and services ecosystems provide to society, causing significant impacts on human welfare (MEA 2005). Households in natural resource-dependent communities respond to disturbance through their livelihoods. Livelihoods are shaped by past decisions and access to natural resource commons, steering households along particular livelihood trajectories (Bagchi et al. 1998). Despite growing interest in understanding how human activities are altering ecosystem goods and services, little research has studied how the factors shaping past decisions in response to ecosystem change have shaped the livelihood trajectories of households in mangrove resource-dependent communities (MRDC). To address this gap, this paper examines the livelihoods of three MRDC in northern Vietnam. It provides valuable insights into mangrove system change and the interdependencies of human activity and mangrove goods and services, through empirical analysis in an understudied development context of a rapidly growing transition economy.

Vietnam's transition from a highly centralized planned economy to a socialist-orientated market economy was initiated in 1986 (Đổi Mới). Reforms included: devolution of land management from centralized collectives to households; decentralization of land allocation decision-making to local government; and market liberalization, while the Communist Party has retained political control. Although transition has fostered significant economic gains, a set of intractable social and environmental issues remain, such as growing inequality and natural resource degradation (Adger 2000). Studying the implications of this transition for the livelihoods of MRDC can extend our understanding of the interdependencies between human activity and mangrove goods and services.

This paper presents the current use of mangrove goods and services. It also identifies the factors influencing past livelihood decisions and the vulnerability and resilience of livelihood trajectories within MRDC. The objectives are to: (1) identify the key aspects impacting the dynamics of each mangrove system; (2) establish the range of mangrove system-related livelihood activities households currently engage in; (3) identify household characteristics related to differing levels of mangrove system dependency and use; and (4) establish which aspects of change have influenced livelihood trajectories of households with different levels of mangrove system dependency and use. Quantitative and qualitative methods are integrated within a case study approach. Integrating the sustainable livelihood framework (SLF) with ecosystem service and livelihood trajectory approaches permits understanding of the links between mangrove system change and livelihood responses. The ecosystem service approach allows categorization of the goods and services that households receive from mangrove systems, allowing comparison of current livelihood strategies and activities across the studied MRDC. Livelihood trajectory analysis allows the charting of mangrove system change over time and exploration of how processes of change have shaped the livelihoods of households with different levels of mangrove system dependency.

The research enhances knowledge on livelihoods and natural resource management and provides insights for future mangrove system planning, through identification of key livelihood vulnerabilities and their drivers. We find that in the context of rapid social and environmental change across Southeast Asia, intensive large-scale commercial aquaculture is: reducing the resilience of livelihoods through losses in buffering capacity and livelihood diversity in response to disturbance; and increasing vulnerability through reduced access to mangrove system commons and increasing livelihood sensitivity by removing a crucial source of livelihood response. Households with greater dependence on a significantly declining and degraded mangrove commons find their livelihood options reduced, locking them into livelihood trajectories that can leave them more vulnerable.

The next sections outline central concepts related to livelihoods, ecosystem services and the research process. Narratives of mangrove system dynamics within each context are provided, exploring the political, socio-economic and environmental aspects contributing to contemporary livelihood strategies. Quantitative analysis then offers insights into the contribution of mangrove systems to households' current livelihood portfolios. Livelihood trajectories of individual households are then examined, providing an essential temporal dimension. Lessons from such insights are then discussed.

Central Concepts

A livelihood comprises the capabilities, assets (including both material and social resources) and activities required to make a living (Chambers and Conway 1992). Assets are the tangible and intangible resources, categorized by Scoones (1998) as natural, social, financial, physical and human capitals. The mix of assets and activities a household employs denotes the "livelihood strategy" (Scoones 1998) and also encapsulates cultural and social elements (Ellis 2000). Livelihoods are shaped by the changing natural environment and form within complex social, economic and political contexts; the "shocks", "trends" and "seasonality" of which combine to determine the livelihood vulnerability context (Chambers and Conway 1992; Scoones 1998). Throughout this paper, shocks, trends and seasonality are collectively referred to as disturbance.

A livelihood is considered sustainable when "it can cope with and recover from stresses and shocks and maintain or enhance its capabilities and assets both now and in the future, while not undermining the natural resource base" (Scoones 1998: p 2). Consideration of sustainability within a livelihood approach resonates with concepts of resilience theory (Marschke and Berkes 2006). Resilience refers to "the capacity of a system to experience shocks while retaining essentially the same function, structure, feedbacks and therefore identity" (Walker et al. 2004: p 2). Nevertheless, the application of concepts relating to ecological resilience in relation to social systems has been questioned because of their inattention to power (Duit et al. 2010), while a resilient system does not indicate whether or not it is in a socially preferred ecological or socio-economic state, or who decides this (Armitage et al. 2012). Both sustainable livelihood and resilience approaches have lacked integration with both livelihood and natural resource system history (Vilardy et al. 2011).

The millennium ecosystem assessment (MEA) broadly defines ecosystem services as the benefits people obtain from natural ecosystems (MEA 2005). The MEA provides a framework that considers feedbacks at multiple scales among direct and indirect aspects of change, ecosystem services and human well-being, organizing ecosystem services under provisioning, regulating, cultural and supporting service categories. The MEA framework nevertheless fails to consider the full ensemble of process and feedbacks required to fully understand complex and dynamic human–ecosystem relations (Carpenter et al. 2009). Ecosystem service approaches tend to focus on current ecosystem benefits and economic value while neglecting political aspects and lack consideration of time (Norgaard 2010).

With the aforementioned limitations in mind, we combine ecosystem service and livelihood approaches to explore livelihood trajectories, i.e. the directions that livelihoods follow over time (Sallu et al. 2010). Such an approach allows exploration of the life histories of individual households and their strategic behavior, and the changing use of mangrove system goods and services in their livelihoods. It enables close examination of dynamic human–environment links between mangrove systems and dependent communities and fosters understanding of the political, socio-economic and environmental aspects underpinning the provision of mangrove system goods and services and the ways people use and manage these services over time (Vilardy et al. 2011).

MATERIALS AND METHODS

Data were collected using mixed methods during February–August 2012, in mangrove systems of three coastal communities in northern Vietnam: Giao Xuan (Nam Dinh Province); Da Loc (Thanh Hoa Province); and Dong Rui (Quang Ninh Province) (Figure 1). A community is considered a subset of a commune and defined as a socio-economic impact area of a given mangrove system (Glaser 2003). Communities represented mangrove systems that are distinct, geographically separate and with different histories. The mangrove systems comprised the interdependent components of trees and intertidal mudflat areas from which communities acquired provisioning goods. In all three communities, mangrove system provisioning goods (MSPG) refer to the wild fish, clam, shrimp, crab and other shoreline animals collected from mangrove system commons.

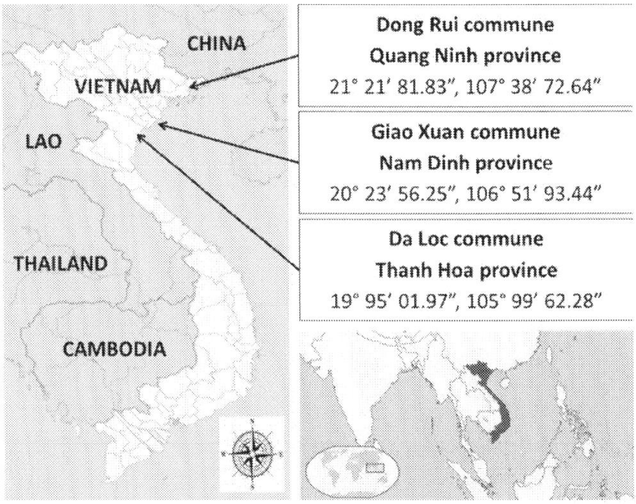

Figure 1: Study site locations and coordinates. *Source* Wikipedia (2014).

Transect walks with local authority and NGO representatives through mangrove system corridors were conducted in each community during scoping studies at the beginning of field work. These aided familiarization with local context, history and issues relating to each mangrove system and enabled more targeted questioning during the main period of data collection. Participant observation of households collecting MSPG and informal conversations with wider community members provided further useful initial information.

Household surveys (n = 248) were conducted with household heads to identify current livelihood strategies and mangrove system use patterns (Giao Xuan, n = 79; Da Loc, n = 70; Dong Rui, n = 99). In this study, household heads are those individuals contributing the largest amount to household income. Semi-structured interviews provided in-depth historical and current perspectives on livelihood strategies and trajectories (n = 10 in each community; total n = 30). Semi-structured interviews elucidated: (a) how households use mangrove systems, (b) how changes in the mangrove system (degradation, storm damage, etc.) affect livelihood decisions, and (c) how aspects identified in (a) and (b) interact with broader political, socio-economic and environmental aspects to determine outcomes and subsequent livelihood impacts.

To achieve objectives 1–3, surveys requested general household information (age, gender, education, etc.) and data on all subsistence and income generating activities. Respondents were selected with the help of local partners and key informants to identify an initial set of households that used the mangrove system for their livelihoods. These households then identified further respondents in a snowball sampling approach (cf. Luttrell 2006). Sampling continued in a respondent-driven way (Heckathorn 1997) until saturation had been reached (i.e. the same names reoccurred in the lists of names provided by respondents). As the composition of the total sample was wholly independent of the initial respondents, it yielded unbiased samples (Heckathorn 1997). This avoided spending time talking to respondents who were not mangrove-dependent. To enable assessment of the relative importance of mangrove resources to household livelihoods, specific information was collected on seasonality, effort, yield and income from mangrove goods. To achieve objective 4, livelihood trajectory data were collected through semi-structured interviews, which covered key events that altered mangrove systems and the subsequent changes in livelihood activities (cf. Sallu et al. 2010). Semi-structured interview participants were selected from survey respondents, maintaining a balance between broad categories of household based on wealth, gender, age and ethnicity (Luttrell 2006). The time covered by the semi-structured interviews was limited to 1975–2012. This covers the period from Vietnam's reunification to the present, encapsulating the collectivized farming era and subsequent political and economic changes: significant events in setting the boundaries of the livelihood context.

Data analysis was iterative and initially involved descriptive analysis to log trends and patterns in preliminary data collected during scoping studies. More detailed analysis was conducted as quantitative and qualitative data accumulated. Quantitative data analysis first explored frequencies of MSPG use in livelihoods using SPSS (IBM SPSS 19). Data were then analyzed using Kruskal–Wallis and Mann–Whitney tests (cf. Cox et al. 2010), with independent livelihood variables categorized using cluster analysis and tested against the dependent variable of percentage of household income derived from MSPG (Brouwer et al. 2007). Qualitative data were coded under emerging themes (Kaplowitz 2001). During livelihood trajectory analysis, tendencies towards resilience or vulnerability were determined by an increase or decrease in access to financial, human, physical, social or natural capitals. This facilitated identification of aspects of change that had played a major role in shaping livelihood strategies. Contradictions and similarities were exposed through repetitive triangulation of findings. Continual iterative reflections were carried out jointly with research participants as further data and results emerged, to determine how and why any conflicts in information may have occurred. This was done via continual dialogue through follow-up interviews and focus groups discussions to validate findings. This resulted in a cyclical process cumulating in inductive interpretation and explanation of results as livelihood system data were positioned within the dynamic socio-economic and political context.

RESULTS

Key Aspects of Change in the Mangrove System

Results in this section are from analysis of transect walk data, semi-structured interviews and focus groups. Mangrove system change is shaped by a complex set of political, socio-economic and environmental aspects identified by research participants (Table 1). Large-scale and intensive commercial aquaculture, privatization of mangrove system tenure rights and the role of local authorities were important in all three communities, along with market liberalization, participation and pollution in Giao Xuan, Da Loc and Dong Rui, respectively. In Giao Xuan and Da Loc, aquaculture comprises large-scale and intensive commercial clam cultivation along the shoreline, while in Dong Rui it refers to large-scale, intensive commercial shrimp cultivation. The unique interactions of political, socio-economic and environmental aspects have defined the contours of the local livelihood context, creating both opportunities and threats to livelihoods over time.

Table 1: Factors of mangrove change

Factors	**Giao Xuan (%)**	**Da Loc (%)**	**Dong Rui (%)**
Aquaculture	73	71	87
Property rights	47	29	67
Local authorities	53	29	60

Pollution	40	18	54
Markets	54	29	14
Participation	–	41	40
Household use	40	6	40
Regulation	30	18	30
Awareness/education	33	6	27
Ecological processes	20	18	–
Population	20	23	–
Severe weather	20	6	7
Infrastructure	20	6	–
Finance	20	–	–
National policy	13	–	–
Knowledge/skills	13	–	–
War	7	–	–

The percentage of interview respondents identifying specific factors in respective communities is provided, while (–) indicates that a factor was not identified (sample sizes: Giao Xuan, n = 15; Da Loc, n = 17; Dong Rui, n = 15)

In all three communities, MSPG played a key role in livelihoods during collectivized farming (1975–1986), when agricultural land was allocated to households and mangrove systems were considered common property. Community rules and traditions determined what people could and could not do. MSPG were crucial for livelihoods, but with no markets for them, there was little incentive to overexploit. In Dong Rui, following the Sino-Vietnamese War (1979), extensive sections of mangrove system land formerly settled by ethnic Chinese groups (approximately 100 km from the Chinese border) were resettled by ethnic "Kinh" Vietnamese from nearby Hai Phong City. Đổi Mới political and economic reforms (1986) fostered lucrative domestic and international markets for MSPG, while changes to the Land Law (1993) devolved land management from the central state to individual households, privatizing tenure rights, and decentralized land allocation authority to local government. Subsequently, regulatory frameworks have struggled to keep pace with changing social, political, economic and environmental conditions. In Dong Rui, subsequent to economic reforms, huge swathes of mangrove system land were sold to shrimp aquaculture investors from Hai Phong

City and surrounding coastal provinces, who had connections to the newly established local authorities. This was done without community consultation, often illegally, by signing land-use contracts using the names of friends, family and community members to circumvent restrictions on the amount of land any one person could own.

During the early 1990s, aquaculture was established on the intertidal mudflat areas in Giao Xuan, with some households benefiting from strong trade links with China and access to lucrative clam markets. Aquaculture took longer to establish in Da Loc due to a lack of trade links and relatively little knowledge of aquaculture farming techniques. However, in the late 2000s, locals observed the financial benefits gained through clam farming in neighboring provinces and aquaculture increased. Growing numbers of people in Giao Xuan and Da Loc claimed land in mangrove areas to establish clam farms, and this caused conflicts within communities. Local authorities intervened by dividing mangrove systems into plots which were auctioned to local households. The authorities benefitted financially through auction processes and subsequent land taxes, while wealthy households and those closely connected to local authorities gained disproportionately.

In all three communities, in addition to poorer households having a vastly reduced area from which to collect MSPG due to the privatization of mangrove system tenure rights, the quantity and quality of MSPG reduced due to the environmental impacts of increased aquaculture. In Giao Xuan and Da Loc, there was concern about disease outbreaks (i.e. infections caused by viral, bacterial and parasitic agents) from aquaculture. Fears were also raised regarding the combined ecological impact of importing vast amounts of alien clam species and associated sand varieties to prepare land for intensive cultivation. Alien species can outcompete and reduce abundance of naturally occurring local species, while imported sand varieties alter the local environmental conditions that local species require to thrive. In Dong Rui, the community's negative experience of mangrove system conversion to aquaculture raised concerns about the subsequent alteration of the hydrological flows that regulate and support the ecosystem. These, along with pollution from the waste discharge of the growing aquaculture industry, were households' main concerns. Dong Rui experienced the complete collapse of the aquaculture industry in the early 1990s (which was owned by external investors) due to mismanagement by local authorities. Such collapse caused severe and widespread degradation in surrounding mangrove system areas.

Current Livelihood Strategies and Activities at the Aggregate Community Level

Results in this section are from the analysis of quantitative data from household surveys. In the context of the above changes, and considering divergent historical perspectives, each community exhibits a distinct set of livelihood strategies and corresponding activities today. High levels of aquaculture activity in Giao Xuan are associated with higher incomes, represented by significantly higher average annual income per capita than in Da Loc and Dong Rui. In Giao Xuan, although aquaculture has significantly contributed to higher incomes, especially for aquaculture farm owners, inequality is reflected in the range of average total household incomes.

Despite comparable average total household incomes between Da Loc and Dong Rui, the almost three times greater income range in Da Loc suggests that the growing aquaculture industry is associated with higher income inequality. Households with the highest incomes are clam farm owners and employees. Although aquaculture has increased the total number of livelihood options available to households in Giao Xuan and Da Loc, the proportion of total livelihood activities each household engages in is higher in Dong Rui. This indicates that households in communities with aquaculture have less diverse livelihoods.

In all three communities, mangrove-based livelihood activities contribute a significant proportion to total income (Figure 2). Conversely, even though 100 % of households in each community are engaged in on-farm activities, these tend to be for household consumption and contribute only a small proportion to total income. Although a small number of households are engaged in off-farm livelihood activities in all communities, income from these activities contribute a larger proportion to total income compared to on-farm activities. Households with more income from on-farm and off-farm activities tend to have lower amounts of income from mangrove systems.

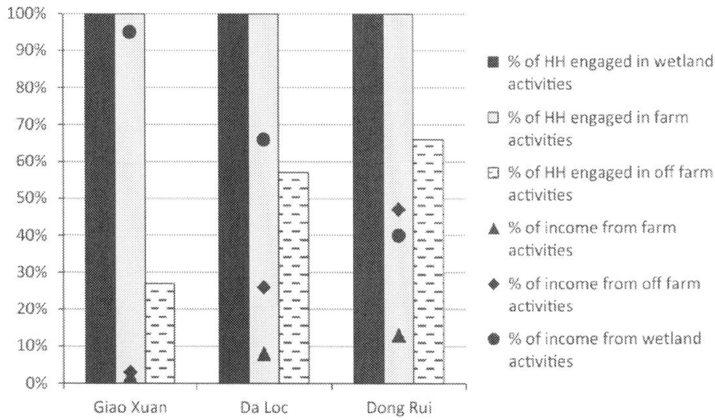

Figure 2: Community livelihood categories (wetland, on-farm and off-farm) and percentage contribution to total income. Wetland activities comprise aquaculture farming, aquaculture employment and wild foraging. On-farm activities comprise crop cultivation (sweet potato, peanut, maize, bean, chilli, sugar cane and fruit) and livestock tending (buffalo, pig, chicken and duck). Off-farm activities include fishing, industry, service, migration. Sample sizes: Giao Xuan, *n* = 79; Da Loc, *n* = 70; Dong Rui, *n* = 99.

Specific livelihood activities relating to mangrove systems in all three communities include the ownership of aquaculture farms, employment on aquaculture farms and collection of wild fish, clam, shrimp and crab from the mangroves for household consumption and/or sale (Figure 3). A number of households are engaged in more than one of these activities. In Giao Xuan, while only 37 % of households own aquaculture farms it contributes 85 % of total income for that community. Although a higher percentage of households are engaged in aquaculture employment (74 %), the proportion of income gained through this activity is low (8 %). Even though 39 % of households are engaged in MSPG collection, it constitutes only 2 % of total income. The unequal distribution of income in Giao Xuan is further apparent because of the 95 % of total income generated through mangrove system activities, and only 6 % comes from aquaculture employment and 1 % from MSPG. The remaining 88 % is derived from the ownership of mangrove system tenure rights that enable income through profits made on aquaculture farms.

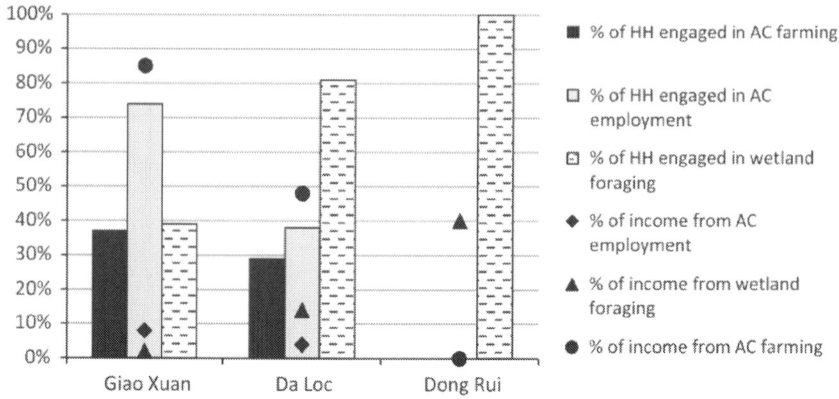

Figure 3: Breakdown of wetland livelihood activities (aquaculture farming, aquaculture employment and wild foraging) and per cent of total income. Sample sizes: Giao Xuan, $n = 79$; Da Loc, $n = 70$; Dong Rui, $n = 99$. *NB: ACaquaculture.*

In Da Loc, where aquaculture farming is in its infancy, a notable portion of households in the community are engaged in it (29 %), and it represents a large portion of aggregate community income (48 %). Over twice as many households in the community are engaged in MSPG collection (81 %) compared to aquaculture employment (38 %), MSPG collection represents over three times the proportion of aggregate community income (14 %) compared to aquaculture employment (4 %). A larger number of households are engaged in non-wetland-related livelihood activities than those observed in Giao Xuan (Figure 2). In Dong Rui, which experienced aquaculture industry collapse, no households engage in aquaculture farming or employment, and 100 % engage in collecting MSPG, representing 40 % of total income, with the remaining 60 % coming from non-wetland-related activities.

These results indicate that when the commercial aquaculture industry is strong: aggregate community income tends to be unequally distributed and concentrated among aquaculture farmers; average household livelihood diversification is lower; and marginalized households remain dependent on collecting MSPG as a livelihood activity.

Table 2: Characteristics of households in Giao Xuan most dependent on MSPG for income

	Giao Xuan				
	Test statistic	**Degrees of freedom**	**Sig.**	*z***score**	**Post hoc** *r***score**
Age	10.961 Ω	4	0.027**	−3.219	0.001
Gender	352 β	–	0.006***	−3.00	0.3
Education	–	–	–	–	–
Years lived in commune	–	–	–	–	–
Household members	–	–	–	–	–
Livelihood diversity	13.344 Ω	2	0.001***	−3.454	0.001
Income	5.935 Ω	2	0.05**	−2.426	0.015
Land user rights	15.416 Ω	2	0.000***	−3.603	0.000

Table 3: Characteristics of households in Da Loc most dependent on MSPG for their income

	Da Loc				
	Test statistic	**Degrees of freedom**	**Sig.**	*z***score**	**Post hoc** *r***score**
Age	–	–	–	–	–
Gender	442.5 β	–	0.087*	−1.710	−0.2
Education	375 β	–	0.026**	−2.221	−0.3
Years lived in commune	5.489 Ω	2	0.064*	−2.228	0.026
Household members	–	–	–	–	–
Livelihood diversity	–	–	–	–	–
Income	–	–	–	–	–
Land user rights	10.459 Ω	2	0.005***	−3.122	0.002

Table 4: Characteristics of households in Dong Rui most dependent on MSPG for their income

	Dong Rui				
	Test statistic	**Degrees of freedom**	**Sig.**	***z*score**	**Post hoc *r*score**
Age	–	–	–	–	–
Gender	685 β	–	0.005***	−2.786	0.3
Education	18.642 Ω	4	0.001***	−2.656	0.008
Years lived in commune	13.409 Ω	2	0.001***	−3.430	0.001
Household members	7.698 Ω	2	0.021**	−2.101	0.036
Livelihood diversity	24.459 Ω	2	0.000***	−2.656	0.008
Income	11.649 Ω	2	0.003***	−3.475	0.001
Land user rights	–	–	–	–	–

* $p = 0.05–0.1$; ** $p = 0.049–0.011$; *** $p = 0.01–0$

β Mann–Whitney test, *Ω* Kruskal–Wallis test

Characteristics of Households Most Dependent On MSPG and Their Use

Results in this section come from the analysis of quantitative data from household surveys and qualitative data from semi-structured interviews and focus groups. In each community, households most dependent on MSPG for their livelihoods were characterized (Tables 2, 3, 4; Table S2 shows the breakdown of variable groups). Female-headed households were more dependent on MSPG than male-headed households in all three communities. In Giao Xuan and Da Loc, where aquaculture prevails, households with less secure mangrove system tenure rights were more dependent on MSPG than those with more secure tenure rights. Where commercial aquaculture is in its infancy or has collapsed (Da Loc and Dong Rui, respectively), households with low education levels were more dependent on MSPG than those with higher education levels. In Giao Xuan, there was more dependence on MSPG among households with high livelihood diversity, while in Dong Rui, higher MSPG dependency was found among households with low livelihood diversity. As low-income households were found to be more mangrove system-dependent in both these communities, this indicates that low-income households in Giao Xuan are using mangroves to diversify their livelihoods, while low-income households in Dong Rui are not. This could be because Dong Rui does not have a commercial aquaculture industry, hence lacking options for diversification.

Table 5: Species, and estimated effort, weight and price of provisional services from mangroves

Catch	Season (height)	Est. effort (hrs)	Est. weight (kg)	Est. $/kg
Giao Xuan				
Fish	All year (March–July)	5–6	5–10	1–1.5
Crab	All year (March–July)	5–6	3–4	1.5
Clam	All year	5–6	2–10	1–3.5
Shrimp	All year (March–July)	5–6	5–30	3–5
Da Loc				
Fish	All year (February–April)	5–6	3	5
Crab	All year (January–August)	5–6	4–5	1
Clam	All year (February–May)	4–6	6–7	0.5
Dong Rui				
Fish	All year (April–June)	8	10	1–5
Crab	All year (March–August)	3	6–8	1–1.5
Clam	All year (May–September)	6	5–7	3–4
Worm	All year (September–February)	8	2–4	2–3
Octopus	All year (June–August)	6	0.5–1	20–50
Shrimp	All year (September–December)	6	4–6	5
Jellyfish	All year (February–March)	6	–	5–10.5

All three communities benefit greatly from mangrove ecosystem services (Table S3). Provisioning and regulating services were the most identified services among all communities, representing more direct benefits. However, perceptions differed between communities. In Giao Xuan, provisioning services were identified more frequently by the highest number of households, with regulating services largely corresponding to storm protection benefits. In Da Loc, regulating services were identified more frequently and by more households. This could be due to experiences of extensive storm damage and saline intrusion from typhoon Damrey in 2005, with resulting damage to arable farm land still fresh in respondents' memories. In Dong Rui, a higher percentage of statements were made regarding regulating services compared with the other communities, although several households identified provisioning services. In Dong Rui, which has no protective community dike, respondents highlighted that moderately intense storms can have severe negative impacts on their crops, and mangroves are seen as crucial for storm protection. With soil quality already poor, saline intrusion resulting from storms is a large community concern. Supporting services were the next most identified service across all communities, particularly soil retention, nutrient cycling, oxygen production and habitat provision. Cultural services were consistently ranked lowest in importance across all three communities, although Dong Rui respondents identified aesthetic qualities and heightened sense of well-being as important ecosystem benefits.

Semi-structured interviews and focus groups revealed that households use diverse strategies to respond to disturbance, such as increased collection of MSPG for sale, drawing on savings, bank loans, social and kinship networks, and sale of assets and labor. Sale of MSPG was the most important safety net in the face of economic shocks because it demanded less reliance on other people: the extended family are commonly poor so cannot offer support, and no repayments are incurred. The MSPG differ in each community due to specific biophysical and geographic mangrove system characteristics (Table 5). Interviews indicated that in all three communities, households with higher dependence on MSPG rely on these goods to sell. Sale of MSPG helps them to cope with disturbances such as crop failures, seasonal fluctuations in the weather, unemployment, celebrations and the start of the new school year when money for schooling and equipment is needed. Semi-structured interviews and focus groups in Giao Xuan revealed that during August and September, when MSPG are at their least available, MSPG-dependent households find it most difficult to meet their subsistence needs. Interviews indicated that although prices offered for MSPG were relatively stable, this was due to wholesalers giving consistently low prices in order to maximize their profits.

Table 6: Livelihood trajectories of households most reflective of the impact of social-ecological dynamics

Case study household	**Factors leading to resilience (*R*) and vulnerability (*V*)**
Case 1—Male, age 54, Giao Xuan Before economic reform, there was no market for mangrove system goods, so he and others in the community foraged MSPG for household consumption. Following economic reforms in 1986, he was employed on a trading boat, which took him to China where he first became aware of the lucrative clam market. He invited a specialist from China to Giao Xuan to teach him clam aquaculture techniques to produce clams for export to China. Subsequent to the success of this, other locals were attracted to mangrove system land to cultivate clams as a commodity and began to assert claims over sections of the land. By 1991, overexploitation resulted in the complete collapse of the native clam population. In 1992, however, he was able to draw on accumulated capital and trading links with neighbouring provinces to import clam seed varieties to cultivate before selling on to China. This was highly successful and the market peaked in 1995. By 1997, however, the imported clams began to die due to incompatibility with local environmental conditions. Many clam fields were abandoned and became available for him to buy as a result. Undeterred, he decided to search further afield, to provinces in the south of Vietnam, to find clam species more suited to local conditions. In 1999, he combined the import of new clam species with new sand varieties to accommodate them. Although there was initial scepticism due to past failures, the enterprise was a success and markets developed both domestically and internationally. Furthermore, due to the stabilizing effect on the environment from mangrove restoration efforts, he no longer needs to import clam seeds from the south. Clam farming is now the major industry in Giao Xuan, which is now one of the biggest producers in Vietnam. Although the industry is more stable now, he still has to make periodic alterations to his fields in order to maintain productivity. He is aware that importing clam and sand varieties is unsustainable and is trying to reintroduce native species	R1. Access to natural resources R2. Salaried employment R3. Access to markets R4. Access to knowledge V1. Loss of natural capital R5. Draws on financial capital R6. Access to social networks V2. Loss of productive capacity R7. Accumulates land R8. Draws on financial capital R9. Rise in demand for aquaculture products R10. Regulating ecosystem service V3. Uncertainty due to suppression of ecosystem functions and processes

Case 2—Female, age 51, Giao Xuan She has lived in Giao Xuan all her life, where she lives with her 21-year-old son. When she was young, her family was poor and life was difficult; often there would not be enough food to eat, and they would have to rely more heavily on foraged MSPG for subsistence. There was no state support at the time, so when her family found themselves in hardship they would have to ask for loans from rich households which they had to pay back with interest. Following the birth of her son, she lost her husband and had to rely on her husband's family to support her and her new born baby. When the clam aquaculture sector began, she could not get access to any land for farming because she was not strong enough to claim land, and she was not rich enough to buy any. Even if she had the money, because she is a woman she cannot own land. When the clam aquaculture sector expanded, she was young and healthy and able to find employment through family contacts. She established a reputation for being a good worker and was respected by her work colleagues, and so aquaculture owners began to ask her to manage work teams on their behalf. She has been able to develop such a wide network of contacts within the industry that she can even find employment in neighboring districts. Her son is now old enough to contribute to household income, and he is also employed on aquaculture farms. However, clam aquaculture does not provide stable employment, particularly during the winter, so she still has to forage MSPG to supplement her income. In recent years, the rains have been less predictable and this has affected her rice crop, so she has to depend heavily on foraging MSPG when this happens. She cannot make as much money from this as she did in the past, as there is less space to forage MSPG and fewer animals available to harvest, even though there are lots more animals in the aquaculture fields. She believes that clam aquaculture is eradicating the natural species and is worried that eventually there will be no MSPG to forage	V1. Lack of subsistence V2. Lack of financial capital V3. Lack of state support R1. Access to ecosystem provisions V4. Debt accumulation V5. Loss of labour R2. Family support network V6. Lack of access to land R3. Salaried employment R4. Applied human capital R5. Extended social networks R6. Gain in human capital V7. Unstable income R7. Provisional ecosystem service V8. Climatic impact on crops V9. Loss of access to ecosystem services V10. Altering ecosystem causes increased livelihood uncertainty

Case 3—Female, age 46, Giao Xuan When she was young she would forage MSPG with her family for household consumption. When the mangrove system area her family had traditionally collected from was divided up and turned into clam aquaculture fields, her husband joined a collective that pooled all their savings together to buy a field. Combined with the income she received from laboring on clam aquaculture fields, they earned enough income for food and to send their son to school. When her husband became terminally ill, he could no longer work, and she had to work fewer hours to tend to him. They received no state support, and with hospital bills mounting, they were forced to sell everything they owned and move into a smaller house next to the dike. The land near the dike is low quality and not suitable for growing crops, and household assets, such as livestock, are often stolen by groups of thieves that target vulnerable households. The community that lives near the dike, made up largely of elderly, disabled and (often illegal) migrant households, are supportive and pool their resources together in order to help each other. In addition, due to the growing clam aquaculture industry she has been able to receive loans from rich owners (usually with interest payable). Although she feels that the rich owners look down on the dike community, they will still employ them to work on their fields, but she still relies heavily on foraged MSPG for food for subsistence. This space, however, has vastly reduced and she must travel through the clam aquaculture fields to get there, and although she can make extra income from collecting the litter thrown from the clam field watchtowers, she must be careful not to stray too close to the fields otherwise the owners will attack her. In addition, because people can make money from foraging MSPG now, they will commit more time and effort which means there are less animals to catch. Although she is aware of some livelihood opportunities available through various NGO projects, she is unable to get to the Women's Union meetings where opportunities are distributed, and she believes that she does not have the adequate level of skills and knowledge required to participate in the projects. Not only that, but these opportunities are usually shared among the families of union leaders	R1. Access to ecosystem services V1. Loss of access to ecosystem services R2. Diversification of income V2. Loss of human capital V3. Loss of income V4. Accumulation of debt V5. Selling of assets V6. Low quality land for arable crops V7. Target of crime R3. Social support networks R4. Access to loans V8. Discrimination R5. Ecosystem service V9. Loss of access to land V10. Overexploitation of resources V11. Lack of access to village meetings V12. Lack of awareness V13. Elite capture

Case 4—Male, age 37, Dong Rui In 1979, he moved to Dog Rui from Hai Phong City as part of the resettlement programme. Life was difficult in the city with little work, and resettlement offered a house with land to cultivate and 6-month worth of rice from the state to help with the transition. The abundance of natural resources meant that food was easy to obtain and life was good. In 1986, encouraged by the local People's Committee, he took out a substantial loan to invest in a wetland boundary pond to allow more effective capture of marine creatures. This was very productive for the first 2–3 years, but then productivity sharply declined due to the impact the ponds had on the natural flow of water and the environment. Many residents raised this as an issue at village meetings at the time, but their concerns were not acted upon by the authorities. As the bank loan repayments were mounting, he took out further loans in the hope that the pond would become productive again. This did not happen and eventually he gave up on the pond. For a while he could still make a living foraging MSPG in the vast wetland area, but when huge areas started to be sold to investors from other provinces this reduced the commons area. Furthermore, pollution from the clam aquaculture fields destroyed the surrounding area, which drove him to destitution. He was the victim of unscrupulous human traffickers to whom he paid money, provided to him by his wife's family in Hai Phong, on the understanding that he would gain well-paid employment in China. On arrival the hours were long, conditions were terrible and the pay was very low, so he fled back to Dong Rui, putting his life in danger and swimming across dangerous waters in order to cross the border from China to Vietnam. Additionally, his rice, peanut and sweet potato crops have been impacted by rapidly changing and unpredictable weather in recent years, with the winters becoming colder and the summer hotter. The irrigation system is inadequate, the quality of local soil is sandy, salty and of poor quality, and this restricts the options for changing crops, planting times and varieties. If people do not get enough rice, they go hungry, but he is lucky that he is still strong and can sell labour to a nearby paper factory and use his earnings to buy rice	R1. Access to ecosystem services R2. Access to loans V1. Loss of ecosystem function and process V2. Accumulating debt V3. Loss of income R3. Access to ecosystem service V4. Loss of access to ecosystem service V5. Onset of poverty V6. Vulnerable target of human trafficking syndicates V7. Negative climatic impact on crops V8. Poor infrastructure V9. Poor quality V10. Lack of diverse cropping options R4. Human capital

Case 5—Female, age 33, Dong Rui She is from the Dao ethnic minority, originally from the mountainous region of the province, and has lived in Dong Rui for 12 years since they were resettled here by the government. The Dao community were promised a better life in Dong Rui, but since arriving she has wanted to return to her home. The district authorities, however, have already converted the land they left for another purpose so she cannot return. She arrived with a small number of other Dao families, but as they did not speak, Vietnamese were not familiar with the environment, and because they have different customs, beliefs and traditions to the ethnic "kinh" Vietnamese, they struggled to integrate into the local community. They soon became isolated and were pushed into the area with lower-quality land where it is difficult to grow crops. Almost all of her income comes from foraging MSPG, and this has been so since she and her family arrived, but she is given a lower price than the ethnic "kinh" wholesaler. Some Dao go to forage MSPG in groups and have developed effective techniques for catching animals, but she is not involved in any of these groups. These groups have become rich, but she remains poor. She has to pay community fees, but she is unsure exactly what this is for as she is very poor but receives no state support	V1. Loss previous support mechanisms V2. Communication difficulties V3. Alienation from wider community V4. Lack of income diversity V5. Discrimination V6. Lack of skills V7. Lack of state support

Livelihood Trajectory Analysis

An in-depth analysis of the livelihood trajectories of illustrative households between 1975 and 2012 ascertained the influence of multiple and interacting key aspects of change that resulted in current livelihood strategies. The livelihood trajectories of five representative households illustrate the wider communities' experiences in using the mangrove system, encompassing all wealth groups (Table 6).

Three distinct types of livelihood trajectory exist across the three communities. The first group, (consolidators) start from a point of power and influence prior to political and economic reform and are currently successful aquaculture farm owners. These households are characterized by high incomes, a middle-aged male head, low livelihood diversity and more secure mangrove system tenure rights. They have typically been able to access MSPG through land grabbing and social influence and prosper through a combination of access to emerging external markets, capital, social networks and knowledge following market liberalization. These households were typically influential during the collective farming era and have leveraged this position during market liberalization to reinforce and increase access to land and resources. Their aquaculture ventures have been consolidated through acquisition of mangrove system tenure rights from struggling aquaculture farmers who lack the skills and knowledge to be successful. These households have also modified the environment in response to ecosystem feedbacks which undermine aquaculture productivity, e.g. by importing alien species of shrimp or clam, and related varieties of sand to accommodate them. Successful aquaculture farmers have also been able to form lobby groups to challenge the local authorities on decisions which impede their aquaculture activities and are able to profit by providing loans to poor households charged at interest. This trajectory of prosperity has been reinforced by some households by leveraging social influence and networks to their advantage.

The second group, (accumulators) were typically from poor backgrounds prior to political and economic reforms, lacking the influence, power and access to resources of the consolidators group. They have increased their incomes largely through aquaculture employment or collective ownership of small-scale aquaculture farms. Households are characterized by mid- to low-level incomes, male or female heads, mid-livelihood diversity to low livelihood diversity and moderately secure land-use rights. These households have been able to improve their livelihood trajectory and gain employment mainly on aquaculture farms (and in construction or manufacturing) through a combination of human capital, social networks and forging reputations as good workers. However, as aquaculture has increased the number of livelihood strategies available to the community, these households have reduced the number of livelihood activities that they engage in, becoming more specialized in aquaculture employment. Some households have been able to either gain access to bank loans or pool resources with family or friends to obtain formal tenure rights and develop aquaculture farms. The local economy provides a sufficient living for these households, and they can overcome livelihood disturbance by seeking alternative employment opportunities beyond the locality if they have an adequate level of human capital. Many of these households still use MSPG to supplement their income, for household use, or in times of livelihood shock and stress.

The third group, (marginalized) were typically from poor backgrounds prior to political and economic reforms and are currently marginalized due to a lack of power, influence and severely limited access to livelihood resources (e.g. labor, skills, networks, capital). These households are struggling to survive and are characterized by low incomes, young, female heads, high livelihood diversity and insecure land-use rights. These households lacked social influence or access to employment opportunities and relied heavily on MSPG for income and subsistence. Usually, they were unable to take advantage of opportunities to access land and resources following political and economic reform due to a lack of social networks and human capital. Increased pressure on mangrove systems from aquaculture disproportionately affects households in this group, who are least able to defend livelihoods or take advantage of market opportunities. Many households shifted from the accumulator group to this group following sickness or death of household members which greatly increased livelihood vulnerability. Other households shifted from the accumulator group to this group subsequent to mounting debts due to failed aquaculture ventures as a result of lack of adequate skills or knowledge. Recently, settled migrants lacking social networks and local knowledge also form a large proportion of this group. To cope with shocks, these households increase livelihood diversity, rely on support from family and close friends and increase their use of MSPG for subsistence and income. However, the impact of aquaculture means they face reduced mangrove system common areas from which to collect MSPG. Feedbacks from rapid economic development (i.e. aquaculture) have also exacerbated negative impacts such as biodiversity loss and water cycle disruption, which has led to reductions in the quantity and quality of MSPG collected. Subsequently, marginalized households often rely on loans, some have pre-existing debt from failed aquaculture ventures, and households often resort to asset selling and move to cheaper, unproductive land. Here, households are susceptible to alienation from the community and often become targets of crime. Other households may have fallen out with local authorities who then use their power and influence to make life difficult for them, and if these households do not have sufficiently strong social networks or human capital to fall back on, they can quickly fall into this group. Households in this group are likely to experience trajectory lock-ins due to lack of access to resources, networks and a greatly reduced and degraded mangrove system commons.

Results highlight the factors that have shaped the resilience and vulnerability of household livelihoods. Factors that increase resilience of consolidator and accumulator livelihood trajectories were: access to aquaculture market opportunities (e.g. employment, knowledge, networks, finance); access to more secure tenure rights; high levels of human capital; and access to support networks and MSPG to buffer disturbance. Factors found to increase vulnerability within the marginalized livelihood trajectory were: loss of access to MSPG reducing buffer to disturbance; low incomes and susceptibility to poverty (e.g. sickness, debt, asset selling); negative environmental impacts from aquaculture (e.g. alteration in ecological processes, exposure to climatic shocks and stresses); discrimination; and elite capture reducing response options.

DISCUSSION

Integrating ecosystem service and livelihood trajectory approaches provided the opportunity to analyse current uses of MSPG in livelihoods between and within communities, and the aspects influencing past livelihood decisions and uses of mangrove system goods and services in response to disturbance. Understanding how this has shaped livelihood trajectories is crucial for understanding the dynamics and interdependencies of mangrove systems and livelihood strategies (Trabucchi et al. 2012; Sallu et al. 2010).

Key Aspects of Mangrove System Change

Aquaculture was found to be the primary driver of mangrove system change, consistent with findings across Southeast Asia (Kirui et al. 2013). Aquaculture encapsulates the key aspects of mangrove system change identified by communities, i.e. changes in mangrove system tenure arrangements, market liberalization and elite capture. Increased intensification and specialization of mangrove system land subsequent to the rapid increase in aquaculture is severely undermining ecosystem functions and processes, in line with findings from Power (2010). Consistent with Gunderson and Holling (2002), feedbacks from rapidly growing aquaculture have exacerbated negative environmental impacts (e.g. biodiversity loss, water cycle disruption). Hence, mangrove system declines, and degradation from conversion to aquaculture has: (1) reduced livelihood resilience through vegetation and biodiversity loss, lowering livelihood options to buffer the impacts of disturbance (cf. Sonwa et al. 2012) and (2) increased the livelihood vulnerability of marginalized households through greater exposure to disturbance and increased sensitivity due to increasingly negative effects from disturbance (cf. Walker et al. 2006).

Livelihood Strategies and Activities

High levels of aquaculture are associated with: greater livelihood vulnerability due to the concentration of wealth among aquaculture farmers and lower livelihood resilience due to lower levels of household livelihood diversity. The distribution of MSPG is becoming increasingly unequal and determined by market forces that serve the interests of elites, while marginalized households face reduced access due to the degradation and loss of mangrove system commons. In line with Adger et al. (2006), income inequality has created vulnerable groups through the concentration of resources among a small number of individuals, thus reducing the livelihood options of marginalized households. Income from MSPG can reduce income inequality. Results show that communities with high levels of aquaculture activity are associated with lower average household livelihood diversity, although livelihood diversity is greater among marginalized groups in these communities. Pomeroy et al. (2006) argue that greater livelihood diversity increases resilience by reducing household dependency on any single income source and provides flexibility by spreading risk among a number of livelihood activities to buffer disturbance. Hence, MSPG can effectively increase livelihood resilience, offering diversification into a low-cost activity requiring little capital outlay.

Household Characteristics and Mangrove System Dependency and Use

In communities with high levels of aquaculture activity, marginalized groups are increasingly vulnerable due to high dependence on MSPG from degraded and declining mangrove system commons. Results indicate that MRDC are differentiated socio-economically with households exhibiting a variation of dependency on MSPG. Results support other findings from Vietnam (Van Hue 2006) and Ethiopia (Asfaw and

Satterfield 2010) where entrenched customary norms and patriarchal cultures constrain women's access to land. Subsequently, female-headed households depend more on MSPG from mangrove system commons. In communities with high levels of aquaculture activity, households earning high incomes from aquaculture have less need to diversify, while low-income households diversify to reduce risk from external disturbances. Conversely, in communities with less aquaculture activity, households able to access a diverse range of livelihood activities, particularly off-farm activities, have less dependence on mangroves. Barrett et al. (2005) state that inter-household heterogeneity plays a fundamental role in determining livelihood diversification patterns, and the constraints and incentives facing various groups must factor prominently in analyses of livelihood diversification. MSPG collection was considered by marginalized households to be the most important response to cope with disturbance. Consistent with findings in Zambia (Kalaba et al. 2013), the sale of forest products was more important than support from kinship ties due to a lack of economic prosperity among kinship networks. It is crucial to consider the impacts that changes in mangrove systems have on household coping strategies and livelihoods (Trabucchi et al. 2012).

Livelihood Trajectories

A household's current livelihood portfolio is largely shaped by factors affecting past livelihood actions, particularly access to ecosystem goods and services (Nyamwanza 2012). Pre-existing conditions (e.g. political power and resource access) shape current livelihoods by influencing access to resources (i.e. natural resources, finance, secure tenure rights, knowledge, technology, social networks, markets). Consolidator households were able to appropriate mangrove system goods and services by using their access to resources to establish successful aquaculture farms, with the subsequent degradation and loss of mangrove system commons depriving marginalized households of MSPG. Pellow (2000) argues that exploring such issues of power helps us to understand how and why people suffer from reduced access to natural resources and subsequent vulnerability. Aquaculture has increased the intensification and specialization of mangrove system cultivation, severely undermining system functions and processes, and disproportionately affecting the livelihoods of those households most dependent on MSPG. This corresponds with findings from Gunawardena and Rowan (2005) in their study of the Rekawa Lagoon system of Sri Lanka, where the negative environmental impacts from aquaculture had disproportionately large impacts on traditional livelihoods. Changes in national-level structures combine with local-level responses to shape multiple livelihood trajectories, such that development gains at the national level ignore crucial environmental justice issues at the local level (Sikor 2001). However, while different, the livelihood trajectories of various groups are interconnected.

Harnessing the knowledge gained through dialogue with the households involved in this study, it was observed that perceptions of potential solutions to enhance livelihood trajectories differed significantly between consolidator and accumulator/marginalized groups. The consolidator group typically expressed their desire for less government interference in the form of tax on their profits and regulations that restrict their management decisions. This, they argued, would enable them to further develop and manage their aquaculture farms as they see fit, which, they believe, would benefit the wider community through increased employment opportunities. Another opinion from consolidators was that successful aquaculture farmers should invest in sustainable development by considering the environmental impact of aquaculture practices. Owners should focus on developing native varieties of aquaculture goods instead of importing alien species and modifying the environment to accommodate them. This was viewed as unsustainable due to the damage caused to the surrounding area. Conversely, respondents from accumulator and marginalized groups expressed the need for more accountable and participatory governance processes. This was due to the perceived unjust distribution of the benefits derived from mangrove system resources which limits their livelihood options. These groups believe that local authority figures are typically appointed by high-ranking Communist Party officials, rather than electing through local elections, which they perceived as a façade. These appointed authority figures typically have little concern for the local communities or environment that they are supposed to represent. Accumulator and marginalized groups suggest that communities should be better able to access higher levels of government in order to mitigate the power of local authorities and elite capture, which they viewed as one some of the most restricting factors in shaping household livelihood trajectories.

CONCLUSION

By analyzing aspects of mangrove system change and the factors that shape household responses, our findings highlight the importance of considering how these interacting elements have shaped livelihoods in three MRDC in northern Vietnam. By using a framework that provides a time dimension to the analyses of household mangrove system use, we have shown how the context within which aquaculture develops, as well as the socio-economic characteristics of households, shapes the vulnerability and resilience of household livelihoods. While calls for efforts to increase mangrove system conservation and restoration in order to increase livelihood resilience are welcomed, it must be acknowledged that MRDC do not use and respond homogenously to mangrove system change.

Findings illustrate how transition processes have altered the governance of mangrove systems through the increasing influence of market mechanisms. For example, households with access to finance, skills, networks and markets have been able to take advantage of the opportunities presented by transition to develop successful aquaculture farms and increase their livelihood resilience. However, aquaculture negatively impacts marginalized households by restricting access and degrading mangrove system resources crucial for households with limited or no access to market opportunities. We have shown how the livelihoods of the marginalized are becoming increasingly vulnerable through: (1) income inequality and the concentration of wealth among a small number of households which diverts resources away from the most marginalized and (2) restricted options for livelihood diversity through limited access and degradation of mangrove systems. Therefore, the impact of aquaculture must be considered if policies to sustainably manage mangrove systems are to be successful. Identifying the characteristics of those households most dependent on MSPG and vulnerable to change is crucial in order to provide targeted livelihood support to those who need it most. Further research will be necessary to identify the specific kinds of support vulnerable groups require, in addition to their desire for greater voice in decision-making. Policies aimed at increasing access of marginalized households to mangrove system resources can reduce income inequalities while increasing livelihood diversification opportunities, thereby increasing resilience. Further research is necessary to understand the institutional structures and processes within which the limits and constraints for increasing access to mangrove system resources exist.

Findings presented here highlight important features of communities that should be considered within environmental governance more widely. For example, the increased influence of market mechanisms on mangrove system governance, income inequality and subsequent constraints on livelihood diversity create path dependencies that shape future options in response to mangrove system change, locking marginalized households into vulnerable livelihood trajectories. How mangrove system change affects livelihoods depends on household characteristics and local context and will manifest differently depending on the equitable and just distribution of resources necessary for the sustainable governance and use of mangrove systems.

ACKNOWLEDGMENTS

This work was funded through the Economic and Social Research Council Centre for Climate Change Economics and Policy (CCCEP). We are grateful to the communities of Giao Xuan, Da Loc and Dong Rui, Marinelife Conservation and Community Development (MCD), the Centre for Natural Resource and Environment Studies (CRES) and CARE Vietnam.

REFERENCES

1. Adger WN (2000) Institutional adaptation to environmental risk under the transition in Vietnam. Ann Assoc Am Geogr 90(4):738–758. doi:10.1111/0004-5608.00220
2. Adger WN, Paavola J, Huq S, Mace MJ (2006) Fairness in adaptation to climate change. MIT Press, Cambridge
3. Armitage D, Béné AT, Charles D, Johnson D, Allison EH (2012) The interplay of well-being and resilience in applying a social-ecological perspective. Ecol Soc 17(4):15. doi:10.5751/ES-04940-170415

4. Asfaw T, Satterfield T (2010) Gender relations in local-level dispute settlement in Ethiopia's Zeghie Peninsula. Hum Ecol Rev 17(2):160–174
5. Bagchi DK, Blaikie P, Cameron J, Chattopadhyay M, Gyawali N, Seddon D (1998) Conceptual and methodological challenges in the study of livelihood trajectories: case-studies in Eastern India and Western Nepal. J Int De Plan Rev 10:453–468. doi:10.1002/(SICI)1099-1328(199806)10:4<453:AID-JID538>3.0.CO;2-Q
6. Barrett CB, Bezuneh M, Clay D, Reardon T (2005) Heterogeneous constraints, incentives and income diversification strategies in rural Africa. Q J Int Agric 44(1):37–60
7. Brouwer R, Sonia A, Luke B, Enamul H (2007) Socioeconomic vulnerability and adaptation to environmental risk: a case study of climate change and flooding in Bangladesh. Risk Anal 27(2):313–326. doi:10.1111/j.1539-6924.2007.00884.x
8. Carpenter SR, Mooney HA, Agard J et al (2009) Science for managing ecosystem services: beyond the millennium ecosystem assessment. Proc Natl Acad Sci USA 106(5):1305–1312. doi:10.1073/pnas.0808772106
9. Chambers R, Conway G (1992) Sustainable rural livelihoods: practical concepts for the 21st century. Institute for development studies, Discussion paper **296**
10. Cox M, Arnold G, Tomas SV (2010) A review of design principles for community-based natural resource management. Ecol Soc 15(4):38
11. Duit A, Galaz V, Eckerberg K, Ebbesson J (2010) Governance, complexity and resilience. Glob Environ Change 20:363–368. doi:10.1016/j.gloenvcha.2010.04.006
12. Ellis F (2000) The determinants of rural livelihood diversification in developing countries. J Agric Econ 51:289–302
13. Friess DA, Phelps J, Leong RC et al (2012) Mandai mangrove, Singapore: lessons for the conservation of Southeast Asia's mangroves. Raffles Bull Zool (25):55–65
14. Glaser M (2003) Interrelations between mangrove ecosystem, local economy and social sustainability in Caete Estuary, North Brazil.

Wetl Ecol Manag 11(4):265–272. doi:10.1023/A:1025015600125

15. Gopal B (2013) Future of wetlands in tropical and subtropical Asia, especially in the face of climate change. Aquat Sci 75(1):39–61. doi:10.1007/s00027-011-0247-y
16. Gunawardena M, Rowan JS (2005) Economic valuation of a mangrove ecosystem threatened by shrimp aquaculture in Sri Lanka. Environ Manag 36(4):535–550. doi:10.1007/s00267-003-0286-9
17. Gunderson L, Holling CS (2002) Panarchy: understanding transformations in human and natural systems. Island Press, Washington DC
18. Heckathorn D (1997) Respondent-driven sampling: a new approach to the study of hidden populations. Soc Probl 44(2):174–199. doi:10.1525/sp.1997.44.2.03x0221m
19. Kalaba FK, Quinn CH, Dougill AJ (2013) Contribution of forest provisioning ecosystem services to rural livelihoods in the Miombo woodlands of Zambia. Popul Environ 35(2):159–182. doi:10.1007/s11111-013-0189-5
20. Kaplowitz D (2001) Assessing mangrove products and services at the local level: the use of focus groups and individual interviews. Landsc Urban Plan 56(1–2):53–60. doi:10.1016/S0169-2046(01)00170-0
21. Kirui KB, Kairo JG, Bosire J (2013) Mapping of mangrove forest land cover change along the Kenya coastline using Landsat imagery. Ocean Coast Manag 83:19–24. doi:10.1016/j.ocecoaman.2011.12.004
22. Luttrell C (2006) Adapting to aquaculture in Vietnam: Securing livelihoods in a context of change in two coastal communities. In: Hoanh CT et al (eds) Environment and livelihoods in tropical coastal zones: managing agriculture-fishery-aquaculture conflicts. CAB International, Wallingford, UK, pp 17–29
23. Marschke MJ, Berkes F (2006) Exploring strategies that build livelihood resilience: a case from Cambodia. Ecol Soc 11(1):42
24. MEA (2005) Millennium ecosystem assessment: ecosystems and human well-being: general synthesis. Island Press and World Resources Institute, Washington DC
25. Norgaard RB (2010) Ecosystem services: from eye-opening

metaphor to complexity blinder. Ecol Econ 69(6):1219–1227. doi:10.1016/j.ecolecon.2009.11.009

26. Nyamwanza AM (2012) Livelihood resilience and adaptive capacity: a critical conceptual review. J Disaster Risk Stud 4(1):55. doi:10.4102/jamba.v4i1.55
27. Pellow DN (2000) Environmental Inequality Formation—Toward a theory of environmental injustice. Am Behav Sci 43:581–601. doi:10.1177/0002764200043004004
28. Pomeroy RS, Ratner BD, Hall SJ, Pimoljinda J (2006) Coping with disaster: rehabilitating coastal livelihoods and communities. Mar Policy 30(6):786–793. doi:10.1016/j.marpol.2006.02.003
29. Power AG (2010) Ecosystem services and agriculture: tradeoffs and synergies. Philos Trans R Soc B 365:2959–2971. doi:10.1098/rstb.2010.0143
30. Sallu SM, Twyman C, Stringer LC (2010) Resilient or vulnerable livelihoods? Assessing livelihood dynamics and trajectories in rural Botswana. Ecol Soc 15(4):3
31. Scoones I (1998) Sustainable rural livelihoods: a framework. IDS Working Paper 72
32. Sikor T (2001) Agrarian differentiation in post-socialist societies: evidence from three upland villages in north-western Vietnam. Dev Change 32(5):923–949. doi:10.1111/1467-7660.00232
33. Sonwa DJ, Somorin OA, Jum C, Bele MY, Nkem JN (2012) Vulnerability, forest-related sectors and climate change adaptation: the case of Cameroon. For Policy Econ 23:1–9. doi:10.1016/j.forpol.2012.06.009
34. Spalding M, Kainuma M, Collins L (2010) World atlas of mangroves. Earthscan, London
35. To PX, Dressler WH, Mahanty S, Thu TP, Zingerli C (2012) The prospects for payment for ecosystem services (PES) in Vietnam: a look at three payment schemes. Hum Ecol 40(2):237–249. doi:10.1007/s10745-012-9480-9
36. Trabucchi M, Ntshotsho P, O'Farrell P, Comin FA (2012) Ecosystem service trends in basin-scale restoration initiatives: a review. J Environ Manag 111:18–23. doi:10.1016/j.jenvman.2012.06.040

37. Tri NH, Adger NW, Kelly NP (1998) Natural resource management in mitigating climate impacts: the example of mangrove restoration in Vietnam. Glob Environ Change 8(1):49–61. doi:10.1016/S0959-3780(97)00023-X
38. Van Hue LT (2006) Gender, Doi Moi and Mangrove management in Northern Vietnam. Gend Technol Dev. doi:10.1177/097185240501000103
39. Van Hue LT, Scott S (2008) Coastal livelihood transitions: socio-economic consequences of changing mangrove forest management and land allocation in a commune of Central Vietnam. Geogr Res 46(1):62–73. doi:10.1111/j.1745-5871.2007.00492.x
40. Vilardy SP, González JA, Martín-López B, Montes C (2011) Relationships between hydrological regime and ecosystem services supply in a Caribbean coastal wetland: a social-ecological approach. Hydrol Sci J 56:1423–1435. doi:10.1080/02626667.2011.631497
41. Walker B, Holling CS, Carpenter SR, Kinzig A (2004) Resilience, adaptability and transformability in social-ecological systems. Ecol Soc 9(2):5
42. Walker BH, Anderies JM, Kinzig AP, Ryan P (2006) Exploring resilience in social-ecological systems through comparative studies and theory development: Introduction to the special issue. Ecol Soc 11(1):12

SECTION V

IMPACTS AND MANAGEMENT MEASURES

CHAPTER

15

ECOLOGICAL IMPACTS OF NON-NATIVE PACIFIC OYSTERS (CRASSOSTREA GIGAS) AND MANAGEMENT MEASURES FOR PROTECTED AREAS IN EUROPE

Roger J. H. Herbert[1], John Humphreys[1], Clare. J. Davies[1], Caroline Roberts[2], Steve Fletcher[3,4], Tasman. P. Crowe[5]

[1] Faculty of Science and Technology, Department of Life and Environmental Sciences, Bournemouth University, Talbot Campus, Fern Barrow, Poole, Dorset BH12 5BB, UK

[2] ABP Marine Environmental Research Ltd., Quayside Suite, Medina Chambers, Town Quay, Southampton, Hampshire SO14 2AQ, UK

[3] Centre for Marine and Coastal Policy Research, Plymouth University, Plymouth PL4 8AA, UK

[4] United Nations Environment Programme-World Conservation Monitoring Centre, Huntingdon Road, Cambridge CB3 0DL, UK

[5] Earth Institute and School of Biology and Environmental Science, Science West, University College Dublin, Belfield, Dublin 4, Ireland

ABSTRACT

Pacific oysters are now one of the most 'globalized' marine invertebrates. They dominate bivalve aquaculture production in many regions and wild populations are increasingly becoming established, with potential to displace native species and modify habitats and ecosystems. While some fishing communities may benefit from wild populations, there is now a tension between the continued production of Pacific oysters and risk to biodiversity, which is of particular concern within protected sites. The issue of the Pacific oyster therefore locates at the intersection between two policy areas: one concerning the conservation of protected habitats, the other relating to livelihoods and the socio-economics of coastal aquaculture and fishing communities. To help provide an informed basis for management decisions, we first summarize evidence for ecological impacts of wild Pacific oysters in representative coastal habitats. At local scales, it is clear that establishment of Pacific oysters can significantly alter diversity, community structure and ecosystem processes, with effects varying among habitats and locations and with the density of oysters. Less evidence is available to evaluate regional-scale impacts. A range of management measures have been applied to mitigate negative impacts of wild Pacific oysters and we develop recommendations which are consistent with the scientific evidence and believe compatible with multiple interests. We conclude that all stakeholders must engage in regional decision making to help minimize negative environmental impacts, and promote sustainable industry development.

Keywords: Invasive species, Non-indigenous species, Environmental risk assessment, Aquaculture, Fisheries, Marine protected areas

INTRODUCTION

The proliferation of non-native species around the globe is considered one of the most important biosecurity concerns of our modern age (IUCN 2000). Although to date, and to the best of knowledge no marine taxon has become extinct as a result of the introduction of non-native species (Rilov 2009) many native species decline when they interact directly or indirectly with non-native species—some have declined considerably and there have been local (site specific) species extinctions as a result of competition (Byers 2009). Invasive species that have the greatest impact are often 'ecosystem engineers' that affect organisms via changes to the physical and chemical environment (Jones et al. 1994, 1997; Jones and Gutierrez 2007; Crooks 2009). These species may create, destroy or modify habitats (Crooks 2009; Sousa et al. 2009; Padilla 2010; Markert et al. 2010; Van der Zee et al. 2012).

Pacific oysters (*Crassostrea gigas*) are now one of the most 'globalised' marine invertebrates and dominate bivalve production in many regions (Ruesink et al. 2005; FAO 2016a, b). The oysters have been introduced to 66 countries outside their native range, mainly for aquaculture, and there are now established self-sustaining populations in at least 17 countries (Ruesink et al. 2005; Smaal et al. 2006; Cardoso et al. 2007; Wrange et al. 2010). Although of considerable importance for coastal economies around the world, the introduction of *C. gigas* has also been very significant in maintaining the oyster fishing and cultivation culture and traditions of communities that have previously relied on native oysters, which in many regions are now declining (Goulletquer and Heral 1997; Zu Ermgassen et al. 2012; Humphreys et al. 2014).

Crassostrea gigas is native in the NW Pacific and Sea of Japan and occurs primarily in warm temperate regions between latitudes 30°N–48°N. It is an estuarine species, generally attached to firm bottom substrates, rocks, debris and shells from the lower intertidal zone to depths of 40 m (FAO 2016a). In Europe and elsewhere, there has been confusion with the introduction of the Portuguese oyster *C. angulata* (Humphreys et al. 2014). Although they are currently considered separate species they may yet be shown to be conspecifics (Gofas 2013).

Pacific oysters have a pelagic larval duration of 2–4 weeks, depending on temperature and nutrition (Rico-Villa et al. 2006; Syvret et al. 2008). The final larval stage will settle on the shore or seabed and develop a hard shell that in time will be recognisable as a juvenile oyster (Arakawa 1990; Reise 1998; Troost 2010). In this review, 'wild settlement' refers to the point when the oysters are first observed on the shore or subtidally either as juvenile or adult stages. Oysters become 'established' when reproduction is at a level sufficient to ensure continued survival. It is likely that a combination of factors enable wild establishment, including a lack of natural predators within receiving systems, beneficial traits such as rapid growth and rising air and sea temperatures as a result of global warming (see Troost 2010 for review). High temperatures appear to have caused an increased spawning frequency in parts of coastal Europe and wild settlement has occurred along much of the continental shoreline (Drinkwaard 1999; Reise et al. 2005; Troost 2010; Wrange et al. 2010; Lejart and Hily 2005, 2011; Herbert et al. 2012; Dolmer 2014). Oyster larvae are known to settle gregariously and to be attracted to conspecifics

(Arakawa 1990; Tamburri and Zimmer 2007). Having initially settled on a shell or a small stone, clumps of oysters may merge to form dense aggregations and potentially a reef. A reef is formed when oyster densities are so high that little space may exist for subsequent oyster settlements or other species on the substrate surface. However, most Pacific oyster reefs in the Dutch and German Wadden Sea do not cover 100 % of the substrate and contain bare patches where soft-sediment communities are still present and shrimps and small fish may be found in shallow pools (Troost 2010).

Reefs have generally been described in terms of the density of oysters per square metre, as opposed to the percentage coverage on the substratum surface. However where densities of live, mature oysters approach or exceed ~200 ind. m^{-2} there is generally little underlying natural substratum visible, especially as an amount of empty shell is also always present (personal observation). In some regions, significant areas of intertidal Pacific oyster 'reef' have developed with densities of 700 ind. m^{-2} (Wehrmann et al. 2006 cited in Markert et al. 2010) creating a hard substratum upon other habitats. Where the oysters have colonised mussel beds (*Mytilus edulis*) in the Wadden Sea, the density of live and dead oysters can be 2000 ind. m^{-2} (Markert et al. 2013). Subsequent settlements will often be on existing oysters, and over time a hard concretion of live and dead oysters develops (Walles et al. 2015), although the area, height and thickness of the reef often vary throughout a site. In the Oosterschelde estuary (Netherlands) wild settlement was first observed in 1975 and the first reefs were mapped in 1980 (Drinkwaard 1999; Smaal et al. 2009). However, in south-east England, although spawning had been observed intermittently over several decades during favourable summers (Spencer et al. 1994; Herbert et al. 2012), dense wild settlement and reefs first became noticed in the mid 2000s (Herbert et al. 2012).

Conservation and Management Concerns

With naturalization of non-native aquaculture species, there is a dilemma between encouraging the beneficial services provided by extensive fisheries and aquaculture and the potential damage caused by their proliferation and potentially invasive traits (Goulletquer 2009; Humphreys 2010; Herbert et al. 2012; Humphreys et al. 2014). Although wild settlement can be beneficial for coastal fishing communities, in protected areas there is now a tension between the continued production of the Pacific oyster and risk to biodiversity associated with the growth of wild populations (Goulletquer 2009; Herbert et al. 2012; Humphreys et al. 2014). Many coastal species and habitats have been designated as protected areas under national and international conservation agreements e.g. European Habitats Directive (92/43/EEC) and Ramsar Sites. Conservation agencies and regulators are concerned that habitats and species of conservation interest are at risk from competition, displacement and proximity to non-indigenous species. If following assessment, sites protected under European law are deemed to be in poor (unfavourable) condition then the European Commission can initiate formal infringement proceedings that can result in financial sanctions for Member States if non-compliance is unresolved (European Commission 2016a).

In parts of continental Europe and in temperate regions elsewhere, the proliferation of wild *C. gigas* is now regarded as an 'invasion' as *C. gigas* is spreading rapidly and is displacing native species and habitats (Diederich et al. 2005; Ruesink et al. 2005; Smaal et al. 2005; Cognie et al. 2006; Lejart and Hily 2005). Moreover the species is listed as 'one of the worst 100 alien species in Europe' (DAISIE 2016). There are also negative socio-economic impacts of wild settlement that range from potential injury to the public from 'razor sharp oyster shells', the maintenance of navigational channels for recreational craft, aesthetic issues (personal communications to authors) and trophic competition with commercial mussel farming (Wijsman et al. 2008). In the southwest Atlantic coast of France, wild populations of *C. gigas* were destroyed because they were competing with cultivated Pacific oysters and therefore limiting growth (Goulletquer 2009).

The overall economic importance of *C. gigas* in many of these regions is extremely high (FAO 2016a, b). However in the UK and parts of Ireland, where the industry is yet to realise its full potential, the future of Pacific oyster aquaculture within designated protected sites is at risk due to conservation concerns about the potential impact of wild oysters (Herbert et al. 2012; Humphreys et al. 2014). The issue of the Pacific oyster therefore locates at the intersection between two policy areas: one concerning the conservation of protected habitats, the other relating to livelihoods and the socio-economics of coastal aquaculture and fishing communities. Other affected stakeholders include port authorities and recreational bodies.

We aim to find sustainable solutions for the aquaculture industry, fisheries, conservation agencies and regulators. Our first objective is to review the evidence for negative impacts of wild Pacific oysters on selected, yet representative, broad-scale coastal habitats and species. Our second objective is to assess the risks and review potential management measures that have been trialled or suggested as a way of containing the impact of the oysters.

METHODS AND APPROACH

This review considers evidence of potential direct impacts of wild settlement and establishment of non-native *C. gigas* on intertidal and subtidal habitats often found within protected areas. The exclusion of some habitats does not necessarily imply that the species will have no impact in them, although deep sea habitats are beyond the species range. We have not specifically considered the impact of *C. gigas* introductions on the spread and ecological and economic damage caused by the 'hitch-hiking' of other non-native species, parasites and pathogens. Neither have we considered in detail the direct environmental impact of cultivation of *C. gigas*.

Although the focus is on European seas, information on ecological impacts was obtained from temperate regions around the world where *C. gigas* is cultivated and where habitats are broadly similar. The published literature was searched using the terms shown in Online Resource Appendix A. Information was obtained from areas with warm or cold temperate climate and within similar 'Biogeographical Realms' (Spalding et al. 2007). This was primarily undertaken through searching online research databases and catalogues (ISI Web of Science, JSTOR, ScienceDirect, Scopus,). To ensure that relevant 'grey literature' was incorporated, an internet search using the same search terms was conducted and professional networks and organizations likely to hold grey literature and information on unpublished and on-going studies were contacted via their websites and libraries (e.g. non-governmental conservation organizations, fisheries research institutes). Scientific experts with specific knowledge about the Pacific oyster and its ecology and habitat were contacted by phone and email and experts and stakeholders professionally engaged in aquaculture were consulted by email with a short questionnaire (UK only), by phone, face-to-face interviews or through visits to cultivation and production businesses. Many of these organizations and experts are listed within the Acknowledgements section of this review. The evidence collected in this way was used as the basis for a narrative review addressing the objectives stated above.

RESULTS

Impacts on Species Diversity and Ecosystem Functioning

Although studies of ecological impact encompass a range of broad-scale intertidal habitats, the number of replicated studies within some habitats and across different regions is relatively low (Table 1). Ecological impacts as a result of reef formation are almost entirely reported from Europe; in the Pacific North-west of the USA, reef formation appears not as extensive. Habitats that are causing most concern are intertidal rocky reefs (Figure 1), muddy intertidal habitats (Figure 2) and biogenic reefs, such as those formed by the honeycomb worm *Sabellaria alveolata*(Figure 3). Studies on ecological impacts are generally local in scale and there has been little work so far on impacts on subtidal species as the extent of recruitment and colonization of sublittoral habitats is currently unclear. A study on the micro-tidal coast of Sweden (Hollander et al. 2015) was technically sublittoral, however the species and communities investigated (*M. edulis* beds and soft-sediments) were comparable with those in the lower intertidal habitats of macro-tidal shores. Studies of impacts on seagrass have only been carried out in the Pacific North-west, whereas impacts on *M. edulis* beds have to date only been studied in Europe, which may reflect differences in the susceptibility of habitats to succumb to wild settlement in different temperate realms. It is uncertain to what extent *C. gigas* has had an influence on the decline of intertidal mussel beds. It is possible that *C. gigas* has colonized shell debris associated with former mussel beds and interfered with their re-colonization through occupation of their former habitat, although there is no experimental evidence for this. Variability in mussel survival is likely to be related to regional levels of invertebrate predation (Nehls et al. 2006; K Reise pers. comm). Reduced fitness may also have facilitated the decline of mussels, as experiments have shown that mussels migrate downwards through the oyster reef to avoid crab predation, yet at the expense of reduced food supply and growth (Eschweiler and Christensen 2011). Yet, in some parts of the central Wadden Sea mussels are now increasing in the shelter of *C. gigas*reefs (Nehls et al. 2009; Markert et al. 2010).

Table 1: Ecological impacts of Pacific oysters in temperate regions

Habitat	Region	Impact	References
Littoral rock	Southern England, Kent	Patches of *C. gigas* reef (200 ind. m^{-2}) present on lower shore of chalk reef (Figure 1)	Herbert et al. (2012)
	France, Bay of Brest (moderately exposed shore)	*C. gigas* reef at all tidal levels (Mean High Water to Mean Low Water); Biomass and species richness significantly higher on *C. gigas* reefs compared to adjacent rock; Deposit and detritus feeders occurred only on oyster reefs and not on adjacent rock	Lejart and Hily (2011)
	Ireland	Experimental addition of living or dead *C. gigas*. Effects varied with state and cover of oysters. Boulders with lowest cover of living *C. gigas* supported greatest diversity	Green and Crowe (2013)
	USA, Pacific Northwest	*C. gigas* common on sheltered rocky shores (low energy littoral rock) and rare (<10 % cover) on exposed shores	Ruesink (2007)
	Canada, Strait of Georgia	Reef formation not reported from British Columbia, though higher densities are present in areas where warmer waters cause more frequent settlement	J. Ruesink (Pers. comm)

		C. gigas settles within the barnacle zone where they may provide a greater surface area for settlement. In experimental manipulations, seastars and crabs reduced monthly survival rates of *C. gigas* by 25 % relative to caged oysters Some neighbouring species on exposed rocky sites might facilitate survival of *C. gigas* by reducing physical stresses	Ruesink (2007); Ruesink et al. (2005)
	Canada, British Columbia	*C. gigas* was able to modify the thermal regime of its habitat and provide refugia for those species that might otherwise suffer from desiccation	Padilla (2010)
	Argentina (1982)	Among eight epifaunal species, three occurred at higher densities within oyster beds and three were more abundant outside these areas	Escapa et al. (2004)
Littoral sediments	Southern England, North Sea, English Channel	Reef formation since 2007 (Figure 2)	Herbert et al. (2012)
	Wadden Sea	Reef formation present on lower shore	Reise (1998)
	Netherlands, Oosterschelde estuary	First natural recruitment in 1975. Reefs mapped in 1980	Drinkwarrd (1999); Smaal et al. (2009)
	France, Bay of Brest, Brittany	Invertebrate species richness on mud beneath *C. gigas* reef was twice that of adjacent mudflats and dominated by carnivores, compared to suspension feeders in mudflats	Lejart and Hily (2005, 2011)

	Ireland	Experimental addition of different covers of oysters in small plots in two estuaries. Diversity and abundance of species increased with cover of oysters. Effects on microbial communities and ecosystem processes varied with cover. Sediment–water fluxes and turnover of ammonium and silicate were greatest at medium cover and decreased with greatest cover	Green and Crowe (2014); Green et al. (2012, 2013)
Saltmarshes and saline reed beds	Southern England, Kent	No settlement observed, however stabilization of sediment by oyster shells may both facilitate further colonization of non-native *Spartina anglica* and potentially create a firm habitat for oyster settlement	McKnight (2011)
	Argentina	Colonization of *C. gigas* on the stems of the saltmarsh cord grass *Spartina alterniflora*	Escapa et al. (2004)
Saline lagoons	Southern England, Fleet Lagoon (1988)	Little settlement. The special flushing characteristics of the lagoon and crab predation may provide resilience to wild settlement	Eno (1994)
	France	*C. gigas* is cultivated in micro-tidal lagoons and has established wild populations in some areas	Miossec et al. (2009)
Blue mussel beds (*Mytilus edulis*)	Netherlands, Wadden Sea From Mean Tide Level (MTL) to the shallow subtidal	*Mytilus*-beds have changed to mixed reefs dominated by 95 % *C. gigas*. Mussels recruit frequently and settle amongst the oysters, migrating to lower regions in the interspaces between the oysters to evade predation	Nehls and Buttger (2007); Nehring et al. (2009); Fey et al. (2010); Eschweiler and Christensen (2011)
	Germany, Wadden Sea	Mixed mussel-oyster reef showed increased species richness, abundance, biomass, diversity and deposit feeding species, compared to mussel beds	Markert et al. (2010)

	Sweden, Skagerrak, shallow subtidal	No significant differences in macrofaunal species richness compared to oyster beds, however species abundance in oyster beds was statistically higher in two out of three sites. Differences in macrofaunal composition were inconsistent	Hollander et al. (2015)
	Ireland	Experimental addition of different covers of oysters in small plots in two estuaries. No effects on diversity or abundance of associated fauna, except a decrease on one sampling occasion at one site. Ecosystem processes including respiration, sediment–water fluxes and turnover of ammonium and silicate increased with increasing cover of oysters	Green and Crowe (2013); Green et al. (2012)
Polychaete worm reefs (*Sabellaria alveolata*)	France, Bay of Mont-Saint Michel	Oysters are colonising some *S. alveolata*reefs with densities >100 ind. m^{-2}	Dubois et al. (2006)
	France, Bay of Mont-Saint Michel	Higher species richness recorded on *Sabellaria* reefs colonised with oysters (and with oysters and algae). Colonization has led to damage of *Sabellaria* by recreational oyster harvesters. aquaculture is also thought to have contributed to habitat deterioration	Desroy et al. (2011); (Figure 3)
	France, Bourgneuf Bay, Brittany	Growth and settlement of wild *C. gigas* has transformed areas where former *S. alveolata* beds had previously been recorded, so recolonization is now unlikely	Cognie et al. (2006)

	Ireland	Oysters attached experimentally to topsides of boulders inhibited settlement of *S.alveolata* on undersides	Green and Crowe (2013)
Sabellaria spinulosa	Southern England, Kent extreme lower shore	An area of intertidal *S. spinulosa* in reef formation is being overgrown by *C. gigas*	McKnight (2011, 2012)
Lanice conchilega	Southern England, Kent Extreme lower shore	Chalk reef colonised by 50 % cover of *L.conchilega* worm reef is partially displaced by Pacific oysters (maximum density of *C. gigas* 14 ind. m^{-2})	McKnight (2011)
Seagrass beds (*Zostera*spp.)	France, Thau lagoon, Mediterranean	Increased water clarity caused by the uptake of particulate material and phytoplankton by *C. gigas* and mussel aquaculture, is thought to have enabled *Zostera* to grow in deeper areas of the lagoon	Deslous-Paoli et al. (1998)
	USA, Washington, Willapa Bay	General pattern of reduced density and shoot size of the native seagrass *Z. marina*on cultured *C. gigas* beds	Tallis et al. (2009)
	USA, Washington, Willapa Bay	Shoot density and cover of *Z. marina*declined with increasing oyster density, attributed to space competition; this competition can generate impacts above thresholds of 20 % oyster cover. At low densities, *C. gigas* has little impact, however oyster cover >50 % is impenetrable to seagrass	Wagner et al. (2012)
	USA, Washington, Willapa Bay	Immediately seaward of the *C. gigas* zone and amongst adjacent *Z. marina* beds, benthic diversity was greatest below the *C. gigas* beds, yet fish and pelagic invertebrates were more abundant within seagrass	Kelly et al. (2007)
Subtidal sediments	Southern England, Thames estuary	Seen at least 3 m below Chart Datum on subtidal sediments	Herbert et al. (2012)

	Northern Ireland, Lough Foyle	Present on subtidal sediments	Herbert et al. (2012)
	Ireland, Lough Swilly	Present on subtidal sediments	Herbert et al. (2012)
	Netherlands, Oosterschelde estuary	Settlement observed in cultivated subtidal Pacific oyster beds (depth 2–3 m) and on adult oysters at 10 m depth	Wijsman, pers. comm.
	Germany, Wadden Sea	Adults found at 10 m below low water, however no juveniles or recruitment observed. Are often large individuals or clusters. Most likely broken off intertidal reef structures	(Reise, pers. comm.)
Subtidal sediments contd.	Wadden Sea	Sublittoral stocks estimated as occupying 700 ha however it is unclear whether this is as a result of recruitment	Kater et al. (2002); Smaal et al. (2005)
	Sweden	Pacific oysters found at depths from 1–9 m	Dolmer et al. (2014)
European native flat oyster *Ostrea edulis*	Wadden Sea	Overlap with native oyster not expected as it is sublittoral	Reise (1998)
	Southern England, Poole Harbour	Found to settle on shells and living *C. gigas*	Authors observation
Fish	USA, Washington, Willapa Bay	Immediately seaward of the *C. gigas* zone and amongst adjacent *Z. marina* beds, fish were much more abundant within the seagrass	Kelly et al. (2008)
Birds	Wadden Sea	Species that have previously relied on mussels (e.g. Eider duck, *Somateria mollissima*), may not be able to feed on the oysters due to their size, shell thickness and cementation	Nehring et al. (2009); Scheiffarth et al. (2007)
	Netherlands, Oosterschelde	Herring gull (*Larus argentatus*) and Eurasian oystercatcher (*Haematopus ostralegus*) are reported to feed on *C. gigas*	Cadée (2008a, b); Troost (2010)

	Wadden Sea	Colonization of *M. edulis* beds by *C. gigas*had a positive impact on feeding rates of the Eurasian oystercatcher and the Eurasian Curlew (*Numenius arquata*)	Markert et al. (2013)
	Argentina	Number of birds (two gulls and four wading bird species) was greater amongst *C. gigas*compared to control areas. Foraging rate of two species was higher amongst oysters, whereas in other two species there was no difference with control plots	Escapa et al. (2004)

Impacts from cultivated stocks are included where considered relevant

Figure 1: *C. gigas* reef establishing on a chalk rocky shore in Kent, south east England (Photo: W. McKnight).

Figure 2: Wild *C. gigas* reef that has established on intertidal mud on the Blackwater estuary at Brightlingsea (UK) in 2008 (Photo: M Gray).

Figure 3: Wild settlement of *C. gigas* on reef of *Sabellaria alveolta.* Bay of Mont-Saint Michel (France) (Photo: N. Desroy).

Small-scale experimental manipulations (0.25 m^{-2} plots) using transplanted oysters on mudflats at two different locations in Ireland showed that invertebrate species density and diversity increased between 5 % cover and 50 % cover of *C. gigas*, but then plateaued with no further increase at 100 % cover (Green and Crowe 2014). The responses of many species differed between locations and over time, suggesting that some effects are context-dependent. A study on mussel beds have found increases in macro-invertebrate diversity with high settlement of wild *C.*

gigas (Markert et al. 2010), yet increasingly unsuitable habitat chemistry and anoxia associated with high rates of decomposition of biodeposits and microbial respiration is likely to be responsible for the suppression of species diversity at very high levels of *C. gigas* cover (Green et al. 2012; Green and Crowe 2013, 2014). Due to a sediment-free upper part of the reef and more turbulent current flow, species richness was significantly greater amongst the oyster beds compared to mussels and some faunal species were exclusively found on oyster beds, particularly anemones and suspension feeders (Markert et al. 2010). Higher species survival amongst the more complex oyster reef structure and bio-deposition of sediments was also thought to explain differences in species richness (Kochmann et al. 2008).

Several studies on soft-sediment and rocky intertidal habitats have shown that species diversity can be greater amongst aggregations of wild Pacific oysters compared to the native habitat in which the oysters settle (Table 1). Although there appear to be significant negative impacts on some native species of conservation concern (e.g. *Sabellaria* reefs) to date there is no evidence for total displacement of any species in Europe. Although there may be beneficial effects of wild settlement on some coastal bird species (Cadée 2008a, b; Scheiffarth et al. 2007; Markert et al. 2013), there remains uncertainty on negative effects on others. The nature and scale of impact and engineering of Pacific oysters is dependent on the type of habitat that is colonised (Padilla 2010) and on the stage of invasion; a low density of scattered individual oysters may have little or no impact on biodiversity at regional scales. Yet at local-scales, the oysters can facilitate grazers (Ruesink 2007) and can modify the thermal regime of the habitat (Padilla 2010). Species interactions also influence community resistance to invasion. While there is evidence that a higher native diversity of sessile invertebrates can suppress a potential invader through competition for space (Olyarnik et al. 2009), this can be alleviated by facilitation, e.g. by the provision of space for secondary settlement on oyster shells.

Crassostrea gigas reefs may compensate for the loss of ecological function of mussel beds in the Wadden Sea (Markert et al. 2010; Troost 2010). Yet comparatively little evidence is available for impacts of wild settlement of *C. gigas* on ecosystem functioning. In field experiments in Lough Swilly, Ireland, (Green et al. 2012; Green and Crowe 2013)

*C. gigas*significantly altered several biogeochemical properties and processes, including fluxes of important limiting nutrients, and microbial assemblages and activity. Sediment–water fluxes of NH4+ and Si(OH)4 and benthic turnover rates increased with increasing cover of oysters in mudflats but decreased at the greatest cover of oysters in mussel-beds (Green and Crowe 2013). Community respiration (CO_2 flux) increased with the greatest cover of oysters on mudflats and among mussels. At 100 % cover compared to 0 % cover, there was significantly greater total microbial activity, chlorophyll content and CO_2 and CH_4 emission from sediments (Green et al. 2012). At 10 % cover, *C. gigas* increased the concentration of total oxidized nitrogen and altered assemblages of ammonia oxidizers and methanogens. At any cover of *C. gigas*, concentrations of pore-water NH4+ were greater than in areas of mudflat without *C. gigas*. Thus *C. gigas* may alter ecosystem functioning not only directly, but also indirectly by affecting the microbial communities that underpin ecosystem processes.

The water-filtering capacity of native suspension feeding benthic species is known to have a significant controlling effect on phytoplankton and nutrient levels in estuarine waters (Hily 1991). It is considered that the high filtration rate of *C. gigas* may also have potential to affect trophic dynamics within ecosystems by consuming high quantities of suspended particles and plankton (Ruesink et al. 2005; Troost et al. 2009; Troost 2010) that could also affect water quality. Cultivated *C. gigas* can reduce the carrying capacity and compete trophically with commercial mussel production (Wijsman et al. 2008). Modelled simulations of nutrients, oyster growth and phytoplankton in the Baie des Veys estuary in northern France showed significant depletion of phytoplankton above areas where *C. gigas* is cultivated, with consequences for the spatial distribution of plankton across the bay (Grangeré et al. 2010) and potential impacts on native suspension feeding species. Field investigations of the impact of extensive cultivated *C. gigas* on chlorophyll concentrations in Wallapa Bay, USA, supported modelled scenarios and showed that although oyster filtration rates were lower than laboratory measurements, the oysters can exert top-down control on phytoplankton production within estuaries (Wheat and Ruesink 2013). Yet isotopic analysis on the diet of wild Pacific oysters on rocky shores along an estuarine gradient in the Bay of Brest (Marchais et al. 2013) found that benthic biofilms and resuspended macro algal detritus, rather than phytoplankton, constituted the greatest proportion of the diet. The 1 m height difference between oysters cultivated on trestles and benthic wild oysters may explain variance in the proportion of benthic v pelagic sources (Marchais et al. 2013), so caution is necessary at extrapolating impacts of aquaculture to the issue of wild settlement. A higher near-bed turbulence caused by the roughness of Pacific oyster reefs, together with the high water filtration capacity of the oysters, may increase food intake rate (Troost et al. 2009; Troost 2010).

Local coastal typology and hydrodynamics are also likely to be influential on water filtration. However, on a regional and landscape-scale it remains unclear what if any impact wild settlement of *C. gigas* will have on ecosystem function. It has been argued that, unlike other coastal developments and stressors, bivalve aquaculture on the Pacific coast of North America does not remove significant habitat area, result in a decline in water quality, nor is there any evidence for causing a shift to alternate states (Dumbauld et al. 2009). Wild settlement and reef formation of Pacific oysters may yet be considered beneficial in some contexts; providing supporting processes such as erosion control (Borsje et al. 2011) and ecosystem services in addition to food provision, such as the use of shells in construction and coastal defence (Piazza et al. 2005; Scyphers et al. 2011; Herbert et al. 2012).

We have not specifically considered in detail the impact of *C. gigas* introductions on the spread and ecological and economic damage caused by the 'hitch-hiking' of other non-native species, parasites and pathogens. Potentially, *C. gigas* reef formation might facilitate their colonization by creating a suitable habitat for their establishment. In the list of species colonising *C. gigas*reefs on rocky and muddy habitats in Brittany (Lejart and Hily 2011) there are only two non-native species that are not present in un-colonised habitat beyond the reefs; the barnacle *Austrominius modestus* (as *Elminius modestus*), that had colonised the shells of *C. gigas* on the muddy shore and the ascidian *Steyla clava* that had colonized the reefs but were not found in either natural habitat. However there are concerns that the invasive brush-clawed crab *Hemigrapsus takanoi* has that has recently been recorded in European *C. gigas* reefs (Wood et al. 2015) will prey on native *Carcinus maenas* (Dauvin et al. 2009; van den Brink et al. 2012). As *C. maenas* is an important prey species for some coastal bird species, these interactions would benefit from further investigation.

Habitat Transformation and Homogenization

There is a risk that the introduction of invasive non-native species might result in taxonomic homogenization of the biota of habitats, as similarity of species composition increases following the combined effects of invasion and extinction of native species (McKinney and Lockwood 1999). Elton (1958) considered the business of culture and transportation of oysters, including *C. gigas*, around the globe was the greatest agency for the spread of non-native species, including oyster pests. This, he asserted, would result in faunas becoming similar across regions.

It is possible that because *C. gigas* can colonize large areas of a wide variety of intertidal habitats, it will result in the biotic homogenization of intertidal habitats at local and regional scales. In the Oosterschelde estuary (Netherlands), *C. gigas* reef has colonized the lower shore and currently represents approximately 8 % of the entire intertidal habitat (Smaal et al. 2009). Of the 115 taxa that were recorded within *C. gigas* reefs that had formed on rock and mud habitats in Brittany (Lejart and Hily 2011) only 11 were common to both reefs. However the reef that formed on the muddy habitat became dominated by carnivores at the expense of suspension feeders. In terms of the proportion of different trophic groups, this reef became comparable to the *C. gigas* reef that colonized the rocky habitat and the rock itself. These findings support those of Markert et al. (2010) who also observed the increasing dominance of carnivores (e.g. crab *Carcinus maenas*) on *C. gigas* reefs that had colonized mussel beds. Therefore although local taxonomic homogenization might be currently considered low (~10 %), functional homogenization (Olden 2006) may become greater.

Large areas of mudflats in the Wadden Sea and rocky shore on the Atlantic coast of France have been transformed to non-native oyster reefs. The spatial dominance of large filter feeding organisms and higher filtration capacity of the oyster reef differs considerably to shores characterized by algae, grazers (e.g. limpets) and other filter feeders (e.g. barnacles) on a rocky reef. Dense wild settlement of *C. gigas* could therefore transform some protected habitats from a functional state in which they were originally designated, to a new functional state.

Although the functional state of some rocky shores might be significantly transformed through the colonization of *C. gigas*, the same may not necessarily be true for habitats consisting primarily of filter-feeders, such as mussel beds, as they may simply be replaced by an equivalent filter feeding species that can accommodate a similar associated fauna. On soft-sediment shores in the Wadden Sea, the distribution of *C. gigas* can overlap with native filter-feeding bivalve species such as *Macoma balthica*, *Scrobicularia plana* and *Cerastoderma edule* (Troost 2010). Yet although expansion of the oyster reef could be expected to have an impact on the diversity and abundance of native soft-sediment species on the lower shore, these species also occur above the tidal level of *C. gigas* reef development, although the extent of habitat overlap is likely to depend on locality. The spatial extent of the impact of wild settlement on the diversity and functionality of native habitats is therefore difficult to determine.

While there is evidence of local impacts of wild *C. gigas* on species diversity and ecosystem functioning across a range of intertidal habitats there is less evidence of wild settlement on the ecological integrity of protected areas. Scaling-up impacts through spatial modelling may be beneficial; however it is also necessary to undertake field studies due to the context dependency of some impacts. Clearly the impact on sublittoral biodiversity also requires further investigation. Considering that the conservation of migratory birds is of great importance in Europe there has been relatively little investigation on the impact of Pacific oysters on behavior and fitness. This is important as this could influence the type and scale of management interventions in areas affected by extensive wild settlement.

Policy Framework and Management Measures

Although species diversity of a non-native oyster reef might be greater than the native habitat and include higher densities of particular species (Lejart and Hily 2005, 2011; Markert et al. 2010), for areas protected under the EU Habitats Directive, it is the fundamental alteration in type and variety of habitats or biotopes that is important. The threshold level and area of impact whereby a site might be considered to have changed has not been quantitatively determined, and to our knowledge has not been legally tested or previously considered for a non-native species.

However, as far as the EU Habitats Directive is concerned, of critical importance to whether a site is classified as being in favorable condition is whether the 'integrity' of the whole designated site is transformed. The integrity of the site has been defined as 'the coherence of its ecological structure and function, across its whole area, that enables it to sustain the habitat, complex of habitats and/or the levels of populations of the species for which it was classified' (European Commission 2000). Broad-scale changes and the transformation of species communities or biotopes might be interpreted as compromising the integrity of the designated site. In Britain, conservation agencies have concluded that even the loss of considerably less than 1 % of designated sites could be significant and in some cases would adversely affect site integrity (Hoskin and Tyldesley 2006), though not specifically for non-native species. The risk of ecological impacts of wild settlement in a warming world has unnerved conservation agencies and the aquaculture industry. For example, in the UK and Ireland, plans to develop a new Pacific oyster farm within or in the vicinity of a protected area may need to satisfy agencies and authorities that the proposal will not have an adverse effect on habitats and species. In Europe, there are various stages in the 'Environmental (Appropriate) Assessment' process; importantly however, the *Likelihood of Significance of the Impact* and the *Impact on the Integrity of the Site* will need to be determined. Examples of 'Significant Impacts' provided by the European Commission and UK agencies are shown in Table 2.

Table 2: Examples of 'significant impacts' on European Natura 2000 sites (English Nature 1999; European Commission 2001)

Alteration of community structure
Reduction in area of habitat/biotope or species for which the site was originally notified
Causes on-going disturbance to species or habitats
Presents a barrier between isolated fragments of native habitats, or reduces the ability of the site to act as a source of new native colonizers
Causes direct or indirect change to the physical quality of the environment (including the hydrology) or habitat within the site
Causes direct or indirect damage to the size, characteristics or reproductive ability of populations on the site
Alter the vulnerability of populations/habitats to other impacts

Due to wide ranging impacts of non-native species, provisions are included in EU policies aimed at the protection of ecosystems and sustainable use of natural resources. However, species that have had a long history of aquaculture and which are of economic value are excluded from the scope of the EU Regulation on the prevention and management of the introduction and spread of invasive alien species (EU) No 1143/2014, as they can be listed within Annex IV of Council Regulation of the European Commission concerning use of alien and locally absent species in aquaculture (EC) No 708/2007 (EC-ASR). Article 4 of this Regulation confers a general obligation for Member States to implement measures to avoid adverse effects to biodiversity "*which may be expected to arise from the introduction or translocation of aquatic organisms and non-target species in aquaculture and from the spreading of these species into the wild*". Yet under EC-ASR, EU Member States (European Commission 2016b) have discretion on whether to impose limits under Annex IV, for example if Pacific oysters have not previously been used in aquaculture.

Invasive species are generally recognized as posing a risk to achieving good ecological status under the EU Water Framework Directive (WFD) and may preclude the area from attaining a high status water body designation. Similarly, invasive species are among the indicators to be used to assess environmental status under the EU Marine Strategy Framework Directive (MSFD 2008). Given these constraints, an authority may prohibit the transfer of oysters to an area in order to maintain or improve the status of a particular site. It should be noted, however, that the presence of Pacific oysters per se does not necessarily mean that action will be taken, as it is the impact of the species on the habitat and not the presence that is a concern. As the Pacific oyster is a non-native species in Europe it is subject to Article 22(b) of the Habitats Directive which requires Member States to "*ensure that the deliberate introduction into the wild of any species which is not native to their territory is regulated so as not to prejudice natural habitats within their natural range or the wild native fauna and flora and, if they consider it necessary, prohibit such introduction*". However pathways for introduction of Pacific oysters other than aquaculture have also been implicated. Wild establishment as a direct result of introductions into marinas, harbors and ports from boat traffic as fouling or entrained larvae are as yet unproven but suspected.

In the UK there are coastal regions where wild settlement is occurring that is distant from Pacific oyster production (Herbert et al. 2012; Smith et al. 2015). Moreover in Lough Foyle on the north coast of Ireland, wild oysters were shown to be genetically different from current stock obtained from local oyster farms (Kochman et al. 2012). While this may reflect changes in broodstock through time, it is also possible that there were other pathways of introduction (Kochman et al. 2012).

Pacific oysters appear vulnerable to invertebrate predators (Syvret et al. 2008; Troost 2010), parasites (Troost 2010) cold winters (Buettger et al. 2011) and smothering by estuarine sediments (N. Miezkowska, pers. comm.). Mass mortality events across Europe caused by the ostreid herpesvirus 1 (OsHV-1) (Cotter et al. 2010; Segarra et al. 2010; Morrissey et al. 2015) may result in temporal and spatial fluctuations of wild settlement in some areas; moreover there is moderate confidence that outbreaks of some variants of the virus may increase with rising temperatures (Rowley et al. 2014). Yet the evidence suggests that frequent and dense settlement over extensive areas of certain habitats, if unmanaged, could put at risk the ecological integrity of protected sites. Within the existing European policy framework, widespread eradication and the prohibition of *C. gigas* aquaculture is highly unlikely as it would considerably reduce the economy of large areas of coastal Europe (Figure 5). In some regions of France, wild spat has now become so economically important for the oyster industry (P Goulletquer, pers. comm.) that it is protected and carefully managed by fisheries administrations. Moreover the environmental sensitivity and impracticability of removing large areas of wild settlement has led some countries to adopt the species as naturalised (e.g. Netherlands, (Drinkwaard 1999). Yet under the EU Marine Strategy Framework Directive (MSFD 2008) member states are required to implement a surveillance programme and evaluate a programme of measures to reduce the impact of non-native species. Although *C. gigas* is listed as one of the worst 100 Alien species (DAISIE 2016) there appears to be no technical or political consensus on its environmental impact and management across Europe.

Risk Assessment and Management Measures

Risk assessment protocols for the introduction of non-native species recommend a detailed analysis and review of possible ecological, genetic and disease impacts of the proposed introduction and the likelihood of spread within and beyond the release site (ICES 2005; Copp et al. 2016a, b). In the context of rising biomass and recruitment of Pacific oysters, these can include specific impacts on native species, such as competition for space with other species, modifications to trophic structure of habitats and sediment, impacts on protected sites and endangered habitats and economic considerations. In the UK, *C. gigas* is classified as presenting a 'medium' risk to nature conservation as the species could become invasive and have an impact on sensitive habitats and species (Sewell et al. 2010). In Scandinavia, separate risk assessments have been conducted for different habitats under IPCC climate scenarios (Dolmer et al. 2014). For example under both short-term and long-term climate scenarios, it was considered that Pacific oysters would have a 'moderate' ecological impact on littoral biogenic reefs in low energy areas (little or no tide, little current and low wave exposure) and a high ecological impact in high energy areas (large tidal fluctuations, strong currents and high wave exposure). Degrees of uncertainty associated with these impacts under different scenarios are also given. An application of the European Non-native species in aquaculture Risk Assessment scheme (ENSARS), a modular scoring scheme developed for evaluating the risk of introduction, establishment, dispersal and impact of species under EC-ASR, rated Pacific oyster as having an overall 'medium' risk (Copp et al. 2016a, b). This was carried out by two French assessors with metropolitan France as the risk assessment area, for which the confidence level of assessment for 'introduction' and 'impact' was rated as 'high'.

In circumstances where the eradication of species that are potentially damaging to ecosystems and the economy is not possible then management measures, that are proportionate to the impact on the environment, should be proposed (European Union 2014). The development of management measures to control the spread of invasive species in open marine ecosystems present unique technical, social, economic and political challenges. Yet there has been progress in the development of decision-making frameworks and protocols that help to assess the risk of control programmes and the effectiveness of different options (Bax et al. 2001; Thresher and Kuris 2004). These can range from physical removal of the pest species to commercial utilization, the application of biocides and biological control agents, genetic approaches and efforts to rehabilitate and improve the biological resistance of the environment (Thresher and Kuris 2004). The level of ecological and socio-economic risk of wild settlement and impact may not have reached its maximum in all regions. However, there is some inevitability that, should predictions of continued warming under the IPCC scenarios be realized, the frequency and magnitude of settlement will increase, causing existing populations to rise and new populations to become established. As the level of environmental and ecological risk varies with locality then a local or regional approach to the management of wild Pacific oyster settlement is likely to be more effective than broad-scale measures that in some areas may currently be irrelevant. Figure 5 summarizes the decision making process that supports this assertion and provides examples of management options that could be selected in a specific regional context.

Here we review various approaches to reduce the risk of wild settlement of *C. gigas* in protected areas and negative impacts of its establishment.

Marine Planning and Husbandry

The rate and extent to which *C. gigas* might become established will depend on the 'invasion (or propagule) pressure' and the biological resistance of the receiving system (Williamson 1996; Rilov and Crooks 2009). Invasion pressure and larval supply will be determined by the frequency of introduction, the size and fecundity of introduced breeding stock and the physical characteristics of the water body, including hydrodynamics that will determine larval transport.

It is generally accepted that wild settlement is dependent on the attainment of critical water temperature thresholds for oyster gametogenesis, spawning and larval development (Miossec et al. 2009; Dutertre et al. 2010). In Europe, the frequency at which temperature thresholds are now reached has increased within the past two decades (Dutertre et al. 2010). The use of *'degree days'* (the annual number of days when temperatures meet thresholds for conditioning (gametogenesis), spawning[1] and recruitment[2]), for assessing 'wild settlement risk' is considered a useful initial screening tool when planning for, or re-licensing *C. gigas* aquaculture developments (Syvret et al. 2008). Yet there is uncertainty with respect to acclimatization to local temperatures, physiological adaptation and duration of the larval development phase in response to available nutrition. Notwithstanding site anomalies and the requirement for accurate temperature measurements, this risk-based management has been incorporated into a 'Pacific Oyster Protocol' that has become acceptable to some in the UK industry and regulators (Syvret et al. 2008; Woolmer 2009).

Through spatial planning, it could be possible to manage the scale of aquaculture operations and introduce husbandry restrictions within each water body to minimize larval production, survival and shore settlement. Water bodies at particular risk could be identified using models of environmental characteristics associated with *C. gigas* establishment (Kochmann et al. 2013) and hydrodynamic models (Brandt et al. 2008; North et al. 2008) that could be adapted to investigate larval transport and dispersal. Assessment of the physical characteristics, flushing characteristics, water residence time and temperature regime of associated water bodies could enable the identification of water bodies that, due to their physical characteristics, might be particularly vulnerable to wild settlement should cultivated (reproductively active) biomass exceed a particular threshold. In this way, restrictions on aquaculture could be appropriately targeted rather than widely applied.

Pacific oysters are cultivated in a variety of ways depending on exposure, substratum and degree of siltation (Miossec et al. 2009). Studies on the impact of oyster husbandry have focused on variation in growth and mortality, rather than reproductive condition and development. In a review of research on the French Atlantic coast (Goulletquer et al. 1998) and south-west England (Robbins 2005), Syvret et al. (2008) concluded that 'parc' or on-bottom culture might result in a greater level of reproductive potential compared to off-bottom cultivation in bags on trestles, though the type of cultivation that is appropriate and employed will be dependent on local factors related to the degree of shelter and hydrodynamics. Studies in North Wales have shown that growth rate can be greater at lower tidal levels (Spencer et al. 1978) and there is evidence from the Atlantic coast of France that reproductive condition (gonadosomatic index) increases with immersion time and in animals greater than 2 years old (Goulletquer et al. 1987). This would suggest that high shore cultivation would result in reduced spawning potential. However contradictory findings were obtained in North Wales by King et al. (2006) who found gonad development to be greater at high shore compared to low shore sites, though oysters showed little maturation and spawning. Experiments in British Columbia on the growth rate and mortality of oysters suspended at different depths within the water column (Cassis et al. 2011) showed that these parameters were affected by depth, temperature, freshwater input, phytoplankton abundance and assemblage composition. On the west coast of Ireland differences in the condition of intertidal and subtidal oysters varied between sites and among months (R. Mag Aoidh, unpubl.data). Experimental studies (Chavez-Villalba et al. 2003) showed that *C. gigas* had flexible reproductive patterns depending on food variability. Clearly the issue of reproductive condition and likelihood of spawning is complex and influenced by several site specific factors and temporal variability.

Aggressive and invasive outbreaks of wild settlement can only be effectively managed if good quality data is available and forthcoming. Frequent surveys are important and necessary to assess the distribution and density of wild settlement. In New South Wales (Australia), up to date distribution data has been used to inform Risk-based Pacific Oyster Regulation and movement controls between estuaries have been implemented to minimise the spread of the species in the wild (NSWDPI 2015). As a requirement for licensing in Wales, oyster growers have been asked to remove any signs of wild settlement (Herbert et al. 2012).

Triploidy

One of the only feasible modes of containment for non-native species within the aquaculture industry is reproductive sterility (Allen and Guo 1996). A method of achieving sterility is induced triploidy, a condition in which a cell or organism has three sets of chromosomes as opposed to the normal two sets of chromosomes. The triploid condition can confer a level of sterility through rendering the oysters unable to produce viable gametes and hence preventing spawning and wild settlement. The likelihood of triploid oysters producing viable offspring has been reported to be extremely low (Guo and Allen 1994) and, by all practical measures, zero (Allen and Guo 1996). Although considered by the authors as an over-estimate Suquet et al. (2016) has calculated that the reproductive potential of triploid Pacific oysters is close to 0.06 % that of diploid individuals. A further issue relates to the stability of the triploid condition. In the USA, a trial in which 'certified triploid' oysters were placed in the York River was halted when it was discovered that about 20 % of the oysters had a 'dual cell state', containing both diploid and triploid cells (referred to as 'mosaics') (Gottlieb and Schweighofer 1996; Allen and Guo 1996; Allen et al. 1999). Investigating the chromosomal stability of triploid populations in the USA, Allen et al. (1999) reported that over a period of 2 years, there was a progressive reversion with more diploid cells accumulating over time. The frequency of reversion in chemically induced triploids had been two to three times higher than in mated triploids and the frequency of reversion also varied between grow-out sites, with harsher environments potentially exacerbating the problem of reversion (Allen et al. 1999). Jouaux et al. (2010) showed that in about 25 % of mated triploids, the process of gamete production closely resembled that of diploid oysters. Triploid oysters cannot be considered to be 'non-reproductive' (Normand et al. 2009) and there is evidence that gonad development and spawning in triploid *C. gigas* may be enhanced in unusually hot summers (Normand et al. 2008) which are predicted in current climate change scenarios (IPCC 2014). There is no doubt that there is still a measure of uncertainty concerning the circumstances and risk of reversion, however given the lower reproductive potential of triploid Pacific oysters, they should be considered as a potential measure for biological containment. The relative reproductive potential of triploids is increased when they are crossed with diploids (Gong et al. 2004), so their introduction into regions where there is wild diploid stock is unlikely to be effective at containing outbreaks. However in regions where diploid stocks are zero or very low, there may be merit in using triploid oysters as a practical measure to reduce the probability of wild settlement. It has been shown that there is no significant difference in growth when the growing conditions of the area are poor (Nell 2002). Yet in the UK, some growers are concerned about high growth rates of triploids and the cost of seed (Herbert et al. 2012), so broad acceptance within the industry might be difficult to achieve (Herbert et al. 2012). Nevertheless, triploid *C. gigas* are used widely in Australia (O'Connor and Dove 2009), France (Normand et al. 2008), the USA and Ireland and this approach may become important for the containment of oysters in culture (Guo et al. 2009).

Mechanical Control

Total eradication is not feasible or practical as densities are now so high in some regions that sufficient brood stock is likely to remain and settlement will continue. After extensive establishment of Pacific oysters in the Wadden Sea, there was unanimous agreement that any large scale eradication or control methods would harm other components of the native ecosystem (Reise et al. 2005). In the Oosterschelde estuary, Netherlands, where wild *C. gigas*can reduce the carrying capacity and compete trophically with commercial mussel production, an experimental dredging of 50 ha of intertidal wild oysters and sub-littoral cultivated beds was carried out using mussel dredges (Wijsman et al. 2008). The oysters could be effectively removed and the operation involved 940 boat hours (20 boat hours per ha). At one of the dredging sites there was little impact on subsequent oyster settlement and, at a second site, some hard shell debris remained on the mud surface. Oysters subsequently settled on the cleared areas, though there was no monitoring plan and it was concluded that beds should be cleared every 5–7 years to reduce potential competition between oysters and other harvestable shellfish species (Wijsman et al. 2008). Yet most reefs have remained undisturbed (i.e. unfished) since their first mapped appearance in the 1980s (Smaal et al. 2009; Walles et al. 2015). The general consensus of experts in the Wadden Sea is that large-scale dredging would cause considerable habitat damage (Reise et al. 2005; K. Reise pers. comm.). Since the 1990s, spatfalls of *C. gigas* have increased on the west coast of France and industrial equipment has been used to clear wild beds of *C. gigas* and colonised infrastructure that were competing with cultivated beds (Goulletquer 2009). In shallow (<2 m) areas of Lake Grevelingen (a saltwater lake in Holland) where the sharp shells of wild Pacific oysters have injured swimmers, the oysters are removed by grabs, and the remaining shells are covered by sand (Wijsman pers. comm.).

The effectiveness of removal of wild *C. gigas* (mean density <1 m^{-2}) has been investigated in Strangford Lough, Northern Ireland (Guy and Roberts 2010). Shells encountered on transects were broken with a hammer during the spring. In the year following the cull, although densities at un-culled sites continued to rise, oyster density at culled sites had dropped by nearly 100 %. It is assumed that oysters that were hammered were killed and that there had been no further settlement at these sites. It was concluded that the measure could be beneficial at reducing population expansion in the early stages of invasion. The probability of settlement may be reduced further if shell debris is also removed as this can be attractive to settling larvae (Arakawa 1990; Gutiérrez et al. 2003).

Rocky habitats may present the greatest challenge in terms of management and containment due to difficulties associated with the physical removal of oysters. A pilot trial to hold the advancing line of Pacific oysters has been conducted in Kent, south east England (McKnight and Chudleigh 2015). The objective was to reduce the wild spawning stock that had colonized chalk intertidal platforms and artificial structures and prevent settlement on nearby protected chalk reefs and intertidal mudflats. Using a variety of tools including edging spades, rods, hammer, pliers and safety equipment, over 40,000 oysters were removed during 43 site visits (96 man hours) along three sections of the coast where they had colonized chalk reefs and mussel beds. However it remains to be seen if the rate of wild settlement in this region is reduced. It is anticipated that further work will be undertaken by a volunteer group in the locality (Figure 4).

Figure 4: Hand removal of *C. gigas* by volunteer workers on a protected chalk shore in Kent, south east England (Photo: W. McKnight).

Opportunities through Hand-Collecting and Fishing

Between 1976 and 1981, handpicking was used to reduce the wild stock of Pacific oysters in the Oosterschelde. These attempts failed and the new inhabitant was accepted as belonging to the Dutch fauna (Drinkwaard 1999). Yet recently, local stakeholders have financed a pilot scheme to remove Pacific oysters from selected beaches where their sharp shells cause injury to tourists (P. van Avesaath pers comm). An efficiency study has shown potential for economic exploitation of isolated oysters and profits appear high enough to support future interventions which are being planned in the southern Dutch Delta (P. van Avesaath pers comm). Guided walks to the oyster reefs are also being organized for visitors to collect wild oysters for food (BBC 2015). Depending on a favorable market, it is possible that some effective control on wild Pacific oyster settlement could be possible through regulated fishing and hand-collecting. Both hand-collecting and dredging over soft sediment habitats creates patches of open mud within oyster reefs for bird feeding. Moreover, the density and spatial arrangement of Pacific oysters has been shown to affect the impact of wild settlement upon the native Australasian oyster *Saccostrea glomerata* (Wilkie et al. 2013). Although fishing activity creates different impacts and disturbances on intertidal benthic species and habitats (Hall and Harding 1997; Spencer et al. 1998; Kaiser et al. 2001; Piersma et al. 2001) and hand-collecting may disturb bird feeding at low tide (Goss-Custard et al. 2006), it is possible that the extent and intensity of these activities could be managed by acquiring licences. In the Blackwater estuary on the south-east coast of England, large areas of *C. gigas* 'reef' have been hand-picked 'clean' of wild oysters to create areas for re-laying oyster seed (Herbert et al. 2012). This seed does not grow to maturity to form reef but is either hand-collected or dredged and re-laid in creeks for on-growing. A variety of wading birds can feed in areas where oysters have been removed and amongst newly laid seed.

Notwithstanding marketing and biosecurity challenges (Humphreys et al. 2014; Schrobback et al. 2014; Ronholm et al. 2016) there is an increasing demand for aquaculture and oyster products (FAO 2016b). As in the Oosterschelde it may be possible to provide financial incentives to support and develop a sustainable industry, particularly where dense reefs have not yet formed, as individual oysters have far greater value than those with distorted shells. For example in ports, harbours and marinas, where fouled vessel traffic is suspected of contributing to wild settlement (Herbert et al. 2012; Smith et al. 2015), harvesting oysters may be a viable way of controlling the stock. Business start-up schemes and fisheries and aquaculture support schemes could be appropriate avenues for support (Figure 5).

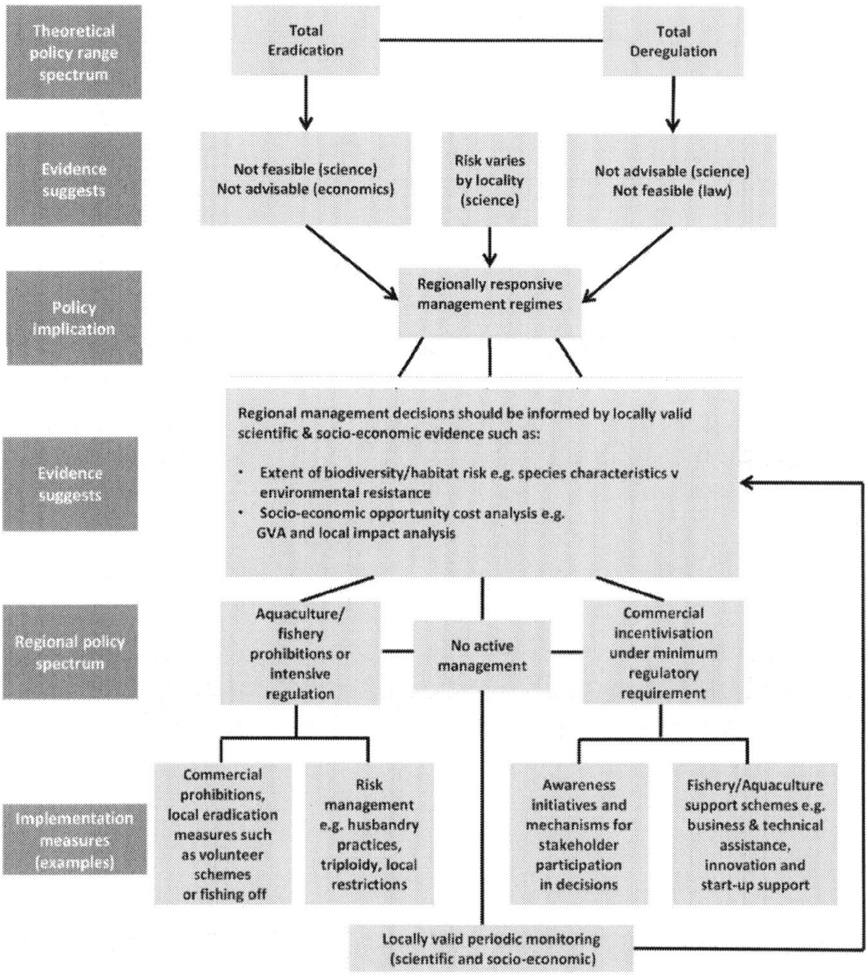

Figure 5: Potential regional approach to management of wild Pacific oyster settlement [Gross Value Added (GVA) is a measure of direct economic contribution and does not represent the full economic impact of an activity].

CONCLUSIONS

Global climate change presents new challenges and risks with respect to the management and conservation of the marine environment (Hawkins

2012). The biosecurity of marine resources, including all cultivated species, must be given a high priority in view of predicted rises in air and sea temperatures and the increased risk of economic and environmental damage caused by invasions of non-native species. Few could have predicted the enhanced fecundity and growth of wild populations of Pacific oysters in Europe as result of higher temperatures, and the potential and actual environmental impacts. We conclude that in view of the potential risks to biodiversity, all stakeholders, including growers, port and harbor authorities and statutory environmental agencies must engage in regional decision making (Bax et al. 2001) to help minimize any negative environmental impacts of wild settlement on features of conservation interest, while at the same time, and within those constraints, maximizing opportunities for sustainable industry development. Without stakeholder co-operation and managed interventions, the ecological impacts of wild settlement on species and habitats are likely to be exacerbated (Herbert et al. 2012). To maintain habitats in good condition and protect features of conservation interest it is important to develop strong partnerships between agencies and fisheries; in the Blackwater estuary, fishing interests, nature conservation organizations and the harbor authority are attempting to ensure aggressive outbreaks of wild settlement on mudflats are controlled through localized dredging (Herbert et al. 2012). However the success and continuity of these partnerships requires a vibrant industry, therefore incentives and assistance with marketing of produce might be required to achieve both commercial and conservation objectives.

It is possible that natural disturbances combined with managed interventions, including some fisheries, could maintain site integrity and functionality in some designated areas. With much uncertainty concerning the impacts on biodiversity features resulting from new aquaculture developments, an 'adaptive management' approach has been applied using trials and essentially 'learning by doing' (Woolmer 2009; Online Resource Appendix B). In terms of specific measures, consideration should be given to establishing regional management plans governing the size of aquaculture operations and number of regional licences. This needs to take account of physical and hydrographic characteristics of water bodies present in the region (Kochmann et al. 2013). In certain circumstances it might be appropriate that a strategy for risk mitigation, such as contributions to the removal of any wild settlement that occurs, could be negotiated as part of the biosecurity aspect of the licensing process. Although there are uncertainties concerning the stability of their sterile condition and effectiveness, in areas where wild settlement is currently absent or where stocks are very low, the use of triploid Pacific oysters within aquaculture should be considered. The spatial extent of any removal of wild settlement would need to be agreed between growers and agencies but a focus on particularly sensitive habitats, such as *Sabellaria* reefs, might be prioritized. In addition, efforts to increase populations of vulnerable or scarce species, such as the restoration of native oyster (*Ostrea edulis*) in the Blackwater estuary (Herbert et al. 2012), could also be encouraged. The economic feasibility of different management options needs to the assessed and capacity building is required in many of these areas to deliver these approaches.

ACKNOWLEDGMENTS

Much of the research was conducted for the Shellfish Association of Great Britain (SAGB) with funding from the European Commission and British Government (Defra). We are particularly grateful to SAGB, Defra, The Marine Management Organisation and Seafish among others for their contributions on the project steering committee. We are also grateful to may acknowledged experts on Pacific oysters including Karsten Reise, Alfred Wagner Institute for Polar and Marine Research, Germany; Jennifer Ruseink, University of Washington, Seattle, USA; John King, University of Wales (Bangor); Nicolas Desroy IFREMER, France; Ximing Guo, Rutgers University, USA; Ian Laing, Mike Gubbins and Rachel Hartnell, Centre for Environment, Fisheries and Aquaculture Science, UK; Martin Syvret, Aquafish Solutions Ltd, UK; John Bayes, Seasalter Shellfish (Whitstable) Ltd, UK; Gary Wordsworth, Othniel Oysters, UK; Richard Haward and Mersea oystermen. We are also grateful to the reviewers of this manuscript for their helpful and constructive comments. Errors remain the exclusive responsibility of the authors.

REFERENCES

1. Allen SK Jr, Guo X (1996) Triploids for biological containment. The risk of heteroploid mosaics. In: Levin MA, Angle JS, McIntyre T, Yu H (eds) Proceedings of the third international conference on

risk assessment methodologies. http://www.nbiap.vt.edu/brarg/brasym96/allen96.htm

2. Allen SK, Howe A, Gallivan T, Guo X, DeBrosse G (1999) Genotype and environmental variations in reversion of triploid *Crassostrea gigas* to the heteroploid mosaic state. J Shellfish Res 18:293
3. Arakawa KY (1990) Natural spat collecting in the Pacific oyster *Crassostrea gigas* (Thunberg). Mar Behav Physiol 17:95–128
4. Bax N, Carlton JT, Mathews-Amos A, Haedrich RL, Howarth FG, Purcell JE, Rieser A, Gray A (2001) The control of biological invasions in the world's oceans. Conserv Biol 15:1234–1246
5. BBC (2015) Reconsider the oyster! Food programme BBC radio 4. http://www.bbc.co.uk/programmes/b054pbb5
6. Borsje BW, van Wesenbeeck BK, Dekker F, Paalvas P, Bouma TJ, van Katwijk MM, de Vries MB (2011) How ecological engineering can serve in coastal protection. Ecol Eng 37:113–122
7. Brandt G, Wehrmann A, Wirtz KW (2008) Rapid invasion of *Crassostrea gigas* into the German Wadden Sea dominated by larval supply. J Sea Res 59:279–296
8. Buettger H, Nehls G, Witte S (2011) High mortality of Pacific oysters in a cold winter in the North-Frisian Wadden Sea. Helgol Mar Res 65:525–532
9. Byers J (2009) Competition in marine invasions. In: Rilov G, Crooks J (eds) Biological invasions in marine ecosystems. Springer, Berlin
10. Cadée GC (2008a) Oystercatchers *Haematopus ostralegus* catching Pacific oysters *Crassostrea gigas*. Basteria 72:25–31
11. Cadée GC (2008b) Herring gulls feeding again on Pacific oysters *Crassostrea gigas* in the Dutch Wadden Sea near Texel. Basteria 72:33–36
12. Cardoso JFM, Langlet D, Loff JF, Martins AR, Witte JIJ, Santos PT, Van de Veer Henk W (2007) Spatial variability in growth and reproduction of the Pacific oyster *Crassostrea gigas*(Thunberg 1793) along the west European coast. J Sea Res 57:303–315
13. Cassis D, Pearce CM, Maldonado MD (2011) Effects of the environment and culture depth on growth and mortality in juvenile

Pacific oysters in the Strait of Georgia, British Columbia. Aquac Environ Interact 1:259–274

14. Chavez-Villalba J, Cochard JC, Le Pennec M, Nbarret J, Enriquez-Ciaz M, Caceres-Martinez C (2003) Effect of temperature and feeding regimes on gametogenesis and larval production in the oyster *Crassostrea gigas*. J Shellfish Res 22:721–731
15. Cognie B, Haure J, Barille L (2006) Spatial distribution in a temperate coastal ecosystem of the wild stock of the farmed oyster *Crassostrea gigas* (Thunberg). Aquaculture 259:249–259
16. Copp GH, Goddard MJ, Russell IC, Peeler EJ, Gherardi F, Tricarico E, Miossec L, Goulletquer P, Almeida D, Britton JR, Vilizzi Mumford J, Williams C, Reading A, Rees EMA, Merino-Aguirre R (2016a) A preliminary evaluation of the European Non-Native Species In Aquaculture Risk Assessment Scheme applied to species listed on Annex IV of the EU Alien Species Regulation. Fish Manag Ecol 23:12–20
17. Copp GH, Russell IC, Peeler EJ, Gherardi F, Tricarico E, MacLeod A, Cowx IC, Nunn Ad, Occhipinti Ambrogi A, Savini D, Mumford J, Britton JR (2016b) The European Non-native Species in Aquaculture Risk Analysis Scheme (ENSARS)—a summary of assessment protocols and decision making tools for use in alien species in aquaculture and stock enhancement. Fish Manag Ecol 23:1–11
18. Cotter E, Malham SK, O'Keeffe S, Lynch SA, Latchford JW, King JW, Beaumont AR, Culloty SC (2010) Summer mortality of the Pacific oyster, *Crassostrea gigas*, in the Irish Sea: the influence of growth, biochemistry and gametogenesis. Aquaculture 303:8–21
19. Crooks JA (2009) The role of exotic marine ecosystem engineers. In: Rilov G, Crooks JA (eds) Biological invasions of marine ecosystems. Springer, Berlin
20. DAISIE (2016) Delivering alien species inventories for Europe. *Crassotrea gigas* species factsheet. http://www.europe-aliens.org/speciesFactsheet.do?speciesId=50156#. Accessed Jan 2016
21. Dauvin JC, Tous Rius A, Ruellet T (2009) Recent expansion of two invasive crabs species *Hemigrapsus sanguineus* (de Haan 1835) and *H. takanoi* Asakura and Watanabe, 2005 along the Opal Coast,

France. Aquat Invasions 4:451–465. doi:10.3391/ai.2009.4.3.3

22. Deslous-Paoli JM, Souchu P, Mazouni N, Juge C, Dagault F (1998) Relationship between environment and resources: impact of shellfish farming on a Mediterranean lagoon (Thau, France). Oceanol Acta 21:831–843
23. Desroy N, Dubois SB, Fournie J, Ricquiers L, Le Mao P, Guerin L, Gerla D, Rougerie M, Legendre A (2011) The conservation status of *Sabellaria alveolata* (L.) (Polychaeta: Sabellaridae) reefs in the Bay of Mont Saint Michel. Aquat Conserv 21:462–471
24. Diederich S, Nehls G, van Beusekom JEE, Reise K (2005) Introduced Pacific oysters (*Crassostrea gigas*) in the northern Wadden Sea: invasion accelerated by warm summers? Helgol Mar Res 59:97–106
25. Dolmer P, Holm MW, Strand A, Lindegarth S, Bodvin T, Norling P, Mortensen S (eds) (2014) The invasive Pacific oyster in Scandinavian coastal waters: a risk assessment on the impact in different habitats and climate conditions. Fisken og havet Report No. 2/2014. Norwegian Institute of Marine Research, Bergen
26. Drinkwaard AC (1999) Introductions and developments of oysters in the North Sea area: a review. Helgol Meeresun 52:301–308
27. Dubois SB, Committo JA, Oliver F, Reti're C (2006) Effects of epibionts on *Sabellaria alveolata*(L.) biogenic reefs and their associated fauna in the Bay of Mont Saint-Michel. Estuar Coast Shelf Sci 68:635–646
28. Dumbauld BR, Ruesink JL, Rumrill SS (2009) The ecological role of bivalve shellfish aquaculture in the estuarine environment: a review with application to oyster and clam culture in West coast (USA) estuaries. Aquaculture 290:196–223
29. Dutertre M, Beninger PG, Barille L, Papin M, Haure J (2010) Rising water temperatures, reproduction and recruitment of an invasive oyster, *Crassostrea gigas*, on the French Atlantic coast. Mar Environ Res 69:1–9
30. Elton CS (1958) The ecology of invasions of animals and plants. Methuen, London, UK
31. English Nature (1999) Habitat regulations guidance note 3. The determination of likely significant effect under the conservation (Natural Habitats &c) Regulations 1994. Peterborough

32. Eno C (1994) Monitoring the temperature regimes of which the Pacific oyster *Crassostrea gigas* is subject to in coastal inlets in Britain (The Fleet lagoon and Teign and Dart estuaries) in relation to their reproductive success. Report to Joint Nature Conservation Committee

33. Escapa M, Isacch JP, Daleo P, Alberti J, Iribarne O, Borges M, Dos Santos EP, Gagliardini DA, Lasta M (2004) The distribution and ecological effects of the introduced Pacific oyster *Crassostrea gigas* (Thunberg 1793) in northern Patagonia. J Shellfish Res 23:765–772

34. Eschweiler N, Christensen HT (2011) Trade-off between increased survival and reduced growth for Blue mussels living on Pacific oyster reefs. J Exp Mar Biol and Ecol 403:90–95

35. European Commission (2000) MANAGING NATURA 2000 SITES: The provisions of Article 6 of the 'Habitats' Directive 92/43/EEC. Brussels

36. European Commission (2001) *Assessment of plans and projects significantly affecting Natura 2000 sites* Methodological guidance on the provisions of Article 6(3) and (4) of the Habitats Directive 92/43/EEC. http://www.ec.europa.eu/environment/nature/natura2000/management/guidance_en.htm. Accessed Jul 2016

37. European Commission (2016a) European commission at work: infringement procedure. http://www.ec.europa.eu/atwork/applying-eu-law/infringements-proceedings/index_en.htm. Accessed Jan 2016

38. European Commission (2016b) Member countries of the European Union. http://www.europa.eu/about-eu/countries/index_en.htm. Accessed Jul 2016

39. European Union (2014) REGULATION (EU) No 1143/2014 OF THE EUROPEAN PARLIAMENT AND OF THE COUNCIL on the prevention and management of the introduction and spread of invasive alien species. Brussels

40. FAO (2016a) Food and Agriculture Organisation of the United Nations. Species Fact Sheets: *Crassostrea gigas* (Thunberg, 1793). http://www.fao.org/fishery/species/3514/en. Accessed Feb 2016

41. FAO (2016b) FAO yearbook. Fishery and aquaculture statistics.

2014/FAO. Rome, Italy. http://www.fao.org/3/a-i5555e.pdf. Accessed Jul 2016

42. Fey F, Dankers N, Steenbergen J, Goudswaard K (2010) Development and distribution of the non-indigenous Pacific oyster (*Crassostrea gigas*) in the Dutch Wadden Sea. Aquacul Int 18:45–59
43. Gofas S (2013). *Crassostrea angulata* (Lamarck 1819). In: MolluscaBase (2016). http://www.molluscabase.org/aphia.php?p=taxdetails&id=146900. Accessed 22 Jul 2016
44. Gong N, Yang H, Zhang G, Landau BJ, Guo X (2004) Chromosome inheritance in triploid Pacific oyster *Crassostrea gigas* Thunberg. Heredity 93:408–415
45. Goss-Custard JD, Triplet P, Suer F, West A (2006) Critical thresholds of disturbance by people and raptors. Biol Conserv 127:88–97
46. Gottlieb SJ, Schweighofer ME (1996) Oysters and the Chesapeake Bay ecosystem: a case for exotic species introduction to improve environmental quality? Estuaries 19:639–650
47. Goulletquer P (2009) History and economic consequences of species invasions on Atlantic coast: "good" & "bad" examples. Total foundation—port cros symposium 2009. http://www.archimer.ifremer.fr/doc/00000/6828/
48. Goulletquer P, Heral M (1997) Marine molluscan production trends in France: from fisheries to aquaculture. In: MacKenzie CL, Burrel VG, Rosenfield A, Hobart WL (eds). The history, present condition and future of the molluscan fisheries of north and central America, and Europe. NOAA Technical Report 129, vol 3, pp. 137–164
49. Goulletquer P, Lombas I, Prou J (1987) Growth and reproduction of the manila clam *Ruditapes philippinarum* and the Pacific oyster *Crassostrea gigas* as influenced by the immersion time. IFREMER Report
50. Goulletquer P, Soletchnik P, Le Moine O, Razet D, Geairon P, Faury N (1998) Summer mortality of the Pacific cupped oyster *Crassostrea gigas* in the Bay of Marennes-Oleron (France). In: Proceedings of Council Meeting of the International Council for the Exploration of the Sea, Cascais (Portugal), 16–19 Sep 1998. Mariculture Committee CM 1998/CC: 14 Theme Session (CC): Population Biology
51. Grangeré K, Lefebvre S, Bacher C, Cugier P, Ménesguen A (2010)

Modelling the spatial heterogeneity of ecological processes in an intertidal estuarine bay: dynamic interactions between bivalves and phytoplankton. Mar Ecol Prog Ser 415:141–158

52. Green DS, Crowe TP (2013) Physical and biological effects of introduced oysters on biodiversity in an intertidal boulder field. Mar Ecol Prog Ser 482:119–132
53. Green DS, Crowe TP (2014) Context-and density-dependent effects of introduced oysters on biodiversity. Biol Invasions 15:1145–1163
54. Green DS, Boots B, Crowe TP (2012) Effects of non-indigenous oysters on microbial diversity and ecosystem functioning. PLoS One 7(10):e48410. doi:10.1371/journal.pone.004810
55. Guo X, Allen SK (1994) Reproductive potential and genetics of triploid Pacific oysters, *Crassostrea gigas* (Thunberg). Biol Bull 187:309–318
56. Guo X, Wang Y, Xu Z, Yang H (2009) Chromosome set manipulation in shellfish. In: Burnell G, Allan G (eds) New technologies in aquaculture: improving production efficiency, quality and environmental management. Woodhead Publishing, Cambridge, pp 165–194
57. Gutiérrez JL, Jones CG, Strayer DL, Iribarne OO (2003) Molluscs as ecosystem engineers: the role of shell production in aquatic habitats. Oikos 101:79–90
58. Guy C, Roberts D (2010) Can the spread of non-native oysters (*Crassostrea gigas*) at the early stages of population expansion be managed? Mar Pollut Bull 60:1059–1064
59. Hall SJ, Harding MJC (1997) Physical disturbance and marine benthic communities: the effects of mechanical harvesting of cockles on non-target benthic infauna. J Appl Ecol 34:497–517
60. Hawkins SJ (2012) Marine Conservation in a rapidly changing world. Aquat Conserv Mar Freshw Ecosyst 22:281–287
61. Herbert RJH, Roberts C, Humphreys J, Fletcher S (2012) The Pacific oyster (*Crassostrea gigas*) in the UK: economic, legal and environmental issues associated with its cultivation, wild establishment and exploitation. Report for the Shellfish Association of Great Britain

62. Hily C (1991) Is the activity of benthic suspension feeders a factor controlling water quality in the Bay of Brest? Mar Ecol Prog Ser 69:179–188
63. Hollander J, Blomfeldt J, Carlsson P, Strand Å (2015) Effects of the alien Pacific oyster (*Crassostrea gigas*) on subtidal macrozoobenthos communities. Mar Biol 162:547–555
64. Hoskin R, Tyldesley D (2006) How the scale of effects on internationally designated nature conservation sites in Britain has been considered in decision making: a review of authoritative decisions. Engl Nat Res Rep, No 704
65. Humphreys J (2010) The introduction of the Manila clam to British coastal waters. Biologist 57:134–138
66. Humphreys J, Herbert RJH, Roberts C, Fletcher S (2014) A reappraisal of the history and economics of the Pacific oyster in Britain. Aquaculture 428–429:117–124
67. ICES (2005) International council for the exploration of the seas (ICES) code of practice on the introductions and transfers of marine organisms. International Council for the Exploration of the Seas, Copenhagen
68. IPCC (2014) Climate change 2014: impacts, adaptation, and vulnerability. Part A: global and sectoral aspects. In: Field CB, Barros VR, Dokken DJ, Mach KJ, Mastrandrea MD, Bilir TE, Chatterjee M, Ebi KL, Estrada YO, Genova RC, Girma B, Kissel ES, Levy AN, MacCracken S, Mastrandrea PR, White LL (eds) Contribution of Working Group II to the fifth assessment report of the intergovernmental panel on climate change. Cambridge University Press, Cambridge
69. IUCN (2000) IUCN Guidelines for the prevention of biodiversity loss caused by alien invasive species. Approved by the 51st meeting of the IUCN council, Gland, Switzerland, February 2000. http://www.iucn.org/themes/ssc/publications/policy/invasivesEng.htm
70. Jones CG, Gutierrez JL (2007) On the purpose, meaning and usage of the physical ecosystem engineering concept. In: Cuddington K, Byers JE, Wilson WG, Hastings A (eds) Ecosystem engineers. Plants to protists. Elsevier, San Diego, pp 3–24

71. Jones CG, Lawton JH, Shachak M (1994) Organisms as ecosystem engineers. Oikos 689:373–386

72. Jones CG, Lawton JH, Shachak M (1997) Positive and negative effects of organisms as physical ecosystem engineers. Ecology 78:1946–1957

73. Jouaux A, Heude-Berthelin C, Sourdaine P, Mathieu M, Kellner K (2010) Gametogenic stages in triploid oysters Crassostrea gigas: irregular locking of gonial proliferation and subsequent reproductive effort. J Exp Mar Biol Ecol 395:162–170

74. Kaiser MJ, Broad G, Hall SJ (2001) Disturbance of intertidal soft-sediment benthic communities by cockle hand raking. J Sea Res 45:119–130

75. Kater B, Baars D, Perdon J, van Riet M (2002) Het inventariseren van sublitorale oesterbestanden in de Oosterschelde mbv side scan sonar. RIVO rapp C058/02 pp26

76. Kelly JR, Volpe JP (2007) Native eelgrass (*Zostera marina* L.) survival and growth adjacent to non-native oysters (*Crassostrea gigas* Thunberg) in the Starit of Georgia, British Columbia. Bot Mar 50:143–150

77. Kelly JR, Proctor E, Volpe JP (2008) Intertidal community structure differs significantly between substrates dominated by native eelgrass (*Zostera marina* L.) and adjacent to the introduced oyster *Crassostrea gigas*. Hydrobiologia 596:57–66

78. King JW, Malham SK, Skov MW, Cotter E, Latchford JW, Culloty SC, Beaumont AR (2006) Growth of *Crassostrea gigas* spat and juveniles under differing environmental conditions at two sites in Wales. Aquat Living Resour 19:289–297

79. Kochman J, Carlsson J, Crowe TP, Mariani S (2012) Genetic evidence for the uncoupling of local aquaculture activities and a population of an invasive species—a case study of Pacific oysters (*Crassostrea gigas*). J Hered 103:661–671

80. Kochmann J, Buschbaum C, Volenborn N, Reise K (2008) Shift from native mussels to alien oysters: differential effects of ecosystem engineers. J Exp Mar Biol Ecol 364:1

81. Kochmann J, O'Beirn F, Yearsley J, Crowe TP (2013) Environmental factors associated with invasion: modelling occurrence data from a

coordinated sampling programme for Pacific oysters. Biol Invasions 15:2265–2279

82. Lejart M, Hily C (2005) Proliferation of *Crassostrea gigas* (Thunberg) in the Bay of Brest: first estimations of the stock and its impact on the global functioning of the ecosystem. 8th international conference on shellfish restoration. Brest
83. Lejart M, Hily C (2011) Differential response of benthic macrofauna to the formation of novel oyster reefs (*Crassostrea gigas*, Thunberg) on soft and rocky substrate in the intertidal Bay of Brest, France. J Sea Res 65:84–93
84. Marchais V, Schaal G, Grall J, Lorrain A, Nerot C, Richard P, Chauvaud L (2013) Spatial variability of stable isotope ratios in oysters (*Crassostrea gigas*) and primary producers along an estuarine gradient (Bay of Brest, France). Estuar Coast 36:808–819
85. Markert A, Wehrmann A, Kronke I (2010) Recently established *Crassostrea* reefs versus native Mytilus beds: differences in ecosystem engineering affects the macrofaunal communities (Wadden Sea of Lower Saxony, southern German Bight). Biol Invasions 12:15–32
86. Markert AW, Esser W, Frank D, Wehrmann A, Exo K-M (2013) Habitat change by the formation of alien Crassostrea-reefs in the Wadden Sea and its role as feeding sites for waterbirds. Estuar Coast Shelf S 131:41–51
87. McKinney ML, Lockwood JL (1999) Biotic homogenization: a few winners replacing many losers in the next mass extinction. Trends Ecol Evol 14:450–453
88. McKnight W (2011) Pacific Oyster distribution within the North East Kent European Marine Sites (intertidal) Phase 4 Report for Natural England. Peterborough
89. McKnight W (2012) Pacific oyster distribution within the North East Kent European Marine Sites (intertidal). Phase 5 Report for Natural England. Peterborough
90. McKnight W, Chudleigh IJ (2015) Pacific oyster *Crassostrea gigas* control within the inter-tidal zone of the North East Kent Marine Protected Areas, UK. Conserv Evid 12:28–32
91. Miossec L, Le Deuff RM, Goulletquer P (2009) Alien species alert:

Crassostrea gigas (Pacific oyster). Copenhagen, International Council for the Exploration of the Sea. ICES Co-operative Research Report No. 299

92. Morrissey T, McClearly S, Collins E, Henshilwood K, Cheslett D (2015) An investigation of ostreid herpes virus microvariants found in *Crassostrea gigas* oyster producing bays in Ireland. Aquaculture 442:86–92
93. MSFD (2008) EU Marine Strategy Framework Directive. Directive 2008/56/EC on establishing a framework for community action in the field of marine environmental policy
94. Nehls G, Buttger H (2007) Spread of the Pacific oyster *Crassostrea gigas* in the Wadden Sea: causes and consequences of a successful invasion. A report for The Common Wadden Sea Secretariat, Wilhelmshaven, p 55
95. Nehls G, Diederich S, Thieltges DW, Strasser M (2006) Wadden Sea mussel beds invaded by oysters and slipper limpets: competition or climate control? Helgol Mar Res 60:135–143
96. Nehls G, WitteS, Büttger H, Dankers N, Jansen J, Millat G, Herlyn M, MarkertA, Kristensen PS, Ruth M, Buschbaum C, Wehrmann A (2009) Beds of blue mussels and Pacific oysters. Thematic Report No. 11. In: Marencic H, De Vlas J. (eds). Quality status report 2009. Wadden Sea Ecosystem No. 25. Trilateral Monitoring and Assessment Group, Common Wadden Sea Secretariat, Wilhelmshaven, p 29
97. Nehring S, Reise K, Dankers N, Kristensen PS (2009) Quality status report 2009 Thematic Report No.7. Wadden Sea Ecosystem Report 25
98. Nell JA (2002) Farming triploid oysters. Aquaculture 210(1–4):69–88
99. Normand J, Le Pennec M, Boudry P (2008) Comparative histological study of gametogenesis in diploid and triploid Pacific oysters (*Crassostrea gigas*) reared in an estuarine farming site in France during the 2003 heatwave. Aquaculture 282:124–129
100. Normand J, Ernande B, Haure J, McCombie H, Boudry P (2009) Reproductive effort and growth in *Crassostrea gigas*: comparison of young diploid and triploid oysters issued from natural crosses or chemical induction. Aquat Biol 7:229–241

101. North EW, Schlag Z, Hood RR, Li M, Zhong L, Gross T, Kennedy VS (2008) Vertical swimming behaviour influences the dispersal of simulated oyster larvae in a coupled particle-tracking and hydrodynamic model of Chesapeake Bay. Mar Ecol Prog Ser 359:99–115

102. NSWDPI (2015) New South Wales Department of Primary Industries. Fishing and Aquaculture. Pacific oyster (*Crassostrea gigas)* http://www.dpi.nsw.gov.au/fisheries/pests-diseases/marine-pests/nsw/pacific-oyster. Accessed 25 Mar 2015

103. O'Connor WA, Dove MC (2009) The changing face of oyster culture in New South Wales, Australia. J Shellfish Res 28:803–811

104. Olden JD (2006) Biotic homogenization: a new research agenda for conservation biogeography. J Biogeog 33:2027–2039

105. Olyarnik SV, Bracken MES, Byrnes JE, Hughes AR, Hultgren KM, Stachowicz JJ (2009) Ecological factors affecting community invisibility. In: Rilov G, Crooks JA (eds) Biological invasions in marine ecosystems. Springer, Berlin, pp 215–238

106. Padilla DK (2010) Context-dependent impacts of a non-native ecosystem engineer, the Pacific oyster *Crassostrea gigas*. Integr Comp Biol 50:213–225

107. Piazza BP, Banks PD, La Peyre MK (2005) The potential for created oyster shell reefs as a sustainable shoreline protection strategy in Louisiana. Restor Ecol 13:499–506

108. Piersma T, Koolhaas A, Dekinga A, Beukema JJ, Dekker R, Essink K (2001) Long-term indirect effects of mechanical cockle-dredging on intertidal bivalve stocks in the Wadden Sea. J Appl Ecol 38:976–990

109. Reise K (1998) Pacific oyster invade mussel beds in the European Wadden Sea. Senkenbergiana Marit 28:167–175

110. Reise K, Dankers N, Essink K (2005) Introduced Species. In: Essink K, Dettmann C, Farke H, Laursen K, Luersen G, Marencic H, Wiersinga W (eds) Wadden Sea quality status Report 2004. Wadden Sea Ecosystem vol 19, pp 155–161

111. Rico-Villa B, Le Coza JR, Minganta C, Robert R (2006) Influence of phytoplankton diet mixtures on microalgae consumption, larval development and settlement of the Pacific oyster *Crassostrea gigas*

(Thunberg). Aquaculture 256:377–388

112. Rilov G (2009) The integration of invasive species into marine ecosystems. In: Rilov G, Crooks JA (eds) Biological invasions in marine ecosystems. Springer, Berlin, pp 241–244
113. Rilov G, Crooks JA (eds) (2009) Biological invasions in marine ecosystems. Springer, Berlin
114. Robbins TE (2005) Evaluation of Pacific oyster cultivation methods in the South West of England. MSc Thesis, University of Plymouth, Plymouth
115. Ronholm J, Lau F, Banerjee SK (2016) Emerging seafood preservation techniques to extend freshness and minimize *Vibrio* contamination. Front Microbiol. doi:10.3389/fmicb.2016.00350
116. Rowley AF, Cross ME, Culloty SC, Lynch SA, Mackenzie CL, Morgan E, O'Riordan RM, Robins PE, Smith AL, Thrupp TJ, Vogan CL, Wootton EC, Malham SK (2014) The potential impact of climate change on the infectious diseases of commercially important shellfish populations in the Irish Sea—a review. ICES J Mar Sci 71:741–759
117. Ruesink JL (2007) Biotic resistance and facilitation of a non-native oyster on rocky shores. Mar Ecol Prog Ser 331:1–9
118. Ruesink JL, Lenihan HS, Trimble AC, Heiman KW, Micheli F, Byers JE, Kay MC (2005) Introduction of non-native oysters: ecosystem effects and restoration implications. Annu Rev Ecol Evol Syst 36:643–689
119. Scheiffarth G, Ens B, Schmidt A (2007) What will happen to birds when Pacific Oysters take over the mussel beds in the Wadden Sea? Wadden Sea Newsl 1:10–14
120. Schrobback P, Coglan L, Pascoe S (2014) Socio-economic determinants for industry development: the case of Australia's Sydney rock oyster industry. Aquat Living Resour 27:167–175
121. Scyphers SB, Powers SP, Heck KL Jr, Byron D (2011) Oyster reefs as natural breakwaters mitigate shoreline loss and facilitate fisheries. PLoS One 6(8):e22396. doi:10.1371/journal.pone.0022396
122. Segarra A, Pépin JF, Arzul I, Morga B, Faury N, Renault T (2010) Detection and description of a particular *Ostreid herpesvirus 1*

genotype associated with massive mortality outbreaks of Pacific oysters, *Crassostrea gigas*, in France in 2008. Virus Res 153:92–99

123. Sewell J, Lindsley-Leake S, Tyler-Walters H (2010) Non-native species risk assessment for the Pacific oyster *Crassostrea gigas*. Report to Great Britain non-native species secretariat and Defra
124. Smaal AC, van Stralen MR, Craeymeersch J (2005) Does the introduction of the Pacific oyster *Crassostrea gigas* lead to species shifts in the Wadden Sea? In: Dame RF, Olenin S (eds) The comparative roles of suspension feeders in ecosystems. Springer, Berlin, pp 277–289
125. Smaal AC, Wijsman JWM, Poelman M (2006) Proliferation of the Pacific oyster *Crassostrea gigas* in Dutch coastal waters and its consequences for human use. ICES Annual Science Conference 2006, theme session G: Human health risks and marine environmental quality, Maastricht
126. Smaal AC, Kater B, Wijsman J (2009) Introduction, establishment and expansion of the Pacific oyster *Crassostrea gigas* in the Oosterschelde (SW Netherlands). Helgol Mar Res 63:75–83
127. Smith IP, Guy C, Donnan D (2015) Pacific oysters, *Crassostrea gigas*, established in Scotland. Aquat Conserv Mar Freshw Ecosyst 25:733–742
128. Sousa R, Gutiérrez JL, Aldridge DC (2009) Non-indigenous invasive bivalves as ecosystem engineers. Biol Invasions 11:2367–2385
129. Spalding MD, Fox HE, Allen GR, Davidson N, Ferdana ZA, Finlayson M, Halpern BS, Jorge MA, Lombana A, Lourie SA, Martin KD, McManus E, Molnar J, Recchia CA, Robertson J (2007) Marine ecoregions of the World: a bioregionalization of coastal and shelf areas. Bioscience 57:573–583
130. Spencer BE, Key D, Millican PF, Thomas MJ (1978) The effect of intertidal exposure on the growth and survival of hatchery-reared Pacific oysters (*Crassostrea gigas* Thunberg) kept in trays during their first on-growing season. Aquaculture 13:191–203
131. Spencer BE, Edwards DB, Kaiser MJ, Richardson CA (1994) Spatfalls of the non-native Pacific oyster, *Crassostrea gigas*, in British waters. Aquat Conserv Mar Freshw Ecosyst 4:203–217
132. Spencer B, Kaiser MJ, Edwards DB (1998) Intertidal clam

harvesting: benthic community change and recovery. Aquac Res 29:429–437

133. Suquet M, Malo F, Quere C, Ledu C, Le Grand J, Benabdelmouna A (2016) Gamete quality in triploid Pacific oyster (*Crassostrea gigas)*. Aquaculture 451:11–15
134. Syvret M, Fitzgerald A, Hoare P (2008) Development of a Pacific oyster aquaculture protocol for the UK—Technical Report FIFG project no: 07/Eng/46/04. Report for the Sea Fish Industry Authority
135. Tallis HM, Ruesink JL, Dumbauld BR (2009) Oysters and aquaculture practices affect eelgrass density and productivity in a Pacific Northwest estuary. J Shellfish Res 28:251–261
136. Tamburri MN, Zimmer CA (2007) Mechanisms reconciling gregarious larval settlement with adult cannibalism. Ecol. Monog 77:255–268
137. Thresher RE, Kuris AM (2004) Options for managing invasive species. Biol Invasions 6:295–300
138. Troost K (2010) Causes and effects of a highly successful marine invasion: case-study of the introduced Pacific oyster *Crassostrea gigas* in continental NW European estuaries. J Sea Res 64:145–165
139. Troost K, Gelderman E, Kamermans P, Smaal AC, Wolff WJ (2009) Effects of an increasing filter feeder stock on larval abundance in the Oosterschelde estuary (SW Netherlands). J Sea Res 61:153–164
140. van den Brink AM, Wijnhoven S, McLay CL (2012) Competition and niche segregation following the arrival of *Hemigrapsus takanoi* in the formerly *Carcinus maenas* dominated Dutch delta. J Sea Res 73:126–136. doi:10.1016/j.seares.2012.07.006
141. van der Zee EM, van der Heide T, Donadi S, Eklöf JS, Eriksson BK, Olff H, van der Veer HW, Piersma T (2012) Spatially extended habitat modification by intertidal reef-building bivalves has implications for consumer-resource interactions. Ecosystems 15:664–673
142. Wagner EL, Ruesink JL, Dumbauld BR, Hacker SD, Wisehart LM (2012) Density dependent effects of an introduced oyster, *Crassostrea gigas*, on native intertidal eelgrass, *Zostera marina*. Mar Ecol Prog Ser 468:149–160
143. Walles B, Mann R, Ysebaert T, Troost K, Herman PMJ, Small AC

(2015) Demography of the ecosystem engineer *Crassostrea gigas*, related to vertical reef accretion and reef persistence. Estuar Coast Shelf Sci 154:224–233

144. Wehrmann A, Markert A, May P, Schieck P, Schmidt A (2006) Gefährdungspotential der eulitoralen Miesmuschelbänke im Niedersächsischen Wattenmeer durch die Bioinvasion der Pazifischen Auster Crassostrea gigas. Abschlussbericht Projekt 7/02 der Niedersächsischen Wattenmeer- Stiftung, 110 pp
145. Wheat E, Ruesink JL (2013) Commercially-cultured oysters (*Crassostrea gigas*) exert top-down control on intertidal pelagic resources in Willapa Bay, Washington, USA. J Sea Res 81:33–39
146. Wijsman JWM, Dubbeldam M, De Kluijver MJ, Van Zanten E, Van Stralen MR, Smaal AC (2008). Wegvisproef Japanse oesters in de Oosterschelde. Eindrapportage. Wageningen Imares, Institute for Marine Resources and Ecosystem Studies. Report CD63/08. CD63/08. Yerseke
147. Wilkie EM, Bishop MJ, O'Connor WA (2013) The density and spatial arrangement of the invasive oyster *Crassostrea gigas* determines its impact on settlement of native oyster larvae. Ecol Evol 3:4851–4860
148. Williamson M (1996) Biological Invasions. Chapman and Hall, London
149. Wood CA, Bishop JDD, Delduca EL, Davies C, Herbert RJH, Clark PF (2015) *Hemigrapsus takanoi* Asakura and Watanabe, 2005 (Crustacea: Decapoda: Brachyura: Grapsoidea): first records of the brush-clawed shore crab from Great Britain. Bioinvasions Rec 4:109–113
150. Woolmer A (2009). Appropriate assessment and shellfisheries: adaptive management protocol. Report to shellfish industry development strategy. Salacia-Marine Ltd
151. Wrange AL, Valero J, Harkestad LS, Strand Ø, Lindegarth Christensen HT, Dolmer P, Kristensen PS, Mortensen S (2010) Massive settlements of the Pacific oyster, *Crassostrea gigas*, in Scandinavia. Biol Invasions 12:1453–1458

152. zu Ermgassen PSE, Spalding MD, Blake B, Coen LD, Dumbauld B, Geiger S, Grabowski JH, Grizzle R, Luckenbach MW, McGraw K, Rodney W, Ruesink J, Powers SP, Brumbaugh RD (2012) Historical ecology with real numbers: past and present extent and biomass of an imperiled estuarine habitat. Proc R Soc B Biol Sci 279:3393–3400

CHAPTER

16

VULNERABILITY OF CHINA'S NEARSHORE ECOSYSTEMS UNDER INTENSIVE MARICULTURE DEVELOPMENT

Hui Liu[1] and Jilan Su[2]

[1] Yellow Sea Fisheries Research Institute, Chinese Academy of Fishery Sciences, Qingdao, China

[2] Second Institute of Oceanography, State Oceanic Administration, Hangzhou, China

ABSTRACT

Rapid economic development and increasing population in China have exerted tremendous pressures on the coastal ecosystems. In addition to land-based pollutants and reclamation, fast expansion of large-scale intensive mariculture activities has also brought about additional effects. So far, the ecological impact of rapid mariculture development and its large-scale operations has not drawn enough attention. In this paper, the rapid development of mariculture in China is reviewed, China's effort in the application of ecological mariculture is examined, and the vulnerability of marine ecosystem to mariculture impact is evaluated through a number of examples. Removal or reduced large and forage fish, due to both habitat loss to reclamation/mariculture and overfishing for food or fishmeal, may have far-reaching effects on the coastal and shelf ecosystems in the long run. Large-scale intensive mariculture operations carry with them undesirable biological and biochemical characteristics, which may have consequences on natural ecosystems beyond normally perceived spatial and temporal boundaries. As our understanding of possible impacts of large-scale intensive mariculture is lagging far behind its development, much research is urgently needed.

Keywords: Large-scale intensive mariculture, Impact on marine ecosystems, Sustainable development, Ecosystem approach to aquaculture

INTRODUCTION

Since the early 1980s, China's economy has developed rapidly with an average annual GDP growth rate over 9 %, according to its National Bureau of Statistics. The growth rate of the 11 provinces and metropolises along the coasts exceeded 10 %. In fact, with only 13 % of China's land area, in 2011, the 11 coastal provinces and metropolises housed about 42 % of its population and produced over 57 % of the national GDP (CCICED Task Force 2013).

Rapid economic growth and urbanization have brought about tremendous pressure on the coastal and marine ecosystems, resulting in the loss or deterioration of many of their services. Land-based pollutant and coastal reclamation are the two most detrimental factors to the health of the ecosystems (CCICED Task Force 2013). According to the official bulletin, in 2009, over half of China's nearshore waters were classified as below the water quality standard and about 20 % as severely below standard. Occurrence of recorded events of harmful algal blooms (HABs) rose sharply to about 80 yearly, covering a total area exceeding 15,000 km^2 (op cit). Between 1990 and 2008, the average reclamation area in China reached 285 km^2 yearly and a further 5780 km^2 was planned for over the next 10 years: The fast pace of reclamation has caused significant coastal landscape fragmentation and loss of biodiversity, as well as destruction of habitats for fish and loss of feeding grounds for shorebirds, leading to sharp decline in fisheries resources and bird species (Wang et al. 2014). In addition, overfishing, rapid expansion of mariculture, and construction of large number of hydraulic works on land all had compounded the impacts on China's coastal and marine ecosystems (CCICED Task Force 2013).

Spurred by both improved culture technology and rise in domestic demand for high-quality seafood, there has been a revolutionary growth of mariculture in China over the last 30 years. China is credited with 63.3 and 54 %, respectively, of the world's farmed marine food fish and aquatic plant production (FAO 2014), which are usually farmed in large-scale aquaculture operations. For such high production, extensive patches of tidal flats and nearshore waters have been converted for mariculture use. However, so far, the ecological impact of rapid mariculture development and its large-scale operations has not drawn sufficient attention in China. Attention is usually directed at the food provision function of mariculture, while the interference of mariculture with other ecosystem services has not been fully recognized.

In this paper, the rapid development of mariculture in China is reviewed, China's effort in the implementation of ecological mariculture practices is examined, and the vulnerability of marine ecosystems to impacts of mariculture is evaluated, pertaining to large-scale operations. Considering the current scale of mariculture worldwide and its significant upward trend, impacts of the world's intensive mariculture on the ocean ecosystems are far from negligible. Understanding the true state of impact of intensive mariculture on coastal and marine ecosystems will facilitate consideration of precautionary measures to ensure sustainable growth of world mariculture.

MARICULTURE IN CHINA

Along with the world's rapid growth in mariculture over the recent decades, China's mariculture has also undergone unprecedented development (Figure 1). Since the middle 1970s, both the mariculture production and capture fisheries experienced fast growth. However, after the middle 1990s, the capture fisheries became stagnant while mariculture kept on expanding. The ratio between the mariculture production and fishery catch was almost nil in early 1950s, reached 20 % in 1975, 50 % in 1988, and 100 % in 2006, and stood at 140 % in 2013. This is in sharp contrast to the ratio for Europe (18 %) and for other continents (<15 %) (data source: FAO FishStatJ). In 2012 China's mariculture contributed to

54 and 63.3 %, respectively, of the world's farmed aquatic plants (largely seaweeds) and marine food fish production (FAO 2014). In light of the depletion of wild fish stocks and the continuing rise in domestic seafood demand, mariculture will undoubtedly lead the future growth of Chinese aquaculture in the long run.

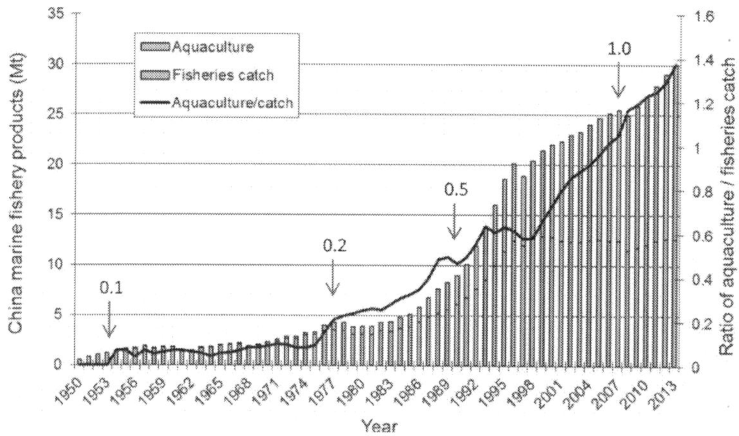

Figure 1: Growth of mariculture production and fishery catch during 1950–2013. The *number above each red arrow* denotes the ratio of "aquaculture/ fishery catch" for that year (data source: MOA 1979–2014).

Throughout the 1950~1980, mariculture species in China were largely confined to seaweeds and bivalves, which are extractive species relying on natural supply of nutrients and plankton. Monoculture on longlines over tidal flats was the main type of operation. Commercial culturing of shrimp, finfish, and other high-value species began in the early 1980s, when land-based hatcheries, earthen ponds, and net cages were rapidly set up. The trend of "fishing down and farming up the food web" has been significant in the world fishery industry for decades (Naylor et al. 2000), and China is no exception. In the last 15 years, mariculture production of fed species (finfish and crustaceans) increased sharply from <5 to >15 %, showing a clear upward trend into the future (Figure 2). This reflects the increased domestic market demand for high-value seafood, corresponding to the economic growth in China. At the same time, this will no doubt bring about increased wastes from mariculture, because, unlike the freshwater finfish which are mostly planktonic filter feeders or omnivorous, marine finfish and crustaceans are generally carnivorous and high in trophic level. Paying attention to this trend is important for rational planning and ecosystem-based management of mariculture.

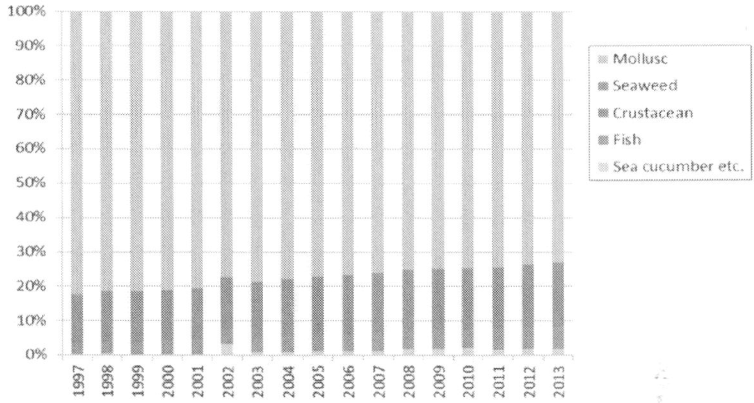

Figure 2: Change of the composition of China's mariculture over time (data source: MOA 1998–2014).

Along with the increase of production, large portions of area from both the intertidal wetlands and nearshore waters of China were used for mariculture (Figure 3). According to the Chinese fishery statistics (MOA 2014), the total mariculture acreage amounted to 2.32 Mha in 2013, among which 0.8 Mha was converted from intertidal wetlands (nearly one third of China's total) and 0.9 Mha from the nearshore waters (nearly 10 % of China's total) was used for mariculture purpose.

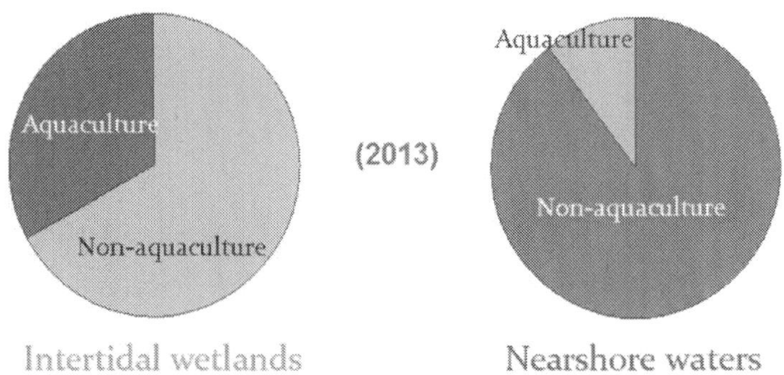

Figure 3: Portion of area used for mariculture in, respectively, intertidal wetlands and nearshore waters of China in 2013. Calculation based on coastal wetland area data from the SFA (2011) and aquaculture area data from MOA (2014).

CHINA'S APPROACH TO ECOLOGICAL MARICULTURE

Aquaculture is basically an economic activity that consumes resources, impacts on biodiversity in different levels and ways, and discharges wastes, hence affecting the environment (e.g., Naylor et al. 2000). With the worldwide rapid development of aquaculture since the 1970s, especially in developing countries, there was first a focus on productivity and benefits, leading to a general scaling up and overstocking. The resulting aquaculture epidemics and mass mortalities raised awareness of ecological carrying capacity of the aquaculture systems (e.g., Grant et al. 2007). This concept soon evolved into ecological aquaculture, with ecosystem approach to aquaculture (EAA) as its implementation strategy (Soto et al. 2008).

Conceptually, EAA embodies the principle of ecosystem-based management by promoting conservation and sustainable use of resources. In implementing EAA, it is necessary to define the ecosystem boundaries in both space and time. For this, three levels of scale have been identified, namely, the farm, the watershed/aquaculture zone, and the global (Soto et al. 2008). In practice, boundaries for the second level, i.e., the watershed/aquaculture zone level, sometimes are not easy to delineate, both managerially and ecologically. For a fjord-like inlet where the physical boundaries are more or less defined, management tools including analysis models for evaluating aquaculture in an ecosystems approach were developed and proved to be useful (Nunes et al. 2011). However, because of the lack of a good understanding of the interaction between a mariculture operation and the marine ecosystems at large, presently, there are no agreed-on guidelines or practical measures for implementing EAA at the level of the watershed/aquaculture zone.

One innovative approach within EAA is the so-called integrated multi-trophic aquaculture (IMTA), in which multi-trophic species with complimentary trophic functions are cultured together. Traditional IMTA has been practiced in the world for thousands of years, including ancient China (Costa-Pierce 2010). Modern polyculture methods in China, as one of the eight techniques embodying China's freshwater aquaculture, were reviewed for the first time in 1959 (CoCFAE 1961).

With the development of intensive aquaculture worldwide since the 1970s, monoculture of omnivorous or carnivorous marine species at high densities, including finfish and crustacean species, was initially carried out in China. The production boom in late 1980s and early 1990s brought about environmental pollution and disease outbreak among the stocks, leading to a crash in harvest. At the same time, technology improvement allowed for restructuring of the existing culture systems. There was then a clear transition in China from intensive monoculture to polyculture. The first successful attempt in marine polyculture was conducted in 1975, when kelp *Laminaria japonica* and mussels *Mytilus edulis* were co-cultured (Xie 1981; Dong 2011). Soon, polyculture in China evolved such that fed (fish and shrimp) and non-fed species (bivalves and seaweeds) were cultured together, where the non-fed species served as a means to recycle wastes and reduce pollution (e.g., Wu et al. 1980; Zhu 1981). Thus, the ecosystem perspective in traditional mulberry-fish pond system from ancient China regained appreciation. Such IMTA mode ecological aquaculture is now commercially practiced in China (Fang 1996; Dong 2011).

A major IMTA demonstration site in China is Sanggou Bay, at the tip of Shandong Peninsula in eastern China, where, according to local statistics, seaweed, shellfish, and finfish are cultured at a ratio of approximately 7:3:1 with an annual seaweed production of about 200 thousand tons (Zhang XJ, Head of Ocean and Fishery Bureau of Rongcheng, private communication). Based on both research findings and technology improvement, the ratio and layout of the seaweed longlines, shellfish lantern nets, and finfish net cages were constantly optimized to minimize the environmental impact from this major IMTA system (Tang et al. 2013, Shi et al. 2013). Although IMTA has become an important component for mariculture in China, there is no official statistics on its scale and total production because it is not yet classified as one standard aquaculture mode.

At the same time, IMTA has been met with worldwide enthusiasm during the last decade, as an effective way to deal with the wastes and improve the output of aquaculture (Barrington et al. 2009; Chopin 2012). In the present practice of IMTA, there are various approaches to achieve the basic goals that feed and wastes are recycled within the system. With different species composition and spatial configuration, IMTA has been applied to both pond and open-water mariculture systems. Although there were attempts to study the nutrient balance within an IMTA system from a modeling point of view (e.g., the study of ratios between fed and extractive species in IMTA farms (Reid et al. 2013)), lack of thorough understanding of environmental impact of IMTA remains one major concern (e.g., IMTA White Paper 2010).

From the point of view of EAA, boundary delineation of an IMTA system at the watershed/aquaculture zone level is often not straightforward. This results in rather complicated interaction between the culture system and the natural habitats. For example, large-scale IMTA in shallow bay will significantly alter the tidal currents, resulting in a reduced flushing capability of the bay, as well as insufficient nutrient supply to the bay from the offshore water (Shi et al. 2011). In addition, organic wastes do accumulate in the sediments of IMTA area and alter both the benthic and water column chemical parameters (Ren et al. 2014). Given the physical conditions of any particular IMTA site, it is extremely difficult, if not impossible, to recycle all the nutrients and wastes within an IMTA system.

VULNERABILITY OF THE NEARSHORE ECOSYSTEM

As commented in the "Introduction" section, over the last 30 years, the nearshore ecosystems of China have been under multiple pressures of human activities, including land-based pollutants, reclamation, port construction, overfishing, and mariculture. Early concern about the impact from mariculture was limited to its waste discharge pollution and escapees. However, as China's mariculture rapidly expands, there are evidences now that large-scale intensive mariculture may bring about undesirable ecological consequence far beyond the normally perceived boundaries of its operation. The combined impact from mariculture and other human activities has an amplification effect that significantly increases the vulnerability of China's coastal and marine ecosystems. In the following, we will use a few examples to illustrate this point.

Degradation and Destruction of Natural Habitats

As is well known, intertidal wetlands and nearshore waters are critical habitats for many marine organisms, including both large and forage fish. As extensive tidal flats and wetlands are transformed to aquaculture ponds, the natural ebb and flow of tides disappear, along with the local benthic organisms and migrating shore birds (Yang et al. 2004, Chen and Ouyang 2014). Mariculture not only introduces pollutants into nearshore material cycle, but also interferes with local hydrologic processes which can disrupt the success of recruitment in the early life of fisheries (e.g., Dong et al. 2007). In open-water aquaculture, the environmental impact is heavily influenced by local circulation (Han and Liu 2014). It is noteworthy that mariculture impact on a sheltered bay is usually more significant due to its limited environmental carrying capacity. Large-scale net cages and longline facilities in shallow bays significantly slow down the tidal currents, resulting in weakened flushing capability of the bay, reduced nutrient supply to the inner bay, increased deposition, and reduced productivity of phytoplankton (Ji et al. 1998; Liu et al. 2004; Shi et al. 2011). In addition, accumulation of waste deposits under shellfish longlines or finfish net cages may severely affect the macro-benthos and change their biodiversity, composition, and distribution (Gao et al. 2003; Yang et al. 2007; Wang et al. 2011). In the long run, the establishment of a mariculture system may result in fragmentation of the original natural ecosystems and disappearance or significant change in some of the original biotic communities (Liu and Fang 2014).

Furthermore, mariculture also imposes deleterious effects on the natural fauna and flora through its chemical or pharmaceutical discharge. To eliminate harmful or undesirable organisms in many of the extensive pond mariculture, pellouxite and other disinfectants have been used in high concentrations before or during the culture period (Luo 2012). When the treated water is discharged without further treatment, these chemicals will invariably cause high mortality to the nearby wild fauna and flora.

Although such impacts are well known, in the case of China, the magnitude of these impacts on the nearshore ecosystems needs to be assessed against the scope of the human activities. As discussed in the "Introduction" section, natural wetlands in China suffered great loss and degradation over decades. More than 50 % of the coastal wetlands were lost due to reclamation for various uses, including mariculture (An et al. 2007). The loss of associated ecosystem services due to reduced purification capacity, failure in recruitment of fishery stock, and reduced biodiversity was estimated to amount to 46 billion USD per year (op cit).

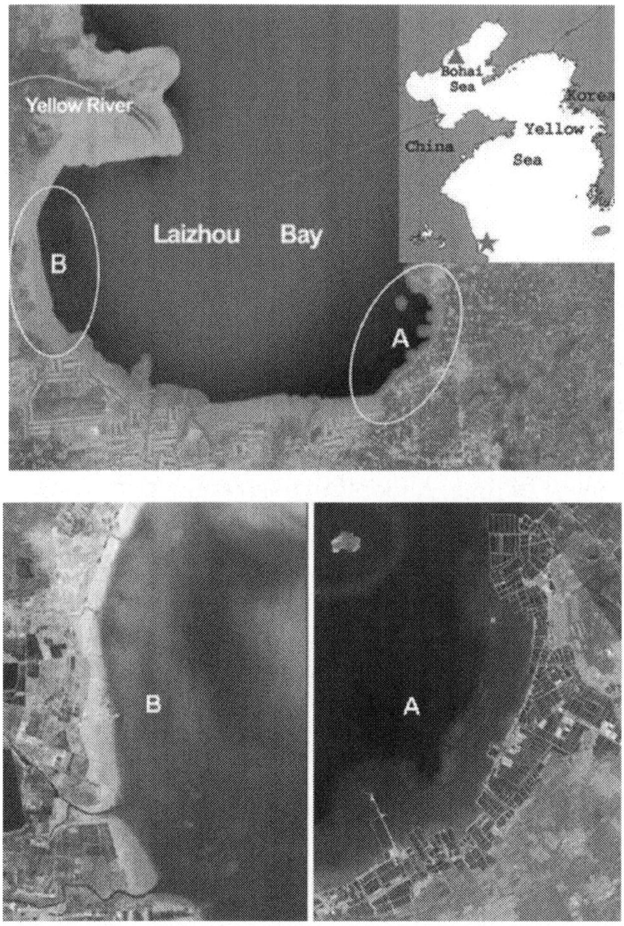

Figure 4: Extensive mariculture facilities along the shores of the Laizhou Bay. In the inserted map on the *top figure*, the *red triangle* and *star* denote, respectively, the location of the large-scale bivalve culture and the extensive Su-Bei tidal shoals.

For example, the Laizhou Bay of Bohai Sea has always been used as spawning grounds for many fishery species of the Bohai Sea and Yellow Sea (Deng and Jin 2000; Xu and Chen 2010). However, presently, most of the estuaries and wetlands around the bay are severely affected by mariculture activities (Figure 4). Furthermore, not shown in the figure, there are also large areas of longline mariculture over the shallow waters in the bay. It is possible that most of the nursery grounds for the fisheries are either lost or severely degraded. Thus, in addition to overfishing, reclamation and mariculture activities may also contribute to the collapse of regional fishery (Zuo and Lin 2008; Wang et al. 2014).

The loss of coastal wetlands along with its biodiversity is not unique in China. As a matter of fact, there has been extensive loss of tidal flats around the whole Yellow Sea region, associated with economic development in China and the two Koreas. It was found that 28 % of the Yellow Sea tidal flats existing in the 1980s had been reclaimed by the late 2000s (Murray et al. 2014).

Pressure on Wild Fisheries

There have been very few observations on the direct aquaculture impact on wild fishery, especially regarding a particular species and over a significant length of time. According to a 7-year study in Canada (Loucks et al. 2014), net cage finfish aquaculture produced odors and changed benthic habitat and water quality, which may affect the behavior and movement of wild fishery species (e.g., lobster). It was found that, within 2 years of a finfish aquaculture operation, ovigerous lobsters abandoned the area and overall lobster abundance decreased; the female lobsters returned to this area only when fish farming was discontinued. On the other hand, aquaculture may provide food for wild species. For example, aggregations of wild fish communities were repeatedly observed in the vicinity of aquaculture farms in the Mediterranean (Grigorakis and Rigos 2011).

In addition to habitats, marine capture fisheries and mariculture also compete for other ecosystem services, such as nutrients and feed, broodstock, and larval supply (Naylor et al. 2000). In China, the catch of wild broodstock for hatchery (such as Chinese shrimp *Fenneropenaeus orientalis*) and of young fish for grow-out or fattening (such as sea bass *Lateolabrax maculatus* and freshwater eel *Anguilla japonica*) has been going on for many years (Niu 1982; Gao and Zhang 2002; Zhi et al. 2013). The above-mentioned human activities together imposed devastating effect on the wild stock and also defeated large-scale restocking efforts conducted in

China, including artificial reefs and annual release of tens of billions of fishery seedlings (Shan et al. 2012).

Another impact of mariculture on wild fishery is the harvest of forage fish, which is either directly fed to aquaculture or processed into fishmeal. Generally speaking, in China, about one third of forage fish (about 4 Mt) are fed directly to aquaculture species (Mai 2010), although there is large site-specific variability. Annual production of these small forage fish amounted to several million tons, though not all such productions are recorded in the official fishery statistics (op cit). Statistics of specific forage fisheries varied widely over the decades, reflecting overfishing pressure on these species (Figure 5).

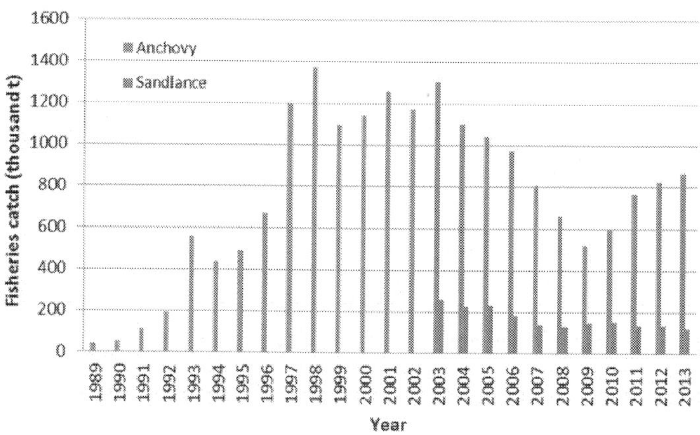

Figure 5: Fishery catch of anchovy and sandlance in China during 1989–2013 (data source: MOA (1990–2014)).

Jackson et al. (2001) have shown that overfishing of large fish fundamentally altered coastal marine ecosystems during each of the cultural periods examined. Forage fish is also an important part of the regional marine ecosystems. With pressure from both habitat loss/ degradation and overfishing, it is hard to imagine how the regional coastal and nearshore ecosystems will evolve when both the large and forage fisheries are destroyed.

Proliferation of Opportunistic Species

Since 2007, there has been yearly outbreak of large-scale green tides over the Yellow Sea in summer. In 2013, at its peak, it covered a total area 790 km^2 distributed over a wide area of 29,700 km^2, while, in 2014, the numbers were, respectively, 540 and 50,000 km^2 (SOA 2014, 2015). Figure 6 shows the distribution pattern of the green tides on 30 June 2013 and 14 July 2014. The green tides are caused by *Ulva prolifera* (*Enteromorpha prolifera*), a macro-algae species often found along the coasts of temperate Asia, including China. Proliferation of *U. prolifera* is associated with the large-scale mariculture system of *Porphyra yezoensis* over the extensive Su-Bei tidal shoals of China (Figs. 4 and 6) off the west boundary of the Yellow Sea (Liu et al. 2009).

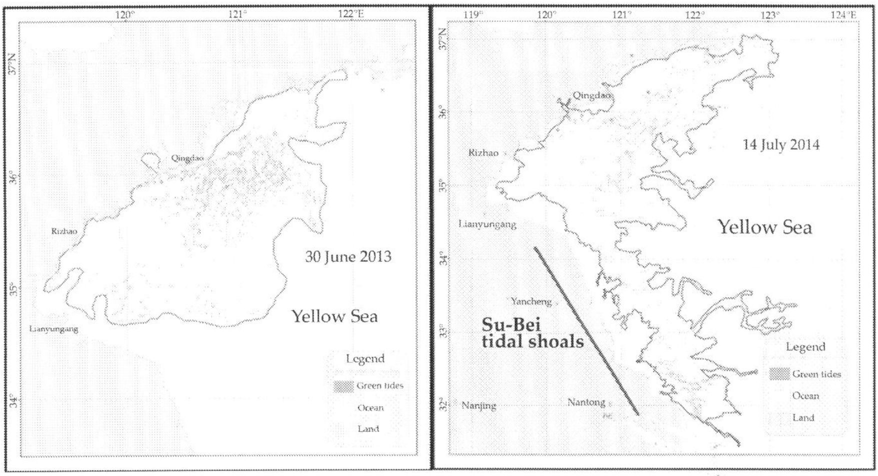

Figure 6: Distribution pattern of the green tides over Yellow Sea on 30 June 2013 and 14 July 2014 (figure source: SOA 2014, 2015).

Micropropagules of *U. prolifera* generally exist in the nearshore water of Yellow Sea. The micropropagule will grow only after attaching itself to a periodically exposed surface, including the rafts used for the *P. yezoensis* culture. A variety of macro-algae species would attach and grow on these rafts from late autumn, including *U. prolifera*, *Ulva linza*, *Ulva compressa*, *Ulva intestinalis*, and *Ulva clathrata*. *U. prolifera* became one of the most dominant species only in late March to mid-May (Fan et al. 2015). At the harvest of *P. yezoensis* in April, the attached macro-algae were stripped off and left on the shoals. Most of the macro-algae left on the shoals were drowned except *U. prolifera* which can float during photosynthesis. Once leaving the shoals, the floating *U. prolifera* would drift in an N-NE direction driven by the prevailing winds. The floating *U. prolifera* grew at an initial daily rate of 20–30 % and soon evolved to green tides (op cit). Some of the drifting green tides landed on the shores of Shangdong Province in June/July while the rest eventually sank offshore.

Figure 7 gives the yearly production, culture area, and productivity of the *P. yezoensis* culture in Jiangsu Province between 1988 and 2013 (most of this culture is over the Su-Bei Shoals). It is noted that the yearly production of *P. yezoensis* reached a plateau after 2007 even though the cultured area kept increasing. The productivity underwent a sharp drop then. The year 2007 was when the green tides over the Yellow Sea first became noticeable. From 2008 to the present, there is a yearly outbreak of green tides covering 250–2100 km^2 distributed over 20,000–60,000 km^2. It is possible that certain bloom-state threshold for the opportunistic species *U. prolifera* was reached after 2007, likely related to the size of the culture area of *P. yezoensis* (Figure 7).

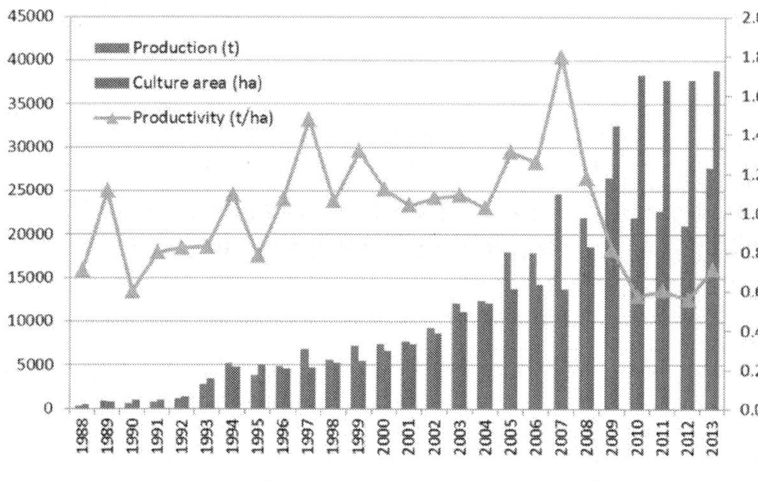

Figure 7: Yearly *P. yezoensis* culture production, culture area, and productivity in Jiangsu Province in 1988–2013 (source of data: MOA 1989–2014).

There is no reported estimate on the portion of the floating mass of *U. prolifera* that sink offshore, nor is there any study on the fate of these submerged macro-algae. It is well known that, in summer, the Yellow Sea Cold Water (YSCW), a water mass, occupies the central part of Yellow Sea deeper than 40 m below a sharp thermocline (e.g., Su 1998). Figures 7-8 and 7-6 in Hong (2012), respectively, showed that the bottom dissolved oxygen in YSCW is highest in spring and lowest in autumn, and in addition, over the past 30 years, it has a continuous decreasing trend with an average drop of about 0.77 mg l^{-1}. A recent ocean acidification survey found that in late autumn of 2012, large area of YSCW had a saturation state of aragonite below 1.0 (SOA 2013), indicating a corrosive environment for many marine organisms (Fabry et al. 2008). It is known that YSCW serves as important oversummer grounds for *Calanus sinicus*, an important copepod providing a wide size spectrum of food for fish stocks (Wang et al. 2003). How the annual outbreak of *U. prolifera* blooms will exert further pressure on the ecosystem services of YSCW is yet to be studied.

Alteration of Biochemistry in Water/Sediment

Although mariculture ponds and enclosures and nearshore longline culture areas are usually counted as artificial wetlands, they are fundamentally different from natural coastal wetlands because their ecosystem services are significantly different. Mariculture is also a major driver on changes in the biogeochemistry of its neighboring water and sediment.

The deleterious environmental effects of the intensive and industrialized mariculture have been well recognized (e.g., National Research Council 2010). Finfish cage culture can cause significant rise in dissolved nitrogen in water column, particulate phosphorous in sediment, and chemical oxygen demand (COD) in both the water column and sediment (e.g., Holby and Hall 1991; Wang 1998). Large-scale oyster (*Crassostrea gigas*) culture may alter the sediment geochemistry by raising NH_4^+ and HPO_4^{2-} by 7.7–11.5 and 1.8–8.0 times and sulfate reduction rates by 2.4–5.2 times (Hyun et al. 2013). Given the scale of bivalve culture in China (total annual production of 12.7 Mt over a total area of 1.6 Mha) (MOA 2014), the biochemical alteration of inshore sediments is considered significant enough to induce eutrophication, change in P/N ratio, and outbreaks of HABs (Bouwman et al. 2013). This was echoed by Hyun et al. (2013), who observed enhanced benthic microalgal biomass and primary production in the water column at the bivalve farms. As discussed in "China's approach to ecological mariculture" section, it is difficult even for an IMTA system to recycle all the nutrients and wastes within its system. Indeed, eutrophication effect of polyculture (with finfish, shellfish, and seaweed) can be significant in a sheltered bay (Zhang et al. 2003).

Nutrients in the reduced form in the ocean are of particular concern. For example, reduced N forms were preferentially taken up during the harmful microalgae blooms of *Karenia mikimotoi* and *Prorocentrum donghaiense* in nearshore waters of the East China Sea (Li et al. 2009). Both of these are common HAB species in coastal China, and the blooms of *K. mikimotoi*usually cause great losses in mariculture (SOA 2005–2015).

Among all forms of mariculture pollutions, direct release of nutrients in the reduced form into the ocean is of more concern (Bouwman et al. 2013). Laboratory experiments showed that ammonia and urea excreted by the cultured animals were readily taken up by phytoplankton and could stimulate their growth (Arzul et al. 1996; Yang et al. 2004). When the level of such nutrient is higher than the assimilative capacity of the marine area, it may fuel the production of algae on which mixotrophic HABs may feed (op cit). In turn, such mixotrophic HABs can kill or intoxicate the mariculture organisms and increase risks to human health. The scallop *Argopecten irradians* culture at the nearshore region of the northwestern Bohai Sea (Figure 4) is number one in China by area scale (Zheng et al. 2011). Coincidentally, large-scale *Aureococcus anophagefferens* brown tides, a form of mixotrophic HABs, have occurred in the northwestern Bohai Sea in early summer for consecutive years since 2009 and have caused significant negative impacts on the scallop culture there (Zhang et al. 2012). China is the third country in the world hit by brown tides, but at a much greater scale. At its peak, the brown tides covered an area of over 3000 km^2 (op cit).

Particulate wastes from aquaculture usually accumulate in nearby sediments, driving an increase in COD value. As organic matter is decomposed by microbes through respiration, the redox potential will decrease, leading to bottom dissolved oxygen (DO) depletion (Wang 1998). Accumulation of wastes and microbial/phytoplankton biomass in proximity to mariculture sites may give rise to hypoxia conditions resulting in an anoxic layer of sediment and bottom waters (Yang et al. 2004), and this develops very often in stratified water. In fact, seasonal bottom oxygen depletion in western Bohai Sea, especially in the vicinity of large-scale bivalve culture sites, has now become normality and is getting worse (Zhai et al. 2012). The progressive depletion in the summer bottom dissolved oxygen in Bohai is rather alarming and raises further concern about Bohai Sea's ecosystems. Observation showed that its minimum bottom oxygen level in August steadily dropped over the years, being 6.3 mg l^{-1} in 1982, 5.1 mg l^{-1} in 1992, 4.5 mg l^{-1} in 2006, 3.5 mg l^{-1} in 2011, and below 3 mg l^{-1} in 2014 (Zhai Weidong, private communication). Analysis by Zhai et al. (2012) indicated that both the bottom DO depletion and acidification in summer were caused by remineralization of biogenic particles, which were supplied by either coastal HABs or regional nearshore mariculture.

CONCLUSION

As the world's population keeps on increasing with high aspiration for better life, rapid intensive reclamation and large-scale mariculture practices will no doubt be replicated in developing countries around the world's coasts. Such human activities are conducted over intertidal wetlands and nearshore waters, which lie in a narrow strip along the coast. These habitats are critical in the life history of many species of large and forage fish. At the same time, overfishing, either for food or for fishmeal, exerts tremendous pressure on the stock of both large and forage fish. During past cultural periods, overfishing of large fish was shown to have fundamentally altered coastal marine ecosystems (Jackson et al. 2001). If the large and forage fish are sharply reduced, both the coastal and shelf ecosystems will likely be significantly altered.

Mariculture replaces intertidal wetlands with artificial wetlands and brings artificial ecosystems into nearshore waters. These artificial ecosystems bring with them different but undesirable biological and biochemical characteristics. Spatial and temporal extents of the effects of these artificial ecosystems, i.e., the boundaries for the watershed/aquaculture zone level, are often hard to define. The speed of mariculture development far outpaces the progress of our knowledge about its impacts. Examples given above show that the coastal and nearshore ecosystems are rather vulnerable when facing intensive large-scale mariculture development.

The fast pace of the worldwide coastal development will likely cause the coastal and nearshore ecosystems to evolve into states unfriendly to either the coastal communities or the ocean ecosystems at large. As our understanding of such possible changes is lagging far behind the economic and social development, much research is urgently needed, including those related to large-scale intensive mariculture.

ACKNOWLEDGMENTS

This work arises from two task force studies, the Problems Facing China's Sea Enclosing and Land Reclamation (2008–2009) and the Scientific Questions and Policy Recommendations for Ecosystem Eveluation and Management of Coastal Waters of China (2014–2016), both funded by the Chinese Academy of Sciences.

REFERENCES

1. An S, Li H, Guan B, Zhou C, Wang Z, Deng Z, Zhi Y, Liu Y, Xu C, Fang S, Jiang J, Li H (2007) China's natural wetlands: past problems, current status, and future challenges. Ambio 4:335–342
2. Arzul G, Clément A, Pinier A (1996) Effects on phytoplankton growth of dissolved substances produced by fish farming. Aquat Living Resour 9:95–102
3. Barrington K, Chopin T, Robinson S (2009) Integrated multi-trophic aquaculture (IMTA) in marine temperate waters. In: Soto D (ed) Integrated mariculture: a global review, vol 529, FAO Fish Aquac Tech Paper., pp 7–46
4. Bouwman L, Beusen A, Glibert PM, Overbeek C, Pawlowski M, Herrera J, Mulsow S, Yu RC, Zhou MJ (2013) Mariculture: significant and expanding cause of coastal nutrient enrichment. Environ Res Lett 8. doi:10.1088/1748-9326/8/4/044026
5. Chen JH, Ouyang L (2014) Let the migrating birds fly freely. Ecol Econ 30(11):10–13 (in Chinese)
6. Chopin T (2012) Integrated multi-trophic aquaculture (IMTA). In: Meyers R (ed) Encyclopedia of sustainability science and technology. Springer Reference, N.Y., pp 184–205
7. Committee on China's Freshwater Aquaculture Experience (CoCFAE) (1961) China freshwater aquaculture. Science Press, Beijing (in Chinese)

8. Costa-Pierce BA (2010) Sustainable ecological aquaculture systems: the need for a new social contract for aquaculture development. Mar Technol Soc J 44:88–112
9. Deng JY, Jin XS (2000) Study on fishery biodiversity and its conservation in Laizhou Bay and Yellow River Estuary. Zool Res 21:76–82 (in Chinese)
10. Dong SL (2011) History, principles, and classification of integrated aquaculture in China. J Fishery Sci China 18:1202–1209 (in Chinese)
11. Dong LX, Su JL, Deng JY, Chen Q (2007) The importance of estuarine gravitational circulation in the early life of the Bohai Penaeid Prawn. J Mar Syst 67:253–262
12. Fabry VJ, Seibel BA, Feely RA, Orr JC (2008) Impacts of ocean acidification on marine fauna and ecosystem processes. ICES J Mar Sci 65:414–432
13. Fan SL, Fu MZ, Wang ZL, Zhang XL, Song W, Li Y, Liu GX, Shi XY, Wang XN, Zhu MY (2015) Temporal variation of green macroalgal assemblage on Porphyra aquaculture rafts in the Subei Shoal, China. Estuar, Coastal Shelf Sci. doi:10.1016/j.ecss.2015.03.016
14. Fang JG, Sun HL, Yan JP, Kuang SH, Li F, Gary F, Grant J (1996) Polyculture of scallop *Chlamys farreri* and kelp *Laminria japonica* in Sungo Bay. Chin J Oceanol Limnol 14(4):322–329 (in Chinese)
15. FAO (2014) The state of world fisheries and aquaculture 2014. Rome. 223 pp
16. Gao HD, Zhang SR (2002) Discussions on sea bass seedling resource management measures. Hebei Fish 122:52 (in Chinese)
17. Gao AG, Yang JY, Chen QZ, Wang ZP, Zhang J, Dong YT, Ning XR (2003) Comparative studies on macrobenthos between cultured and non-cultured areas in Xiangshan Bay. J Fish China 27(1):25–31 (in Chinese)
18. Grant J, Curran KJ, Guyondet TL, Tita G, Bacher C, Koutitonsky V, Dowd M (2007) A box model of carrying capacity for suspended mussel aquaculture in Lagune de la Grande-Entree, Iles-de-la-Madeleine, Quebec. Ecol Model 200:193–206
19. Grigorakis K, Rigos G (2011) Aquaculture effects on environmental and public welfare—the case of Mediterranean mariculture.

Chemosphere 855:899–919

20. Han QX, Liu DY (2014) Review on the effect of coastal bivalve raft mariculture on macrobenthic assemblages. Mar Sci Bull 33(3):352–359 (in Chinese)
21. Holby O, Hall PO (1991) Chemical fluxes and mass balances in a marine fish cage farm II. Phosphorus. Mar Ecol Prog Ser 70:263–272
22. Hong HS (ed) (2012) Regional oceanography of china seas-chemical oceanography. Ocean Press, Beijing, 374pp (in Chinese)
23. Hyun JH, Kim SH, Mok JS, Lee JS, An SU, Lee WC, Jung RH (2013) Impacts of long-line aquaculture of Pacific oysters (*Crassostrea gigas*) on sulfate reduction and diffusive nutrient flux in the coastal sediments of Jinhae–Tongyeong, Korea. Mar Pollut Bull 74:187–198
24. IMTA White Paper (2010) White paper integrated multi-trophic aquaculture: a workshop. September 14–15, 2010. Port Angeles, Washington http://www.nmfs.noaa.gov/aquaculture/docs/imta/imta_white_paper.pdf Accessed on February 27, 2015.
25. Jackson JBC, Kirby MX, Wolfgang HB, Bjorndal KA, Botsford LW, Bourque BJ, Bradbury RH, Cooke R, Erlandson J, Estes JA, Hughes TP, Kidwell S, Lange CB, Lenihan HS, Pandolfi JM, Peterson CH, Steneck RS, Tegner MJ, Warner RR (2001) Historical overfishing and the recent collapse of coastal ecosystems. Science 293:629–638
26. Ji RB, Mao XH, Zhu MY (1998) Impacts of coastal shellfish aquaculture on bay ecosystem. J Oceanogr Huanghai Bohai Seas 16(1):21–27 (in Chinese)
27. Li J, Glibert PM, Zhou MJ, Lu SH, Lu DD (2009) Relationships between nitrogen and phosphorus forms and ratios and the development of dinoflagellate blooms in the East China Sea. Mar Ecol Prog Ser 383:11–26
28. Liu H, Fang J, Zhu J, Dong S, Wang F, Liang X, Zhang J, Lian Y, Wang L, Jiang W (2004) Study on limiting nutrients and phytoplankton at long-line-culture areas in Laizhou Bay and Sanggou Bay, northeastern China. Aquat Conserv Mar Freshwat Ecosyst 14(6):551–574
29. Liu DY, Keesing JK, Xing QG, Shi P (2009) World's largest

macroalgal bloom caused by expansion of seaweed aquaculture in China. Mar Pollut Bull 58:888–895

30. Liu H, Fang JG et al (2014) Environmental impact of large scale mariculture. In: Su JL (ed) Scientific issues and policy recommendations on ecosystem safety of ocean and coastal engineering in China. Science Press, Beijing, In Chinese
31. Loucks RH, Smith RE, Fisher EB (2014) Interactions between finfish aquaculture and lobster catches in a sheltered bay. Mar Pollut Bull 88:255–259
32. Luo R (2012) Pay attention to disappearing coastal wetlands. China Nat 9:8–11 (in Chinese)
33. Mai KS (2010) Achievements and prospects for China aquaculture and feed industry. Sci Fish Aquac 11:1–2 (in Chinese)
34. Ministry of Agriculture (MOA) (1979–2014) China fishery statistical yearbook. China Agriculture Press, Beijing (in Chinese)
35. Murray NJ, Clemens RS, Phinn SR, Possingham HP, Fuller RA (2014) Tracking the rapid loss of tidal wetlands in the Yellow Sea. Front Ecol Environ 12:267–272. doi:10.1890/130260
36. National Research Council (2010) Ecosystem concepts for sustainable bivalve mariculture. National Academies Press, Washington DC. http://www.nap.edu/catalog/12802.htmlAccessed on March 5, 2015
37. Naylor RL, Goldburg RJ, Primavera JH, Kautsky N, Beveridge MCM, Clay J, Folke C, Lubchenco J, Mooney H, Troel M (2000) Effect of aquaculture on world fish supplies. Nature 405:1017–1024
38. Niu ZZ (1982) Several technical issues on gathering and transporting shrimp broodstock. Hebei Fisheries Sci Technol 5:17–19 (in Chinese)
39. Nunes JP, Ferreira JG, Bricker SB, O'Loan B, Dabrowski T, Dallaghan B, Hawkins AJS, O'Connor B, O'Carroll T (2011) Towards an ecosystem approach to aquaculture: assessment of sustainable shellfish cultivation at different scales of space, time and complexity. Aquaculture 315:369–383
40. Reid GK, Chopin T, Robinson SMC, Azevedo P, Quinton M, Belyea E (2013) Weight ratios of the kelps, *Alaria esculenta* and *Saccharina*

latissima, required to sequester dissolved inorganic nutrients and supply oxygen for Atlantic salmon, *Salmo salar*, in integrated multi-trophic aquaculture systems. Aquaculture 408–409:34–46

41. Ren L, Zhang J, Fang J, Tang Q, Zhang M, Du M (2014) Impact of shellfish biodeposits and rotten seaweed on the sediments of Ailian Bay, China. Aquacult Int 22:811–819
42. Shan XJ, Jin XS, Li ZY, Chen YL, Dai FQ (2012) Fish community structure and stock dynamics of main releasing species in the Bohai Sea. Progress Fish Sci 33(6):1–9 (in Chinese)
43. Shi J, Wei H, Zhao L, Yuan Y, Fang JG, Zhang JH (2011) A physical-biological coupled aquaculture model for a suspended aquaculture area of China. Aquaculture 318:412–424
44. Shi HH, Zheng W, Zhang XL, Zhu MY, Ding DW (2013) Ecological-economic assessment of monoculture and integrated multi-trophic aquaculture in Sanggou Bay of China. Aquaculture 410–411:172–178
45. Soto D, Aguilar-Manjarrez J, Hishamunda N (eds) (2008) Building an ecosystem approach to aquaculture. FAO/Universitat de les Illes Balears Expert Workshop. 7–11 May 2007, Palma de Mallorca, Spain. FAO Fisheries and Aquaculture Proceedings No. 14. FAO, Rome, p 221
46. State Forestry Administration (SFA) (2011) The basic situation of forestry in China in 2011. http://www.forestry.gov.cn/ Accessed on October 28, 2014
47. State Oceanic Administration (SOA) (2005–2015). China marine disaster bulletin, 2004–2014. http://www.soa.gov.cn/zwgk/hygb/zghyzhgb/ Accessed on March 16, 2015
48. State Oceanic Administration (SOA) (2013). China marine environment bulletin, 2012. http://www.soa.gov.cn/zwgk/hygb/zghyhjzlgb/ Accessed on June 10, 2015
49. Su J (1998) Circulation dynamics of the China Seas: north of 18°N, In: Robinson AR and Brink K [eds] "The sea," vol. 11, The global coastal ocean: regional studies and syntheses, Wiley, 483–506
50. Tang QS, Fang JG, Zhang JH, Jiang ZJ, Liu HM (2013) Impacts of multiple stressors on coastal ocean ecosystems and integrated multi-

trophic aquaculture. Progress Fish Sci 34(1):1–11 (in Chinese)

51. Task Force CCICED (2013) Ecosystem issues and policy options addressing the sustainable development of China's ocean and coasts. China Environmental Science Press, Beijing, p 493 (in both Chinese and English)
52. Wang XP (1998) The nitrogen and phosphorus in sediment interstitial water of cage cultural waters in the Depth Bay of Niutou Island. J Fish Sci China 5(3):126–128 (in Chinese)
53. Wang R, Zuo T, Wang K (2003) The Yellow Sea Cold Bottom Water—an oversummering site for *Calanus sinicus* (Copepoda, Crustacea). J Plankton Res 25:169–183
54. Wang ZX, Sun PX, Liu CX, Liu XJ, Lian Y, Xu RS (2011) The species diversity of macrobenthic fauna in Sanggou Bay, China. Period Ocean Univ China 41(7/8):79–84 (in Chinese)
55. Wang W, Liu H, Li YQ, Su JL (2014) Development and management of land reclamation in China. Ocean Coastal Manag 102:415–425
56. Wu CD, Tang TD, Lin ZX, Li WJ (1980) Co-culture of shrimp and mullet. 1980 (3): 11 (in Chinese)
57. Xie CL (1981) High economic return generated by bivalve and seaweed co-culture. China Fish 1981(5):20 (in Chinese)
58. Xu ZL, Chen JJ (2010) Analysis to population division and migratory routine of populations and migratory routines of *Argyrosomus argentatus* in the north China waters. Acta Ecol Sin 30:6442–6450 (in Chinese)
59. Yang YF, Hou LC, Nie XP, Tang DL, Chung IK (2004) Development of mariculture and its impacts in Chinese coastal waters. Rev Fish Biol Fish 14:1–10
60. Yang JY, Gao AG, Ning XR, Zhang DS (2007) Characteristics on macrofauna and the responses on aquiculture in Yueqing Bay. Acta Ecolog Ica Sinica 1:34–41 (in Chinese)
61. Zhai WD, Zhao HD, Zheng N, Xu Y (2012) Coastal acidification in summer bottom oxygen-depleted waters in northwestern-northern Bohai Sea from June to August in 2011. Chin Sci Bull 57(9):1062–1068
62. Zhang J, Wu AY, Shi QS (2003) Mariculture and its environment

effect in the Xiangshangang Bay. Donghai Mar Sci 21(4):54–59 (in Chinese)

63. Zhang QC, Qiu LM, Yu RC, Kong FZ, Wang YF, Yan T, Gobler CJ, Zhou MJ (2012) Emergence of brown tides caused by *Aureococcus anophagefferens* Hargraves et Sieburth in China. Harmful Algae 19:117–124
64. Zheng XR, Zhang FC, Wu XM, Zhang B, Wang LS (2011) Nutrients in bay scallop culture areas in Changli County. Hebei Fish 205:35–38 (in Chinese)
65. Zhi YL, Hou JL, Zhang T, Zhang LZ, Zhao F, Yang G, Zhuang P (2013) Spatiotemporal distribution patterns of elver (*Anguilla japonica*) in the Yangtze River Estuary of China. Chin J Ecol 32(10):2750–2755 (in Chinese)
66. Zhu YG (1981) Co-culture of clam *Meretrix meretrix* in a shrimp pond. Mar Fish 1981(2):25 (in Chinese)
67. Zuo YH, Lin GL (2008) Resource utilization and environment protection in the coastal zone. Science Press, Beijing, p 169 (in Chinese)

CITATION

CHAPTER 1

Guangyu Wang, Shari Mang, Haisheng Cai, Shirong Liu, Zhiqiang Zhang, Liguo Wang, John L. Innes, "Integrated watershed management: evolution, development and emerging trends," DOI 10.1007/s11676-016-0293-3.

CHAPTER 2

Guangyu Wang, Shari Mang, Haisheng Cai, Shirong Liu, Zhiqiang Zhang, Liguo Wang, John L. Innes, "Integrated watershed management: evolution, development and emerging trends," DOI 10.1007/s11676-016-0293-3.

CHAPTER 3

Rebecca M. Quiñones, Theodore E. Grantham, Brett N. Harvey, Joseph D. Kiernan, Mick Klasson, Alpa P. Wintzer, Peter B. Moyle, "Dam removal and anadromous salmonid (*Oncorhynchus*spp.) conservation in California," DOI 10.1007/s11160-014-9359-5.

CHAPTER 4

Trina Rytwinski, Dirk A. Algera, Jessica J. Taylor, Karen E. Smokorowski, Joseph R. Bennett, Philip M. Harrison and Steven J. Cooke, "What are the consequences of fish entrainment and impingement associated with hydroelectric dams on fish productivity? A systematic review protocol," DOI 10.1186/s13750-017-0087-x.

CHAPTER 5

Wollebæk et al.: Population connectivity: dam migration mitigations and contemporary site fidelity in arctic char. BMC Evolutionary Biology 2011 11:207. doi:10.1186/1471-2148-11-207.

CHAPTER 6

Ki Won Jung, Z. Daniel Deng, Jayson J. Martinez, David R. Geist, Geoffrey A. McMichael, John R. Stephenson and Peter J. Graf, "Performance of an acoustic telemetry system in a large fishway," DOI 10.1186/s40317-015-0052-9.

CHAPTER 7

Gulab D Khedkar, Rahul Jamdade, Amol Kalyankar, Anita Tiknaik, Tetsuzan Benny Ron and David Haymer, "Genetic fragmentation in India's third longest river system, the Narmada," doi:10.1186/2193-1801-3-385.

CHAPTER 8

Kari Stange, Judith van Leeuwen and Jan van Tatenhove, "Boundary spaces, objects and activities in mixed-actor knowledge production: making fishery management plans in collaboration," DOI 10.1186/s40152-016-0053-1.

CHAPTER 9

Lina M. Saavedra-Díaz, Robert Pomeroy and Andrew A. Rosenberg, "Managing small-scale fisheries in Colombia," DOI 10.1186/s40152-016-0047-z.

CHAPTER 10

Tatenda Dalu and P William Froneman, "Diatom-based water quality monitoring in southern Africa: challenges and future prospects," http://dx.doi.org/10.4314/wsa.v42i4.05.

CHAPTER 11

Magdalena Bełdowska, Agnieszka Jędruch, Joanna Słupkowska, Dominka Saniewska and Michał Saniewski, "Macrophyta as a vector of contemporary and historical mercury from the marine environment to the trophic web," DOI 10.1007/s11356-014-4003-4.

CHAPTER 12

Thaıs Dalzochio, Gabriela Zimmermann Prado Rodrigues, Ismael Evandro Petry, Gunther Gehlen, Luciano Basso da Silva, "The use of biomarkers to assess the health of aquatic ecosystems in Brazil: a review," DOI 10.1007/s40071-016-0147-9.

CHAPTER 13

Cooke et al.: Tracking animals in freshwater with electronic tags: past, present and future. Animal Biotelemetry 2013 1:5. doi:10.1186/2050-3385-1-5.

CHAPTER 14

Steven Emmerson Orchard, Lindsay Carman Stringer and Claire Helen Quinn, "Mangrove system dynamics in Southeast Asia: linking livelihoods and ecosystem services in Vietnam," DOI 10.1007/s10113-015-0802-5.

CHAPTER 15

Roger J. H. Herbert, John Humphreys, Clare. J. Davies, Caroline Roberts, Steve Fletcher, Tasman. P. Crowe, "Ecological impacts of non-native Pacific oysters (*Crassostrea gigas*) and management measures for protected areas in Europe," DOI 10.1007/s10531-016-1209-4.

CHAPTER 16

Hui Liu and Jilan Su, "Vulnerability of China's nearshore ecosystems under intensive mariculture development," DOI 10.1007/s11356-015-5239-3.

INDEX

N

P

R

S

T

W

X

Y